Remote Sensing

Remote Sensing

The Image Chain Approach

JOHN R. SCHOTT
Rochester Institute of Technology

New York Oxford
OXFORD UNIVERSITY PRESS
1997

Oxford University Press

Oxford New York
Athens Auckland Bangkok Bogota Bombay
Buenos Aires Calcutta Cape Town Dar es Salaam Delhi
Florence Hong Kong Istanbul Karachi
Kuala Lumpur Madras Madrid Melbourne
Mexico City Nairobi Paris Singapore
Taipei Tokyo Toronto

and associated companies in
Berlin Ibadan

Copyright © 1997 by Oxford University Press, Inc.

Published by Oxford University Press, Inc.,
198 Madison Avenue, New York, New York 10016

Oxford is a registered trademark of Oxford University Press

All rights reserved. No part of this publication may be reproduced,
stored in a retrieval system, or transmitted, in any form or by any
means, electronic, mechanical, photocopying, recording, or otherwise,
without the prior permission of Oxford University Press.

Library of Congress Cataloging-in-Publication Data

Schott, John R. (John Robert), 1951-
Remote sensing : the image chain approach / John R. Schott.
p. cm.
Includes bibliographical references and index.
ISBN-13: 978-0-19-508726-0
ISBN 0-19-508726-7
1. Remote sensing. I. Title. 95-47216
G70.4.S36 1996 CIP
621.36 ' 78--dc20

9 8 7
Printed in the United States of America
on acid-free paper

Dedicated with love and respect to

Pam, Stephen, Jesse, and Jack

PREFACE

Much of the remote sensing literature is written by and for the applications specialists who are users of remotely sensed data. The remote sensing "science" is often neglected or given only cursory treatment because of the need to stress the principles of the application area (*i.e.* geography, geology, forestry etc.). Those books that more directly address remote sensing as a discipline have tended to heavily emphasize either the optics and physics of remote sensing or the digital image processing aspects.

This book treats remote sensing as a continuous process, including energy matter interaction, radiation propagation, sensor characteristics and effects, image processing, data fusion and data dissemination. The emphasis is on the tools and procedures required to extract information from remotely sensed data using the image chain approach.

This approach to remote sensing has evolved from over a decade of teaching remote sensing to undergraduate and graduate students and two decades of research and consulting on remote sensing problems for government and industry. That experience has often shown that individuals or organizations all too often focus on one aspect of the problem before considering the entire process. Usually this results in a great deal of time, effort, and expense to achieve only a small improvement, because all the effort was placed somewhere other than the weak link in the chain. As a result, the perspective on remote sensing presented here is to treat the process as a continuous flow and to study the underlying science to a level sufficient to understand the many constrictions that limit that flow of information to the eventual user.

Because the field of remote sensing is so large, I have chosen to limit the treatment to aerial and satellite imaging for earth observation. In addition, because the vast majority of remote sensing is done passively in the visible through the thermal infrared region, I have emphasized this area. Within this spectral region, the underlying science and techniques of quantitative radiometric image acquisition, image analysis, and multispectral image processing are emphasized. The details of specific sensors and software packages are downplayed because of their ephemeral nature and photo interpretation and photogrammetry are only briefly introduced because of their thorough treatment elsewhere.

In writing I've always had two audiences in mind. The first is the traditional student. As a text, this book is aimed at graduate students in the physical or engineering sciences taking a first course in remote sensing. It would also be appropriate for advanced undergraduates or as a second course for students in applications disciplines. In several cases, where the mathematical principles may be beyond what students in a particular discipline are required to know (e.g., 2-D linear systems theory), I have attempted to use more extensive graphical and image examples to provide a conceptual understanding. I have assumed a working knowledge of university physics and calculus. In addition, parts of Chapters 7 and 9 draw on linear systems theory, although these sections can be treated in a more descriptive fashion when necessary.

The second audience I had in mind when writing is the large number of scientists and engineers who were trained in the traditional disciplines and find themselves working in the remote sensing field. Having worked extensively with many of these scientists in government and industry, I wanted to compile a book that could be used to understand and begin to address many of the questions these individuals must first ask of the remote sensing field. I hope both of these groups will find this a useful tool box for working the problems ahead.

J.R.S.

Wyldewood Beach
Port Colbourne, Ontario
December 29, 1995

ACKNOWLEDGMENTS

When I took on this book, I expected it would be daunting. When I look back at all the effort that others have made to bring this project to completion, I wonder that any work of this type ever reaches completion. I could never have begun to do it alone. I want to gratefully acknowledge a team of people who made this happen. Margôt Delage produced the majority of the illustrations from my very poor hen-scratches, Nina Raqueño filled in on several of the more technical illustrations. Paul Barnes produced most of the plots and performed the related calculations. Scott Brown prepared most of the images, including most of the image processing examples, and together with Emmett Ientilucci, was responsible for most of the final layout and data organization. Rolando Raqueño and Tim Gallagher provided software and hardware support, respectively, to fill in any gaps the rest of us couldn't fill. As with any project we've undertaken in recent years, Carolyn Kitchen provided the word-processing, overall organizational framework, and communications links that made this possible. As you all know so well, I could not have begun to do this without you, and I am deeply grateful.

I am also pleased to acknowledge the research staff and numerous students who have worked with me over the years in DIRS. Their inquiries, insights, energy, and enthusiasm for this field have kept me as intrigued with it today as when I was first captivated by it at the age of twenty. You have collectively given me a gift that few people ever experience, and I truly cherish it.

It is always gratifying when you reach out to busy professionals and receive their generous response. In trying to gather data and illustrations for this text, I was met over and over, not by obstacles, but by generous individuals who went the extra mile to help out. The individual organizations are referenced with each figure, but I want to specifically recognize and thank some of the individuals who were so generous. To Carroll Lucas of Autometric, Steve Stewart, Dan Lowe, and Robert Horvath of ERIM, Paul Lowman, Jim Irons, and Darrel Williams of NASA Goddard, Larry Maver of ITEK, Al Curran of Georgia Tech, Phil Slater and Bob Schowengerdt of the University of Arizona, Tom Chrien and Robert Crippen of NASA JPL, Tom Steck of the Army Night Vision Lab, Keith Johnson of KRC, Ron Beck and Jeffrey Eidenshink of the EROS Data Center, Phil Teillet of Canada Centre for Remote Sensing, Mike Duggin of SUNY CESF, Jim Barrow of GDE, and Lenny Lafier and Mike Richardson of Eastman Kodak, my respect and thanks.

I also want to thank the many individuals who, through their careful reviews, so often pulled my foot out of my mouth by catching the numerous blunders in the early versions of the text. Their insight and guidance added greatly to the final version. Much of the credit for the clarity and coherence go to these individuals. What errors remain, rest with me for this crew did a fine job of a thankless task. I salute Bernie Brower, John Mason, Rulon Simmons, Jon Arney, Mark Fairchild, Jim Jakobowski, John Francis, Bill Philpot, Roger Easton, Carl Salvaggio, Kurt Thome, Carolyn Kitchen, Perti Jarvelin, and Zoran Ninkov.

Finally, I'd like to thank RIT for supporting me in this effort, and the editorial staff at Oxford for encouraging me to start this project and their patience in awaiting its completion.

CONTENTS

CHAPTER

1

Introduction

The standard opening for any reference book is to define the subject: *Remote sensing is the field of study associated with extracting information about an object without coming into physical contact with it.* Most readers having read that vague definition look out the nearest window and remotely sense the weather conditions to help them decide whether to head out into the sunshine or muddle forward through this introduction. You have just engaged in a *chain* of events that represents remote sensing in its broadest context. You used your eyes to *acquire* data and your visual cognitive system to *process* the data, and you *output* a decision to read on and/or return to this later in the day. As we will see, this sequence of acquisition, processing, and output is characteristic of remote sensing systems. In most cases, we will find that the general definition of remote sensing given above is too broad for a manageable treatment. As a result, we will spend most of this chapter limiting the field to that segment we intend to address and providing a perspective for the approach taken in the succeeding chapters.

1.1 WHAT IS REMOTE SENSING (AS FAR AS WE'RE CONCERNED)?

The broad definition of remote sensing would encompass vision, astronomy, space probes, most of medical imaging, nondestructive testing, sonar, observing the earth from a distance, as well as many other areas. For our purposes in this text, we are going to restrict our discussion to earth observation from overhead. Within this restricted context, we could trace the origins of the field to prehistoric hunters or explorers who climbed the nearest hill to get the lay of the land. However, our main interest will be in the principles behind overhead earth observation using aircraft and satellite (aerospace) remote sensing systems. Given this restriction, we recognize that we are talking about a field of study that for the most part is only a few decades old. This is an important perspective for the reader to keep in mind. In such a relatively young field, many of the principles are still being formulated, and in many areas the consistent structure and terminology we expect in more mature fields may be lacking. On the positive side, young fields, such as remote sensing, offer a myriad of opportunities for exploring unanswered (and often as yet unasked) questions. These questions address what is to be learned about the earth's surface and about the earth's land, water, and atmosphere. They can be further extended to include the condition of the water quality, the vegetation health, the pollutant levels, and how these conditions are changing with time. The tools used in addressing these questions will be our major concern.

1.2 WHY REMOTE SENSING?

As we continue to try to narrow our definition of remote sensing (at least as far as what we will cover in this book), it is important to keep in mind why we use remote sensing in the first place. The reason most often cited is that remote sensing literally and figuratively gives us a different way of looking at the world. This different view often adds a significant amount of incremental data that are useful to a host of applications. Some of these incremental data are due to the synoptic perspective provided by overhead images. Many more traditional approaches to earth observation (i.e., surface studies) can be limited by too much detail on too few samples or by only having data from a very restricted locale. This is the classic "can't see the forest for the trees" problem. The synoptic perspective offered by remote sensing lets us look at whole forests, regions, continents, or even the world and yet, at appropriate scales, can let us see not only the whole forest but also the individual trees as well (cf. Figs. 1.1 through 1.4). This perspective lets us look for large-scale patterns, trends, and interactions and serves as an excellent means of guiding efforts to interpolate or extrapolate parametric values from extensively studied ground sites (ground truth sites). Coupled to this synoptic perspective is the opportunity to view the world in ways our visual system cannot. Even if we ignore sound waves, magnetic fields, nuclear radiation, etc., and restrict ourselves to remote sensing of electromagnetic (EM) radiation (as we will in this treatment), there is a great deal of the EM spectrum our eyes can't see. If we look at a transmission spectrum of the earth's atmosphere (cf. Fig. 1.5), we see a number of transmission bands or windows through which we could peek if we could find appropriate sensing technology. As we will see in later chapters, sensing through these windows lets us look at vegetation stress, surface temperature, atmospheric water content, and a host of other parameters that our visual system couldn't begin to see.

Another view offered by remote sensing that is hard to achieve in other ways is a temporal perspective over large areas. This lets us look at changes over time spans of minutes to decades (cf. Figs. 1.6 and 1.7). Not only can this be used to study changes in the condition of the earth's surface and the atmosphere, but it can also be used to effectively "see through clouds" in atmospheric windows where this would normally be impossible. This is accomplished by taking a sequence of images (say on five consecutive days) and cutting and pasting together an image that represents a cloud-free composite as shown in Figure 1.8.

1.3 WHAT KINDS OF REMOTE SENSING?

Remote sensing of EM radiation is often described in terms of what spectral windows one is peeking through. We are going to restrict ourselves primarily to remote sensing through the windows between 0.4 and 15 μm. Figure 1.9 shows how different the earth can appear when viewed simultaneously in four different spectral regions. As we will see, it is these differences that facilitate many of the information extraction techniques we will discuss in later chapters.

Much of remote sensing is done in the visible and near infrared (VNIR) in the daytime. However, nighttime imaging in the VNIR window can yield information on the location,

Figure 1.1 Photographic image of the earth acquired by Skylab astronauts. See colotr plate 1.1. (Image courtesy of NASA Goddard.)

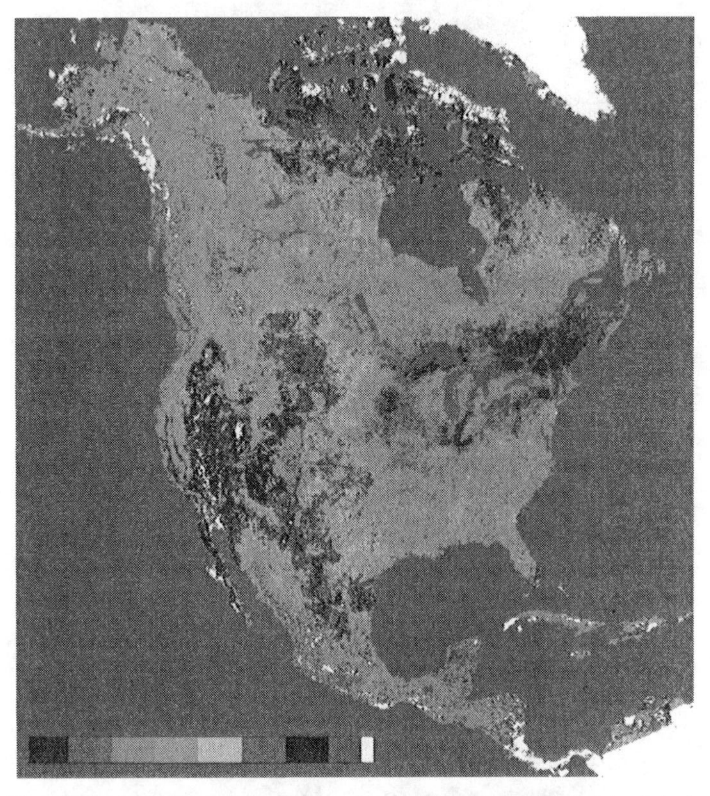

Figure 1.2 Map of the normalized difference vegetative index of North America derived from AVHRR data. See color plate 1.2. (Image produced jointly by the Canada Center for Remote Sensing and the EROS Data Center.)

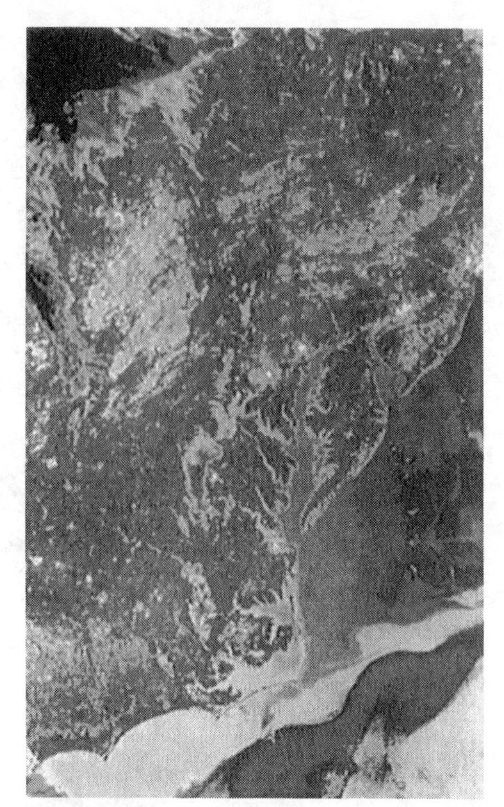

Figure 1.3 B/W image of Eastern U.S. produced from the thermal channel of the Heat Capacity Mapping Mission's (HCMM) radiometer. Note the warm urban areas and the warmer water in the Gulf Stream. See color plate 1.3. (Image courtesy NASA Goddard.)

Figure 1.4 B/W infrared aerial photograph of a forested area. See color plate 1.4.

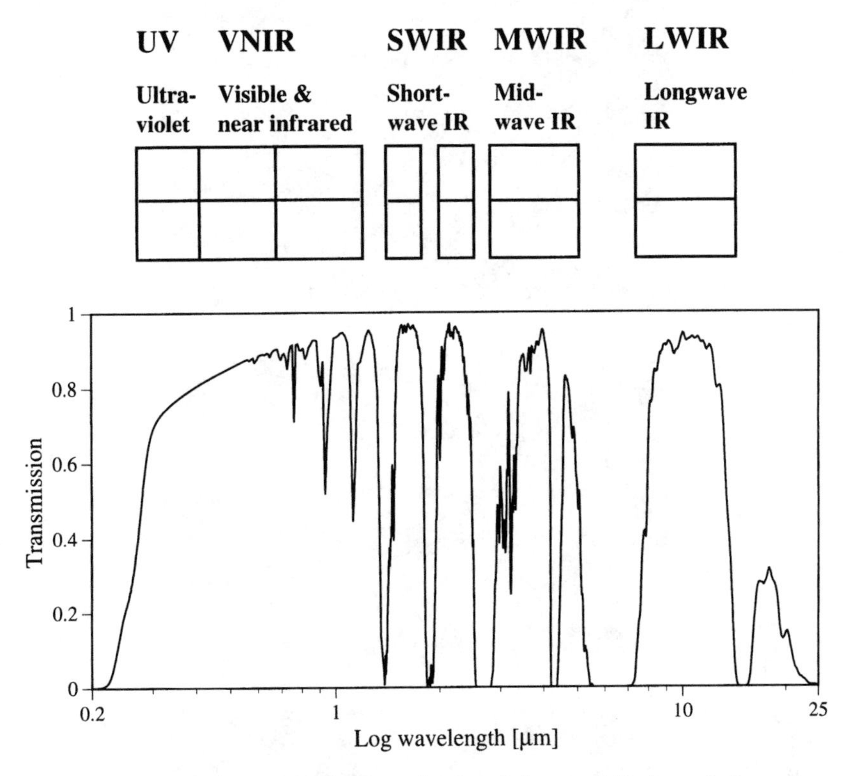

Figure 1.5 Atmospheric transmission spectra showing windows available for earth observations.

extent, and activity level of population centers, as illustrated in Figure 1.10. By selecting a window in the thermal infrared region dominated by self-emission due to the temperature of objects, we can even "see" in the dark making 24-hour remote sensing a possibility. The thermal infrared windows let us image a great deal more detail at night, ranging from terrain features to detailed inspection of facilities (cf. Fig. 1.11). By choosing to restrict ourselves to passive remote sensing, we will limit our discussion to sensors that collect energy that is either emitted directly by the objects viewed (e.g., thermal self-emission) or reflected from natural sources (typically the sun). We are imposing this limit on the topics covered merely to reduce the volume of material.

There are many very useful active remote sensing systems whose treatment is simply beyond our scope. These active systems employ an active source that illuminates the scene. In some cases, the sensed energy is reflected or scattered from the source as in the case with the synthetic aperture radar (SAR) image shown in Figure 1.12. Radar images the reflected microwave energy emitted by the sensor itself. By selecting suitably long wavelengths, it is possible to use radar to image through clouds, providing an all-weather capability not available with visible or IR systems.

In other cases the active source energy is absorbed by the scene elements and reradiated in other wavelength regions. This is the case, for example, in laser-induced fluorescence imaging. The surface is irradiated by a laser that stimulates some of the materials to fluoresce. The fluorescent energy is emitted at longer wavelengths and sensed by the imaging system. The amount of fluorescence is a function of material type and condition. Many of the principles discussed in this volume are applicable to active remote sensing systems. However, covering the specifics of these systems would make this volume prohibitively long. The reader should consider Elachi (1987) and/or Colwell (1983) for a thorough introductory

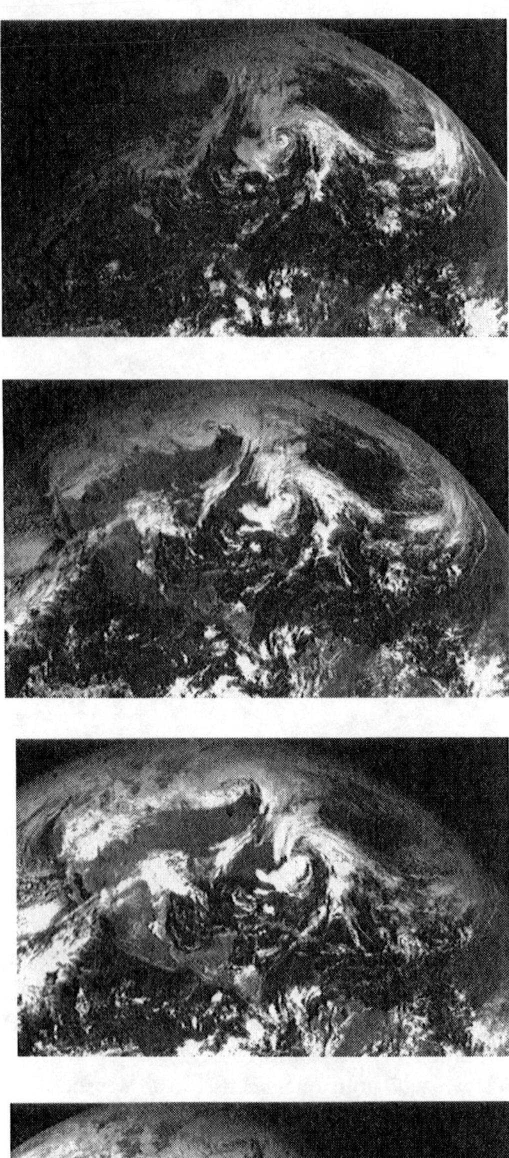

Figure 1.6 Time sequence of GOES images showing cloud dynamics. These images are acquired every 30 minutes. The images selected here are more widely spaced over the course of a day to emphasize the movement of the clouds. (Image courtesy of NOAA.)

(a) 

(b) 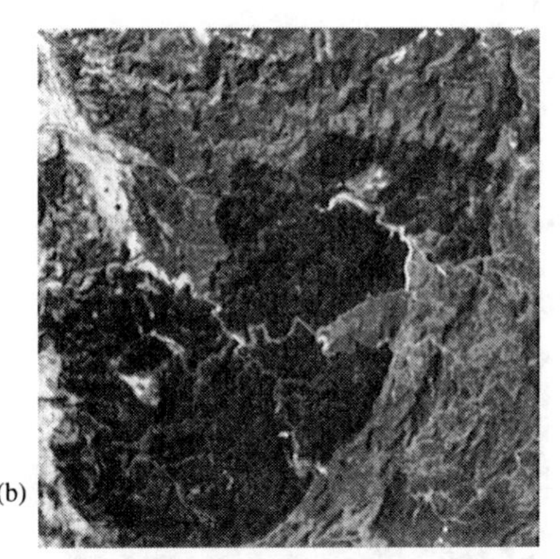

Figure 1.7 Images from portions of Landsat TM images. The region is a section of the Custer National Forest (a) before and (b) after a forest fire. See color plage 1.7. (Image courtesy of EROS Data Center.)

treatment of radar principles and Measures (1984) for a treatment of active remote sensing with laser systems.

So far, we have restricted our scope to passive aerospace remote sensing of the earth with emphasis on the 0.4-to-15 μm region. Within that scope, we are going to place a strong emphasis on quantitative analysis, particularly digital image processing for target identification and radiometric analysis for condition assessment. The more classical analysis referred to as *photo* or *image interpretation* will be briefly discussed in Chapter 2. This is the process of extraction of information by a human analyst viewing the photographic or electro-optical image to determine the location and condition of features. It is based largely on spatial patterns and is a critical part of most remote sensing activities. Expertise in this field is, however, largely a function of the application area of interest (e.g., geology, forestry). More detailed treatment of photo interpretation approaches and tools can be found in Lillesand and Kiefer (1987), Colwell (1960), and Liang et al. (1951).

In contrast to photo interpretation, which relies heavily on the skills of an individual analyst, quantitative analysis attempts to develop analytical tools often based on brightness levels in one or more spectral bands. From an operational standpoint, a major goal of quantitative analysis is to reduce the burden on human analysts by performing tasks that are difficult or tedious for an image interpreter. This is largely accomplished using two approaches. The first approach uses quantitative measurement of radiance levels that are very difficult to differentiate visually. The second approach uses computer-based algorithms to combine many sources of data (e.g., spectral bands, texture metrics, spectral feature vectors, etc.) that can overwhelm a human analyst. From a scientific perspective, a major goal of quantitative analysis is to develop functional relationships between remotely sensed data and parameters of interest (e.g., land cover or material type, location, extent, concentration, orientation, condition, change in condition).

An additional advantage of quantitative analysis over image interpretation is the ability to perform error analysis. Functional relationships are quantitative such that errors can be assigned to estimates of parameter values and confidence levels to classification parameters.

Another often overlooked goal of quantitative analysis is to make the remote sensing data as useful as possible to the interpreter or decisionmaker. This can take the form of improving the visual appearance of an image or a data set for human analysis or transforming a maze of quantitative analytical results to a simplified form (e.g., crop yield is expected to be X metric tons with an error of $\pm Y$ metric tons). From this standpoint, the whole field of

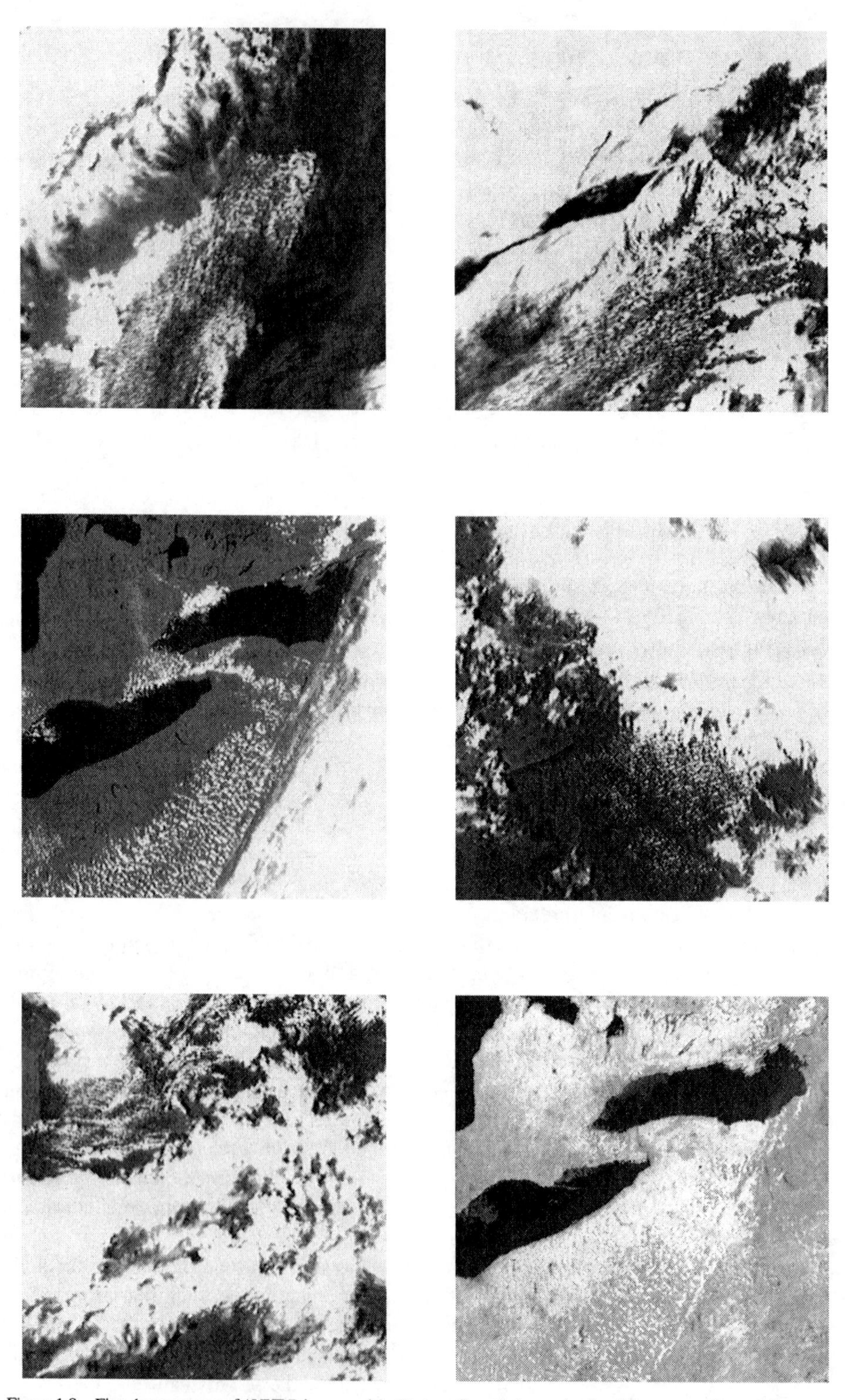

Figure 1.8 Five-day sequence of AVHRR images of the Eastern Great Lakes and a cloud-free composite made by combining the five-day sequence. See color plate 1.8.

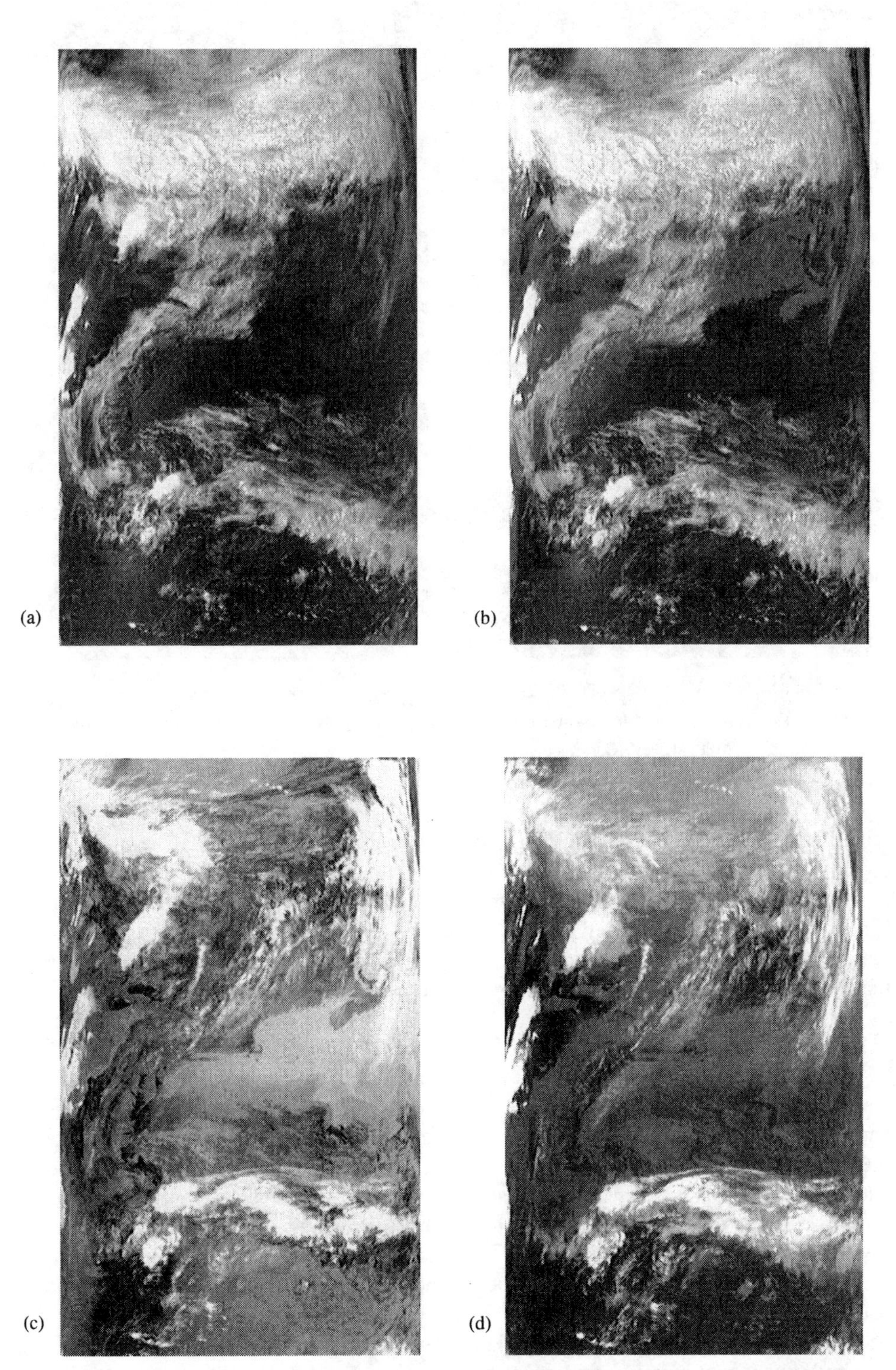

Figure 1.9 Simultaneously acquired AVHRR images showing the East coast of North America and the Great Lakes region as viewed through the (a) red, (b) near infrared, (c) midwave infrared, and (d) long-wave infrared windows. (Note that the MWIR and LWIR images are inverted, so white is cold to preserve the white shades for the clouds.)

Figure 1.10 Nighttime visible image of the U.S. street lighting clearly demonstrates major population centers and traffic corridors, but no terrain or feature detail is available. This image is a composite made up from Defense Meteorological Satellite Program (DMSP) images. (Copyright Hansen Planetarium.)

scientific visualization becomes closely coupled to the output side of the remote sensing process.

In looking at quantitative analysis from the scientific perspective, it is easy to delineate some of the capabilities, as well as clear limitations of the remote sensing approach. In an idealized functional form, quantitative analysis would yield an objective, reproducible, quantitative, error-free functional relationship for each parameter of interest of the form

$$Y = f(X1, X2, \cdots, XN) \tag{1.1}$$

where Y is one of several parameters of interest (e.g., material type, concentration of pollutant), and $X1$, $X2...XN$ are the remotely sensed independent variables needed to characterize the Y parameter(s). These variables ($X1$, $X2...XN$) would ideally be in the form of reproducible material characteristics such as spectral reflectance, temperature, orientation, emissivity, scattering function, or polarization response. However, in many cases the variables might be expressed in terms of observed values such as digital counts in each spectral band in an image. The tools used in trying to define the inputs to Eq. (1.1) (i.e., XI values) and the nature and form of the functional relationships (f) with the parameters (Y) of interest will occupy most of the following chapters. However, the critical assumption of remote sensing (for both image interpretation and quantitative analysis) should be addressed here. That is, that a relationship exists between the parameter of interest and some combination of EM signals from the earth or its atmosphere. For example, the concentration of suspended solids in surface waters normally impacts the reflectivity of the water. Therefore, we could hope to use remote sensing to observe and maybe quantify the concentration of suspended solids. On

(a)

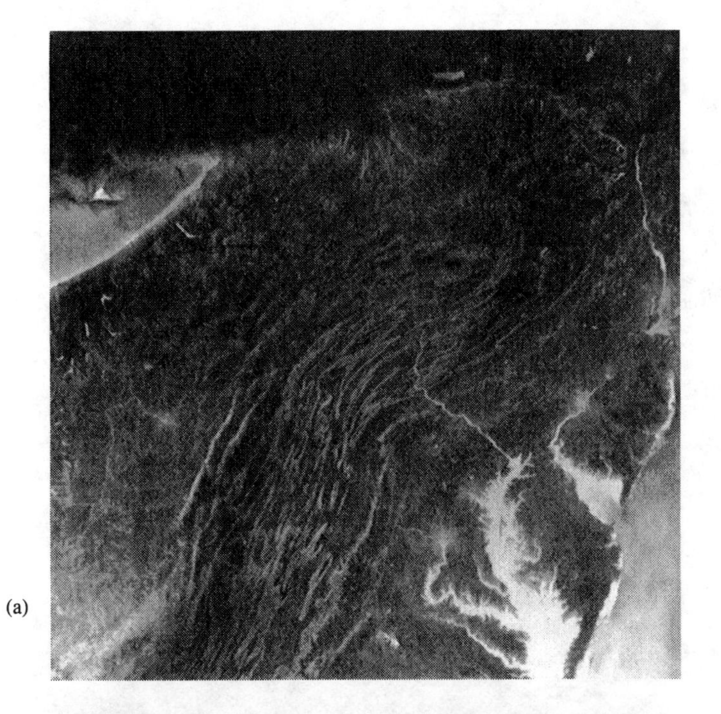

(b)

Figure 1.11 Nighttime thermal infrared image of (a) a portion of a HCMM image of the Appalachian Mountains in Central Pennsylvania showing terrain features, and (b) an airborne thermal infrared image of a government office complex in Albany, New York, acquired in the winter as part of a heat-loss study. Note in particular how the service floors in the office towers are bright (hot). The egg-shaped object includes a large auditorium whose heated shape shows through the roof structure. (Image courtesy of RIT's DIRS Laboratory.)

(a)

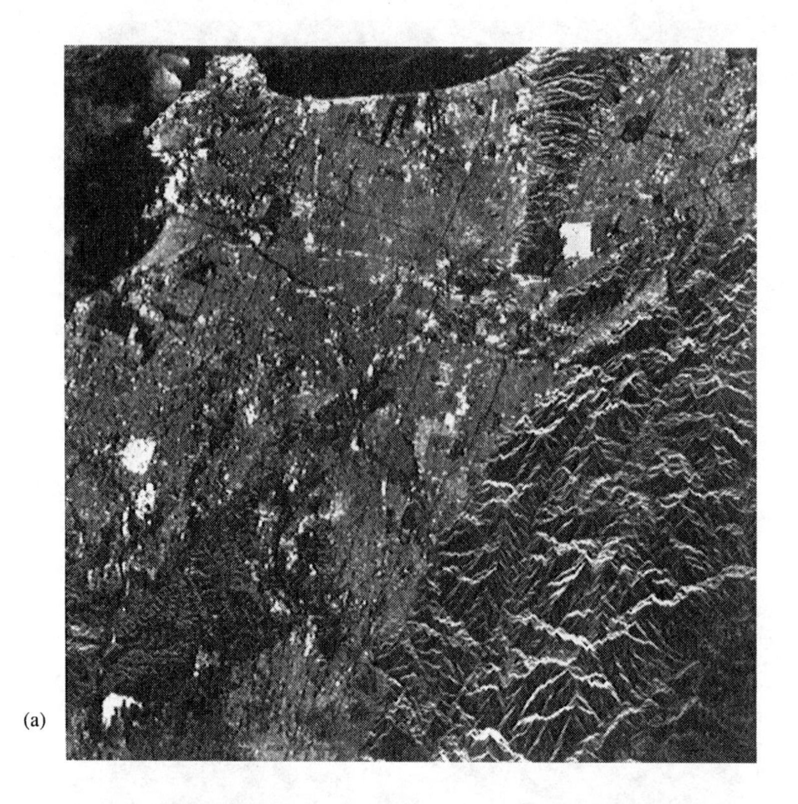

(b)

Figure 1.12 (a) Synthetic aperture radar (SAR) image of Los Angeles, and (b) a portion of a Landsat image of the same region shown for reference. (Images courtesy of NASA Goddard.)

the other hand, the concentration of D_2O (heavy water) in a body of water has no obvious EM manifestations and could not be observed using the remote sensing methods we are considering. There is no functional relationship of the form of Eq. (1.1), and EM remote sensing is not an available option for addressing this latter problem. On the other hand, there are times when a parameter may be indirectly measured, or at least inferred. For example, if a factory is releasing a colorless chemical into the surface waters, there may be no direct functional relationship of the form of Eq. (1.1). On the other hand, if the chemical is toxic to phytoplankton (algae), which do have an optical manifestation, then it may be possible to infer the presence and even the concentration of the chemical (though other means would clearly be required for corroboration). Part of the remote sensing process is the search (both theoretical and experimental) for these EM manifestations and the definition, development, and application of tools to observe and measure them.

Our emphasis here will be on characterizing the properties of materials that affect the EM radiation emitted, reflected, or scattered, the processes that govern the propagation of EM radiation to the sensing platform, and the types of sensors that can be used to collect and record the relevant EM energy levels. Then we will look at tools that can be used to derive the inputs to functions of the form of Eq. (1.1) and methods to improve the quality of both the analytical and visual data available to image analysts. This emphasis on theoretical principles and analytical tools stems from the assumption that, in most cases, the remote sensing scientist will be working closely with experts trained in the application areas of interest. In such teams, the remote sensing scientist must learn enough of each application area to understand how remote sensing can be of use and must teach the corresponding application expert enough remote sensing fundamentals so they can help define the problems and expectations in a common format. Our emphasis here is on the tools the remote sensing scientist needs and the principles with which the applications expert needs to be familiar. In the context used throughout this treatment, a tool is an instrument, algorithm, or often a combination of instruments and algorithms used to construct a solution to a problem. Tools in this context are most often thought of as general and reusable (e.g., an algorithm for atmospheric correction or edge detection). The tools would then be used to help build a solution to an application-specific problem. Readers uncomfortable with their backgrounds in the relevant physics related to electromagnetic radiation or basic optics should consult a university physics text such as Halliday and Resnick (1988). For the more ambitious reader, a complete treatment of some of the optical principles drawn on here is contained in Hecht (1987).

Applications of remote sensing will be treated only to the extent that they illustrate how certain analytical tools can be used. With remote sensing extensively used in fields including meteorology, oceanography, forestry, agriculture, archeology, reconnaissance, geology, geography, range management, hydrology, and atmospheric and soils science, we could not begin to adequately address the domain-specific remote sensing issues. Many of these applications areas are addressed in Volume II of the *Manual of Remote Sensing* [cf. Estes and Thorley 1983], as well as in numerous remote sensing texts that approach the field from the applications perspective, e.g., Barrett and Curtis (1976) and Lo (1986).

1.4 THE IMAGE CHAIN APPROACH

In order to take advantage of the synoptic perspective that remote sensing offers, we will emphasize imaging systems throughout our treatment. An imaging system is a system where the input is a scene and the output is an image. A simple imaging system would involve image capture, image processing, and image display. This is not intended to exclude point radiometers (such as those used in atmospheric sounders) or other nonimaging devices, but merely to indicate that most of our treatment of remote sensing will be interwoven with elements of imaging science. We will often find it convenient to think of the remote sensing process as a chain of events or steps that lead to a final output. This output will often be an image, but it might as easily be a map or a table of figures or a recommendation to a decisionmaker (e.g., harvest the trees in an infested forest before the mortality rates increase the fire hazard and reduce the timber value). From a conceptual or philosophic point of view, we will refer to the study or characterization of the sequence of steps or chain of events that lead to that final output as the *image chain approach.*

The events along the way may be very different and apparently unrelated. For example, some events will be natural processes, such as sunlight striking the ground. Some will be the result of simple choices or procedures, such as what type of sensor to use or what time of day (TOD) to acquire the image. Some of the events may be very intense activities directed at the imaging process, such as calibration of the image data or digital image processing. Finally, some events involve the inclusion of external data (e.g., ground truth, previous maps of the area, or an inclusion of an expert's knowledge of a target). All of these events can be linked together by the fact that they impact the final output.

This perspective of viewing remote sensing as a series of events is called the *image chain approach.* Each event or step can be thought of as a link in the chain. The image chain approach is based on the premise that if we study and understand the chain of events associated with a particular output image or product, we will be better able to understand what we have (i.e., what the product tells us or means), what we don't have (i.e., the limitations or confidence levels associated with the output product), and where those limitations were introduced (i.e., where are the weak links in the image chain). The image chain approach is often useful when we are trying to understand the fidelity of the image or information output from the remote sensing process or designing remote sensing systems, or procedures to ensure fidelity. Recognizing that each link in the chain results from the relationship to previous links, we begin to recognize that the chain, and hence the image fidelity (or output product), is only as strong as the weakest link.

The image chain approach treats the entire remote sensing process as an interrelated system. It generally includes an imaging system, and the components or transfer relationships in the imaging system will be referred to as links in the *imaging chain.* The *image chain* combines the links in the imaging chain with any nonimaging components or transfer relationships. The image chain describes the components and transfer mechanisms that characterize the entire remote sensing process as a system. The systems nature of the image chain approach lets us also think about the image chain as being comprised of many interwoven strands. These subchains or subsystems, which we will often refer to as the *strands* of the image chain, are often separated out and treated independently for convenience and then recombined to form the complete image chain.

Our approach will be to try to define (in most cases quantitatively) the links in the image chain. This approach serves several purposes. First, and most fundamentally, by fully characterizing the image chain from end to end, we ensure an understanding of the process. Besides the academic value and the warm fuzzy feeling associated with understanding the process, the image chain approach lets us analyze the process and address questions related to what we have, what information we can extract from the data, and (often as helpful) what information is not available from the data. The image chain may be modeled quantitatively. Such a model can then be used as the basis for reverse engineering to help in analyzing image data. If we know all the links along the chain, then it is often possible to work backwards to extract critical information from a particular link (e.g., backing out atmospheric effects to compute surface reflectance values). A quantitative analysis of the image chain also helps the analyst identify weak links in the system. This points to where one can most effectively expend efforts to improve the chain. This might take the form of drawing on other data to reinforce a weak link or identifying where one might want to try estimation, modeling, or iterative techniques to strengthen a link in the chain. The image chain approach can be particularly useful in many aspects of image analysis because it points to where a process can

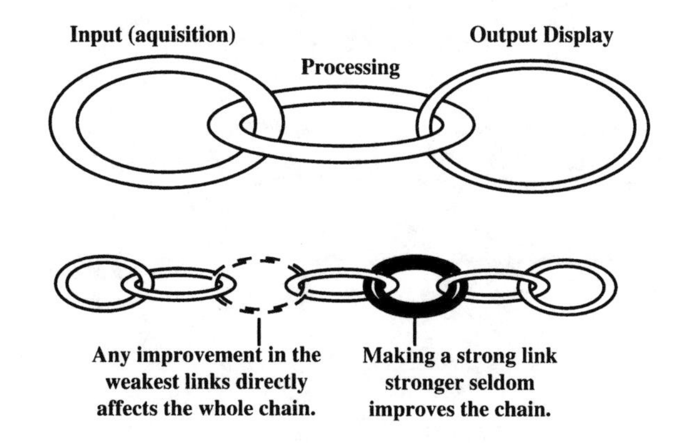

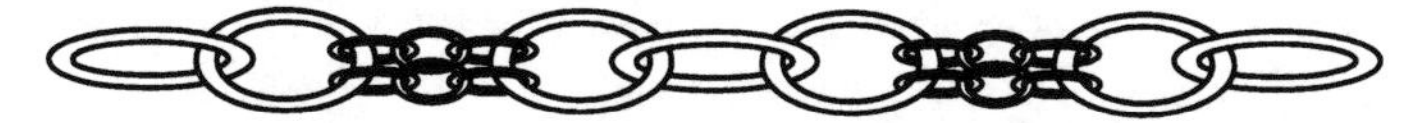

Figure 1.13 Image chain analogy. The chain is only as strong as its weakest link.

be improved. This is particularly useful for system design and upgrade studies, but can also be useful as pointed out above in focusing our image analysis efforts. By focusing our attention on the weak links, we can often achieve significant end-to-end improvements in the overall system. On the other hand, as the chain analogy suggests, we could spend a great deal of time, money, and effort trying to improve the system by working on a link that was already strong and achieve little or no net improvement in the overall system because we had ignored weaker links (cf. Fig. 1.13).

Throughout this treatment, we will be concerned with how the image chain impacts image fidelity. We have carefully chosen to use the term *fidelity* rather than *quality*. Image quality is usually associated with the visual appearance of images and, as a result, will be greatly influenced by visual and psychophysical response functions. Our concern here emphasizes ways to make measurements about the earth using remotely sensed data. As a result, our concern focuses on how well the remote sensing system reproduces the characteristics of interest. We will use the term *fidelity* to refer to how well the image chain reproduces these characteristics.

In general, there is good correlation between fidelity and perceived image quality. An image with high fidelity will, in general, be perceived as being of high quality. However, it is not uncommon for images with poorer fidelity to be perceived to be of higher quality. This is because the visual processing system applies different criteria (or at least different weights to similar criteria) in evaluating quality (cf. Chap. 9). By emphasizing the importance of fidelity, we seek to ensure the integrity of the final image for measurement purposes. In addition, the perceived image quality will usually be maintained. However, if a system is being designed solely or predominantly for visual image quality, the fidelity criteria emphasized here should be adjusted for their importance to perceived image quality.

The image chain approach can often be segmented into different aspects or parameters for convenience. For example, we sometimes will find it convenient just to look at certain parameters of a system such as spatial resolution, radiometric fidelity (including signal-to-

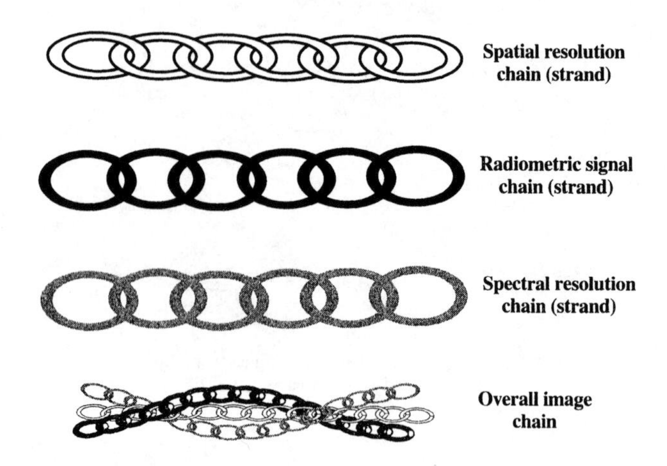

Figure 1.14 Aspects of the image chain. The overall remote sensing system can often be divided into several chains for convenience, but the final system performance is a function of the interplay of all of the chains.

noise issues), spectral resolution, or temporal fidelity. We can think of each of these parameters as having its own image chain, with the appropriate characteristics of the final output (e.g., spatial image fidelity) a function of the interplay along the links in the chain. We must, of course, keep in mind that these various chains are seldom independent of each other, and the final system is the result of the interplay and collective strength of all of the strands in the chain (cf. Fig. 1.14).

In most cases, we will look separately at different strands in the image chain (e.g., spatial resolution) and track some measure of fidelity through the links in the chain. The spatial fidelity of the image will be measured based on how well the image reproduces the radiance variations in a scene. As discussed in Chapter 9, we will use the modulation transfer function (MTF) to measure spatial image fidelity, with an ideal system having an MTF of one at all possible spatial frequencies. The radiometric fidelity is usually characterized in two ways. The first method uses the noise level defined as the signal variation about what should be a constant level (i.e., when the sensor is sensing a uniform surface). An ideal system would have zero noise at all spatial frequencies. The second is the difference between a measured radiometric value (e.g., radiance or reflectance) and the actual value. An ideal system would be perfectly calibrated such that there is no error between measured and actual values. These and other measures of fidelity will be introduced in the following chapters. In Chapter 9 we will look at how these measures can be used to evaluate the performance of individual links in the image chain, as well as end-to-end performance.

We will often use a "systems theory" approach to modeling the imaging chain. This approach is treated in considerable detail in Gaskill (1978) and Dainty and Shaw (1974). We will, for the most part, make a number of simplifying assumptions. In particular, we will usually assume that the systems are linear and shift (space) invariant. The linearity assumption requires that a change in an input variable will result in a proportional change in an output parameter. The space invariance assumption requires that an input at any spatial location will produce the same output function (e.g., the image of a point source will produce a brightness distribution with the same shape whether imaged at point A or point B). Most systems violate our simplifying assumptions to some extent, and in some cases a more rigorous treatment will be required. However, experience has shown that most reasonably well-behaved imaging systems can be approximated with good success using the linear-space invariant approximations. Thus, unless otherwise noted, we will assume that a linear space invariant system model is being assumed for our imaging chain.

While we don't believe it is useful to push any analogy too far, we think the reader may find the image chain a useful and rewarding perspective from which to study remote sensing. As a result, this book is organized around the image chain approach. Chapter 2 contains mostly historical and conceptual background material and terminology. Chapters 3 and 4 treat the early links in the image chain by tracing EM flux from its origins to the sensor. The emphasis in these chapters is on the strand of the image chain associated with the radiometric signal. Chapter 5 deals with spatial, spectral, and radiometric links in the image chain associated with imaging sensors. Chapter 6 deals with reverse engineering back along the radiometric links in the image chain from the sensor to the ground to compute target-specific values such as reflectance and temperature. The links in the chain associated with image processing and data output are covered in Chapters 7 and 8, respectively. By the end of Chapter 8 we've introduced the basic principles associated with the entire image chain as it applies to remote sensing systems. In Chapter 9, we treat various aspects of the image chain to see how we could identify and potentially correct limitations in the chain. Finally, in Chapter 10 we discuss how synthetic image generation can be used in conjunction with the image chain approach as a tool to understand and analyze remotely sensed images and imaging systems.

1.5 REFERENCES

Barrett, E.C., & Curtis, L.F. (1976). *Introduction to Environmental Remote Sensing*. Chapman & Hall, London.

Colwell, R.N., ed. (1960). *Manual of Photo Interpretation*. American Society of Photogrammetry, Falls Church, VA.

Colwell, R.N., ed. (1983). *Manual of Remote Sensing*. Vol. I, 2d ed., American Society of Photogrammetry, Falls Church, VA.

Dainty, J.C., & Shaw, R. (1974). *Image Science*. Academic, NY.

Elachi, C. (1987). *Introduction to the Physics and Techniques of Remote Sensing*. Wiley, NY.

Estes, J.E., & Thorley, G.A., eds. (1983). *Interpretation and Applications*. In Vol. II of *Manual of Remote Sensing*, R.N. Colwell, ed. American Society of Photogrammetry, Falls Church, VA.

Gaskill, J.D. (1978). *Linear Systems, Fourier Transforms, and Optics*. Wiley, NY.

Halliday, D., & Resnick, R. (1988). *Fundamentals of Physics*. 3d. ed., Wiley, NY.

Hecht, E. (1987). *Optics*. 2d. ed., Addison-Wesley, Reading, MA.

Liang, T., Costello, R.B., Fallon, G.J., Hodge, R.J., Ladenheim, H.C., Lueder, D.R., & Lo, C.P., (1986). *Applied Remote Sensing*. Longman, NY.

Lillesand, T.M., & Kiefer, R.W. (1987). *Remote Sensing and Image Interpretation*. 2d. ed. Wiley, NY.

Mallard, J.D. (1951). "Airphoto Analysis: Cornell University Land Form Series," Vols. 1 - 6, NRL 257-001.

Measures, R.M. (1984). *Laser Remote Sensing: Fundamentals and Applications*. Wiley, NY.

CHAPTER 2

Historical Perspective and Photo Mensuration

Observing the earth and recording those observations as images is a form of remote sensing that has been taking place for nearly a century and a half. The development of the first commonly used form of photography is often attributed to L. J. M. Daguerre, who from 1835 to 1839 perfected the daguerreotype. The first known photographs from an overhead platform were taken about 20 years later by a French portrait photographer, Gaspard Felix Tournachon (a.k.a. Nadar). In 1858 Nadar took photographs from a balloon equipped with a darkroom to process the wet plates used in the collodion process he employed. In addition to their novelty value, Nadar's aerial photographs soon found use in surveying and map making [cf. Stroebel and Zakia (1993)]. For over 100 years, camera photography remained the only tool for remote sensing from overhead platforms. The platforms changed to include kites, pigeons, and airplanes. Eventually, 101 years after Nadar's first flight, the first photographs of earth were made from space. These were acquired in August 1959 by Explorer 6 in a nonorbital space flight. The cameras and photographic processes also improved considerably over that period to include the first color photographs from an unmanned Mercury-Atlas flight (MA-4) in 1960.

The modern era of remote sensing began that same year, 1960, with the launch of the Television Infrared Observation Satellite (TIROS-1). This satellite and its successors carried vidicon cameras into space to begin the systematic process of monitoring our weather and global environment from space [cf. Simonett (1983)]. The modern era is characterized by the availability of satellite platforms, electro-optical (EO) sensor systems, and quantitative analytical tools for processing both photographic and electro-optical images. The major thrust of the following chapters will be on sensors and analytical methods that relate to this more recent era of quantitative radiometric photographic and electro-optical imaging, both for aircraft and satellite systems. In this chapter, we want to lay the foundation for the rest of the book by looking at how this new era of remote sensing evolved, and then by delving into some aspects of the field that are supportive of our major thrust. Section 2.1 deals with photo interpretation and Section 2.2 with photogrammetry (the science of taking spatial measurements from photographs). Both topics are critical to understanding the origins of contemporary remote sensing. The interested reader should consider Estes and Thorley (1983), Wolf (1983), and Slama (1980) for a more complete technical treatment of these topics.

2.1 PHOTO INTERPRETATION

Since its inception, overhead photography has been (to use today's vernacular) a dual-use technology (i.e., it has commercial and military applications). Over the decades following Nadar's 1858 images, advances in the civilian use of aerial photography in the mapping and surveying sciences were adopted by the military. Similarly, technological advances in cameras and film for reconnaissance were rapidly employed in surveying, land cover mapping, and resource assessment. The field of photo interpretation evolved from the joint civilian and military need to learn how to extract more information from aerial images, particularly black and white images. In the early days, black and white images were all that was available. Even as color film became readily available, its spatial resolution, so critical to visual analysis, could not match the high resolution of the black and white films preferred by photo interpreters for most applications.

Estes et al. (1983) describe visual cues that have been identified as part of the psychovisual trigger mechanisms used by photo interpreters in extracting information from overhead images. These cues are interesting to consider since visual analysis of overhead images is still a major aspect of essentially all remote sensing applications. Furthermore, as we look at the more quantitative, machine-oriented methods of analysis discussed in later chapters, it is important to recognize what humans can do well and how they do it. This may, in some cases, help us to mimic what our visual system does with machine-based methods. It is also important to recognize what the human visual and cognitive system does not do well to ensure that, where possible, we develop machine-based methods to cover these limitations.

Figure 2.1 lists some of the cues that Estes et al. (1983) have identified as useful in photo interpretation, along with visual examples of each cue. As we review these cues, it is useful to imagine how you would use them individually or in combination to address a particular problem. Recognize that the utility of a cue depends on the area of application and the analyst's level of knowledge regarding the application area. It is also useful to think of how one might teach a machine (computer) to reach the same conclusion as a human observer. We will begin by briefly looking at the visual cues listed in Figure 2.1, recognizing that these are not exhaustive but merely indicative of keys that might be used by a photo interpreter.

Shape Shape is simply the geometric outline of an object. Often this simple outline carries a surprising amount of information about the nature or function of the object. Even with no additional information, we can tell that objects B and C are manmade, and A is unlikely to be manmade. Thus, we see how shape can provide important information, even with extremely little detail in the image.

Size Size may refer to the area of an object or to a single dimension such as the length of a road or airport runway. As soon as we see an image our eye-brain cognitive system automatically assigns a scale. This is not a conscience logical process, nor is the scale factor quantitative. We simply recognize one or two objects (e.g., car, house, lake, continent), and by knowing their size, we proportionately dimension other objects in the image and recognize the extent or coverage of the entire image. In most cases of earth observation this is an easy process because easily recognized objects are readily found. However, its importance is perhaps best recognized by examining an image of the moon or some landscape with which you are completely unfamiliar. In the example, the same squarish shape shown at D could be a tombstone, a car (E), or house (F).

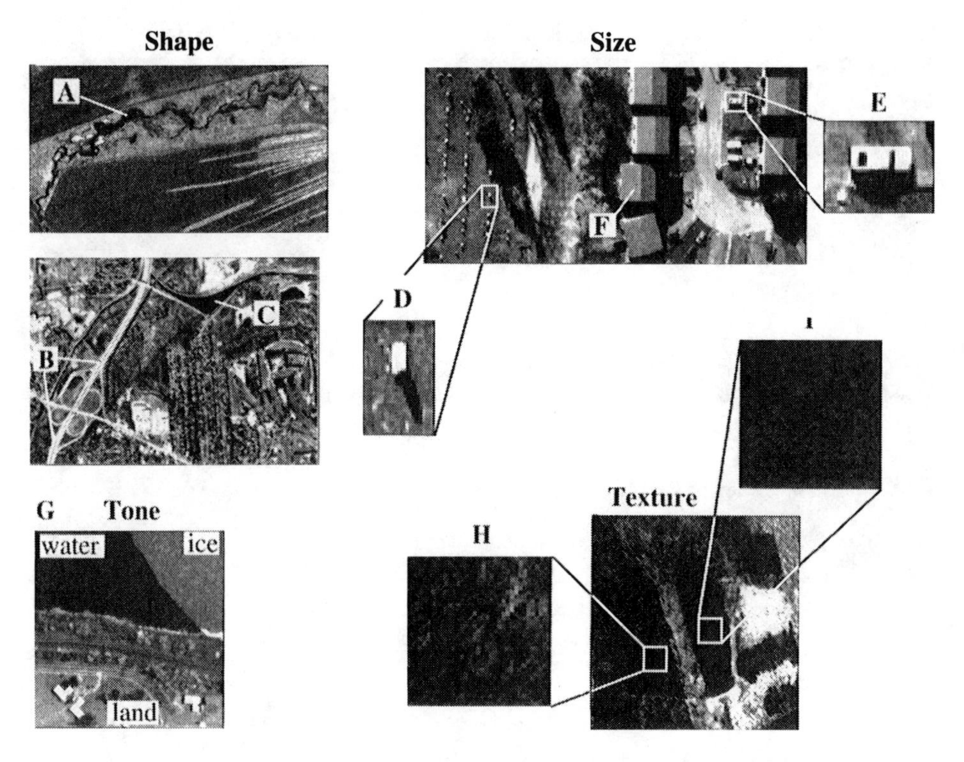

Figure 2.1 Cues used in photo interpretation.

Tone Tone is the brightness level in a monochrome image or the combination of brightnesses (i.e., color), in a color image. In the image at G, tone easily lets us differentiate water from ice. The visual system cannot easily quantify brightness, but as we discuss in Section 8.1, it relies heavily on tonal differences as cues. In multispectral images (color) the human's ability to quantify becomes even more restricted compared to the potential information content. This is one area where machines are much more adept than man and is the basis for much of the multispectral scene classification described in Section 7.2.

Texture Texture describes the structure of the variation in brightness within an object. Forest canopies and water in certain spectral bands have the same mean brightness but very different texture, which allows them to be easily distinguished (e.g., H is forest, I is water).

Pattern Patterns are shapes with identifiable geometric or periodic attributes. The patterns associated with a drive-in movie theater and a baseball diamond are shown as J and K. These are manmade patterns. However, both natural and manmade patterns are common and exist at all scales (e.g., corn rows and drainage patterns). It is important to realize that the process of recognizing the handful of edges associated with objects J and K as a drive-in and a ball diamond involves processing a great deal of cultural knowledge about your environment. This same recognition process would be much more difficult for someone who had never seen these structures before. Thus, the extraction of information from pattern data requires some form of prior learning process. This turns out to be true for both human and machine observers.

Shadow Shadows are often thought of in a negative sense by interpreters because it can be

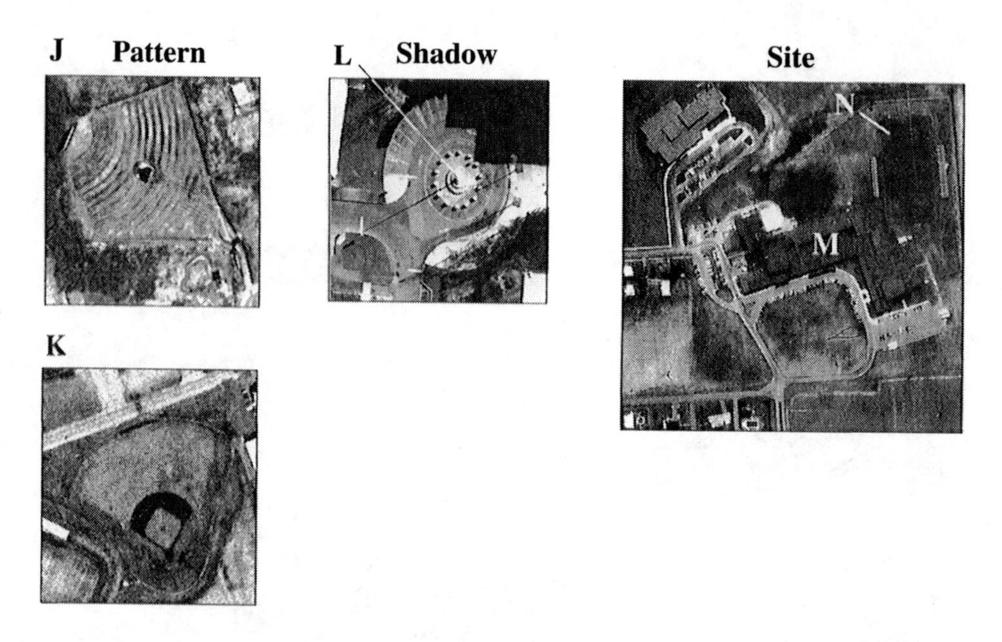

Figure 2.1 Cues used in photo interpretation (continued).

difficult to see objects in deep shadow. On the other hand, shadows can provide insight into the height profile of objects. The flagpole is much easier to identify by its shadow at L than in the actual image.

Site Site refers to the geographic location of a target or the location of one feature relative to another. Some authors differentiate these concepts with the geographic location being referred to as *site* and the relative location of features as *association*. Site information can help us determine that trees are more likely to be conifers because we know we are well up on a mountainside. It can also help us determine that if the building at M is co-located with the pattern at N then it is probably a high school. Use of site as a cue requires a great deal of location-specific geographic knowledge and/or cultural knowledge. For the building at M to become a high school, we first recognize it as a building, second we see it co-located with a pattern we recognize as a playing field (football). This combination of interrelated features, coupled with our cultural and geographic knowledge of construction patterns in North America, lets us conclude that the building is a high school. Implicit in this conclusion is a series of hypothesis tests, additional search procedures, and cues of which we may not even consciously be aware. For example, seeing the playing field we look for indications that this is a sports complex (i.e., stadium-style seating and large parking lots). We reject this hypothesis and suggest a school. Grammar schools are ruled out since they normally don't have football fields. A college is a possibility but would more likely be associated with a cluster of buildings. We reject that hypothesis as unlikely and conclude there is a high probability the building is a high school.

The ability of the human analyst to combine the types of cues listed above with a host of other data acquired over a lifetime makes human analysis a critical component in most remote sensing tasks. We currently have no machine-based equivalent with anywhere near the capabilities necessary to allow us to off-load this task to machines. On the other hand, human analysts require training and experience before they become proficient in many aspects of image analysis. They are also expensive, subject to fatigue, and have a difficult time when quantitative analysis is required or where many image inputs must be combined simultaneously. As a result, we have moved into an age of machine-assisted image analysis where we are attempting to move more of the burden of image analysis to machines. This lets human analysts do only tasks that the machines cannot.

Our ability to analyze and reason based on spatial patterns was what drove most of early remote sensing. The human visual-cognitive system has evolved to perform this task. Applying it to overhead images is a fairly straightforward transition requiring only adaptation for the perspective change. Indeed, much early remote sensing simply involved physically providing the analyst with a synoptic perspective. In reconnaissance, this meant that the pilot or an observer took notes or marked up a map to record visual observations of interest. This procedure is still used today in some applications. The forest service, for example, might have an observer in a light plane mark up a map showing forest type and condition prior to a spray program to control a gypsy moth infestation. These aerial sketch maps, as they are called, are one of the most direct forms of remote sensing. Because of the inherent potential for placement error, lack of detailed analysis, and ease of missing a critical phenomena, direct visual observation was, and is, normally augmented or replaced by photographic records. This allows for more detailed analysis, measurement, and mapping, and in most cases provides higher resolution through the use of increasingly sophisticated film, camera, and mechanical systems. In many cases, the photographic process is augmented by the visual system in terms of pointing the camera at the right target or locating the aircraft over the target. For much of the photography from space on manned missions, this is still the most common approach.

The analysis of aerial photos evolved along two separate but closely coupled paths. The civilian applications to mapping and surveying continued to grow and become more sophisticated and quantitative as the field of photogrammetry evolved. In parallel with a growing civilian interest, particularly in resource inventory and mapping (e.g., forest cover, soil type and condition, geologic patterns), the First and Second World Wars saw a tremendous push by the military for improved collection systems for reconnaissance applications.

Improvements in film and camera systems were quickly adapted for civilian applications. These applications continue to push the evolution of imaging technology in many areas. A good example of this evolution is the history of color infrared film, or camouflage detection (CD) film as it was originally known. As photography evolved, it became possible to produce color images by making multilayered photographic emulsions where each layer, through a combination of inherent spectral sensitivity and filtering, was effectively sensitive to a different spectral region (e.g., red, green, and blue). When the film is developed, different color dyes are chemically generated in the different layers to produce a full-color image. This basic process is illustrated in Figure 2.2 and treated in greater detail by James (1977). The location and concentration of the dye is a function of the exposure at any given point. The resultant color images are inherently registered, and by selecting the dyes appropriately, an image that approximates true color can be produced.

When camouflage nets or paints were imaged with panchromatic or color film, they looked enough like natural vegetation to make detection difficult. However, the camouflage of that era typically did not have a high reflectance in the near infrared. Vegetation does (cf. Fig. 2.3), and camouflage detection film was developed to take advantage of this difference in spectral reflectance. The sensitivities of the emulsion layers are shown in Figure 2.4(a). The *color IR film*, as it is commonly called, is normally flown with an external yellow (minus blue) filter to negate the blue sensitivity of the film. In order to display the images, dyes are coupled to the emulsion layers as shown in Figure 2.4(b). Objects such as vegetation with high infrared reflectance values appear red in the resultant image. Camouflage, which was designed to look like vegetation only in the visible, did not have a high infrared return. Therefore, it had more neutral tones on CD film, making it much easier to detect.

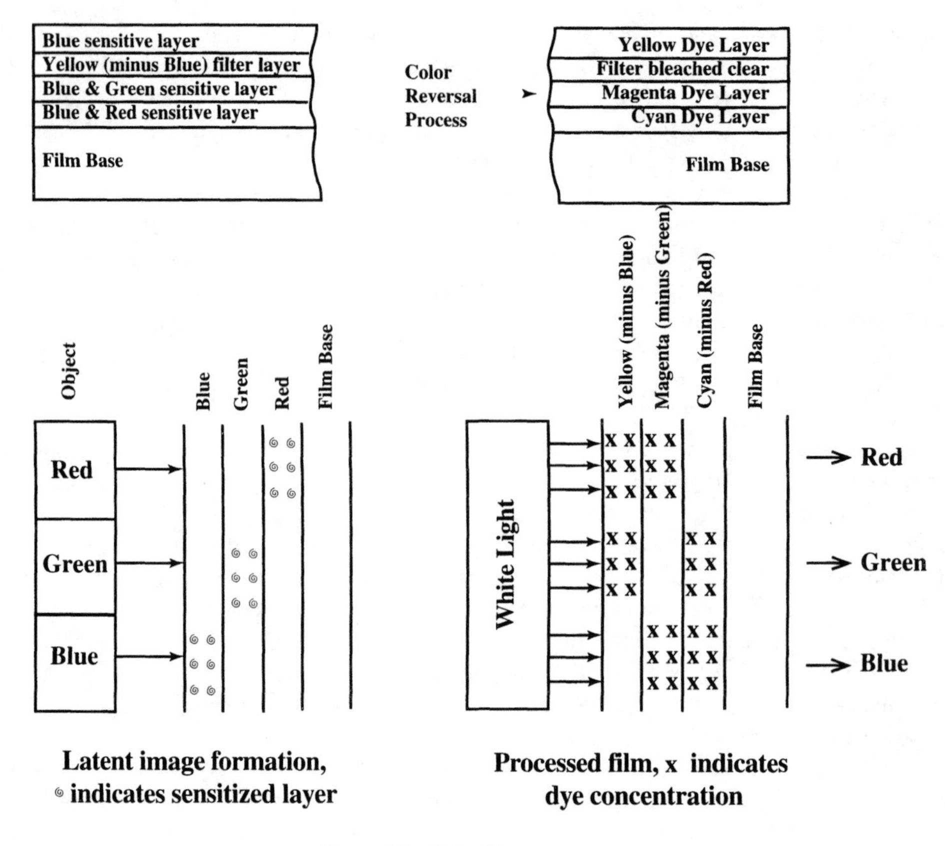

Figure 2.2 Color film concepts.

Over time camouflage became more sophisticated, and materials were developed that matched vegetation over a broader spectrum. Figure 2.5 shows an example of a color infrared image of camouflage. In the meantime, CD film was quickly discovered by the rest of the remote sensing community to have other uses. It is widely used in nearly all photographic studies of vegetation condition, because variations in vegetation density and health are manifest as changes in color tones in the color IR images (cf. Sec. 5.1). Figure 2.6 shows an example of how color infrared images can be used to delineate vegetation condition.

The interest in imaging in multiple spectral bands grew as scientists discovered that changes in the condition or makeup of soils, water, rock formations, and vegetation were often manifested in subtle changes in the reflectance properties in certain spectral bands or by the relative changes between multiple spectral bands. In order to observe these spectral reflectance phenomena, referred to as *spectral signatures*, two approaches were pursued. In some cases, specialized multilayer films were developed. Sprechter et al. (1973) describe a two-layer color film designed for water penetration to aid in the study of water depth and bottom condition in shallow waters. However, this approach is limited to a few spectral bands, and the cost and technical complexity of developing specialized films limits this option to high-volume applications. An alternative, and much more commonly used, approach was found in the form of multilens or multicamera systems. Here spectral filters are used in combination with black-and-white film types to allow the simultaneous acquisition of many images of the same scene in different spectral bands. Figure 2.7 shows an ITEK six-lens system and part of an image set acquired with the camera system. The advantage of this approach over multilayer films is the large number of spectral bands that can be acquired and

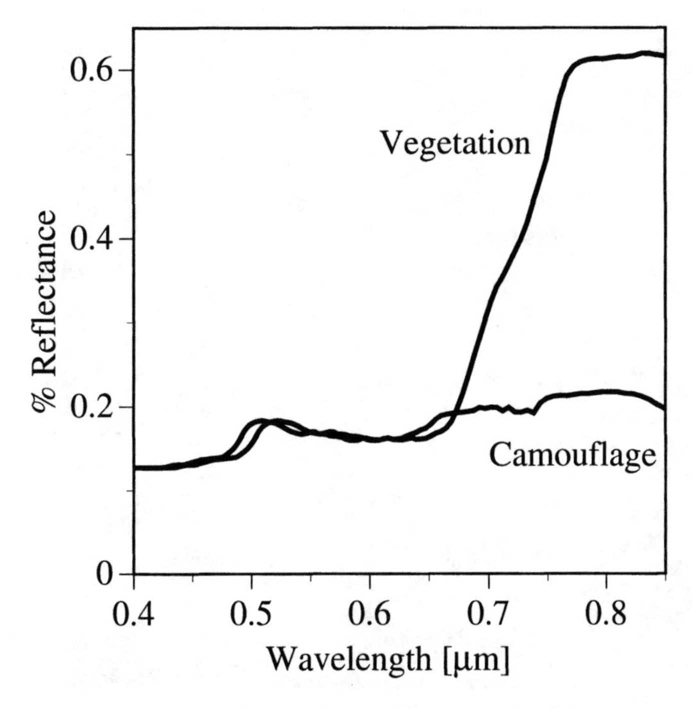

Figure 2.3 Spectral reflectance curves for vegetation and first-generation camouflage material.

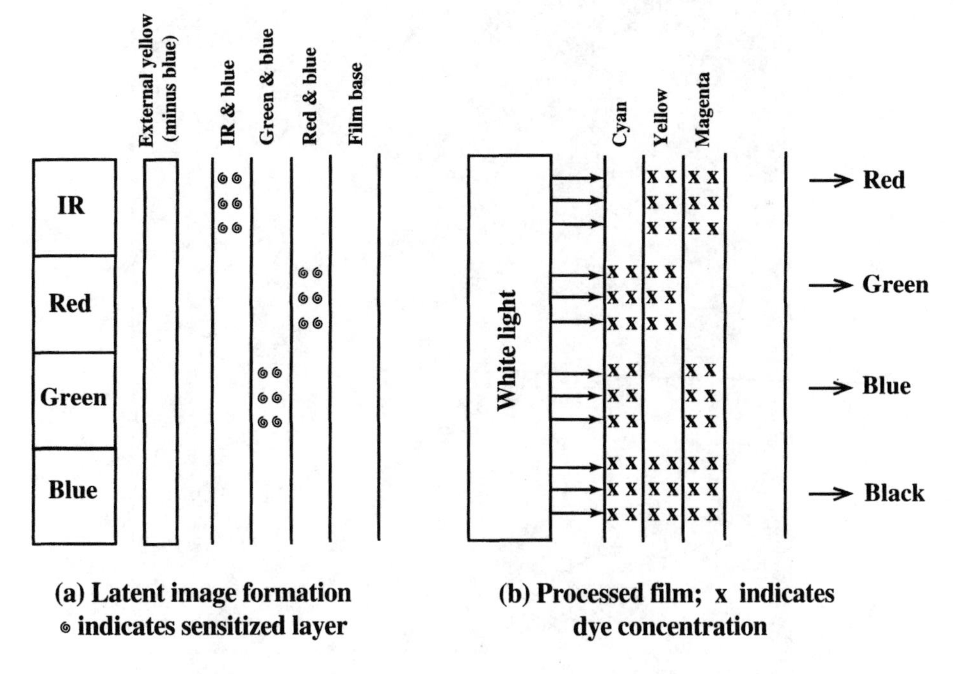

Figure 2.4 Spectral sensitivity of color infrared film and dye coupling concepts. (Spectral sensitivity curves are shown in Figure 5.5.)

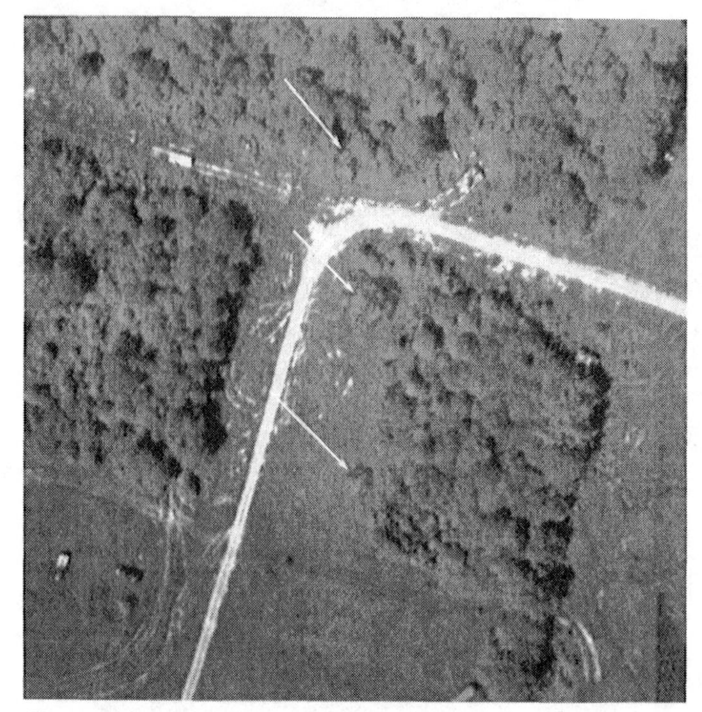

Figure 2.5 B/W infrared image of camouflaged objects. See color plate 2.5.

Figure 2.6 B/W photo showing vegetation condition. A gypsy moth infestation has defoliated most of the trees in the image. “Cosmetic” spray programs along the roadways have prevented severe defoliation in these areas. See color plate 2.6. (Image courtesy of RIT's DIRS laboratory.)

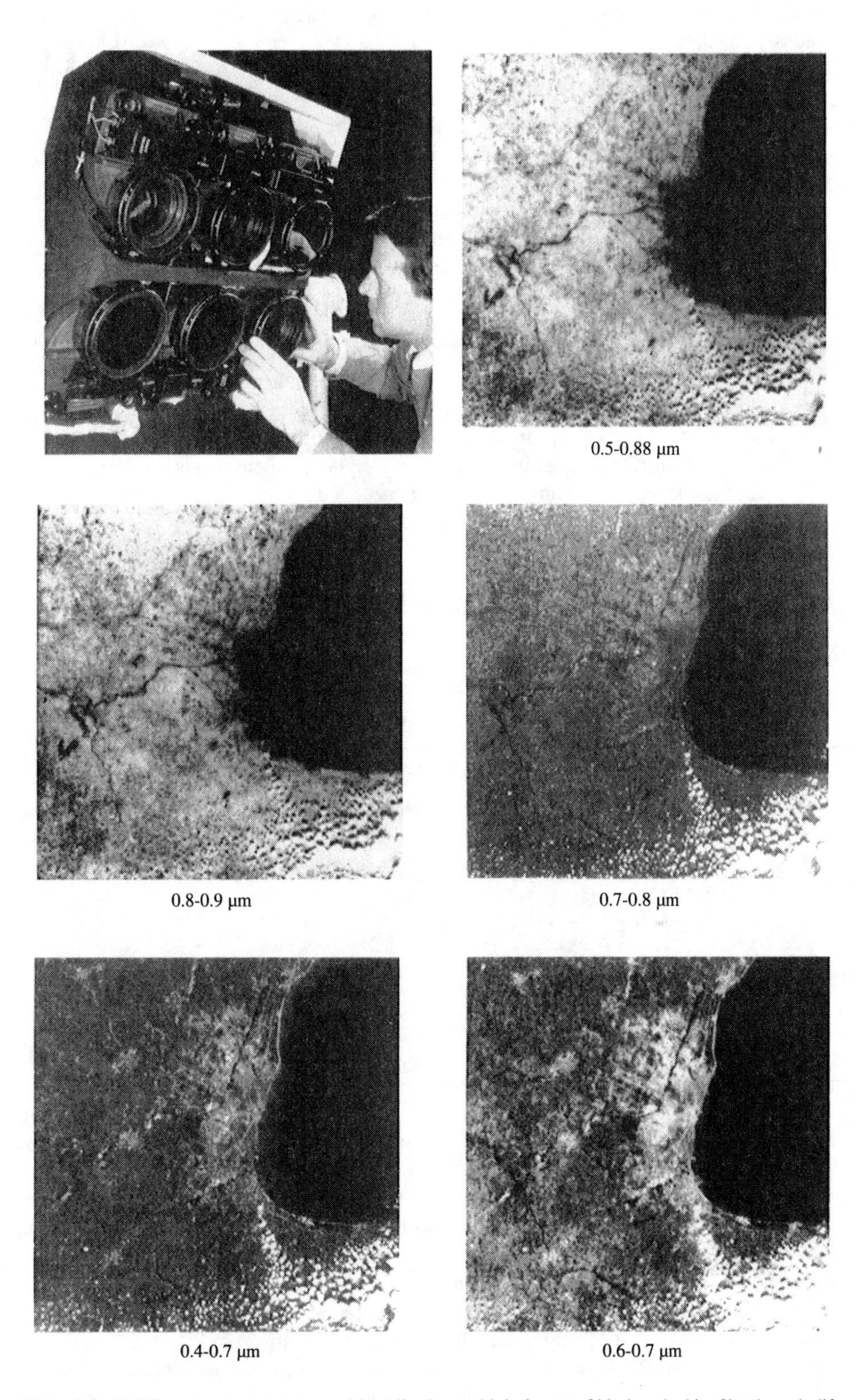

Figure 2.7 Multilens (camera) systems used in collecting multiple frames of black and white film through different spectral filters. (Images courtesy of Itek Corporation.)

the ability to control the spectral region sampled through careful selection of film-filter combinations. The disadvantages of this approach are the increased size and cost of the collection system coupled with the problem that the spectral band images are no longer inherently registered as they are with color film. In the early multiband photographic systems, this problem was solved by projecting the images in groups of three onto a common plane using various forms of optical recombination. The images were then shifted and rotated until they were in visual alignment. By projecting each of three images through a red, green, or blue filter, the spectral information in the resulting color image could be interpreted. With today's technology, it would be more common to digitize all the images into a computer compatible form (cf. Sec. 5.1) and use digital image registration and resampling techniques (cf. Sec. 8.3).

The growth of photographic science and technology, coupled with the synoptic perspective provided by remote sensing, allowed us to look at the earth's surface in new ways. These new perspectives, and hitherto unseen spectral and spatial patterns, revealed to scientists a host of new applications. These applications include archeology, meteorology, marine science, water resources assessment, land use analysis, geology, soils science, civil engineering, agriculture, forestry, range management, and wildlife biology (cf. Estes and Thorley (1983). The application of photographic and electro-optical remote sensing to these various fields not only spurred the rapid advance in image acquisition techniques but also stimulated the evolving field of quantitative imagery analysis.

2.2 QUANTITATIVE ANALYSIS OF AIR PHOTOS

The quantitative analysis of remotely sensed data is normally divided into at least two parts. The first, photogrammetry, is concerned with using photographic images to make measurements of the size, height, and location of objects or land forms. As such, it includes the science of mapping the topography of the earth's surface and of locating and measuring the dimensions of objects on the surface. The second part is based on radiometric analysis of the images. It is this radiometric analysis that will occupy the bulk of our attention in later chapters. However, in this chapter we want to lay the foundation for the latter treatment by introducing some basic photogrammetry concepts, as well as the concept of using a camera as a radiometer.

2.2.1 Photogrammetry

To keep our treatment relatively simple, we are going to restrict our consideration of photogrammetry to measurements from conventional, vertical (nadir) viewing photographic systems. However, the treatment is easily extended to more complex EO systems. Furthermore, we are only going to introduce the most rudimentary photogrammetric concepts that are necessary for general remote sensing purposes. A more complete treatment can be found in Wolf (1983) and Slama (1980). Our interest in photogrammetry parallels that of early photogrammetists who wanted to know, (a) how to convert a distance measurement on a photo to a distance on the ground, (b) how photo coordinates could be related to ground or map coordinates, (c) how height could be determined from photo coordinates, and (d) how air photos could be used to make topographic maps.

To begin, we need to introduce the terminology and a set of parametric conventions to be used in our discussion of photogrammetry. These are illustrated in Figure 2.8. For simplicity we will assume that we are only dealing with vertical images, all measurements are from positive photographic prints, and that no geometric distortions exist due to imperfections in the camera system or spatial instabilities in the film. Note: All these restrictions can be relaxed, but the treatment is beyond our scope. The camera location when the image is

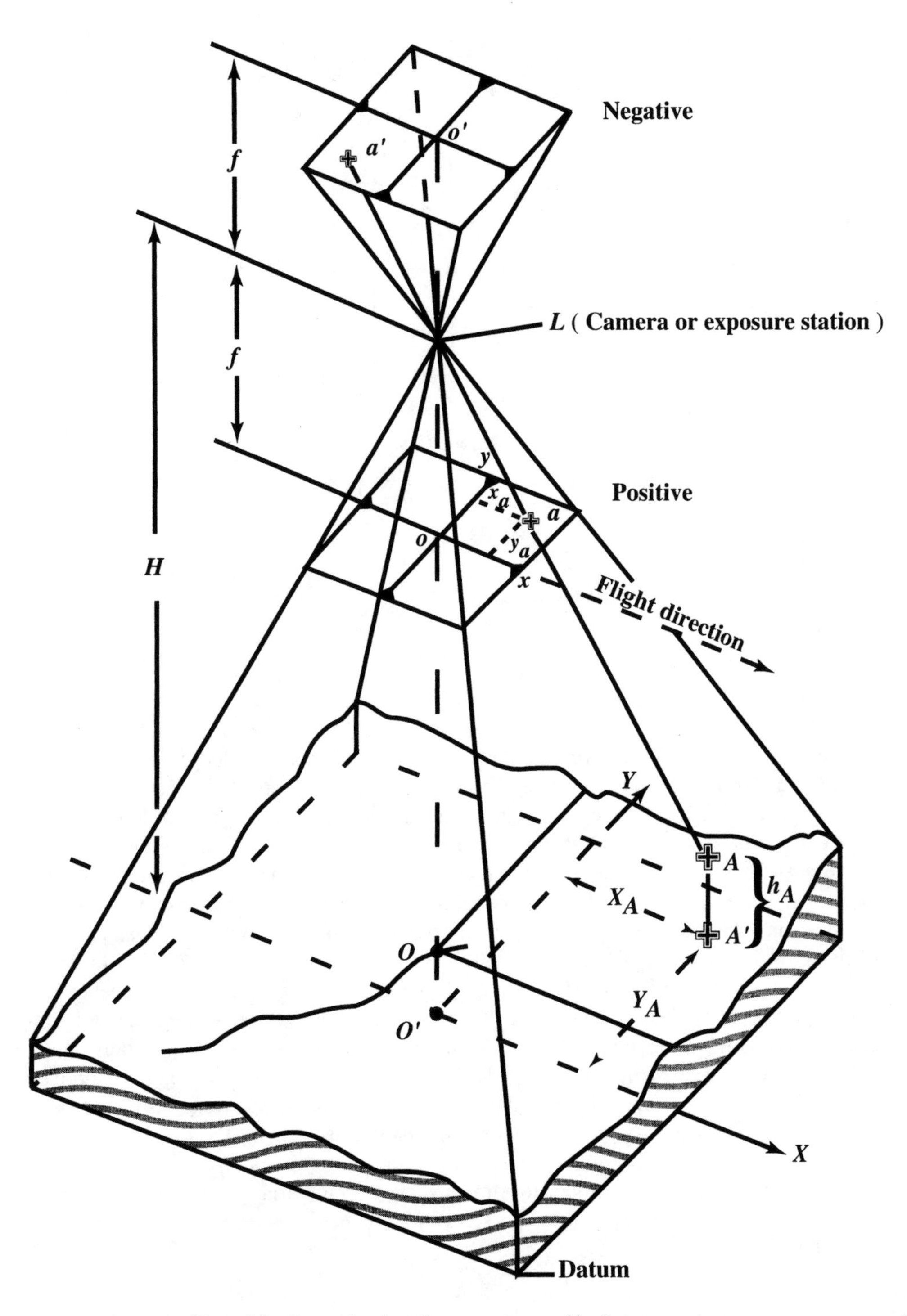

Figure 2.8 Geometric orientation parameters used in photogrammetry.

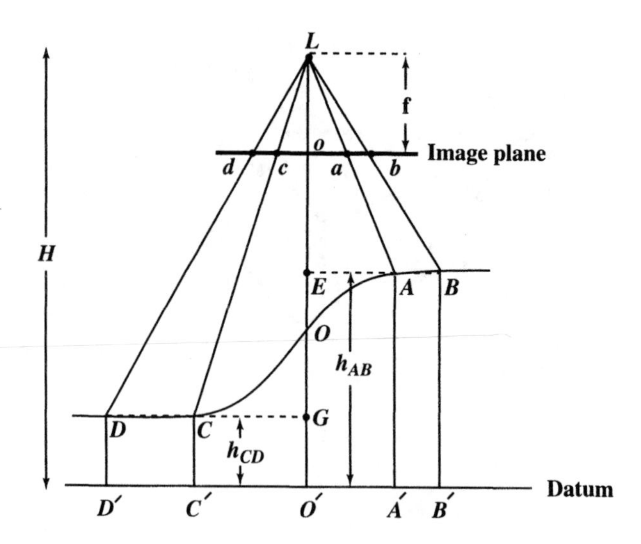

Scale at height h_{AB} is $\quad S_{AB} = \dfrac{\overline{ab}}{\overline{AB}} = \dfrac{\overline{La}}{\overline{LA}} = \dfrac{\overline{Lo}}{\overline{LE}} = \dfrac{f}{H - h_{AB}}$

Scale at height h_{CD} is $\quad S_{CD} = \dfrac{\overline{cd}}{\overline{CD}} = \dfrac{\overline{Lc}}{\overline{LC}} = \dfrac{\overline{Lo}}{\overline{LG}} = \dfrac{f}{H - h_{CD}}$

Figure 2.9 Scale determination.

acquired is referred to as the *exposure station* (L). The camera focal length is f. The positive image is considered to be projected back through the exposure station and located in a plane a distance f below the exposure station (i.e., it is a simple mirror image of the negative at the exact scale such as would be obtained by contact printing). The datum refers to a reference plane to which points on the terrain are projected for mapping or measurement purposes [e.g., mean sea level (MSL) is a common datum plane]. By convention, the plus x dimension in the image is along the flight path, with the origin at the optical axis, and the plus y axis is 90° counterclockwise from the plus x axis. Lowercase letters are used to define image locations on the positive print (primed on the negative), with the uppercase letter representing the corresponding point in the scene (primed uppercase values represent the projection of the point to the datum plane). The points on the optical axis are generally denoted by the letter o and referred to as *principal points*; therefore, the optical axis would pass through the points o', L, o, O, and O'. An arbitrary scene coordinate system is then defined by projecting the image's x and y axes onto the datum plane, with the plus Z axis being straight up.

An analysis based on the proportionality of sides of similar triangles in Figure 2.9 shows how image measurements are related to scene or map measurements by the local image scale(s), i.e.,

$$s = \frac{f}{H - h} \tag{2.1}$$

where s is the image scale, f is the focal length, H is flying height above datum, and h is the height of the measured point above datum. Clearly, the scale of the image will vary with elevation, and this variation will be greatly reduced in high altitude or satellite images.

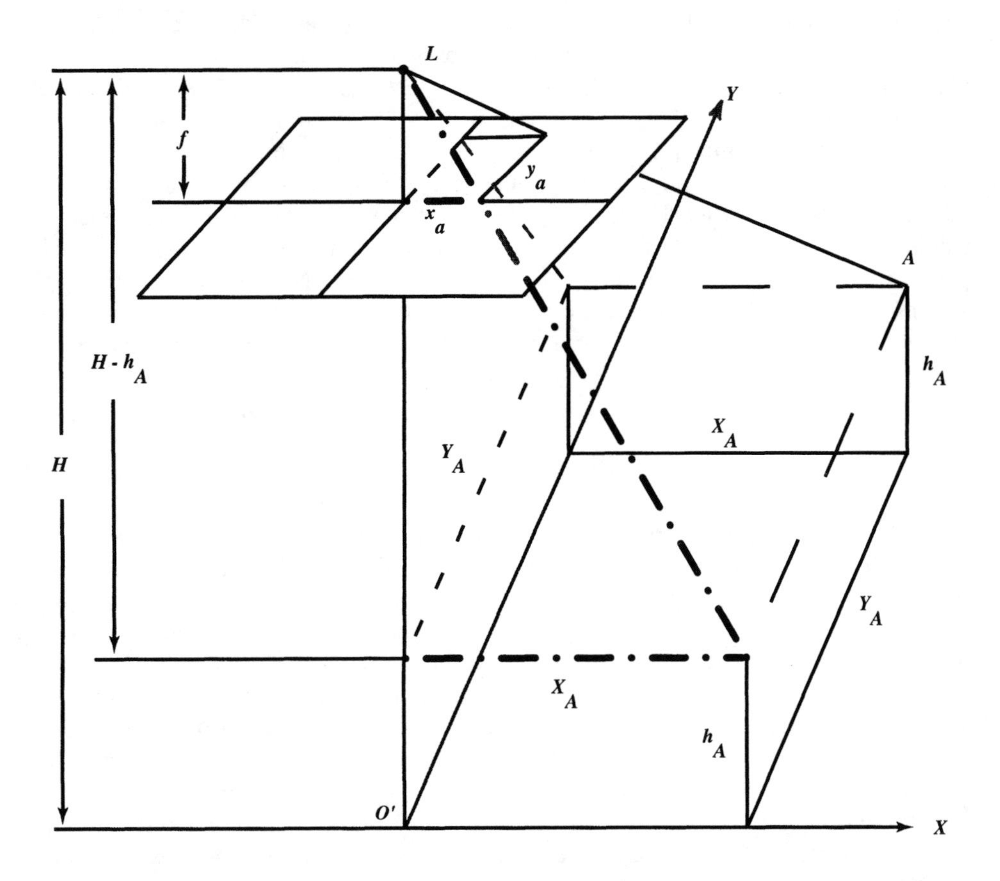

$$\frac{x_a}{X_A} = \frac{f}{H - h_A} \quad \text{from similar triangle } (\text{-·-·-·-·-})$$

$$\frac{y_a}{Y_A} = \frac{f}{H - h_A} \quad \text{from similar triangle } (\text{- - - - -})$$

Figure 2.10 Geometric construction for derivation of ground coordinate transforms.

Figure 2.9 also makes it clear that measurements (distances, angles) made on the image must first be corrected for elevation variation before they are meaningful. In many cases this requires an ability to convert image coordinates into ground coordinates.

The equations of transformation can again be easily derived using the proportionality feature of sides of similar triangles based on the geometric construction shown in Figure 2.10. The resulting transformation equations are:

$$X_A = \frac{H - h_A}{f} x_a \text{ and}$$

$$Y_A = \frac{H - h_A}{f} y_a \qquad (2.2)$$

where X_A and Y_A are the geometric coordinates in the ground coordinate system for point a in the image with image coordinates x_a, y_a. Using Eq. (2.2) we could transform any data in the image to a ground coordinate plane and make quantitative dimensional and angular measurements in the ground coordinate plane. In addition, if the image contained two or more points whose location and height were known in a global ground coordinate system, a simple geometric transform from the arbitrary *X*, *Y* space to a global reference system can easily be derived. The only drawback to this process is that Eqs. (2.1) and (2.2) require the elevation of the target (h_A)

There are two common methods used in photogrammetry for determining elevation. The first is normally used for determining the height of vertical targets such as towers and buildings relative to the local ground plane. This method uses the relief displacement that results when the top of an object viewed off nadir is imaged farther away from the principal point than the bottom of the object, as shown in Figure 2.11. Note that for a photographic system, the relief displacement is radially outward from the center of the format and increases with object height and radial distance. Once again, using cascaded similar triangles, we can derive an expression for the height of a target as:

$$h_A = \frac{Hd}{r} \tag{2.3}$$

where h_A is the height of the target, d is the relief displacement as measured on the photo, H is the flying height above the base of the target, and r is the radial distance from the principal point to the top of the target. Note that many EO systems that don't look fore and aft of nadir generate images with relief displacement only outward from the center line of the image along a line perpendicular to the flight line. In those cases where the side of an object can be seen due to relief displacement, the height can be determined from two simple measurements if the flying height above local datum is known. Regrettably, this method is not applicable to objects near the center of the format or for finding the height of sloped surfaces such as terrain.

For mapping height throughout a scene, images of the scene from two different perspectives are required. Such stereo pairs have long formed the basis for three-dimensional viewing using optical systems designed to effectively place one eye at one camera station and the other at a second. As seen in Figure 2.12, height variations result in differential displacement distances along the x direction for objects in the scene which the visual system interprets as height changes in the same way our normal stereoscopic vision does. By simply forcing one eye to see one image and the other a second, stereo perception can be achieved. Figure 2.13 is a stereo pair that can be viewed using a simple pocket stereo viewer. One can take advantage of this phenomenon by measuring the x axis displacement of scene elements in a stereo pair. This displacement is known as *parallax* (p). The geometric analysis is shown in Figure 2.14. First the location of the first images principle point is found in image 2 (and vice versa) to determine the local line of flight to be used as a common x axis. Next the point of interest (a, a') is located in both images, with the prime used to designate objects in image 2. We define the air base (B) to be the distance between the exposure stations (L and L'). We identify two similar triangles by projecting a, a', and A onto the X, Z plane to form LA_xL' and shifting $L'o'$ to coincide with Lo to form a'_xLa_x. Then we have:

$$\frac{p_a}{f} = \frac{B}{H - h_A} \tag{2.4}$$

where $p_a = x_a - x'_a$ is the parallax. (Note: x'_a would be negative in the example shown.) Rearranging yields an equation for height determination for an arbitrary point A.

$$h_A = H - \frac{Bf}{p_a} \tag{2.5}$$

Equation (2.4) can be rearranged and substituted into Eq. (2.2) to also yield expressions for

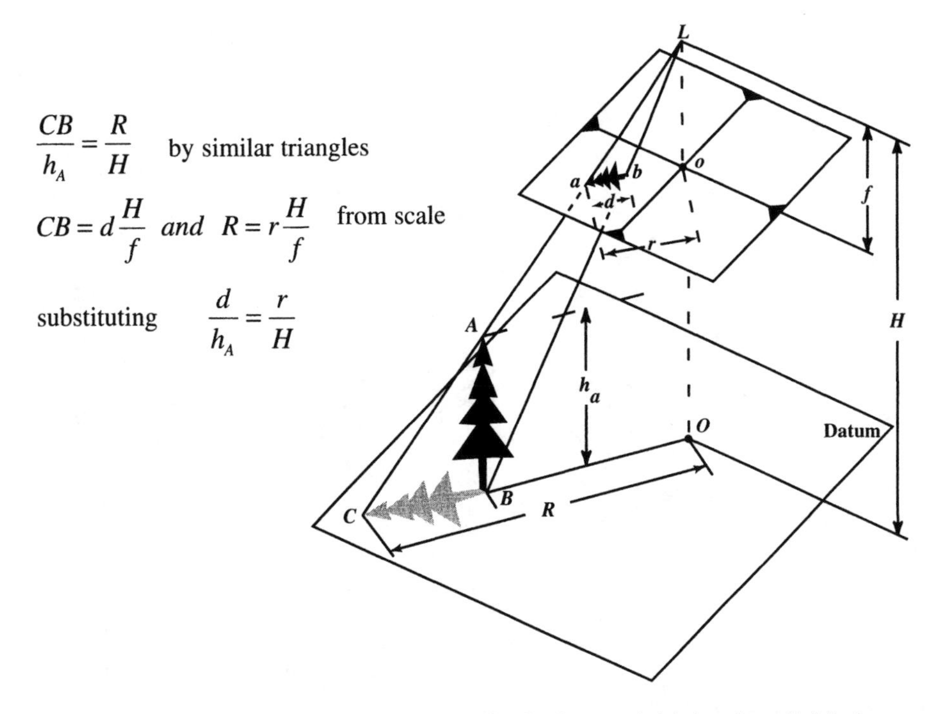

(a) Geometric construction used to derive an expression for the target height based on relief displacement.

(b) Portion of an image showing radial relief displacement.

Figure 2.11 Relief displacement concepts. (Image courtesy of GDE.)

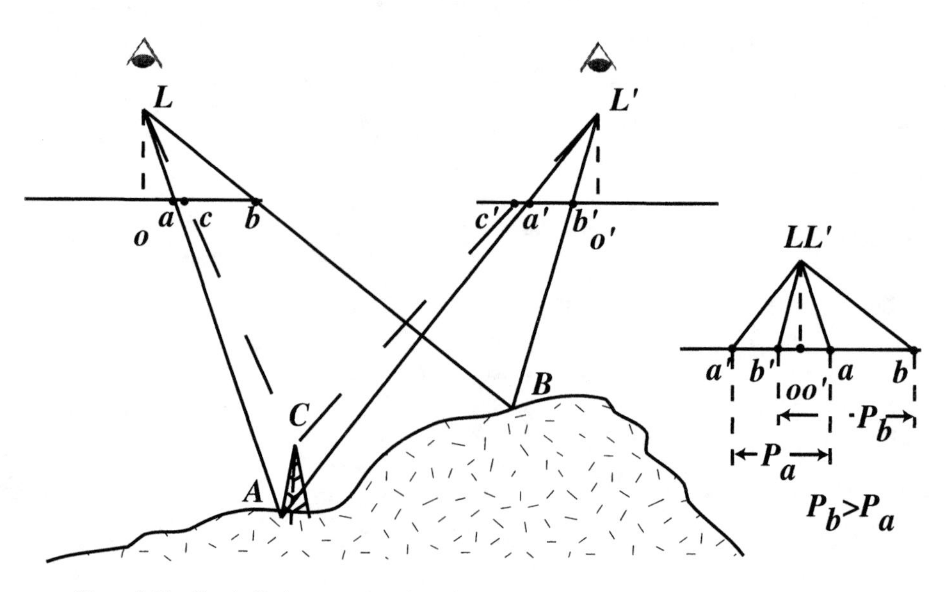

Figure 2.12 X axis displacement (parallax) in stereo pair images is a function of object height.

the ground coordinates in terms of parallax, i.e.,

$$X_A = \frac{Bx_a}{p_a} \tag{2.6}$$

$$Y_A = \frac{By_a}{p_a} \tag{2.7}$$

The parallax measurements can be made by direct measurements from the photos with a ruler or more commonly with any of a host of devices designed to simplify the job of mapping height throughout an image. Many of these employ the human visual system's stereoscopic perception to allow height determination even over uniform terrain where no distinct features exist for reference. These devices, called *stereo plotters*, can be used to compile topographic maps or digital elevation models (DEM) of the scene. A DEM is a representation of the terrain in a grid form where every grid center represents an *x,y* location in the datum plane having a height associated with it, as shown in Figure 2.15. The elevation, slope, and orientation data that can be derived from stereo image pairs are extremely useful in all the obvious engineering and hydrological applications of remote sensing, as well as in many more subtle radiometric applications that will be introduced in later chapters.

2.2.2 Camera as a Radiometer

As the field of remote sensing evolved in the middle of the 20th century, interest grew in trying to quantify some of the information that appeared to be available in the air photos. For example, an image analyst could see that one water body was more turbid than another, but not how much more, or that one soil condition was different than another, but not what caused the difference or what the magnitude of the difference might be. Laboratory and field studies showed that many questions about the condition of objects were related to the reflectance of the object. Since the optical density recorded on the film varied with the radiometric

Figure 2.13 Black-and-white stereo pair reproduced from an early color film. This stereo pair was used to assess bomb damage during World War II. (Images courtesy of Eastman Kodak.)

reflectance of the object, the logical question became how to quantify the relationship between the two, or how to employ the camera as a radiometer. Piech and Walker (1971, 1974) demonstrated that water turbidity could be related quantitatively to film density and that soil texture and moisture could be studied based on the optical density values recorded on color film transparencies. The details of this type of analysis are developed in Chapters 3, 4, 5, and 6. For the present, we simply postulate that the required functional relationships exist, i.e.,

$$r = f'(Cond.) \tag{2.8}$$

$$L = f''(r) \tag{2.9}$$

$$H = f'''\ (L) \tag{2.10}$$

$$D = f''''(H) \tag{2.11}$$

$$D = f(\text{Cond.}) \tag{2.12}$$

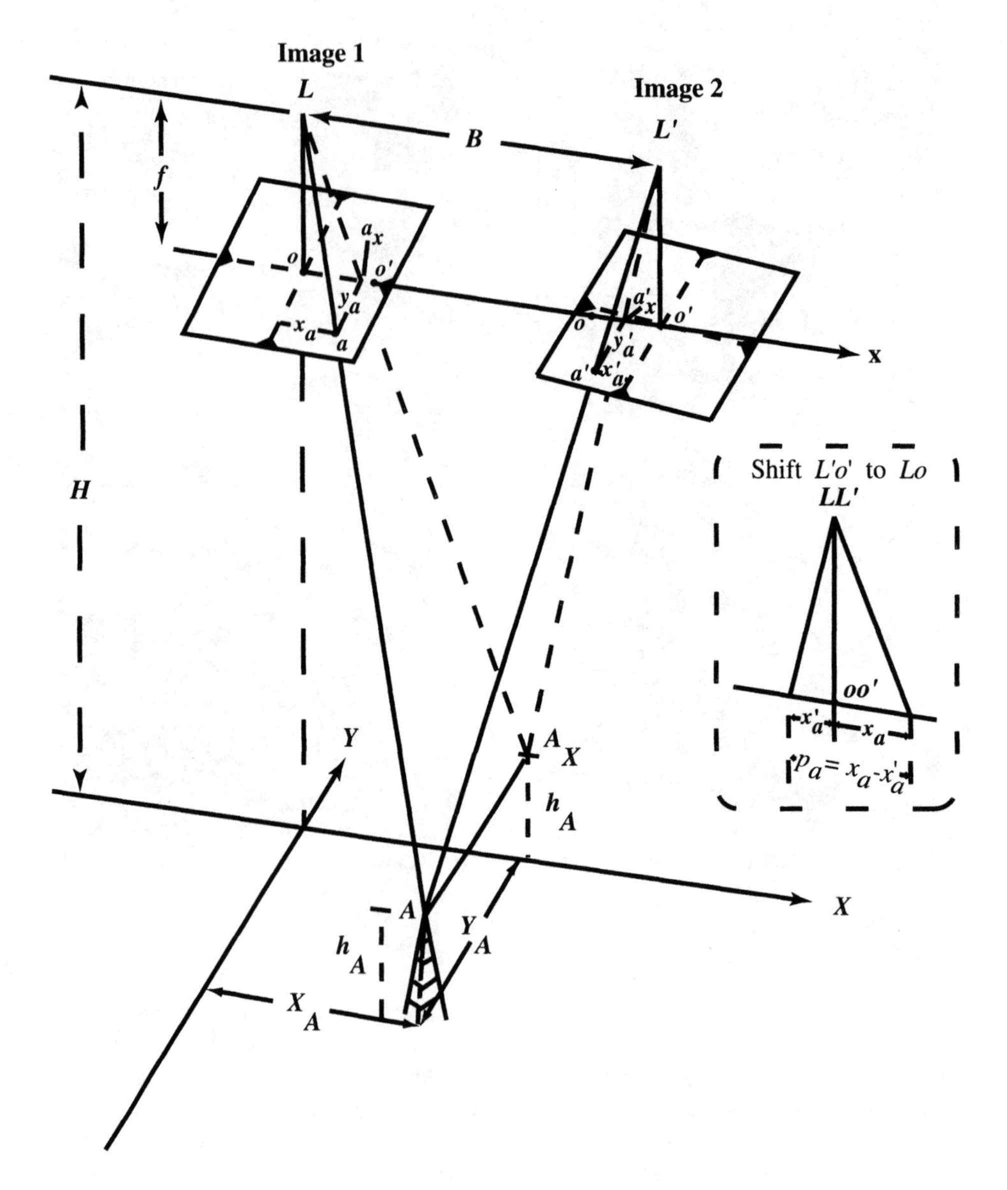

Figure 2.14 Geometric construction for derivation of the parallax equation for height determination.

where Cond. is a parameter representing the type or condition of the target (e.g., turbidity, soil moisture, vegetation health), r is the reflectance of the target (possibly in several spectral regions), L is the radiance reaching the sensor, H is the exposure on the film, D is the density (- log of the transmission) recorded on a photographic transparency, and f, f', etc., represent the functional relationships between the indicated parameters.

If we can derive an expression for each of the primed functions in Eqs. (2.8) through (2.11), or for the aggregate function f in Eq. (2.12), then it is possible to begin to think of the camera as a radiometer. In early applications, this was accomplished empirically by making many field observations of the parameter of interest (Cond.) and film density readings at image locations corresponding to where the "ground truth" observations were acquired.

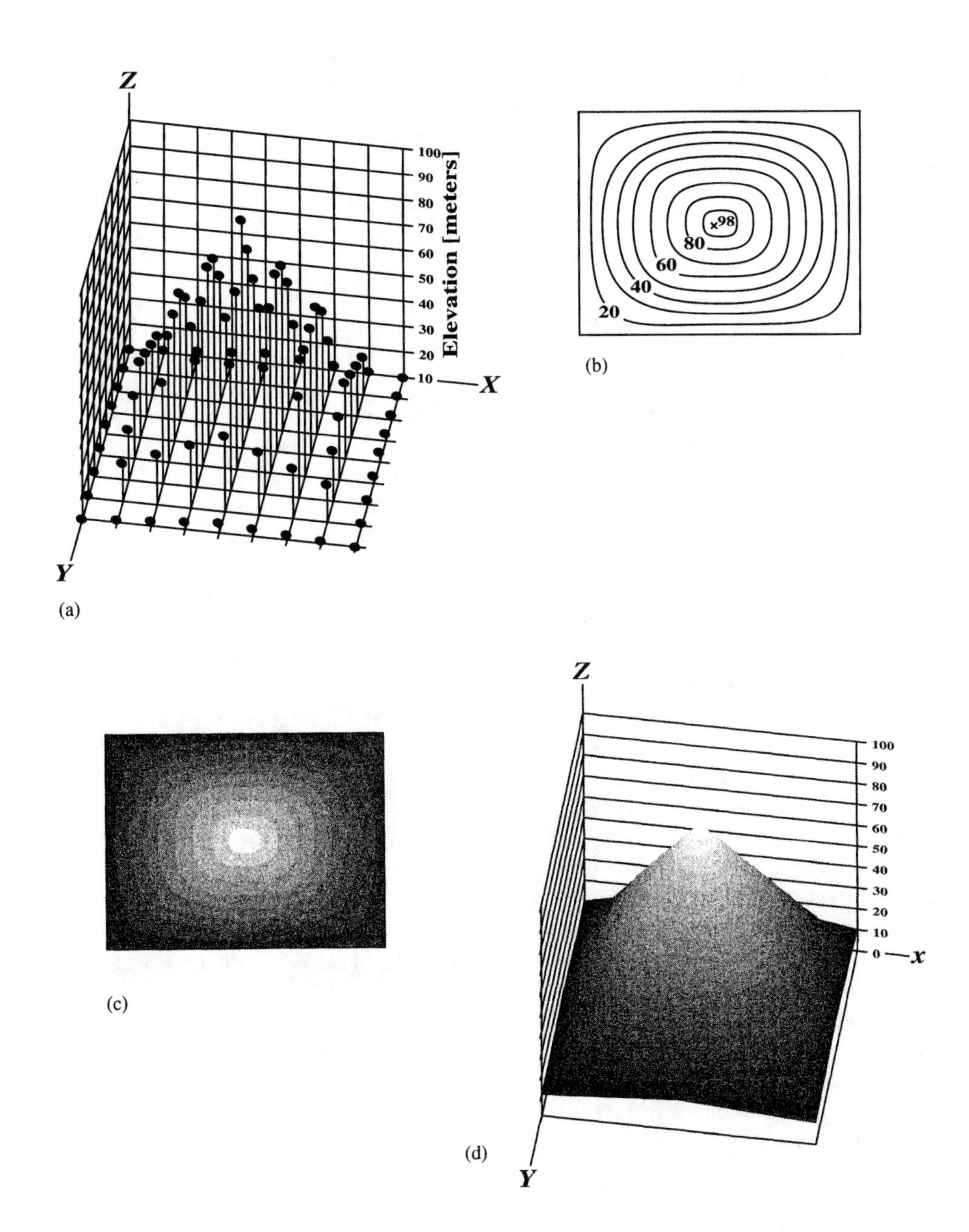

Figure 2.15 Digital elevation model concept. (a) Plot of elevation heights (points), (b) countour plot, (c) gray level coding of elevation, and (d) combination of (a) and (c) to allow surface visualization.

Ground truth is a term used to describe any surface measurements used to assist in the analysis of remotely sensed data. In this case, the ground truth would consist of an extensive set of measurements to quantify the condition parameter (e.g., soil moisture). A curve fit between the two data sets would generate an approximation for the functional form of f in Eq. (2.12). The resulting relationships were invariably nonlinear, required extensive ground truth to develop, required near simultaneous ground truth and overflight data, were not robust, and had errors associated with not knowing the nature of the cascaded functions (f', f'', etc.) [cf. Lillesand et al. (1973)]. Nevertheless, early successes with this "empirical" approach led to efforts to derive theoretical expressions for the various functions and to calibration methods aimed at improving the robustness, accuracy, and utility of quantitative radiometric image analysis. The bulk of our attention in the rest of this volume will be directed at understanding the tools that evolved in response to the need to use the camera as a radiometer.

2.3 EVOLUTION OF EO SYSTEMS

From the late 1950's onward, electro-optical (EO) systems have competed with and/or complemented photographic systems in most applications. Electro-optical systems range from the simple vidicon and solid-state video cameras (familiar to us through the consumer market) to very exotic systems with either complex opto-mechanical systems for image formation or large arrays of tiny electronic detectors. Film cameras have a distinct advantage in that they provide a low-cost, high spatial resolution solution to many problems. However, as imaging acquisition requirements became more demanding, film solutions were not always available. For example, film spectral response could not be pushed much beyond 1 μm. Imaging in the thermal infrared spectral region required, and became a driving force behind, the development of EO sensors. Another force behind the evolution of EO systems was the growing interest in using spectral reflectance variations to help identify and characterize earth surface features. As discussed in Section 2.1, to do this simultaneously in many spectral bands is very difficult with film. These problems can be overcome with EO systems, although only at the cost of geometric integrity and spatial resolution (particularly in early systems). This ability to "see" things the camera couldn't by using a wider range of the electromagnetic spectrum and also to see in many relatively narrow spectral regions, led to the development of a host of airborne EO systems. We will consider these systems in more detail in Chapter 5. The utility of these systems can be seen in the thermal infrared image of a cooling water discharge from a power plant shown in Figure 2.16. The brightness of this image is proportional to temperature. Images from EO IR systems such as this were used to map the magnitude and extent of the environmental impact of cooling waters on the receiving waters and to monitor compliance with environmental regulations. These images let us study the world in ways that had never before been possible. Figure 2.17, for example, shows several images from an airborne EO system representing different regions of the EM spectrum. Note both the dramatic and the more subtle changes between materials in different wavelength regions. In Chapter 7 we will discuss how these variations in the observed spectral brightness can be used to map land cover type and the condition of selected targets (e.g., crop stress, water quality).

2.4 SPACE-BASED EO SYSTEMS

Early photographs from space clearly demonstrated the value of the synoptic perspective from space. This interest in space-based systems helped to spur the evolution of EO systems because of the operational problems associated with the recovery of film from space. The inherent electronic nature of the EO images solved the problem, since the images could be transmitted to ground receiving stations.

In order to understand the performance of some of the space-based sensors described in Chapter 5, we need to understand a little about satellite orbits. Simple scale effects dictate that most systems for earth observation use low earth orbits (LEO) to maintain acceptable spatial resolutions. Furthermore, to facilitate comparable image scales anywhere on the earth, near circular orbits are usually preferred. Orbit selection has to take into account the satellite motion, earth rotation, and the relative sun-earth-satellite orientation angles. In

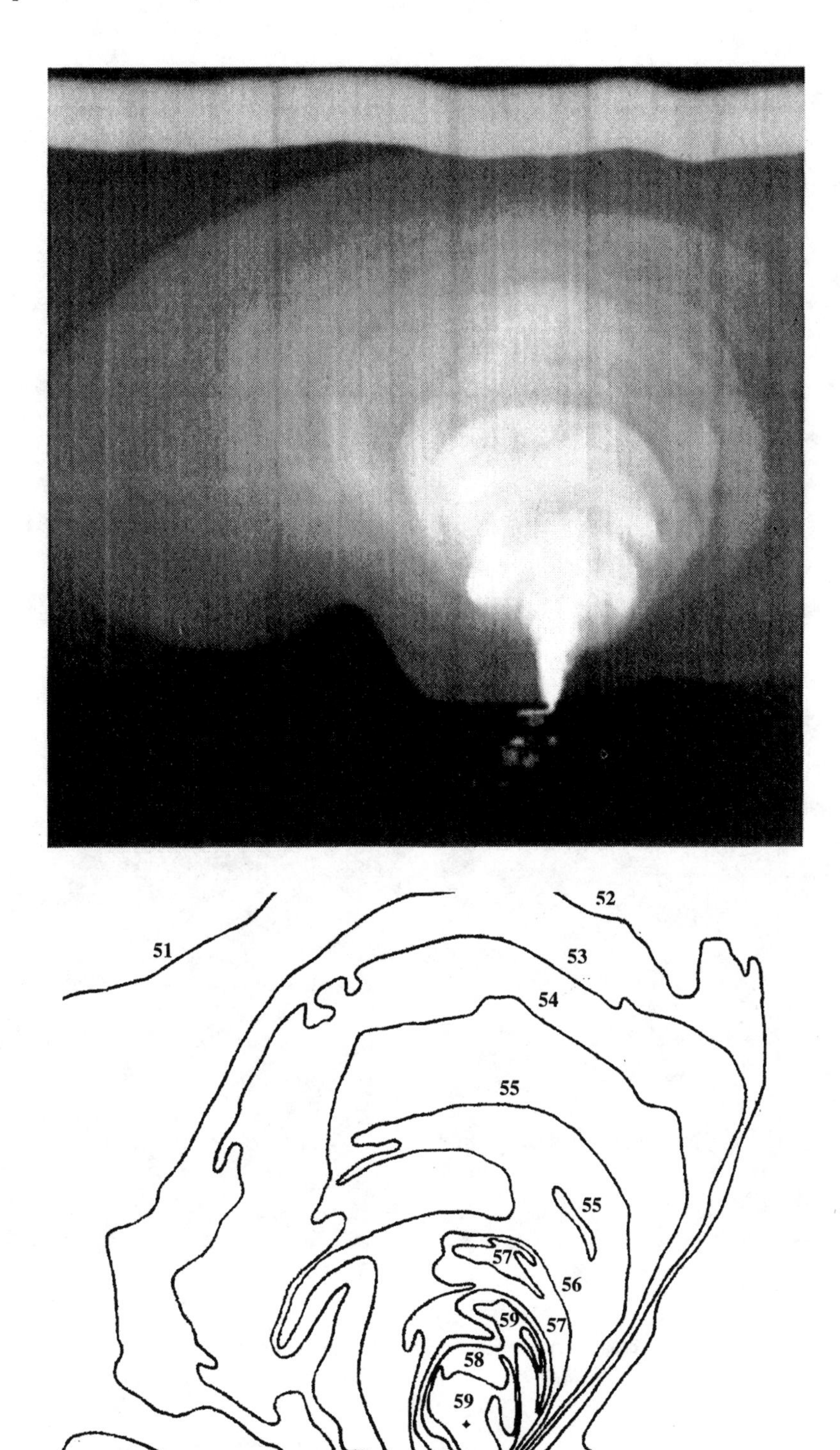

Figure 2.16 Thermal infrared image of a power plant cooling water discharge (white is hot) and an isothermal image derived from the map (degrees Farenheit). (Image courtesy of RIT's DIRS laboratory.)

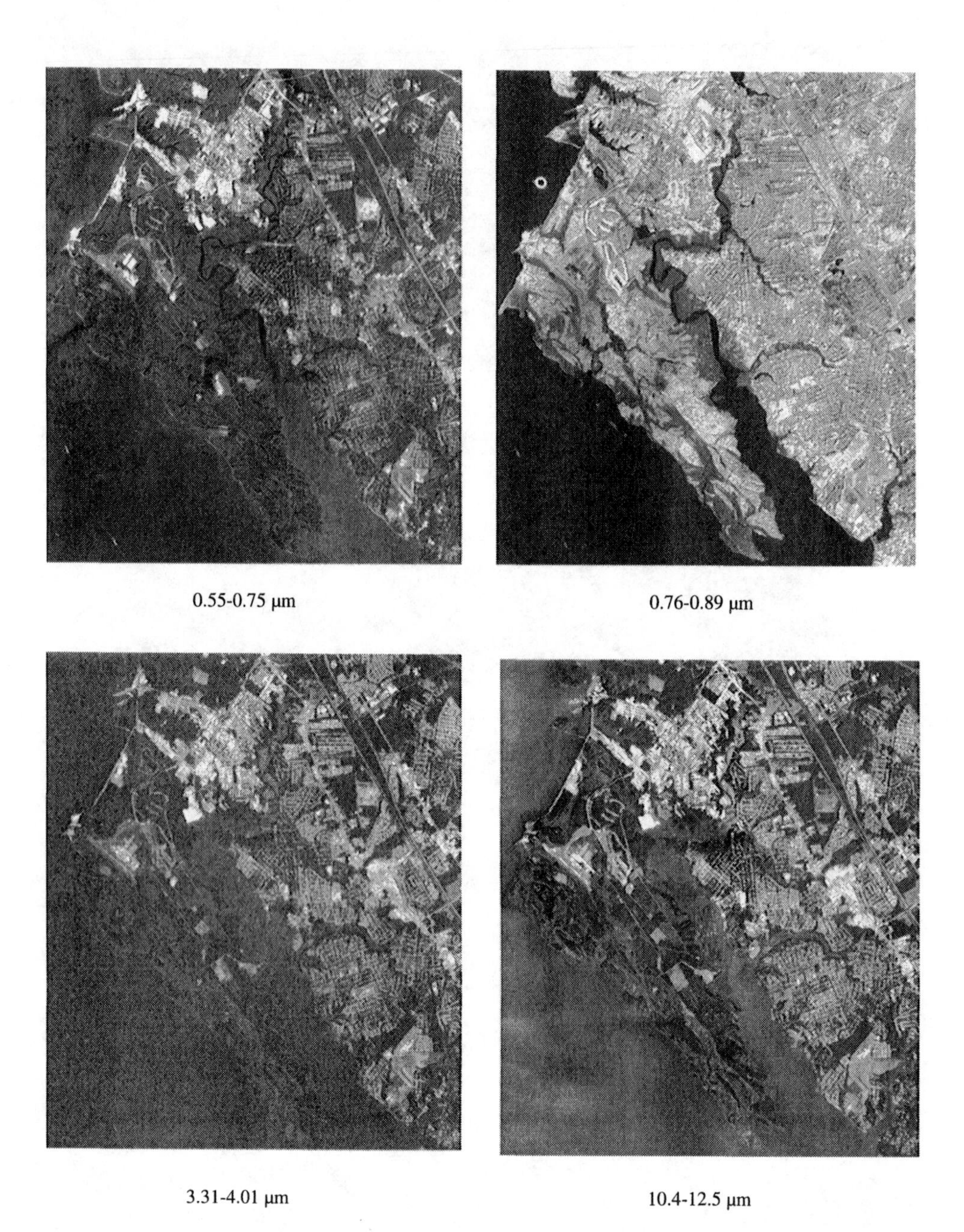

Figure 2.17 Several spectral channels from a multispectral airborne line scanner. Images of Fort Eustis Military Reservation, Virginia. This image is an M-7 mapper mosaic provided by the Environmental Research Institute of Michigan, Ann Arbor, Michigan.

many cases, we desire an imaging system that will periodically pass over nearly all points on the earth (i.e., it would periodically image most of the earth when in nadir viewing mode).

To accomplish this, near polar orbits are used to take advantage of the earth rotation as shown in Figure 2.18. To facilitate analysis of the satellite images, it is often desirable to have the sun-earth-sensor angle reasonably constant over time. This makes images taken on

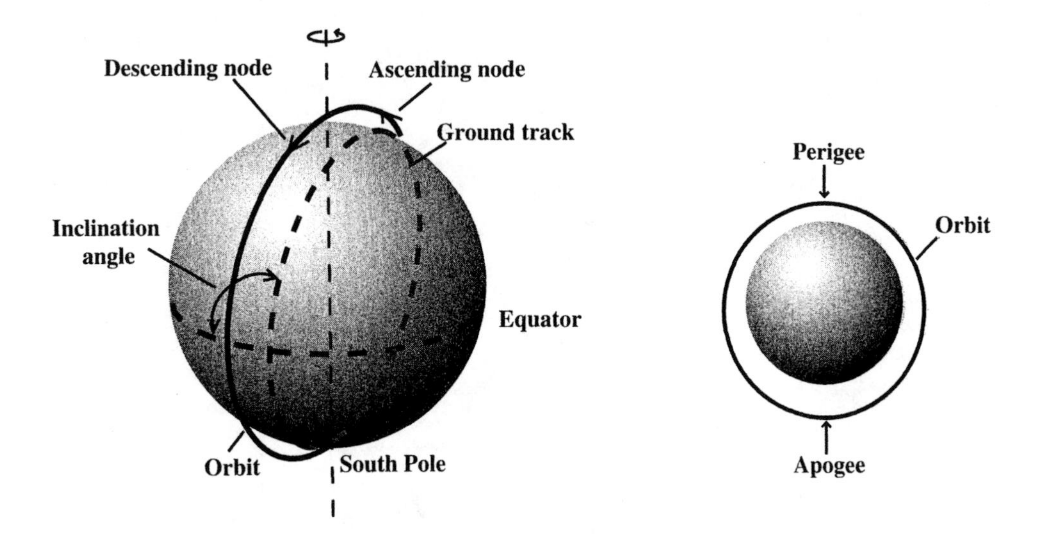

Figure 2.18 Some low earth orbit (LEO) concepts.

multiple overpasses more comparable, since we don't have to correct for illumination angle variations. This can be accomplished to a good approximation using a sun synchronous orbit. In a sun synchronous orbit, the satellite crosses the equator at the same local solar time with each pass, and, since it is traveling relatively fast compared to the earth's rotation, it will pass over points north and south of the equator at approximately constant solar time. The only variations in sun-object-sensor angle will be due to seasonal variations in the solar angle. Repetitive images of the same area at the same time of year will have nearly equal sun-object-sensor angles. All of these factors combine to dictate sun-synchronous, near-polar, low earth orbits for most sensors. These are typically several hundred kilometers above the earth, have orbital periods of about 90 to 100 minutes, and are inclined 5 to 10 degrees from the poles (inclination angles of 95 to 100°). The Landsat 4 and 5 satellites, for example, were placed in approximately 705-km-high sun-synchronous orbits crossing the equator at 9:45 a.m. in the descending node. The 98.2° inclination angle, coupled with the 90-minute orbital period, 185-km swath width, and the earth rotation, results in complete global coverage (except at the poles) every 16 days. This type of orbit is commonly used for earth observation and high-resolution meteorological sensors (e.g., SPOT, Landsat, NOAA).

Another common type of circular orbit is the geosynchronous earth orbit (GEO), which has a 24-hour period and is used to locate a satellite roughly over a fixed longitude. A satellite in GEO revolves around the earth with a period equal to the earth's rotational period. The ground track will tend to be roughly a figure 8 as shown in Figure 2.19, with the latitude limits a function of inclination. If the inclination is near zero, the satellite will effectively remain stationary over a point on the equator in what is called a *geostationary orbit.* These geostationary orbits are convenient when rapid repetitive coverage of an area is required. The moving images of cloud patterns so commonly seen on televised weather reports are acquired from geostationary satellites with sensors that acquire images every half hour (cf. Fig. 1.6). Regrettably, in order to maintain a satellite in GEO it must be at an altitude of approximately $36 \cdot 10^3$ km (roughly 40 times higher than the LEO systems discussed above). As a result, it is very difficult to operate high-resolution systems from GEO. On the other hand, it is a very attractive orbit for communications systems. A communications satellite in geostation-

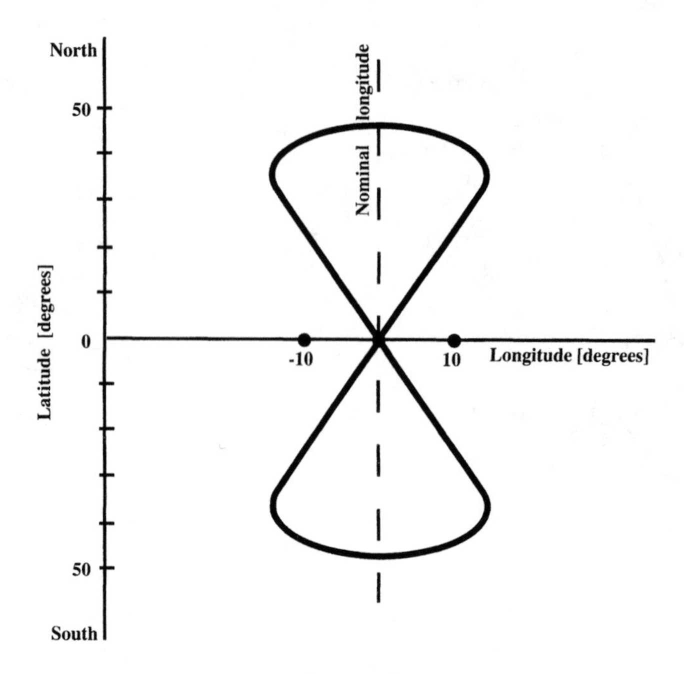

Figure 2.19 Ground track for a satellite in geosynchronous orbit.

ary orbit will always be in line-of-sight communication with the same points on the ground. A constellation made up of three geostationary satellites can be in line of sight of each other and any point on the ground. The Landsat 4 and 5 satellites used the tracking data relay satellite system (TDRSS) to send images from essentially any point in LEO to a single ground receiving station by first relaying the images up to GEO and then down to the receiving station (cf. Fig. 2.20).

Satellite imaging of the earth developed hand in hand with the international space program. Early photos were recovered from V-2 launches in the late 40's and early 50's. Following the first manmade satellite launch of Sputnik 1 on October 4, 1957, photographs and video images were acquired by U.S. Explorer and Mercury programs and the Soviet Union's LUNA series. Just 3 years after Sputnik 1's debut (April 1960), the U.S. initiated its space-based reconnaissance program acquiring high-resolution photographic images from space (cf. McDonald, 1995). Figure 2.21 shows the first photograph from the CORONA Satellite Program of a military target. This program of returning photographic images to earth using reentry vehicles called *film return buckets* continued into the 1970's. That same year (April 1960), the first U.S. orbiting electro-optical system for regular monitoring of earth resources was launched. This first television and infrared observation satellite (TIROS) carried vidicon cameras that generated low-resolution images suitable for meteorological purposes. With time, a series of TIROS launches led to the improved TIROS operational satellites (ITOS 1 was launched in January 1970). These were called the NOAA (National Oceanic and Atmospheric Administration) series as they became operational and continued to evolve. Eventually line scanner systems were added to the NOAA series payload and improved to include the current advanced very high-resolution radiometer (AVHRR) which has a 1-km ground spot at nadir and a wide field of view (cf. Sec. 5.1.1).

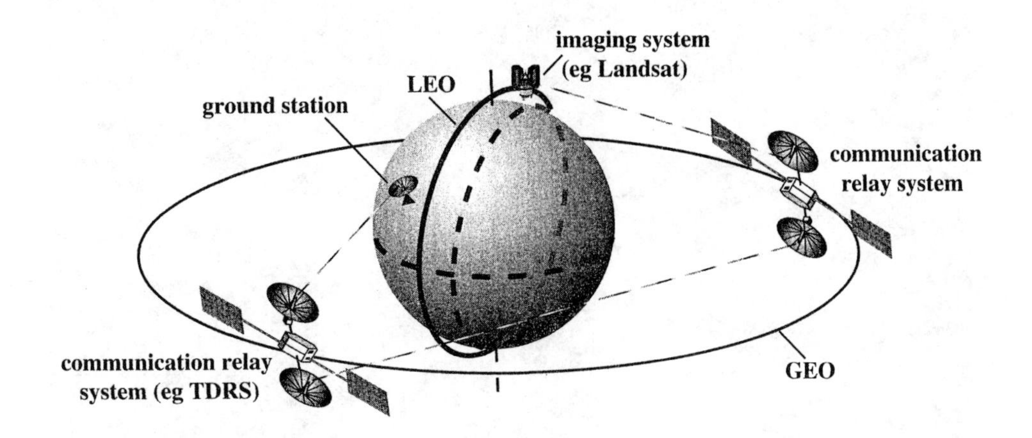

Figure 2.20 Satellites in geostationary orbit can be used to relay image data from LEO to a fixed ground station.

2.5 DIGITAL CONCEPTS

With the evolution of digital computing in the sixties and seventies, we began to think of images in a new way. In its digital representation, an image is divided into a two-dimensional array of picture elements (pixels) with each pixel having an integer value associated with it that represents the image brightness at that location (cf. Fig. 2.22). Color or multiple band images can be thought of as being made up of layers of two-dimensional arrays (or three-dimensional arrays with the layers comprising the third dimension) with each layer (or dimension) representing brightness in a different spectral band.

Digital images would eventually offer great advantages because of their ease of manipulation in readily available digital computers. Even in the late sixties and early seventies when digital image processing was still in its infancy, recording and transmitting of images in digital form had major attractions. One of the foremost attractions was the value of maintaining the quantitative integrity of the signal during transmission over long distances (e.g., from space). Another attraction was that the digital images could be stored, retrieved, and reproduced many times, or after long periods of time, with no loss in content or quality. For these reasons, starting in the early seventies, most satellite systems intended for quantitative analysis or for archival purposes began to convert the image to digital form for transmission and storage. As the user community became more comfortable with digital image concepts, image data were increasingly distributed in digital form.

The Landsat series of satellites typify this evolution to digital image analysis. The first Landsat system originally called the Earth Resources Technology Satellite 1 (ERTS-1) was designed in the late 1960's and launched in 1972. It was the first of a series of operational sensors designed specifically to study the earth's land masses. To accomplish this, it had much better resolution than the operational meteorological systems that preceded it. The multispectral scanner (MSS) on Landsat 1 had an 80-meter spot size on the ground and acquired data in four spectral channels (cf. Sec. 5.4). The Landsat data were digitized onboard, and, after being transmitted and processed on the ground, they were archived in digital form. Much of the earliest use of the Landsat data was classical photo interpretation of conventional photographic prints or transparencies made from the digital data. With the rapid evolution of digital computing, interpretation methods quickly changed with more and more

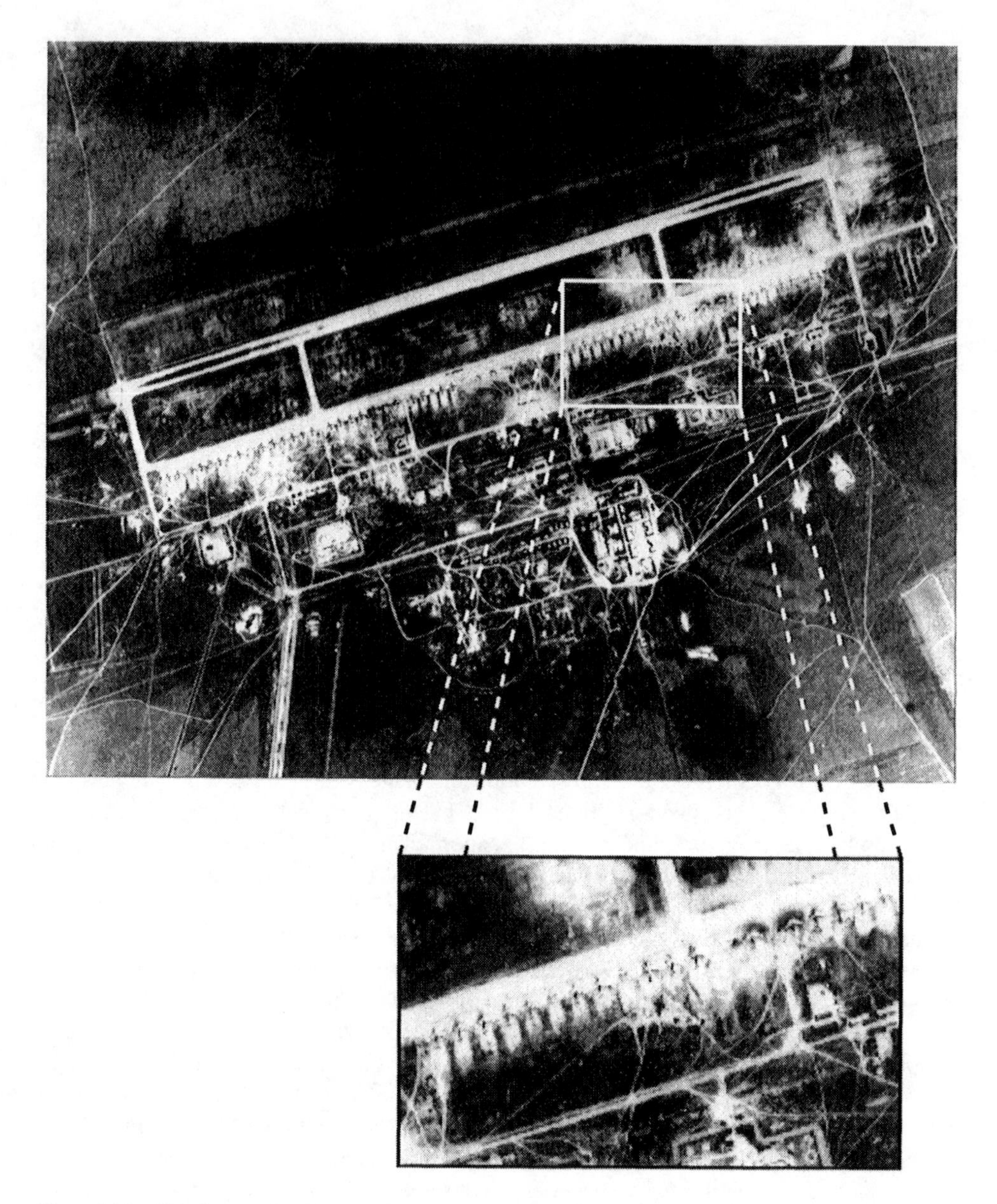

Figure 2.21 This August 18, 1960, photograph of a Soviet Union airfield is the first intelligence target imaged from the first CORONA mission.

users performing their analysis on digital images produced from the archived records.

This archive of Landsat images contains records of the earth surface condition and how it has changed since 1972. The global perspective, repetitive history, and potentially quantifiable nature of these records are characteristic of the opportunity and challenge facing remote sensing today. How can we unlock and make available to scientists, decisionmakers, and the common man the wealth of information available from today's remotely sensed systems? Perhaps more importantly, can we decipher the host of archival data that can help us understand the changes taking place on our planet? The remainder of this volume focuses on the tools needed to acquire and analyze remotely sensed data in a manner designed to unlock this information storehouse.

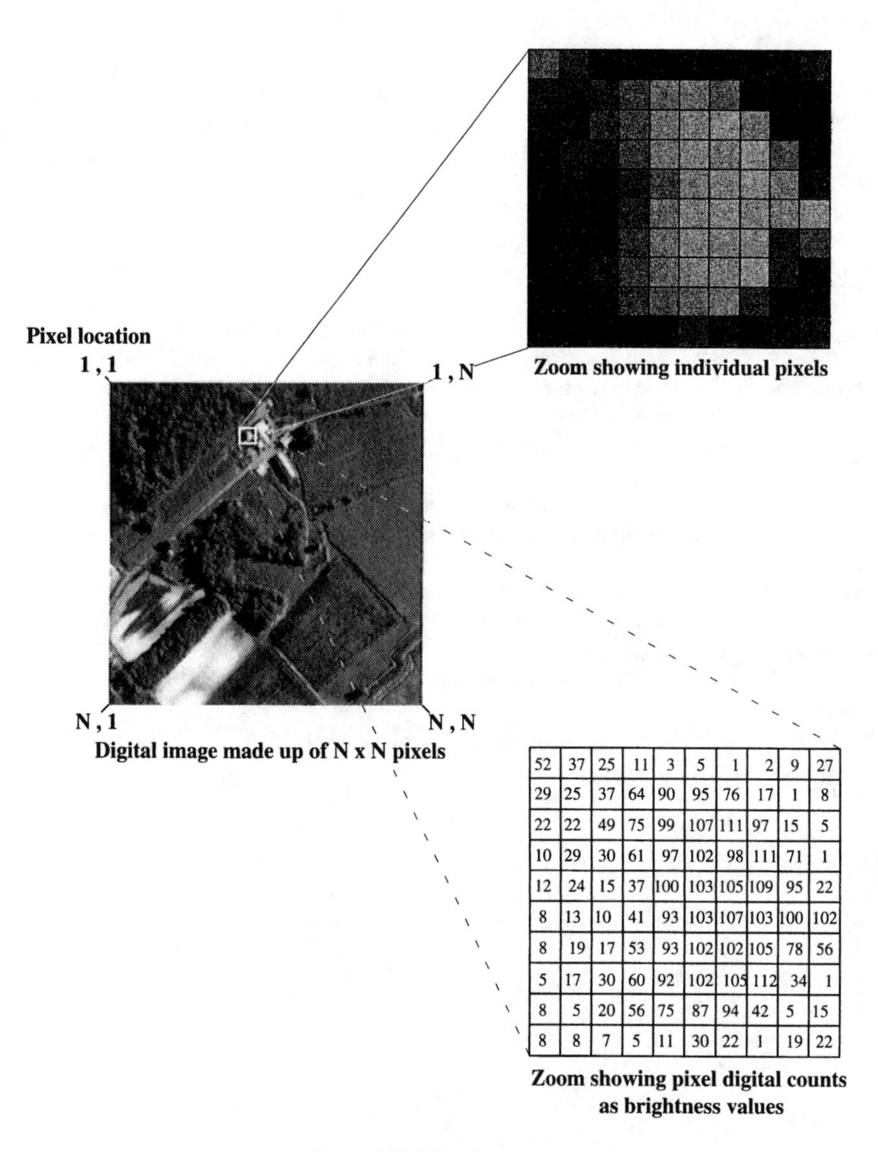

52	37	25	11	3	5	1	2	9	27
29	25	37	64	90	95	76	17	1	8
22	22	49	75	99	107	111	97	15	5
10	29	30	61	97	102	98	111	71	1
12	24	15	37	100	103	105	109	95	22
8	13	10	41	93	103	107	103	100	102
8	19	17	53	93	102	102	105	78	56
5	17	30	60	92	102	105	112	34	1
8	5	20	56	75	87	94	42	5	15
8	8	7	5	11	30	22	1	19	22

Figure 2.22 Digital image concepts.

2.6 REFERENCES

Estes, J.E., & Thorley, G.A., eds. (1983). *Vol. II: Interpretation and Applications.* In Colwell, R.N., ed., *Manual of Remote Sensing*, 2d ed., American Society of Photogrammetry, Falls Church, VA.

Estes, J.E., Hajic, E.J., Tinney, L.R., (1983). *Analysis of Visible and Thermal Infrared Data.* In Colwell, R.N., ed., *Manual of Remote Sensing* Vol. 2, 2d ed., American Society of Photogrammetry, Falls Church, VA.

James, T.H. (1977). *The Theory of the Photographic Process*, 4th ed., Macmillan, NY.

Lillesand, T.M., Scarpace, F.L., & Clapp, J.L. (1973). "Photographic quantification of water quality in mixing zones." NASA-CR-137268.

McDonald, R.A. (1995). "Opening the cold war sky to the public: declassifying satellite reconnaissance imagery." *Photogrammetric Engineering and Remote Sensing*, Vol. 61, No. 4, pp. 385-390.

Piech, K.R., & Walker, J.E. (1971). "Aerial color analysis of water quality." *Journal of Surveying and Mapping Division, American Society of Civil Engineers*, Vol. 97, No. SUZ, pp. 185-197.

Piech, K.R., & Walker, J.E. (1974). "Interpretation of soils." *Photogrammetric Engineering*, Vol. 40, pp. 87-94.

Simonett, D.S., ed. (1983). *The Development and Principles of Remote Sensing.* In Colwell, R.N., ed., *Manual of Remote Sensing*, 2d ed., American Society of Photogrammetry, Falls Church, VA.

Slama, C.G., ed. (1980). *Manual of Photogrammetry.* American Society of Photogrammetry, 4th ed., Falls Church, VA.

Sprechter, M.R., Needler, D., & Fritz, N.L. (1973). "New color film for water-penetration photography." *Photogrammetric Engineering*, Vol. 39, No. 4, pp. 359-369.

Stroebel, L., & Zakia, R., eds. (1993). *The Focal Encyclopedia of Photography*, 3d ed., Focal Press, Boston, MA.

Wolf, P.R. (1983). *Elements of Photogrammetry*, 2d ed., McGraw-Hill, NY.

CHAPTER 3

Radiometry and Radiation Propagation

Radiometric image analysis involves a quantitative assessment of the energy or flux recorded by an imaging system and some form of reverse engineering to determine the parameters that controlled the energy levels observed. For example, we might want to know the reflectance of elements in a scene, their temperature, or the distribution of water vapor in the atmosphere above the scene. This chapter contains the basic radiometric principles needed to describe, and eventually analyze, the brightness levels in an image. Before beginning this quantitative review of the fundamental physics, we will conceptually describe the energy paths we may need to analyze in deciphering an image. This "big picture" look will help to clarify what radiometric principles we need to review and will set the stage for the detailed development of the governing equation for radiometric analysis. This treatment is simplified by neglecting the dependency of radiometric signals on spatial or temporal frequency. These dependencies are introduced in Chapters 5 and 9.

3.1 ENERGY PATHS

If we restrict ourselves to passive remote sensing in the 0.4- to 15-μm spectral region, the energy paths can be conveniently divided into two groups. The first energy paths are those followed by radiation originating from the sun, while the second set of paths are those associated with radiation resulting from the temperature of objects other than the sun.

3.1.1 Solar Energy Paths

In trying to develop an understanding of radiation propagation and eventually an equation to describe the energy reaching an imaging system, it is useful to try to imagine all the ways electromagnetic energy "light" could reach an airborne or satellite camera. To keep this men-

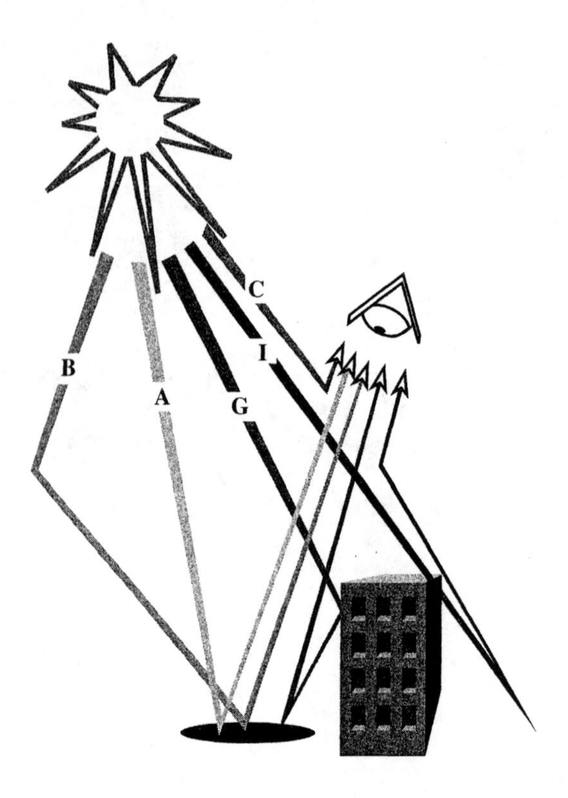

Figure 3.1 Solar energy paths.

tal game somewhat bounded, we should try to place an intuitive estimate on the relative amounts of energy. Terms that are two or three orders of magnitude smaller than the total will not be significant in most applications.

The most significant solar energy paths are illustrated in Figure 3.1. The path we think of first and most commonly is for packets of energy or photons that originate from the sun, pass through the atmosphere, are reflected from the earth's surface, and propagate back through the atmosphere to the sensor (type A photons in Fig. 3.1). This is the path that lets us "see" sunlit objects with variations in reflectance providing the character or information content of the image. Another important path is followed by photons that originate at the sun, are scattered by the atmosphere, and are then reflected by the earth to the sensor (type B photons). These photons make up what we commonly refer to as *skylight* or *sky shine*. If you imagine walking into the shadow of a building on a sunny day (effectively eliminating type A photons), you realize that there is less light, but still plenty by which to see and discriminate objects. Thus, it is clear that at least in the visible region there are typically fewer type B than type A photons, but that the number of type B photons is still significant. Our intuition also correctly tells us that the relative number of type B to A photons will increase as we move from clear to hazy to overcast skies.

Another group of photons (type C) that our experience correctly tell us are important are those that originate at the sun and are scattered into the camera's line of sight without ever reaching the earth. This haze, air light, or flare light, which we will come to call *upwelled radiance*, is what washes out the contrast when we look at air photos or out the window of an aircraft. Clearly it is also a function of how "hazy" the atmosphere is and can be a very large contributor to the overall flux, in extreme cases (fog and thick cirrus clouds) completely overwhelming the flux reflected from the earth. Piech and Schott (1974) indicate that in the visible region under clear-sky conditions, this upwelled radiance term can range from an equivalent reflectance of a few percent at low altitudes to 10% or more at higher altitudes and depends strongly on wavelength and atmospheric conditions. If the upwelled radiance were equivalent to a 10% reflector, then if we actually imaged a 10% reflector, half the energy would come from the upwelled radiance term. More to the point, the contrast between a 10% and a 5% reflector rather than being 2 to 1 would be 1.33 to 1.

Another possible photon path is illustrated with type G photons in Figure 3.1. These originate at the sun, propagate through the atmosphere, reflect from background objects, and are then reflected from the object of interest back through the atmosphere to the sensor. This is clearly a more convoluted path than we have considered thus far, involving multiple reflections or bounces of the photons. To attempt to assess the relative magnitude of this phenomena, imagine a wall painted half black and half white, with grass growing up to it as shown in Figure 3.2(a). Your intuition tells you that the grass will appear the same, even though there must be slightly more photons reflected from the grass near the white wall and fewer photons from the grass by the black wall. The issue here is the number of incremental photons. In this case, it seems as though they would be overwhelmed by the number of type A, B, and C photons (i.e., lost in the noise). On the other hand, consider the same high-contrast background made up of a black and white wall with a dark car parked in front of it. We might expect the photon flux from the background to change the appearance of the car. This is illustrated in Figure 3.2(b) where the combination of the dark car, the high surface reflectance due to the shiny (specular) finish, and the high contrast of the background combine to cause the difference in the number of multiple bounce photons to be large enough to affect the appearance (i.e., they are no longer lost in the noise). Thus, we see that whether multiple bounce photons are important depends on the sensitivity of our measurements (i.e., what is the noise level), the roughness of the surface, and the brightness or contrast of the background. The contrast between the background object and the sky is particularly important because the sky would be the source of photons from the background if the object were not present. For most natural (rough) surfaces in reasonably level terrain, the multiple bounce photons do not make a significant contribution and can and will be neglected. However, if specular targets or targets with a high background contrast are of interest, multiple bounce photons must be considered.

Another source of multiple bounce, or bounce and scatter photons (type I in Fig. 3.1), are caused by what has been termed the *adjacency effect*, [cf. Kaufman (1982)]. These are photons that are reflected from surrounding objects and then scattered into the line of sight of the sensor. In most cases [e.g., where the average background reflectance (albedo) is slowly varying], these photons can be treated as a constant and lumped in with the path radiance (type C photons). However, if a medium-gray object is imaged with a very dark or very bright background, there will be more photons in the bright background case due to the adjacency effect. (Note: Visually it will appear darker due to psychophysical effects.)

In summary, for energy originating at the sun, we need to be concerned with type A, B, and C photons. In general, we will not be concerned with type G and I photons, although we should keep in mind that in special circumstances they may be important.

3.1.2 Thermal Energy Paths

Recall from elementary physics that all objects with temperatures above absolute zero radiate energy, therefore, we must also consider electromagnetic energy or photons from this source. Our intuition is not as useful in understanding these photons, since in the visible region so few photons are emitted by anything other than the sun that they are completely negligible. However, in other portions of the spectrum (e.g., near 10 μm) self-emitted photons become very important, so we will want to consider energy paths that would lead these

(a) Grass with black and white background.

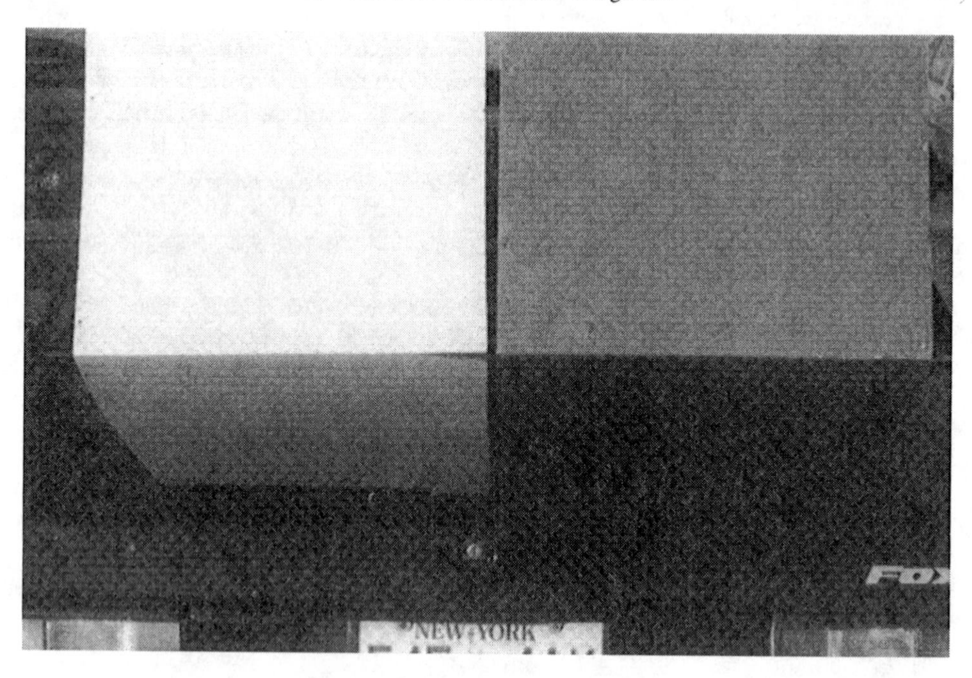

(b) Car hood with black and white background.

Figure 3.2 Illustration of multiple bounce effects.

self-emitted photons to our sensor (cf. Fig. 3.3).

The photons most often of interest are caused by radiation due to the temperature of the target itself. These photons propagate through the atmosphere to the sensor where their num-

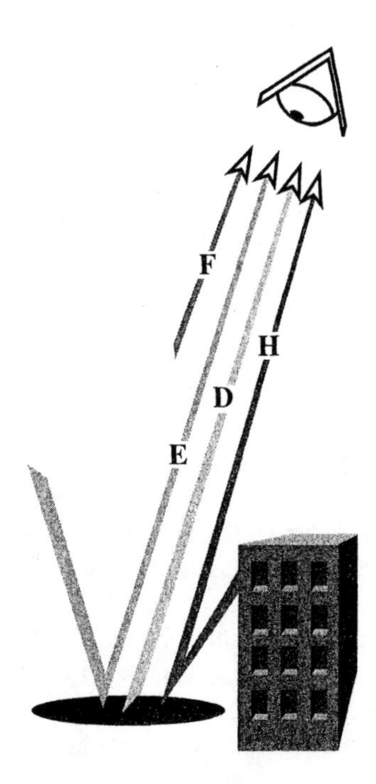

Figure 3.3 Self-emitted thermal energy paths.

bers provide information about the temperature of the target (type D photons in Fig. 3.3). Because the atmosphere above the target has some nonzero temperature, it radiates and scatters self-emitted energy down onto the target which can be reflected and propagate up to the sensor (type E photons in Fig. 3.3). These photons are the thermal equivalent of skylight photons from the sun, and type B and E photons produce what is referred to collectively as *downwelled radiance*. The atmosphere along the line of sight path will also emit and scatter self-emitted photons directly to the sensor independent of the surface (type F photons). They are the thermal equivalent of the type C solar scattered photons, and type C and F photons produce the upwelled radiance or path radiance. Finally, background objects obscuring the sky above a target radiate energy due to their temperature and produce photons that can be reflected by the target and propagate to the sensor (type H photons in Fig. 3.3). These are similar to the multiple-bounce-type G solar photons in that their relative importance will depend on how diffuse (rough) the target is and how bright (hot) the background. For most natural surfaces without large background objects, type H photons are negligible.

In summary, a more quantitative analysis (cf. Sec. 4.6) will show that in the MWIR and LWIR spectral regions there are significant numbers of type D and F photons (more D than F) and smaller numbers of type E and H photons, which, depending on circumstances, may or may not be negligible.

Assessment of the radiation propagation paths shown in Figures 3.1 and 3.3 suggests that in order to describe quantitatively the radiometry, we need to characterize self-radiating sources, propagation of energy from the sun to the earth, propagation through the atmosphere (e.g., transmission losses), scattering in the atmosphere, reflection by the earth, and the radiometric characteristics of radiation propagation within the sensor. The rest of this chapter deals with the tools needed to characterize the radiometric strand in the image chain. These tools are then applied in Chapter 4 to describe quantitatively the first links in the image chain.

3.2 RADIOMETRIC TERMS

Radiometry is formally defined as the science of characterizing or measuring how much EM energy is present at, or associated with, some location or direction in space. It has evolved separately in the fields of physics, illumination or vision, and engineering, and as a result a host of terms are used to describe various radiometric concepts. Often one concept has several different names, and it is also common for the same term (e.g., intensity) to mean different things to different authors. To provide a common framework, we will briefly review the definition of the relevant physical parameters and radiometric terms. An emphasis will be placed on the units of measure in this section and throughout the book to ensure a clearer understanding—units are usually designated with square brackets ([ ]) for clarity, and, where relevant, a unit's cancellation analysis may be performed within square brackets. In reading other authors, particularly older works or work drawn from other disciplines, the reader should carefully evaluate the author's definition and units of measure to determine what term is being applied to each radiometric concept. The definitions used throughout this volume are consistent with those established by the Commission Internationale de l'Eclairage (CIE) and adopted by most international societies [see CIE Publication (1970)]. In addition, to the extent practical, the parameters, nomenclature, and symbology are consistent with the relevant reference material [see Grum and Becherer (1979) and Nicodemus (1976)].

3.2.1 Definition of Terms

For most radiometric considerations, we can use the ray/particle simplification of optics. This approach is based on geometric optics and assumes that light travels in straight lines and transfers energy in discrete packets or quanta. The physical optics effects of diffraction and interference associated with the wave nature of EM energy can be largely ignored in simple radiometric calculations in the visible and thermal infrared. The wave nature of EM energy is important in image formation but need not be considered in most simple radiometric calculations.

Recall that wavelength (λ[μm]), frequency (ν[sec^{-1}]), and the speed of light (c[m/sec]) are related as:

$$c = \lambda \nu \tag{3.1}$$

where wavelength is the distance between two consecutive like elements (same phase) in a wave representation (e.g., peak-to-peak or trough-to-trough). It is commonly referred to in units of microns [μm = 10^{-6} m] or nanometers [nm = 10^{-9} m]. Frequency is the number of waves (cycles) that would travel past a fixed point in 1 second and has units of hertz, cycles per unit time [Hz] or just sec^{-1}. The speed of light clearly has units of distance per second and in vacuum has a constant value of $2.9979 \cdot 10^8$ m/sec. Spectral references will generally be given with respect to wavelength. However, some computations are more readily represented by the wave number ν' [cm^{-1}], which is simply the number of waves that would fit in a 1-centimeter length, i.e.,

$$\nu' = \frac{1}{\lambda}[\text{cm}^{-1}] = \frac{\nu}{c} \tag{3.2}$$

The common spectral regions and the nomenclature used in this volume are delineated in Figure 3.4. In radiometric calculations, it is generally easier to think of energy as being

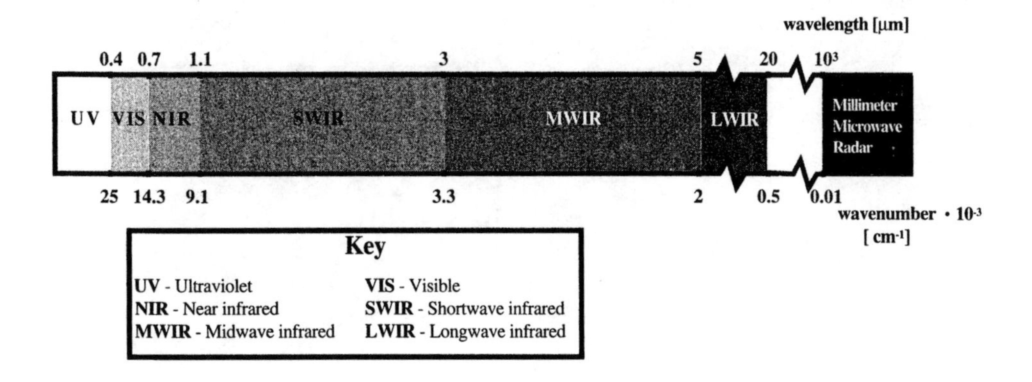

Figure 3.4 Nomenclature for various regions of the electromagnetic spectrum.

transferred in terms of energy packets or quanta in accordance with quantum theory. The particle or energy carrier is called a photon, and each photon carries energy

$$q = h\nu = \frac{hc}{\lambda}[\text{joules}] \tag{3.3}$$

where $h = 6.6256 \cdot 10^{-34}$ joules·sec is Planck's constant, and energy is expressed as joules [j]. Thus we see that shorter wavelength photons carry more energy than longer wavelength photons. This becomes very important when we begin to look at the spectral response of detectors in Chapter 5. The total energy (Q) in a beam or ray is a function of the number and spectral makeup of the photons according to

$$Q = \sum q_i = \sum_{i=1} n_i h\nu_i \tag{3.4}$$

where the sum is over all frequencies present and n_i is the number of photons at each frequency.

It is usually more convenient to think of a beam or bundle of rays, not in terms of the total energy but rather in terms of the rate at which the energy is passing or propagating cf. Fig. 3.5(a). This rate of flow of energy is called the *radiant flux*, or *power* (Φ), and is defined as the first derivative of the radiant energy with respect to time (t), i.e.,

$$\Phi = \frac{dQ}{dt}[\text{watts, w}] \tag{3.5}$$

Often we are interested in the rate at which the radiant flux is delivered to a surface (e.g., the responsive surface of a detector). This concept is given the term *irradiance* (E) and is defined as

$$E = E(x, y) = \frac{d\Phi}{dA}[\text{wm}^{-2}] \tag{3.6}$$

where dA [m^2] is an area element on the surface of interest, and (x,y) are generic spatial location parameters that, for convenience, will generally not be explicitly expressed. [Equation (3.6) is characteristic of a shorthand we will use to indicate a simplification of notation where E and $E(x,y)$ are identical, but the dependence on x and y will only be explicitly stated where

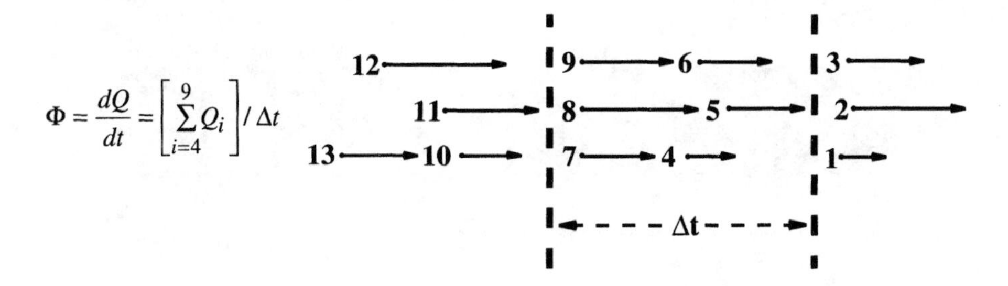

(a) Radiant flux-time rate of energy delivery, production, or propagation.

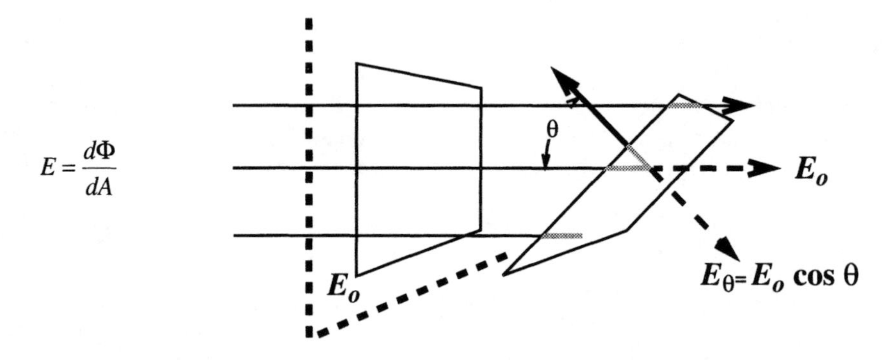

(b) Irradiance flux per unit area onto a surface. The first surface is perpendicular to the incident flux and has irradiance E_0 The second is rotated through the angle θ and has irradiance E_θ.

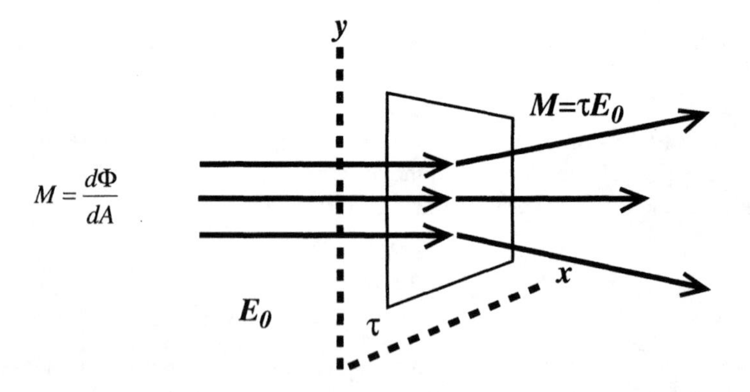

(c) Radiant emittance or radiant exitance - flux per unit area away from the surface. A surface with E_o irradiance from the left and transmission τ would have exitance M away from the right-hand side.

Figure 3.5 Illustration of radiometry definitions.

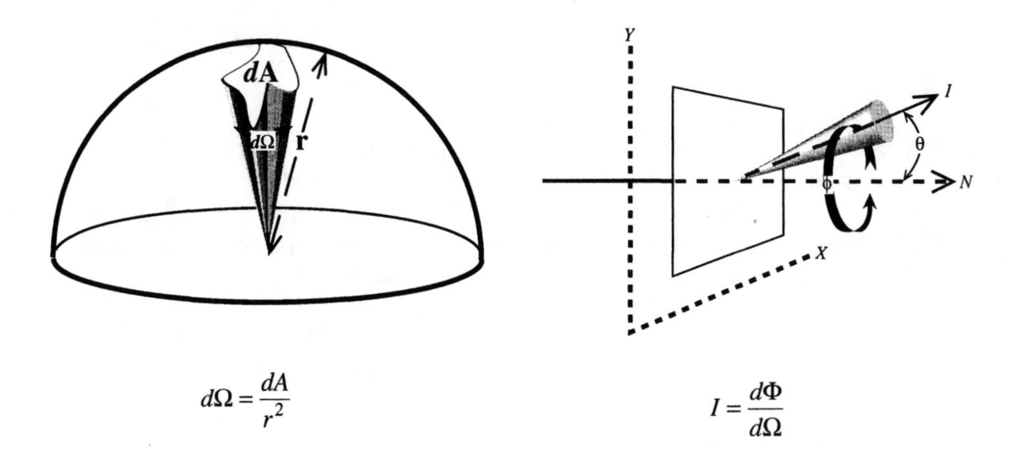

(d) Element of solid angle (steridian).

(e) Radiant intensity = flux per unit solid angle into the direction defined by θ and φ, where θ is the angle from the normal to a reference surface, and φ is an azimuthal angle.

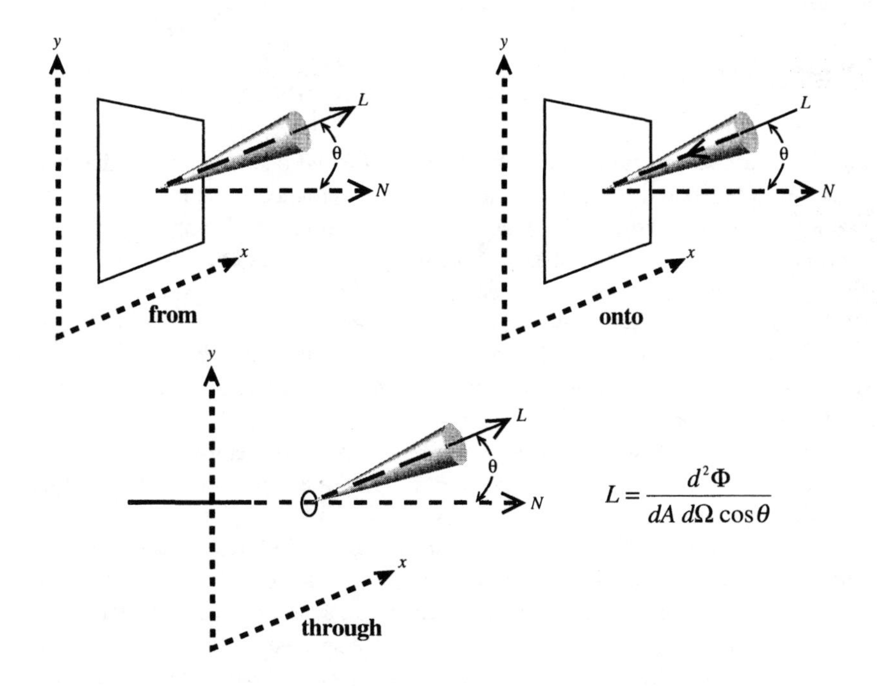

(f) Radiance - flux per unit projected area per unit solid angle from, onto, or through the plane. N is the normal to the plane.

Figure 3.5 Illustration of radiometry definitions (con't.).

it is required for clarity.] Irradiance, as illustrated in Figure 3.5(b), is the flux per unit area *onto* a surface. It is very similar to radiant exitance (or radiant emittance), which is defined as:

$$M = M(x,y) = \frac{d\Phi}{dA}[\text{wm}^{-2}] \tag{3.7}$$

and describes the flux per unit area *away from* a surface [cf. Fig. 3.5(c)]. This term describes the power per unit area radiated by a source or reflected from a surface.

Both the irradiance and the exitance provide spatial information about the flux, but no angular or directional information. The simplest term used to describe directional or dispersive information about the flux is the *radiant intensity* (I), defined as:

$$I = I(\theta,\phi) = \frac{d\Phi}{d\Omega}[\text{wsr}^{-1}] \tag{3.8}$$

where $d\Omega = dA/r^2$ [steradian, sr] is the element of solid angle. The element of solid angle is defined as the conic angle encompassing the area element dA on the surface of a sphere of radius r [cf. Fig. 3.5(d)], and θ and ϕ are generic orientation angles, as illustrated in Figure 3.5(e), which will not be explicitly expressed unless required.

The radiant intensity describes the flux per unit solid angle from a point source into a particular direction. While the intensity provides directional information, it does not provide any spatial information. The use of the radiance term ($L[\text{wm}^{-2}\text{sr}^{-1}]$) to characterize the flux overcomes this limitation. It is the most complex of the radiometric terms we will consider, but also the most useful and ubiquitous. It is defined as:

$$L = L(x,y,\theta,\phi) = \frac{d^2\Phi}{dA\cos\theta d\Omega} = \frac{dE}{d\Omega\cos\theta} = \frac{dI}{dA\cos\theta} = \frac{dM}{d\Omega\cos\theta} \tag{3.9}$$

where x and y define a location in the plane of interest, and θ and ϕ are angles that define the direction of interest relative to the normal to the plane. The radiance is the flux per unit projected area (at the specified location in the plane of interest) per unit solid angle (in the direction specified relative to the reference plane). Note that while radiant exitance and intensity are generally source terms and irradiance is generally associated with receivers or detectors, radiance can be used to characterize the flux from or onto a surface, as well as the flux through any arbitrary surface in space [cf. Fig. 3.5(f)]. In addition, it has some very useful properties of constancy of propagation that make it an attractive parameter to use in most treatments of radiation propagation.

Nicodemus (1976) demonstrates the important concept of the constancy of radiance through an isotropic lossless media (i.e., no transmission losses and unit index of refraction). Referring to Figure 3.6, we assume a beam of energy with constant radiance across the profile of the beam. We select two arbitrary points along the beam and two surfaces with arbitrary orientation containing those points. It may be convenient to conceptualize the first surface as a source (i.e., the earth's surface) and the second as a sensor. If we consider the flux associated with a bundle of rays at surface 1 contained in a surface element dA_1, which are also contained in dA_2 on surface 2, we see that in a lossless media, the flux $d\Phi_1$ through dA_1 must equal the flux $d\Phi_2$ through dA_2. We want to evaluate how the radiance at surface 1(L_1) is related to the radiance at surface 2 (L_2). We see that the radiance L_1 along the primary ray at p_1 toward the surface element dA_2 is, by definition:

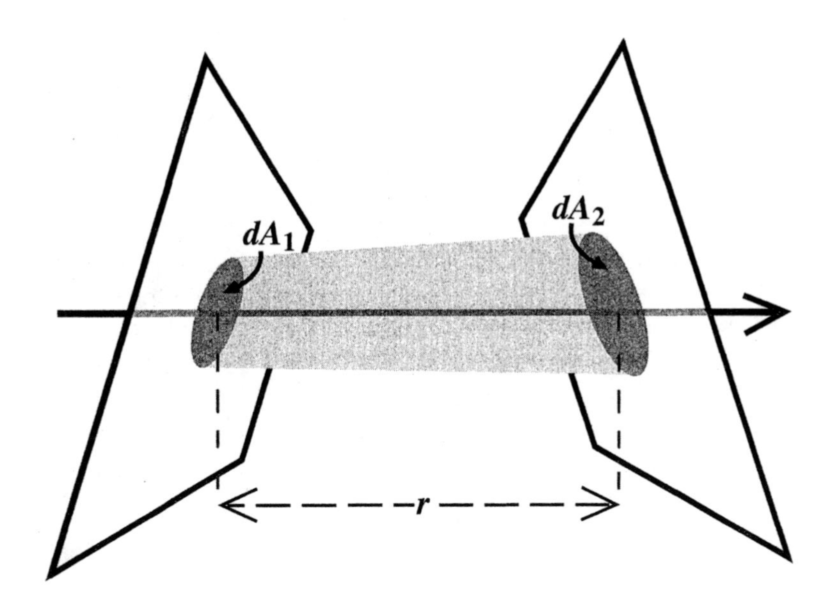

(a) Radiance along a beam.

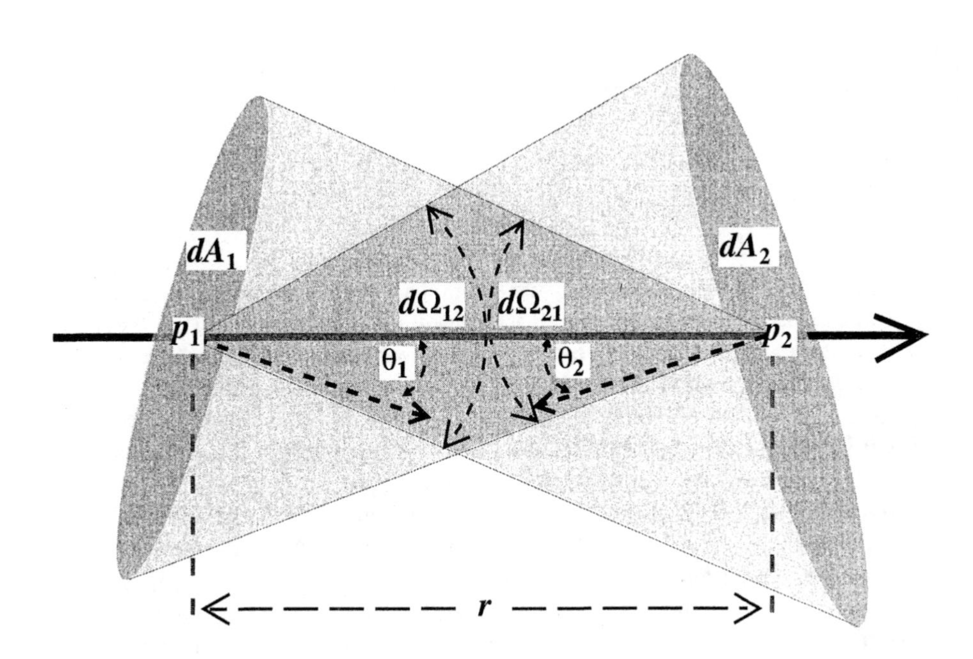

(b) Illustration of the terms used in the definition of radiance at the points p_1 and p_2 along a beam.

Figure 3.6 Constancy of radiance.

$$L_1 = \frac{d^2\Phi_1}{dA_1 \cos\theta_1 d\Omega_{12}} \tag{3.10}$$

and the radiance at p_2 is:

$$L_2 = \frac{d^2\Phi_2}{dA_2 \cos\theta_2 d\Omega_{21}} \tag{3.11}$$

where θ_1 and θ_2 are the angles from the normal to the surface to the primary ray, and $d\Omega_{12}$ is the element of solid angle encompassed by the area element dA_2 at p_1, and similarly $d\Omega_{21}$ is the element of solid angle encompassed by the area element dA_1 at p_2. If we let r represent the arbitrary distance between p_1 and p_2, we see that the throughput $\mathcal{T}$ can be expressed as:

$$\mathcal{T}_1 = dA_1 \cos\theta_1 d\Omega_{12} = dA_1 \cos\theta_1 \frac{dA_2 \cos\theta_2}{r^2}$$

and

$$\mathcal{T}_2 = dA_2 \cos\theta_2 d\Omega_{21} = dA_2 \cos\theta_2 \frac{dA_1 \cos\theta_1}{r^2} \tag{3.12}$$

and that $\mathcal{T}_1 = \mathcal{T}_2 = \mathcal{T}$. Expressing the radiance as:

$$L_1 = \frac{d^2\Phi_1}{\mathcal{T}_1} \tag{3.13}$$

and

$$L_2 = \frac{d^2\Phi_2}{\mathcal{T}_2} \tag{3.14}$$

and recalling that $d\Phi_1 = d\Phi_2 = d\Phi$, we have

$$L_1 = L_2 = \frac{d^2\Phi r^2}{dA_1 \cos\theta_1 dA_2 \cos\theta_2} = \frac{d^2\Phi}{\mathcal{T}} \tag{3.15}$$

Since all the terms in this analysis were completely arbitrary, we see that the radiance along a ray is constant over distance in a lossless media. Thus it is the term most readily used for radiation propagation, since it is independent of geometric considerations and only losses due to the medium need to be considered (e.g., absorption and scattering).

To this point we have ignored the spectral character of the radiometric terms. In fact, the flux is spectrally variable and, therefore, each of the radiometric terms will vary with wavelength. In general we will be interested in spectral density expressed as flux per unit wavelength interval and designated with a wavelength subscript. Thus, the spectral irradiance would be expressed as E_λ [wm^{-2} μm^{-1}]. The responsivity of the detectors is also a function of wavelength and must be cascaded with the spectral flux to generate effective bandpass values for the radiometric terms (i.e., what is the effective magnitude of the radiometric term relative to the spectral response of the detector?). The responsivity at each wavelength is defined as the signal out (S) per unit flux incident on the detector at the wavelength of interest. Therefore the spectral response function is defined as:

$$R(\lambda) = \frac{dS}{d\Phi(\lambda)} \tag{3.16}$$

with units of [amps w^{-1}] or [volts w^{-1}] depending on the signal out of the detector. The unitless peak normalized spectral response function is

$$R'(\lambda) = \frac{R(\lambda)}{R(\lambda)_{max}} \tag{3.17}$$

where $R(\lambda)_{max}$ is the maximum value of the $R(\lambda)$ function. Thus, $R'(\lambda)$ is normalized to a maximum value of unity. The peak normalized effective value of a radiometric term over the detector bandpass is then obtained by weighting the radiometric term by this normalized response value (cf. Fig. 3.7). For example,

$$L = L_{eff} = \int_0^\infty L_\lambda R'(\lambda) d\lambda \ [\text{wm}^{-2}\text{sr}^{-1}] \tag{3.18}$$

where the subscript (eff) is usually implied, rather than explicitly indicated, and a numerical approximation to the integral is used in practice. The signal from a sensor can be computed by integrating the spectral flux weighted by the spectral response function according to:

$$S = \int_0^\infty \Phi_\lambda R(\lambda) d\lambda \tag{3.19}$$

where the output signal (S) has units of amps or watts depending on the type of detector, and the integral or its numerical approximation need only be performed over the nonzero spectral response range. We can also express the effective bandpass responsivity of a detector as:

$$R = \frac{\int R(\lambda)\Phi_\lambda d\lambda}{\int \Phi_\lambda d\lambda} = \int R(\lambda) \frac{\Phi_\lambda}{\int \Phi_\lambda d\lambda} d\lambda \tag{3.20}$$

It is important to recognize that when the bandpass value for responsivity is used, it is calculated for a specific source spectral distribution. The same detector will exhibit different bandpass responsivity values when irradiated by sources with differing spectral distributions.

3.2.2 Blackbody Radiators

One of the cornerstones of modern physics and a critical element of quantitative radiometric remote sensing is the formula for spectral exitance from a blackbody radiator. A blackbody is an idealized surface or cavity that has the property that all incident electromagnetic flux is perfectly absorbed and then re-radiated (i.e., the reflectivity is zero and absorptivity is one). Planck (1901) derived an expression for the spectral radiant exitance from a blackbody based on statistical calculation of the vibrational energy states between the atoms and the assumption that the vibrational resonation between the atoms could only emit or absorb energy in discrete levels proportional to the frequency of the oscillation state. Thus all the energy states are defined by $Q = mh\nu$ where m can only take on integer values, h was an empirically derived value that we now refer to as *Planck's constant*, and ν is the frequency of oscillation. Einstein's later work on the quantum theory of light and the concept of the photon provided the theoretical foundation for Planck's results [cf. Einstein (1905)]. The Planck or blackbody radiation equation for the spectral radiant exitance from a surface is:

$$M_\lambda = 2\pi hc^2 \lambda^{-5} \left(e^{\frac{hc}{\lambda kT}} - 1\right)^{-1} [\text{wm}^{-2}\mu\text{m}^{-1}] \tag{3.21}$$

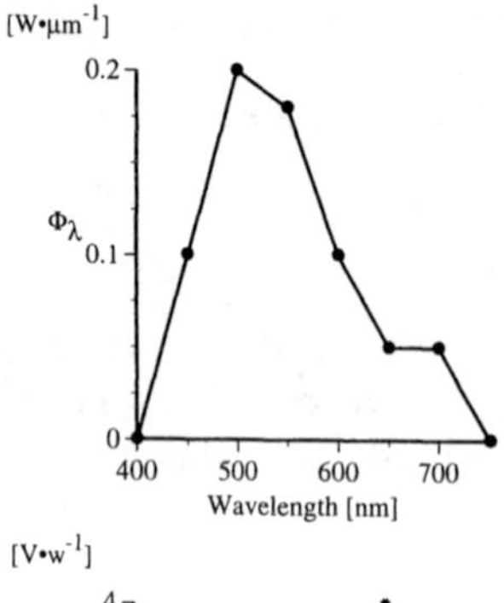

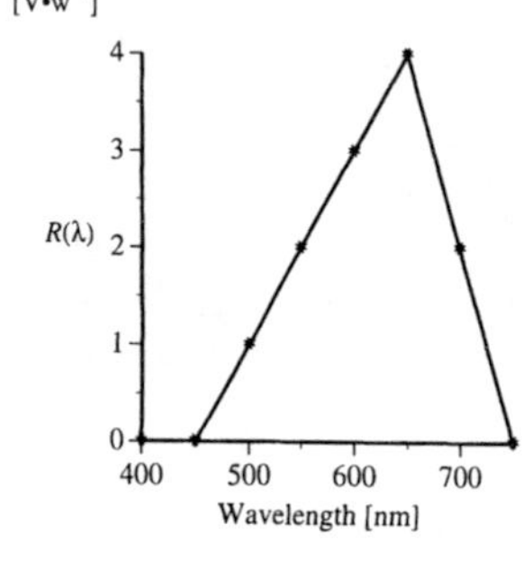

λ[nm]	Φ_λ[wnm^{-1}]	$R(\lambda)$[v/w]	$R'(\lambda)$[v/w]	$\Phi_\lambda / \int \Phi_\lambda d\lambda$
400	0	0	0	0
450	0.1	0	0	2.94 x10^{-3}
500	0.2	1	0.25	5.88 x10^{-3}
550	0.18	2	0.5	5.29 x10^{-3}
600	0.1	3	0.75	2.94 x10^{-3}
650	0.05	4	1	1.47 x10^{-3}
700	0.05	2	0.5	1.47 x10^{-3}
750	0	0	0	0

$$\Phi = \int \Phi_\lambda d\lambda = \Sigma \Phi_\lambda \Delta\lambda = 34[\mathrm{W}]$$

$$\Phi_{eff} = \int \Phi_\lambda R'(\lambda) d\lambda = \Sigma \Phi_\lambda R'(\lambda) \Delta\lambda = 14.5[\mathrm{W}]$$

$$R = \int R(\lambda) \frac{\Phi_\lambda}{\int \Phi_\lambda d\lambda} d\lambda = \Sigma R(\lambda) \frac{\Phi_\lambda}{\Sigma \Phi_\lambda d\lambda} \Delta\lambda = 1.7\ [\mathrm{v/w}]$$

$$S = \int \Phi_\lambda R(\lambda) d\lambda = 58\ [\mathrm{V}]$$

$$S = R\Phi \cong 58[\mathrm{V}]$$

Figure 3.7 Effective flux and responsivity sample calculations.

where T is the temperature in degrees Kelvin, k is the Boltzmann gas constant ($1.38 \cdot 10^{-23}$ jK^{-1}), and h and c are the Planck constant and the speed of light as previously defined. Examination of the Planck equation shows that radiant exitance is a function of both temperature and wavelength. By holding temperature fixed at selected values, a family of blackbody curves can be generated as shown in Figure 3.8 relating spectral exitance to wavelength. These curves show how the exitance increases with temperature and that it is a well-behaved function whose peak value shifts to shorter wavelengths as the temperature increases. In practice, the ideal blackbody can only be approximated by imperfect absorbers. To describe this phenomenon, we introduce the concept of emissivity $\varepsilon(\lambda)$, defined as the ratio of the spectral exitance $M_\lambda(T)$ from an object at temperature T to the exitance from a blackbody at that same temperature $M_{\lambda BB}(T)$, i.e.,

$$\varepsilon(\lambda) = \frac{M_\lambda(T)}{M_{\lambda BB}(T)} \tag{3.22}$$

The emissivity describes how well an object radiates energy compared to the perfect black-

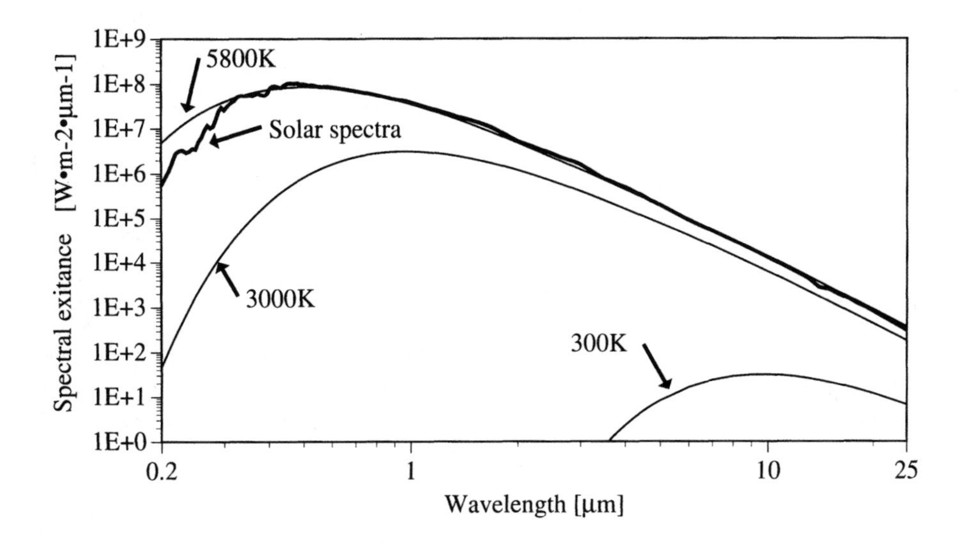

Figure 3.8 Blackbody curves and solar exitance spectra.

body radiator and is a unitless value with a range from 0 to 1. Objects whose emissivity is approximately constant with wavelength are referred to as *gray bodies*, while objects with spectrally varying emissivities are called *selective radiators*. Objects that approximate *gray-body* radiators over all or part of the spectrum are often described or approximated by a blackbody that would produce the equivalent exitance. The exitance from the sun can be approximated by a blackbody at approximately 5800 K (cf. Fig. 3.8).

It is important to recognize that emissivity is a fundamental property of matter just like absorptivity, reflectivity, and transmissivity. In the formalism we have introduced, the transmissivity is the ability of the material to allow the flux to propagate through it. The transmittance or transmission τ can be expressed as the unitless ratio of the exitance on the back of a sample (M_τ) to the irradiance on the front of a sample (E_i), i.e.,

$$\tau = \frac{M_\tau}{E_i} \tag{3.23}$$

Clearly the spectral transmittance $\tau(\lambda)$ is simply the ratio of the spectral exitance and spectral irradiance. Similarly, the reflectivity is the ability of the material to turn incident flux back into the hemisphere above the material, and the reflectance r can be expressed as the ratio of the exitance from the front of a sample (M_r) to the irradiance onto the front of the sample, i.e.,

$$r = \frac{M_r}{E_i} \tag{3.24}$$

Finally, the absorptivity is the ability of the material to remove electromagnetic flux from the system by converting incident flux to another form of energy (e.g., thermal energy). The absorptance α can be expressed as the ratio of the flux per unit area incident on the surface that is converted to another form of energy (M_α) to the irradiance onto the surface, i.e.,

$$\alpha = \frac{M_\alpha}{E_i} \tag{3.25}$$

Since conservation of energy requires all the incident flux to be absorbed, transmitted, or reflected, we have

$$\alpha + \tau + r = 1 \tag{3.26}$$

or in the case of an opaque material, where τ is zero we have

$$\alpha + r = 1 \tag{3.27}$$

Furthermore, according to Grum (1979), Kirchhoff's law states that the emissivity must be numerically equal to the absorptance for surfaces in thermodynamic equilibrium (i.e., good absorbers are good emitters). Therefore, we can also express the conservation of energy relationship as:

$$\varepsilon + \tau + r = 1 \tag{3.28}$$

or for opaque objects

$$\varepsilon + r = 1 \tag{3.29}$$

We can compute the total exitance from a blackbody by integrating the Planck equation over all wavelengths. This yields the familiar Stefan-Boltzmann equation, i.e.,

$$\begin{aligned} M &= \int_0^\infty M_\lambda d\lambda = \int 2\pi hc^2\lambda^{-5}(e^{\frac{hc}{\lambda kT}} - 1)^{-1} d\lambda \\ &= \frac{2\pi^5 k^4 T^4}{15c^2h^3} = \sigma T^4 \end{aligned} \tag{3.30}$$

where σ ($5.67 \cdot 10^{-8} \mathrm{wm^{-2}K^{-4}}$) is the Stefan-Boltzmann constant. It is important to recognize that this fourth-power relationship holds only for the integral over all wavelengths and is mostly of use for energy exchange calculations in thermodynamics. The exitance within a bandpass can be expressed as:

$$M = \int_{\lambda_1}^{\lambda_2} M_\lambda d\lambda \tag{3.31}$$

and must be solved in numerical form since no closed form solution exists.

Another fundamental natural law can be derived from the Planck equation by taking the first derivative with respect to wavelength and setting it equal to zero, i.e.,

$$\frac{dM_\lambda}{d\lambda} = 0 \tag{3.32}$$

Since we have already seen that the Planck equation is well-behaved with a single maximum, the zero point in the first derivative will yield the wavelength of maximum exitance. Solving Eq. (3.32) and rearranging produces the Wien displacement law, i.e.,

$$\lambda_{max} = \frac{A}{T} \tag{3.33}$$

where A (2898 μm K) is the Wien displacement constant. This expression predicts that the peak radiance from the sun at approximately 6000 K will occur in the visible portion of the spectrum at approximately 0.5 μm and that the peak flux for an object near the earth's ambient temperature of 300 K will occur at approximately 10 μm (cf. Fig. 3.8). This is conveniently in the center of an atmospheric transmission window (cf. Fig. 1.5), which is extensively used for studying the thermal characteristics of the earth.

3.3 RADIOMETRIC CONCEPTS

This section introduces some basic radiometry concepts that draw on the definitions and terms we have introduced thus far, and which we need to define the equations governing radiometric remote sensing of the earth.

3.3.1 Inverse-Square Law for Irradiance from a Point Source

It is often useful to know how irradiance varies with distance from a small source (i.e., in the ideal, a point source). Consider two spherical surfaces at distances r_1 and r_2 from a point source located at p as shown in Figure 3.9(a). We construct two area elements dA_1 and dA_2 defined by the extent of two elements of angle $d\theta$ and $d\phi$ as illustrated. If we assume that the irradiance (E_1) at surface element dA_1 is known, then we wish to know how the irradiance (E_2) at surface dA_2 is related to E_1. If we assume a lossless isotropic media, then by construction the element of flux ($d\Phi_1$) through the first surface must be identical to the element of flux ($d\Phi_2$) through the second surface. This is because every ray (or photon) contained within the angle element $d\theta$ and $d\phi$ must pass through dA_1 and dA_2, i.e.,

$$d\Phi_1 \equiv d\Phi_2 [\text{w}] \tag{3.34}$$

By definition the irradiance on each surface can be expressed as:

$$E_1 = \frac{d\Phi_1}{dA_1} [\text{wm}^{-2}]$$

and

$$E_2 = \frac{d\Phi_2}{dA_2} [\text{wm}^{-2}] \tag{3.35}$$

but the area elements dA can be represented as the product of the arc lengths $rd\theta$ and $rd\phi$, i.e.,

$$dA_1 = r_1 d\theta r_1 d\phi [\text{m}^2]$$

and

$$dA_2 = r_2 d\theta r_2 d\phi [\text{m}^2] \tag{3.36}$$

Substituting Eqs. (3.36) and (3.35) into Eq. (3.34) and rearranging, we have

$$E_1 r_1^2 d\theta d\phi = d\Phi_1 = d\Phi_2 = E_2 r_2^2 d\theta d\phi [\text{w}] \tag{3.37}$$

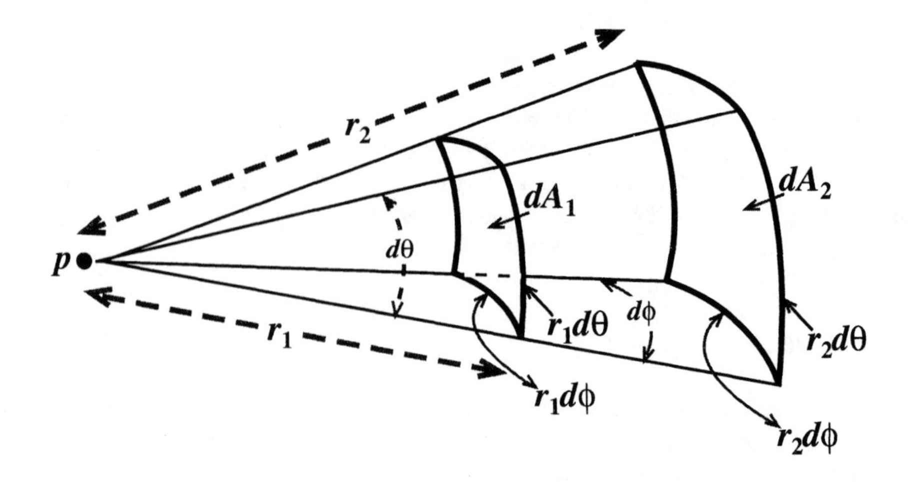

(a) The irradiance E_2 at r_2 compared to the irradiance at r_1 is given by $E_2 = E_1 r_1^2 r_2^{-2}$.

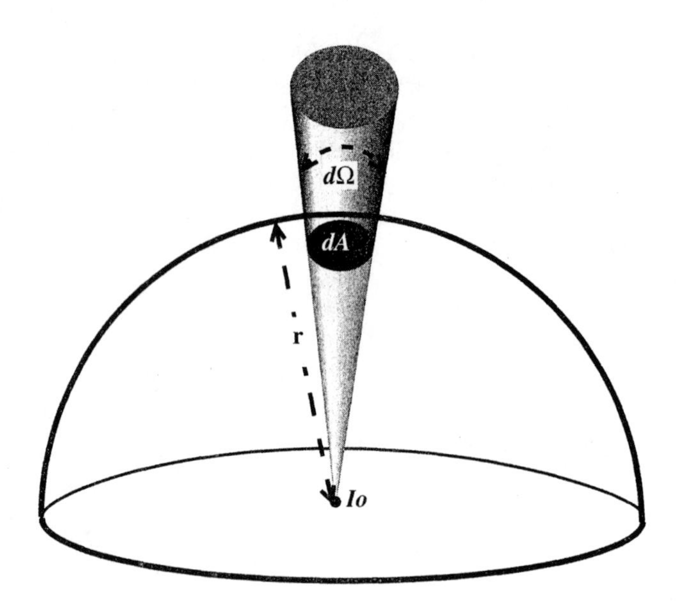

(b) A source of radiant intensity I_o produces an irradiance at a distance r of I_o/r^2.

Figure 3.9 Inverse-square law for irradiance variation from a point source.

or

$$E_2 = \frac{E_1 r_1^2}{r_2^2} \tag{3.38}$$

The irradiance varies inversely with the square of the distance from a point source. A similar relationship can be derived for the irradiance at a distance r[m] from a source of known radiant intensity I_0 [wsr^{-1}] as shown in Figure 3.9(b). If we consider an element of solid angle $d\Omega$ encompassing an area element dA on the surface of interest, the irradiance onto the surface must by definition be:

$$E = \frac{d\Phi_E}{dA} \tag{3.39}$$

where $d\Phi_E$ is the element of flux onto the surface dA. Also by definition the radiant intensity is

$$I_0 = \frac{d\Phi_I}{d\Omega} \tag{3.40}$$

where $d\Phi_I$ is the element of flux encompassed by the element of solid angle $d\Omega$. Clearly, in a lossless isotopic media $d\Phi_E = d\Phi_I$, or substituting

$$EdA[\text{wm}^{-2}\text{m}^2] = d\Phi_E[\text{w}] = d\Phi_I[\text{w}] = Id\Omega[\text{wsr}^{-1}\text{sr}] \tag{3.41}$$

Recall that by definition $d\Omega = dA/r^2$. Substituting and rearranging yields

$$E = \frac{I}{r^2}[\text{wm}^{-2}] \tag{3.42}$$

which again reflects a variation in irradiance from a point source that is inversely proportional to the distance squared.

To see where this concept comes into play, let us consider how much energy from the sun reaches the outside of the earth's atmosphere in terms of irradiance, i.e., the exoatmospheric irradiance. We then wish to know how much of the exoatmospheric irradiance would reach the earth in the absence of an atmosphere. To begin, we will make the simplifying assumption that if we compute the total flux from the sun and then treat the sun as a point source radiating equally in all directions, the irradiance we compute at the earth-sun distance will be a reasonable approximation of the true irradiance. Grum (1979) suggests that if the ratio of the source sensor distance to the source radius is more than a factor of 10, then point source approximations should introduce relatively small error.

To begin, we assume that if the sun is a 5800-K blackbody radiator, then the exitance from the solar surface is:

$$M = \sigma T^4 = 5.6697 \cdot 10^{-8}[\text{wm}^{-2}K^{-4}](5800[K])^4 = 6.42 \cdot 10^7[\text{wm}^{-2}] \tag{3.43}$$

and the total flux from the solar surface is

$$\begin{aligned}\Phi &= M \cdot 4\pi r_s^2 = 6.42 \cdot 10^7[\text{wm}^{-2}] \cdot 4\pi \cdot (695.5 \cdot 10^6[\text{m}])^2 \\ &= 3.9 \cdot 10^{26}[\text{w}]\end{aligned} \tag{3.44}$$

where $r_s = 695.5 \cdot 10^6$ [m] is the mean radius of the sun.

If we assume this flux comes from a point source at the center of the sun such that it would produce the observed exitance at the sun's surface, then it should produce a flux at the mean earth-sun distance $r'_{es} = 149.5 \cdot 10^9$[m] of

$$E_{ex} = \frac{\Phi}{4\pi r_{es}'^2}$$

$$= \frac{3.9 \cdot 10^{26}[w]}{4\pi(149.5 \cdot 10^9[m])^2} = 1390[wm^{-2}] \tag{3.45}$$

These crude calculations are in good agreement with the estimate for the "solar constant" (exoatmospheric irradiance at the mean earth-sun distance) of 1373 ± 20 [wm^{-2}] as put forward by Frohlich (1977) based on a number of measurement programs. If we assume the earth's atmosphere to be approximately 200-km thick, then, according to Eq. (3.38), the variation in irradiance from the exoatmospheric value (E_{ex}) to the value at the earth's surface due strictly to geometric effects will be

$$E_s = \frac{E_{ex} r_{es}'^2}{(r_{es}' + 200 \cdot 10^3[m])^2} = E_{ex}(0.999997) \cong E_{ex} \tag{3.46}$$

The only losses that must be considered in propagating the solar beam are due to atmospheric absorption and scattering, since no geometric effects need to be considered. On the other hand, Chen (1985) points out that the variation in the earth-sun distance due to the ellipticity of the earth's orbit can produce as much as a 3.4% variation in the exoatmospheric irradiance from the value at the mean earth-sun distance. We will often be interested in the exact exoatmospheric spectral irradiance. This value can be computed from tabulated spectral values as found in Thekaekara (1972) for the exoatmospheric irradiance at the mean earth-sun distance $E_{exavg\lambda}$ corrected for the earth-sun distance on the day of interest according to

$$E_{ex\lambda} = E_{exavg\lambda} \frac{r_{es}'^2}{r_{es}^2} \tag{3.47}$$

where $E_{ex\lambda}$ is the exoatmospheric spectral irradiance on a specific day when the earth-sun distance is r_{es}.

3.3.2 Projected Area Effects (cos θ)

Most radiance calculations involve dealing with radiation propagating through or onto a surface that is not perpendicular to the propagating beam. In these cases, it is important to take into account the effects of projected area. Consider the problem illustrated in Figure 3.10(a). In this case, the element of flux $d\Phi$, associated with the ray **P**, is incident on the element of unit area dA_u such that the normal to the surface **N** is coincident with **P** (i.e., the beam is perpendicular to the surface). In this case, the irradiance onto the surface is by definition

$$E_0 = \frac{d\Phi_0}{dA_u} \tag{3.48}$$

where the subscript 0 refers to the angle from the normal to the surface to the incident ray. In case (b) illustrated in Figure 3.10(b), the same amount of incident flux is spread over a larger area, since the surface is rotated through an angle θ. The irradiance in this case is

$$E_\theta = \frac{d\Phi_\theta}{dA_\theta} = \frac{d\Phi_0}{dA_u/\cos\theta} = E_0 \cos\theta \tag{3.49}$$

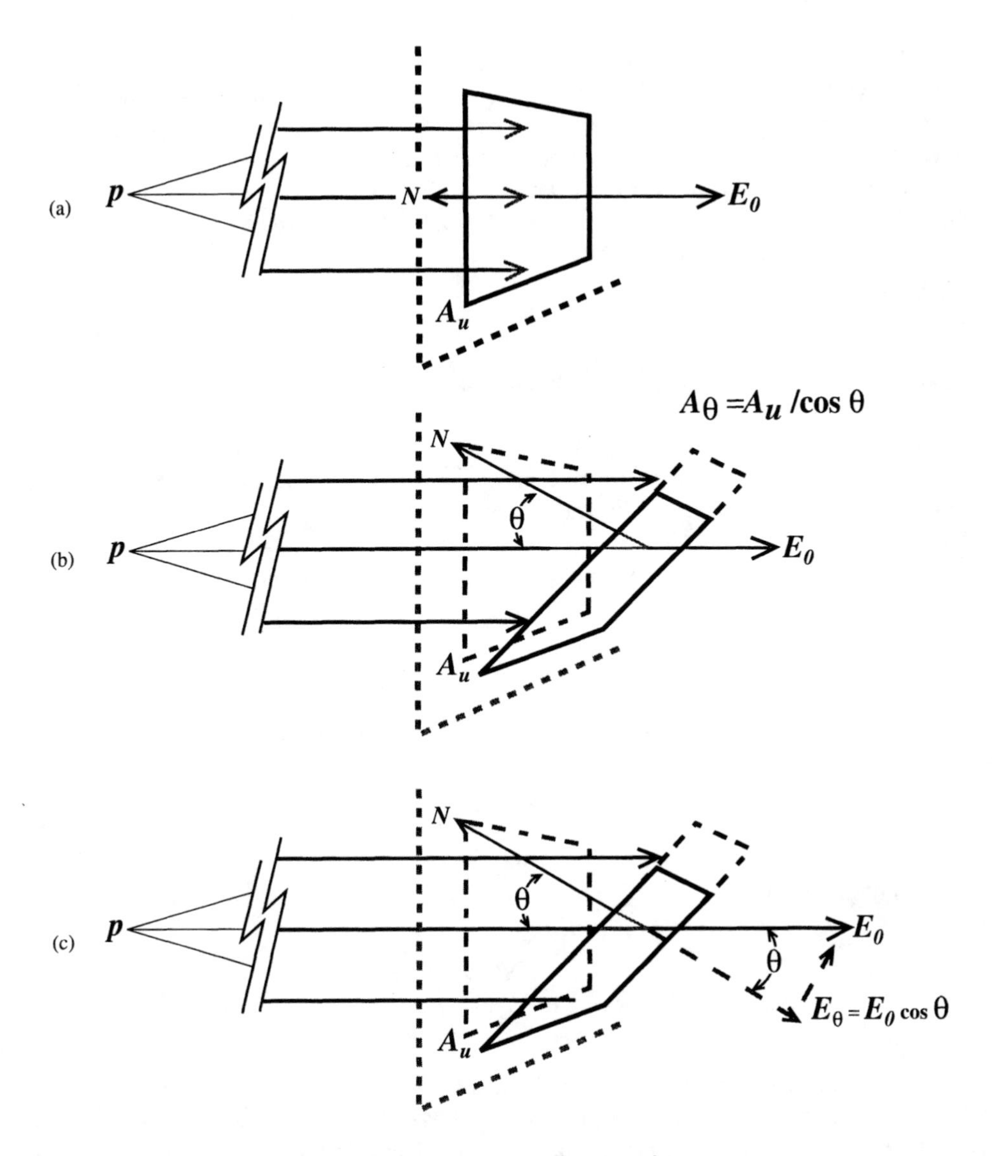

Figure 3.10 Projected area effects [cos (θ)].

where θ is the angle from the normal to the surface to the incident ray, and we have used the geometric construction that, in order for the area element to capture the same amount of flux (i.e., for $d\Phi_\theta = d\Phi_0$), the area must be increased by a factor of cos θ. That is, the area element A_θ is the projection of the area element perpendicular to the ray (A_u) onto the plane of the surface of interest ($A_0/\cos \theta$). Many people find it easier to think of the irradiance onto the surface as the component of the irradiance vector along the ray, which is normal to the surface as illustrated in case (c) in Figure 3.10(c).

3.3.3 Lambertian Surfaces

One of the most difficult questions we will have to deal with in remote sensing is the question of how the energy leaving a surface is angularly distributed into the hemisphere above the surface. Is all of it perpendicular to the surface? Is the radiant intensity the same in all directions? Can the intensity distribution be characterized in any functional form? We will often make use of an idealized surface known as a *Lambertian surface* to help us deal with these questions. A Lambertian radiator, or reflector, is characterized by a well-behaved variation in radiant intensity according to

$$I_\theta = I_0 \cos\theta \tag{3.50}$$

where I_0 is the intensity normal to the surface and θ is the angle from the normal to the surface to the direction of interest. Thus, a Lambertian surface will have a steady decrease in the intensity approaching zero at grazing angles. It is often more intuitive to consider the angular distribution of radiance from a Lambertian surface. Visual response is proportional to radiance, and this will tell us how a Lambertian surface will "look" when viewed at different angles.

By definition the radiance along the normal from a Lambertian surface will be

$$L_0 = \frac{dI_0}{dA\cos(0)} = \frac{dI_0}{dA} \tag{3.51}$$

Similarly, the definition of radiance into any direction θ from the normal is

$$L_\theta = \frac{dI_\theta}{dA\cos\theta} \tag{3.52}$$

Substituting Eqs. (3.50) and (3.51) into Eq. (3.52) yields

$$L_\theta = \frac{dI_\theta}{dA\cos\theta} = \frac{dI_0\cos\theta}{dA\cos\theta} = \frac{dI_0}{dA} = L_0 \tag{3.53}$$

which states that the radiance from a Lambertian surface is the same in all directions (i.e., the decreasing intensity with angle is exactly compensated by a decrease in projected area). Since perceived brightness is proportional to the radiance in the visible region, this means that a Lambertian surface would look the same from all directions. There are many examples in nature of surfaces that are approximately Lambertian, particularly if grazing angles are avoided. Thus, while we cannot assume that all surfaces are Lambertian, we will often use Lambertian assumptions as approximations or as a point of reference for discussion of less well-behaved surfaces.

3.3.4 Magic π

Because of the importance of Lambertian surfaces as reference materials, we will consider the relationship between the exitance from a Lambertian surface and the radiance from that surface. For example, consider the case illustrated in Figure 3.11(a) of a Lambertian surface having reflectance r exposed to a beam having irradiance E_0 onto an imaginary surface perpendicular to the beam at the surface. We know the radiance from the surface will be the same in all directions, but we seek its magnitude. The irradiance onto the surface from vec-

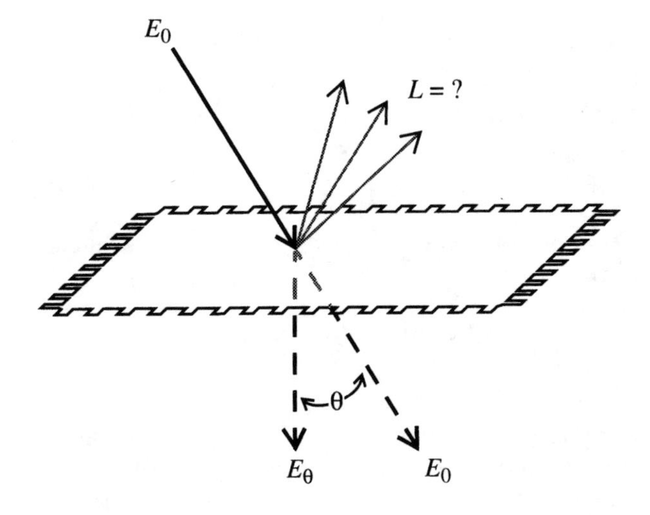

(a) Statement of problem in terms of reflected radiance.

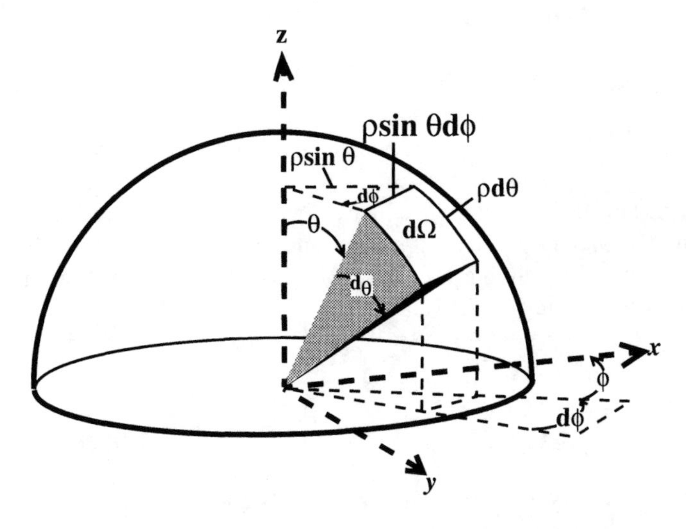

(b) The element of solid angle $d\Omega$ can be expressed as $d\Omega$=sinθdθdϕ.

Figure 3.11 Magic π.

tor analysis of Figure 3.11(a) is

$$E_\theta = E_0 \cos\theta \;[\mathrm{wm}^{-2}] \tag{3.54}$$

where θ is the angle from the normal to the surface to the incident ray. From the definition of reflectance Eq. (3.24), we have the exitance expressed as:

$$M = E_\theta r = E_0 r \cos\theta \;[\mathrm{wm}^{-2}] \tag{3.55}$$

The radiance by definition can then be expressed as:

$$L = \frac{dM}{d\Omega \cos\theta} \; [\text{wm}^{-2}\text{sr}^{-1}] \tag{3.56}$$

or rearranging

$$dM = L d\Omega \cos\theta \; [\text{wm}^{-2}] \tag{3.57}$$

If we take the integral of both sides to find the total exitance into the hemisphere above the reflecting surface, we have

$$M = \int dM = \int L \cos\theta d\Omega [\text{wm}^{-2}] \tag{3.58}$$

where the left-hand side is simply the exitance we already know from Eq. (3.55), and the right-hand side is the integral with respect to solid angle over the hemisphere above the surface. Cos (θ) changes with respect to solid angle, so we need to convert $d\Omega$ into a more tractable form to perform the integration. If we construct the element of solid angle associated with an arbitrary (or unit) sphere of radius ρ, as illustrated in Figure 3.11(b), then we have by definition:

$$d\Omega = dA/\rho^2 \tag{3.59}$$

and by construction:

$$dA = \rho d\theta \rho \sin\theta d\phi [\text{m}^2] \tag{3.60}$$

where the surface is defined by the x,y plane, θ is the angle from the normal to the surface (i.e., from the z axis), and ϕ is the azimuthal angle swept about the z axis in any plane parallel to the x,y plane. Substituting Eq. (3.60) into Eq. (3.59), we see that the dependence on the size of the hemisphere vanishes leaving

$$d\Omega = \frac{\rho^2 d\theta \sin\theta d\phi}{\rho^2} \frac{[\text{m}^2]}{[\text{m}^2]} = \sin\theta d\theta d\phi [\text{sr}] \tag{3.61}$$

Substituting this expression for the element of solid angle in Eq. (3.58), we are left with a double integral on θ and ϕ of the form

$$M = \int\int L \cos\theta \sin\theta d\theta d\phi \tag{3.62}$$

where inspection of Figure 3.11(b) for the limits of integration over the hemisphere yields

$$M = \int_{\phi=0}^{2\pi} \int_{\theta=0}^{\frac{\pi}{2}} L \cos\theta \sin\theta d\theta d\phi \tag{3.63}$$

If *and only if* we have a Lambertian surface, the radiance is a constant and can be taken outside the integral. The integral on θ can then be seen to be

$$M = L \int_{\phi=0}^{2\pi} \left[\int_{\theta=0}^{\frac{\pi}{2}} \cos\theta \sin\theta d\theta \right] d\phi$$

$$= L \int_{\phi=0}^{2\pi} \left[\frac{\sin^2\theta}{2} \right] \Bigg|_0^{\frac{\pi}{2}} d\phi = \frac{L}{2} \int_{\phi=0}^{2\pi} d\phi \tag{3.64}$$

and the integral on ϕ yields

$$M = L[\text{wm}^{-2}\text{sr}^{-1}]\pi[\text{sr}] = L\pi[\text{wm}^{-2}] \tag{3.65}$$

or on rearranging, we see that for a Lambertian surface the radiance and radiant exitance are related by a factor of π, i.e.,

$$L = \frac{M}{\pi} \tag{3.66}$$

or in our case

$$L = \frac{M}{\pi} = \frac{E_0 r \cos\theta}{\pi} \tag{3.67}$$

This factor of π[sr] shows up repeatedly (magically) in radiometric calculations (usually without its derivation and often inappropriately), but it is generally valid only for treatment of Lambertian surfaces or when a comparison to a Lambertian surface is used as is the case with the reflectance factors that are introduced in the next chapter (cf. Sec. 4.2.1). For example, if we assume a blackbody is a Lambertian radiator, the Planck equation can be expressed in terms of spectral radiance as:

$$L_\lambda = \frac{M_\lambda}{\pi} = 2hc^2\lambda^{-5}(e^{\frac{hc}{\lambda kT}} - 1)^{-1}[\text{wm}^{-2}\text{sr}^{-1}\mu\text{m}^{-1}] \tag{3.68}$$

It is important to realize that in order to find the relationship between radiance and exitance [i.e., to solve Eq. (3.63)], the relative variation in radiance into all possible directions in the hemisphere above the surface would have to be known. It is the magnitude of this task that forces us to make the assumption of Lambertian behavior whenever possible.

3.3.5 Lens Falloff

In Chapter 2, we introduced the concept of using a film-based camera as a radiometer, and in the sections that follow we will expand on this theme by addressing the use of a variety of imaging systems as quantitative radiometers. The imaging nature of these radiometers will cause many of them to have a substantial angular field of view. To see what effect this off-axis viewing will have on the radiometry, we will consider the simplest case of a pinhole camera and examine the variation in exposure as a function of the view angle (θ). A pinhole camera is simply a light-tight box (e.g., a shoe box) with a sheet of film at the back, as illustrated in Figure 3.12. The image is acquired by opening a shutter (removing your finger from the hole in the shoe box) and exposing the film for a short period of time. The small aperture acts as a lens by only allowing flux from a very limited direction to hit each point on the film. In our case we want to imagine imaging a scene of uniform radiance (e.g., a uniform

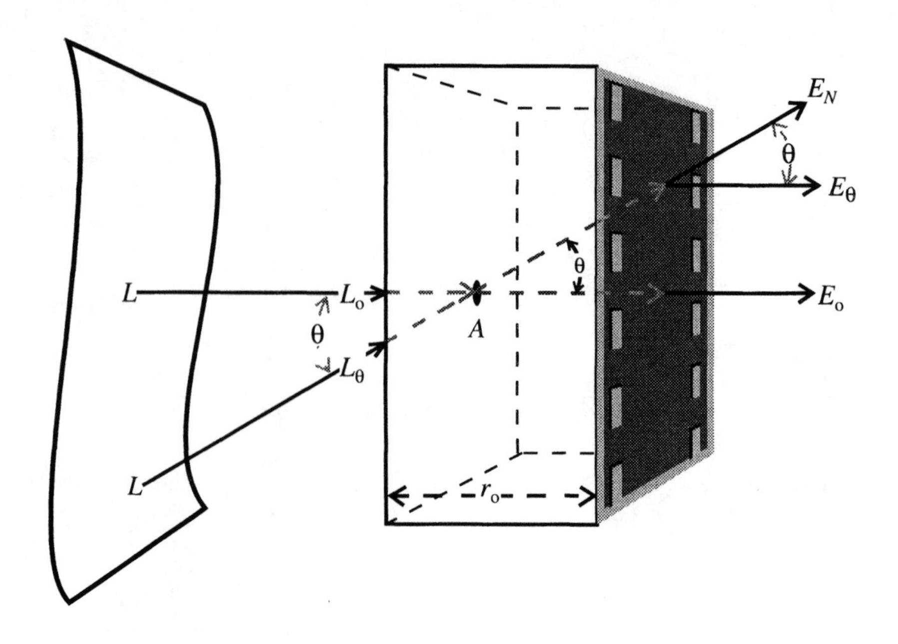

Figure 3.12 Cos^4(θ) lens falloff for a pinhole camera.

gray Lambertian reflector) and determine if the exposure will vary across the image format (i.e., will two identical surfaces appear different due solely to their location in the image?). To begin, we compute the irradiance (E_0) on the focal plane along the primary axis. If the scene has radiance $L[\mathrm{wm}^{-2}\mathrm{sr}^{-1}]$, then in a lossless medium we will have radiance $L_0 = L$ reaching the aperture along a $\theta = 0$ view angle (i.e., along the primary axis). The radiant intensity through the aperture in the $\theta = 0$ direction is

$$dI_0 = L_0 dA \cos\theta \tag{3.69}$$

from Eq. (3.9). For small apertures, we can assume that θ and dI_0 do not vary across the aperture, so

$$I_0 = L_0 A \cos(0)[\mathrm{wm}^{-2} sr^{-1}][\mathrm{m}^2] = L_0 A[\mathrm{wsr}^{-1}] \tag{3.70}$$

where A is the area of the aperture $[\mathrm{m}^2]$. From Eq. (3.42), the irradiance onto the center of the focal plane a distance r_0 away is

$$E_0 = \frac{I_0}{r_0^2}[\mathrm{wm}^{-2}] = \frac{L_0 A}{r_0^2}[\mathrm{wm}^{-2}] \tag{3.71}$$

and recalling that the exposure is simply the irradiance times the time of exposure (t) yields the on-axis exposure of

$$H_0 = E_0 t[\mathrm{wm}^{-2}][\mathrm{sec}] = \frac{L_0 A t}{r_0^2}[\text{joules m}^{-2}] \tag{3.72}$$

Similarly, in a lossless medium the radiance reaching the pinhole headed in a direction defined by the view angle (θ) from the primary axis is

$$L_\theta = L = L_0 \tag{3.73}$$

By analogy to Eq. (3.70), the radiant intensity in the direction defined by the view angle θ radiating from the aperture is

$$I_\theta = L_0 A \cos\theta \tag{3.74}$$

i.e., the radiant intensity is reduced due to the aperture appearing smaller for off-axis viewing. The irradiance E_N that would be observed on a surface normal to the ray at the focal plane (cf. Fig. 3.12) is

$$E_N = \frac{I_\theta}{r_\theta^2} = \frac{I_\theta}{(r_0/\cos\theta)^2} = \frac{L_0 A \cos^3\theta}{r_0^2} \tag{3.75}$$

i.e., the irradiance is reduced by the square of the increased distance from the aperture to the focal plane at the view angle θ. To find the irradiance onto the focal plane, we take the vector component of E_N perpendicular to the focal plane yielding

$$E_\theta = E_N \cos\theta = \frac{L_0 A \cos^4\theta}{r_0^2} \tag{3.76}$$

The resulting exposure would be

$$H_\theta = E_\theta t = \frac{L_0 A t}{r_0^2}\cos^4\theta = H_0 \cos^4\theta \tag{3.77}$$

indicating that the exposure would fall off from the center of the format as the cosine of the view angle to the fourth power. Since most of us don't do radiometric image analysis with pinhole cameras, we need to know how this derivation relates to conventional imaging systems. From the standpoint of radiometry, the major function of a lens is to gather flux more effectively from all directions. As a result, the effective size of the aperture is more nearly constant with view angle—i.e., $\cos^4(\theta)$ becomes approximately $\cos^3(\theta)$. With a conventional camera, the effects due to increase in path and projected area of the sensor in the off-axis direction remain approximately the same as for the pinhole camera. The fact that the aperture may not be a good approximation of a point source results in a more complicated version of Eq. (3.42) that must account for variation in the distance and irradiation angles from the various elements of area across a finite aperture. For a real camera system, the relation between irradiance across the format can usually be approximated as:

$$E_\theta = E_0 \cos^n(\theta) \tag{3.78}$$

where for simple cameras n will usually take on a value of approximately 3, with the exact value normally obtained by curve fitting through experimental data.

To evaluate the magnitude of this effect, consider first a 70-mm format camera flown with an 80-mm focal length lens. The actual image format is approximately 55-mm or 28-mm from center to edge for a field of view of

$$\theta = \tan^{-1}\left(\frac{28\text{ mm}}{80\text{ mm}}\right) \cong 19^\circ \tag{3.79}$$

If an optical bench analysis of the lens yields a value for n in Eq. (3.78) of 3.2, then the falloff in exposure will be

$$H_{19} = H_0 \cos^{3.2}(19^\circ) \cong 0.83\ \mathrm{H}_0 \tag{3.80}$$

i.e., There will be only 83% as much exposure at the edges of the format as at the center for identical objects. This is clearly a nonnegligible effect that must be compensated for in quantitative image analysis. Because it is a well-behaved repeatable effect, it is easily accounted for by measuring the exposure (H_θ) or irradiance at any point of interest on the focal plane and then correcting for what that exposure would have been if the object were imaged at the center of the format. The corrected value H'_0 would be

$$H'_0 = \frac{H_\theta}{\cos^n(\theta)} \tag{3.81}$$

If we consider the exposure falloff from a standard 9-inch format camera (with actual image area 230 mm x 230 mm) with a 6-inch focal length lens, we have a field of view to the edges of the format of

$$\theta = \tan^{-1}\left(\frac{115\text{mm}}{152\text{mm}}\right) \cong 37^\circ \tag{3.82}$$

and 47° to the corners of the format. This results in exposures of 51% and 32% of the on-axis values for $n = 3$. This effect is very dramatic and causes the edges of the format to be underexposed and difficult to interpret when the center of the format is properly exposed. To compensate for this effect, antivignetting filters are often used with this type of camera to partially account for the lens falloff effects. These filters have lower transmission in the center of the format than at the edges where the transmission is near unity. The combined effects of lens falloff and an antivignetting filter are illustrated in Figure 3.13. The $\cos^3(\theta)$ curve shows how exposure would vary with angle with no filter present. The filter transmission with angle $\tau(\theta)$ is also shown, with the actual exposure variation resulting from the product of the $\cos^3(\theta)$ curve and the transmission curve. In general, this will still result in considerable residual variation across the format, which must be measured and corrected. Making an exact antivignetting filter (i.e., $\cos^n(\theta)\tau(\theta)$ = a constant) is seldom attempted. Besides the technical difficulties, the central transmission would be so low that excessively long exposure times would be required to properly expose the film.

3.4 ATMOSPHERIC PROPAGATION

One of the most critical factors affecting radiometric remote sensing is the effect of the atmosphere on the propagating energy. In most cases, the atmosphere is perceived as a hostile entity whose adverse impacts must be neutralized or eliminated before remotely sensed data can be properly analyzed. In this section, we will examine the underlying principles that describe radiation propagation through the atmosphere. These principles will be used in the next chapter to describe the governing equation for the radiance reaching a remote sensing system.

3.4.1 Atmospheric Absorption

We want to examine the effect of the atmosphere on a beam of energy propagating through the atmosphere (e.g., a beam associated with the exoatmospheric solar irradiance as it prop-

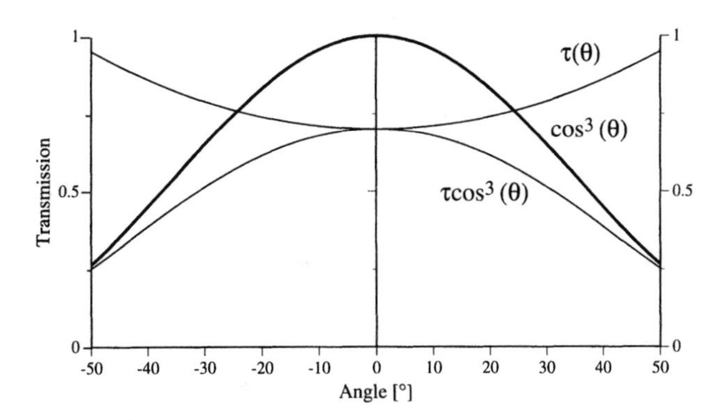

Figure 3.13 Lens falloff and antivignetting filter effects.

agates to the earth's surface). To begin, we will consider an idealized atmosphere that only exhibits absorption effects (i.e., nonscattering). We define absorption as the process of removal of energy (photons) from the beam by conversion of the electromagnetic energy to another form (usually thermal). For atmospheric propagation, this results from absorption of the photons by the constituent molecules in the atmosphere when a photon induces a molecular vibration, rotation, or electron orbital transition to an alternate energy state. Recall from modern physics that these are discrete (i.e., quantized) energy transitions so that only photons with selected energy levels can be absorbed. In the simplest case, we would have absorption occurring in narrow spectral lines associated with those photons having the exact energy ($h\nu$) needed to induce an allowable energy transition. Because of the low densities in the atmospheric media of interest, we can assume that each absorption event is discrete and that the total absorption is simply the sum of the individual events (i.e., the absorption of each photon only impacts the molecule that absorbs it). This would lead to an atmospheric absorption spectra made up of discrete lines as illustrated in Figure 3.14(a). In fact, several factors contribute to the broadening of these discrete lines [cf. USAFGL (1985) Chapter 18 for a general treatment]. The Heisenberg uncertainty principle indicates that the time interval of energy transitions can only be known within set limits, and, therefore, the associated emission, or absorption frequency, will have some uncertainty or width. However, this effect is small compared to Doppler and pressure broadening of the absorption lines. Doppler broadening is due to the fact that a particle moving toward or away from an observer will emit or absorb frequency-shifted photons in accord with conventional Doppler theory. This results in a random distribution of the absorption frequencies about the nominal absorption line caused by the random distribution of particle velocities relative to the incident beam. Since particle motion is increased at higher temperature and dampened by inertia, the extent of Doppler broadening increases with temperature and decreases with molecular mass according to

$$\Delta\nu \propto \sqrt{\frac{T}{M}} \tag{3.83}$$

where $\Delta\nu$ is the width of the absorption frequency due to broadening, T is the temperature in Kelvin, and M is the gram molecular weight.

Pressure or collision broadening of the absorption lines results from the interaction or collision of one molecule with another perturbing the energy states. The degree of broadening ($\Delta\nu$) is proportional to the relative pressure (p). The overall theory of line broadening is

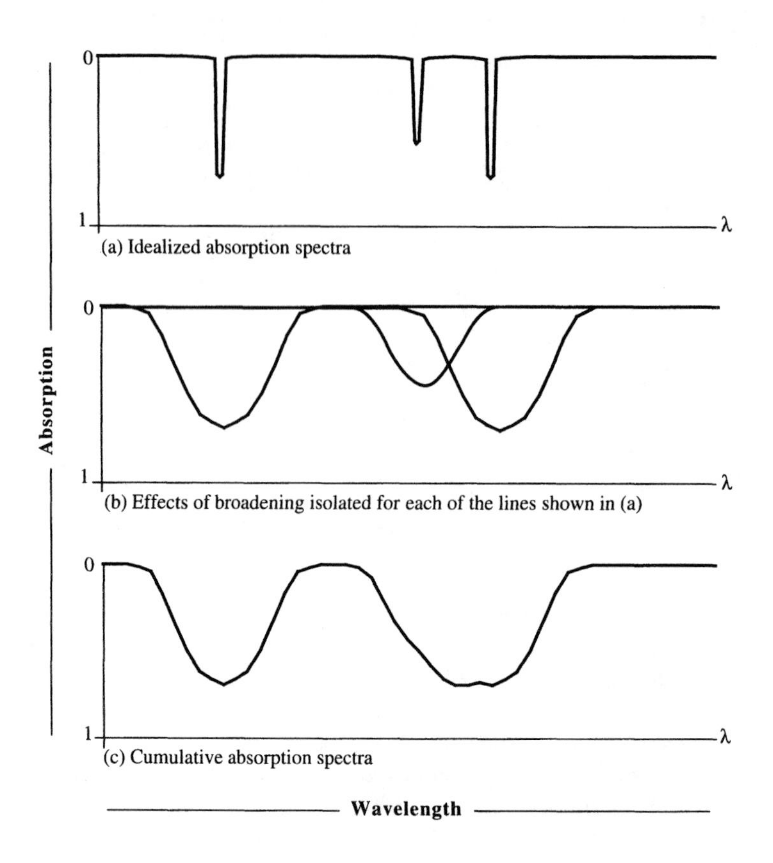

Figure 3.14 Characteristics of absorption spectra.

beyond the scope of this treatment, and we must be satisfied at this point to illustrate the net result in Figure 3.14. What we see is that the absorption spectra is made up of many discrete lines that are broadened to various extents. Furthermore, when many broadened lines are close together, we tend toward a continuum of absorption where the discrete line nature is largely lost. Depending on the spectral region of interest and the spectral resolution of the sensor, both the individual broadened lines and the continuum concept will be important for remote sensing.

In simplified form, we can derive the following relationships between the number and efficiency of absorbers and their effect on the propagating beam. At each wavelength, we define the absorption cross section C_α to be the effective size of a molecule relative to the photon flux at that wavelength. Conceptually, this can be expressed as:

$$C_\alpha = C_g \xi = \pi r^2 \xi [\mathrm{m}^2] \tag{3.84}$$

where C_g [m^2] is the geometric cross section for a molecule of radius r [m] and ξ is a unitless wavelength-dependent efficiency factor that is proportional to the molecule's ability to absorb flux. Values of C_α are available as a function of wavelength for the atmospheric constituents. These values can be derived for particular temperatures and pressures from experimental data or through molecular energy theory, then adjusted for the effects of the temperature and pressure of interest as alluded to above. The molecule is then assumed to be a perfect absorber over that cross-sectional area. To compute the fractional amount of energy lost

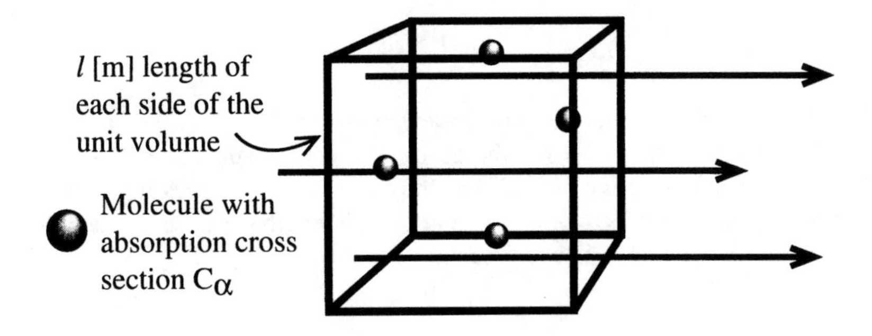

(a) A unit volume containing m' absorption centers. We assume that the medium has a large mean free path such that in a small volume, if we project the molecules onto one face, there will be no overlap, i.e.,

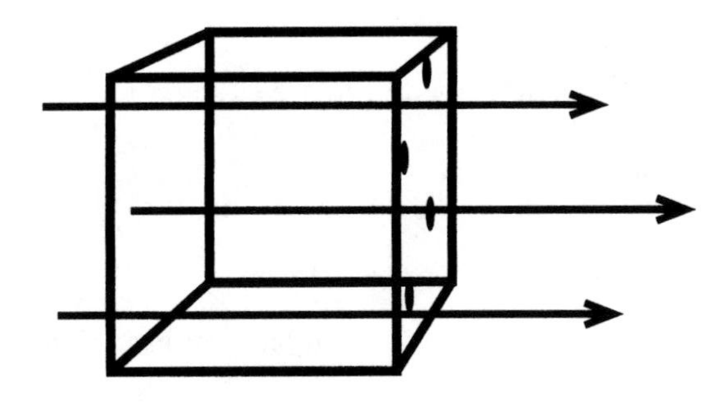

(b) Projection of absorbers onto the face of the volume.

Figure 3.15 Computation of the absorption coefficient.

per unit length of transit in a propagating beam, we need to know the number density of the molecules. Referring to Figure 3.15, we let m' be the number of molecules in a unit volume of side dimension l[m]. Then, the area blocked (A_b) by the molecules is

$$A_b = m' C_\alpha [\mathrm{m}^2] \tag{3.85}$$

The area on the face of the volume (A_f) onto which the molecules were projected is

$$A_f = l^2 [\mathrm{m}^2] \tag{3.86}$$

The fraction of the face blocked by the absorbing molecules (F) is

$$F = \frac{m' C_\alpha}{l^2} \left[\frac{\mathrm{m}^2}{\mathrm{m}^2} \right] \tag{3.87}$$

Therefore, the fraction amount of flux absorbed β_α per unit length of transit (l) is

$$\beta_\alpha = \frac{F}{l} = \frac{m'}{l^3} C_\alpha [\text{m}^{-1}] = \frac{m'}{V} C_\alpha = m C_\alpha [\text{m}^{-1}] \tag{3.88}$$

where V is the unit volume, m is the number density of molecules, defined as the number of molecules per unit volume and β_α is the absorption coefficient, defined as the fractional amount of flux lost to absorption per unit length of transit in a propagating beam.

According to Grum (1979), for an element of path length dz[m] in the medium, the element of fractional flux lost can be expressed as:

$$\frac{d\Phi}{\Phi} = -\beta_\alpha(z)dz \tag{3.89}$$

where we have made the dependence of β_α on location in the media explicit. For propagation along a finite path starting at distance zero where we have initial flux Φ_0 to distance z where we have flux Φ_z, we have

$$\int_{\Phi_0}^{\Phi_z} \frac{d\Phi}{\Phi} = \int_0^z -\beta_\alpha(z)dz = \ln \Phi \Big|_{\Phi_0}^{\Phi_z} = \int_0^z -\beta_\alpha(z)dz = \ln \Phi_z - \ln \Phi_0$$
$$= \ln\left(\frac{\Phi_z}{\Phi_0}\right) = \int_0^z -\beta_\alpha(z)dz \tag{3.90}$$

Making both sides powers of e to simplify the left-hand side yields

$$\frac{\Phi_z}{\Phi_0} = e^{-\int_0^z \beta_\alpha(z)dz} \tag{3.91}$$

Recognizing the left-hand side as a definition of transmission and solving for the simplified case of a homogeneous medium, we have

$$\tau = \frac{\Phi_z}{\Phi_0} = e^{-\beta_\alpha \int_0^z dz} = e^{-\beta_\alpha z} \tag{3.92}$$

which is variously known as Lambert's law or Bouguer's law.

The product $\beta_\alpha z$ is generally referred to as the *optical depth* (δ), i.e.,

$$\delta_\alpha = \beta_\alpha z \tag{3.93}$$

To this point we have implicitly assumed a media containing a single constituent. For a homogeneous atmosphere containing many types of molecules, we introduce the subscript i to denote the particular constituent. If we assume that the molecules interact independently with the propagating flux, we can express the transmission as:

$$\tau = \prod \tau_i = e^{-\sum \delta_i} = e^{-\sum \beta_{\alpha i} z} = e^{-\sum m_i C_{\alpha i} z} = e^{-\beta_\alpha z} = e^{-\delta_\alpha} \tag{3.94}$$

where Π designates the product of the transmission values for each constituent if computed separately, the summation (Σ) is over all constituents, and we redefine $\beta_\alpha = \Sigma \beta_{\alpha i}$ to be the composite absorption coefficient and δ_α to be the composite optical depth due to absorption.

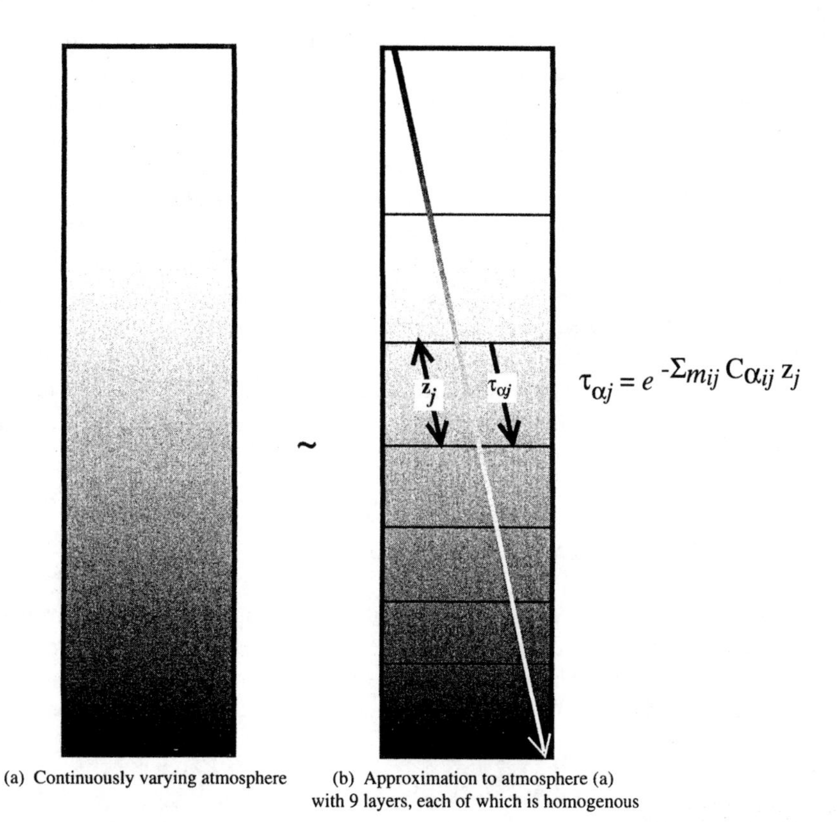

Figure 3.16 Transmission calculations using homogeneous layers to approximate a stratified atmosphere. Each constituent designated by the subscript i has an absorption cross section $C_{\alpha ij}$ and a number density m_{ij} where the subscript j designates the layer.

From Eq. (3.94) it is clear that to find the transmission along a beam, we need to know the number density of each constituent molecule along the path, the absorption cross section as a function of wavelength for each constituent, and how that cross section varies with environmental parameters (e.g., T,P) along the path. Often the atmosphere can be treated as a series of homogeneous layers as shown in Figure 3.16. The overall optical depth can then be found from a numerical integration of Eq. (3.90) to be a simple sum of the optical depths for each of the homogeneous layers. The transmission is then given by

$$\tau = e^{-\sum_j \left(\sum_i m_{ij} C_{\alpha ij} z_j \right)} = e^{-\sum_j \delta_{\alpha j}} = e^{-\delta_a} \tag{3.95}$$

where the subscript i differentiates the constituents in the atmosphere and the terms are as described in Figure 3.16.

3.4.2 Atmospheric Scattering

In addition to absorption losses, energy may leave the beam due to scattering. This process is most easily thought of as a disturbance of the electromagnetic field by the constituents in the atmosphere resulting in a change in the direction and spectral distribution of the energy in the beam. The details of scattering theory and the related radiation propagation are quite complex, and only a limited treatment will be given here. The simplified treatment needed for our purposes has been reduced from the rigorous coverage by Chandrasekhar (1960) and Van de Hulst (1981). There are three primary forms of scattering. Rayleigh scatter is the result of the EM wave interacting with the very small particles or molecules that we think of as making up the atmosphere itself. It occurs when the particles are much smaller than the wavelength of the incident flux. Because it can be described in a closed form, we will treat it in some detail and use the formalism developed as a model for our more general treatment of Mie scatter and nonselective scatter. Mie scattering occurs when the wavelength of the energy is approximately equal to the size of the particles. Aerosols, small dust particles, fossil fuel combustion products, and suspended sea salts are some of the major atmospheric constituents in this size range. Nonselective scattering occurs when the particles are very large compared to the wavelength of the incident energy. Large dust particles, water droplets, and ice crystals are sources of nonselective scattering.

3.4.2.1 Rayleigh Scatter

The fractional amount of energy scattered into a solid angle at an angle θ from the propagation direction per unit length of transit in the medium was first characterized by Lord Rayleigh (1871) in an attempt to explain the blue color of the sky. His expression for randomly polarized flux is still valid and can be written as:

$$\beta_r(\theta) = \frac{2\pi^2}{m\lambda^4}(n(\lambda)-1)^2(1+\cos^2\theta)[\mathrm{m}^{-1}\mathrm{sr}^{-1}] \tag{3.96}$$

where $\beta_r(\theta)$ is the Rayleigh angular scattering coefficient defined as the fractional amount of energy scattered into the direction θ per unit solid angle about θ per unit length of transit, θ is the deflection angle from the beam direction, $n(\lambda)$ is the wavelength dependent index of refraction of the medium, and m is the number density as previously introduced. We explicitly write the functional dependence of the angular scattering coefficient $\beta_r(\theta)$ on θ to differentiate it from the total scattering coefficient β_r to be introduced shortly. Inspection of Eq. (3.96) shows an inverse dependence with wavelength to the fourth power and an inverse dependence on number density. The wavelength dependence can explain why the sky is blue, since more short wavelength energy will be scattered out of the beam. It also explains why the solar disk is red at sunrise and sunset, since over long paths the fractional amount of energy in the beam will shift to longer wavelengths as the shorter wavelengths are disproportionately scattered out of the beam.

The inverse dependence of the Rayleigh law on number density appears to contradict our intuition. Since the number density near the earth's surface is higher, we should expect less scattering because of the inverse relationship. However, recognize that the index of refraction will also become larger as we approach the earth's surface and that the term $(n(\lambda)-1)^2$ will increase more rapidly than m as we approach the earth. So our intuition is satisfied with scattering increasing in the lower atmosphere.

The Rayleigh expression for scattering expressed in Eq. (3.96) does not directly tell us the energy loss from the beam per unit length (cf. Fig. 3.17). In order to find this, we must integrate the angular scattering coefficient over all possible solid angles according to

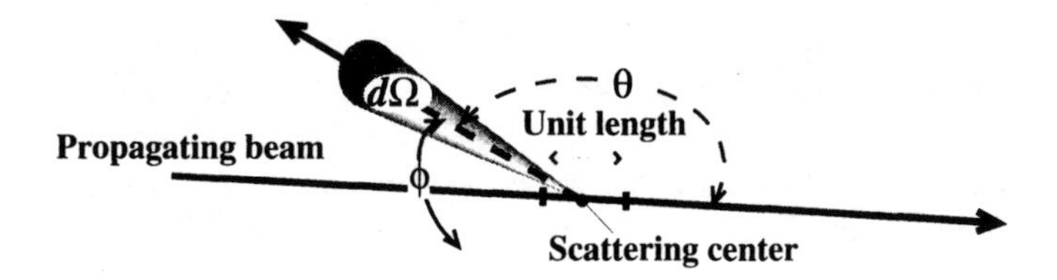

(a) The fractional amount of energy scattered in the direction θ per unit solid angle per unit length through the medium.

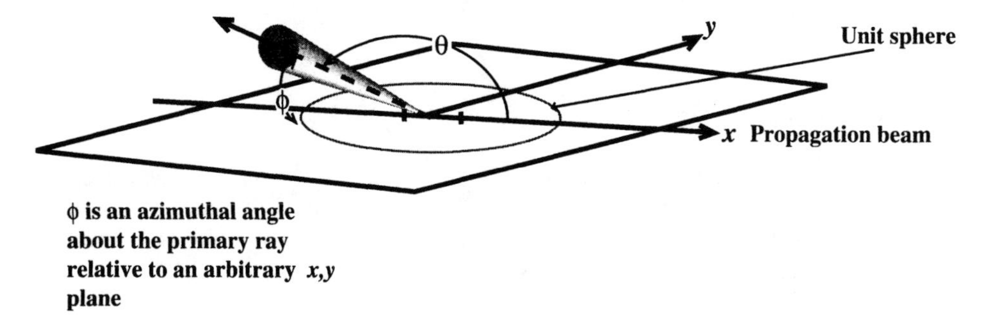

(b) Geometric terms.

Figure 3.17 Angular scattering coefficient.

$$\beta_r[\mathrm{m}^{-1}] = \int \beta_r(\theta)d\Omega = \int_{\phi=0}^{2\pi}\int_{\theta=0}^{\pi} \frac{2\pi^2}{m\lambda^4}(n(\lambda)-1)^2(1+\cos^2\theta)\sin\theta d\theta d\phi$$
$$= \frac{32\pi^3(n(\lambda)-1)^2}{3\lambda^4 m}[\mathrm{m}^{-1}] \tag{3.97}$$

where we have substituted sin $\theta d\theta d\phi$ for $d\Omega$ (cf. Fig. 3.11) and β_r is the Rayleigh scattering coefficient and can be interpreted in a fashion analogous to the absorption coefficient β_α.

Also by analogy to the absorption cross section, we introduce the wavelength-dependent Rayleigh scattering cross section to be

$$C_r[\mathrm{m}^2] = \frac{\beta_r}{m} \tag{3.98}$$

to define the effective size of a scattering center in terms of how efficiently it scatters flux from the beam. The Rayleigh optical depth can then be expressed as:

$$\delta_r = \beta_r z = mC_r z \tag{3.99}$$

which results in a transmission value of

$$\tau = e^{-\delta_r} \tag{3.100}$$

The angular scattering coefficient is often broken down into a product of two terms. The first term defines the magnitude or amount of scattering, and the second, referred to as the scattering phase function, describes how the scattered energy is angularly distributed. Van de Hulst (1981) describes the scattering phase function $p(\theta)$ as the ratio of the energy scattered per unit solid angle into a particular direction to the average energy scattered per unit solid angle in all directions. This can mathematically be represented as:

$$p(\theta) = \frac{\beta_r(\theta)}{\frac{\int \beta_r(\theta) d\Omega}{4\pi}} = \frac{4\pi\beta_r(\theta)}{\beta_r} \tag{3.101}$$

or rearranging we have

$$\beta_r(\theta) = \frac{\beta_r}{4\pi} p(\theta) \tag{3.102}$$

where we see that the magnitude of the angular scattering term is simply the scattering coefficient divided by 4π steradians, and the scattering phase function is normalized such that its integral over all solid angles (i.e., over the sphere about the scattering center) must equal 4π. We could, therefore, express the Rayleigh scattering law of Eq. (3.96) as:

$$\beta_r(\theta) = \frac{8\pi^2(n(\lambda)-1)^2}{3m\lambda^4} p(\theta) = \frac{8\pi^2(n(\lambda)-1)^2}{3m\lambda^4}\left[\frac{3}{4}\left(1+\cos^2\theta\right)\right] \tag{3.103}$$

where $p(\theta) = 3/4\ (1+\cos^2\theta)$ is the Rayleigh scattering phase function and is plotted in Figure 3.18 along with other scattering phase functions.

3.4.2.2 Mie and Nonselective Scatter

It is not as straightforward to write an expression for the angular scattering coefficient associated with Mie scattering as it is for Rayleigh scattering. In general, the magnitude and scattering phase function are dependent on the number density, particle size distribution, and type (which controls the shape and complex index of refraction) of the scatterers. All the input parameters are generally not available for a rigorous solution to Mie scattering theory. As a result, empirical approximations are often used based on general classes of aerosols. The magnitude and scattering phase functions are available for a variety of atmospheric types based on tabulated data derived from empirical measurements [cf. Kneizys (1988)]. In general, Mie-type scattering is loosely referred to as *aerosol scattering* and represented with a subscript (a). Similar to Rayleigh scatter, we can define an angular scattering coefficient, $\beta_a(\theta)$ made up of a magnitude $\beta_a/4\pi$ and phase $p_a(\theta)$. Also by analogy, we can define an aerosol scattering cross section C_a and optical depth δ_a.

Mie scattering differs from Rayleigh in that it is highly forward-scattered as seen in Figure 3.18. The Rayleigh scatter is symmetric with significant and equal amounts of forward- and backscatter. Both types of scattering show relatively little energy scattered at right angles to the propagating beam. Mie scattering is also less a function of wavelength. Curcio (1961) indicates that the magnitude of the scattering term dependency ranges from λ^{-2} to $\lambda^{0.6}$ for various atmospheres.

The term nonselective scattering is used to describe the scattering from particles that are

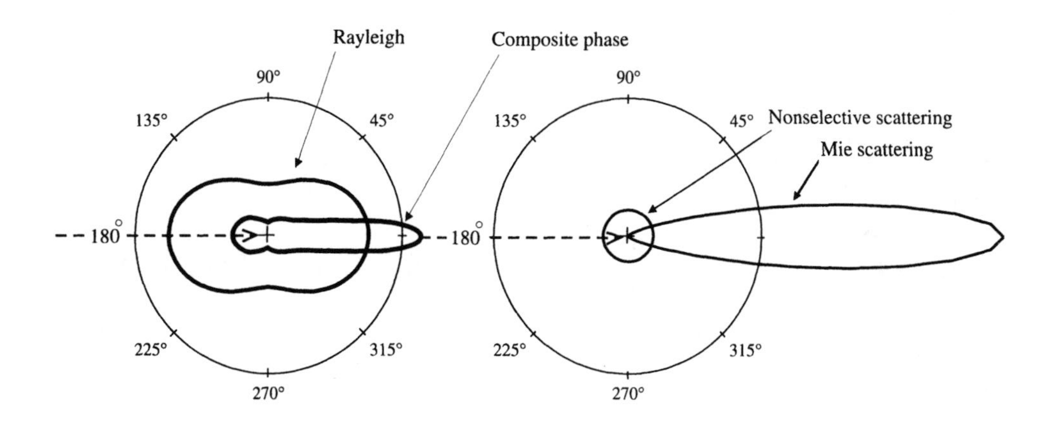

Figure 3.18 Shapes of scattering phase functions.

large compared to the wavelength of the EM energy. Nonselective scattering (denoted with the subscript ns) shows little dependency on wavelength and is approximately the same in all scattering directions (cf. Fig. 3.18). This type of scattering is most often associated with large dust particles, water droplets, ice, and hail (e.g., smoke and fog).

Scattering by all sources can be treated independently and combined with absorption effects to yield the overall transmission along a beam according to

$$\begin{aligned} \tau &= e^{-(\delta_a+\delta_r+\delta_a+\delta_{ns})} = e^{-\delta} \\ &= e^{-(\beta_a+\beta_r+\beta_a+\beta_{ns})z} = e^{-\beta_{ext}z} \end{aligned} \tag{3.104}$$

where the extinction coefficient (β_{ext}) has been introduced as the sum of the absorption and scattering coefficients, and the total optical depth is simply designated δ. It is important to realize that each extinction term or optical depth term in Eq. (3.104) may be made up of contributions from several constituents in the atmosphere. Furthermore, for a nonhomogeneous atmosphere, the atmosphere would be approximated as a series of layers, and the optical depths summed in a piecewise fashion over each layer, as indicated in Figure 3.16.

3.5 CHARACTERISTICS OF THE EM SPECTRUM

In order to evaluate the relative importance of many of the concepts introduced in this chapter, we need to combine them as a function of wavelength. Figure 3.19 is a graphical attempt to combine several of these concepts. The atmospheric transmission is plotted as a function of wavelength for transmission through a single standard atmosphere (i.e., vertical transmission earth to space for the U.S. standard atmosphere). Also plotted is the exoatmospheric spectral irradiance [$wm^{-2}\mu m^{-1}$] from the sun and the radiant exitance [$wm^{-2}\mu m^{-1}$] from a 300 K blackbody. These curves show that from the visible through the short-wave infrared (SWIR), there will be several orders of magnitude more flux from the sun than from self-emission at 300 K. In general, we will refer to this spectral range as the *reflective region* and ignore thermal or self-emitted flux when we are considering this region (i.e., referring to Figure 3.1, only type A, B, and C photons are relevant). In the 8- to 14-μm window just the opposite occurs, and there will be several orders of magnitude more flux from self-emission

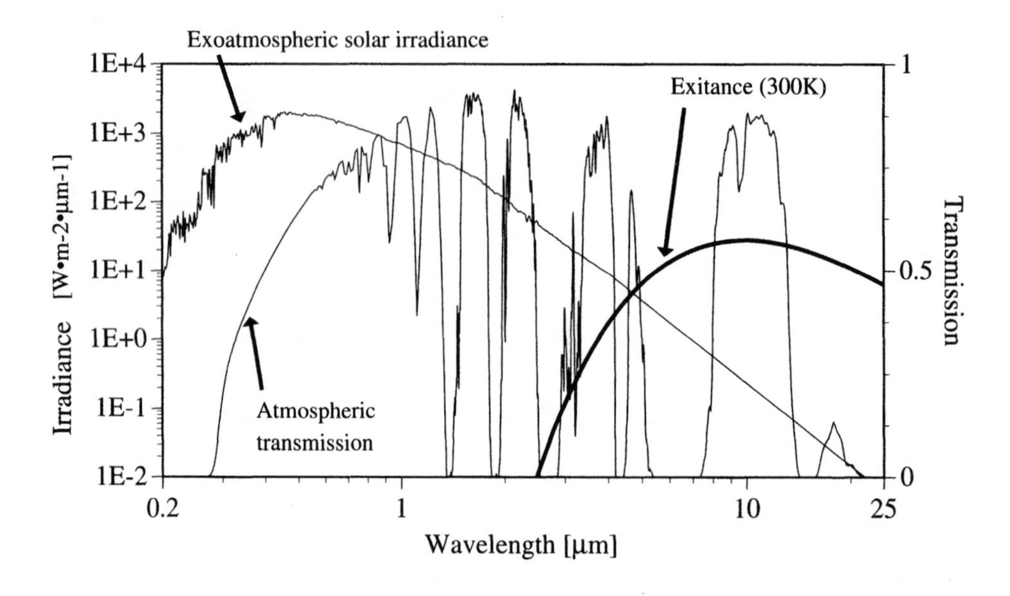

Figure 3.19 Comparison of atmospheric transmission, solar irradiance, and self-emission spectra.

than from reflected solar flux. Only approximately one photon in 4000/5000 (for a 10% reflector at 300 K) will be from the sun. Since we will rarely be able to make measurements where the noise limits are better than 1 part in 1 thousand, we can generally ignore solar photons in the 8- to 14-µm region (i.e., referring to Figure 3.3 only type D, E, and F photons are relevant). In the midwave infrared (MWIR) window between 3 and 5 µm, the situation is considerably more complicated. In this region the number of solar and self-emitted photons are of the same order of magnitude. Therefore, under daylight conditions we must consider both solar and self-emitted photon paths. The photon flux in this spectral region simplifies somewhat at night and becomes similar to the 8- to 14-µm region. It is important to realize that Figure 3.19 represents a simplified case with direct sunlight and a 300 K target. Under low-sun conditions, or for hotter objects, the relative amounts of self-emitted flux can be more important, particularly at the longer wavelength end of the SWIR region.

The wavelength scale of Figure 3.19 makes the solar irradiance and transmission curves deceptively smooth. In Figure 3.20 (derived from the MODTRAN code described by Berk, 1989), we show an expanded wavelength scale which more clearly shows some of the spectral structure. The absorption line highlighted in the irradiance spectra is one of the Fraunhofer lines. These are very narrow deep lines caused by absorption in the solar atmosphere of the near-blackbody radiation from the sun. If we were to image the earth in this spectral region, even on a sunlit day, we would see little or no reflected solar flux. In Section 5.6, we will see how we can take advantage of this to study solar-induced luminescence. The transmission plot shows how the transmission spectra is actually made up of the overlapping absorption characteristics of the various constituents in the atmosphere. Figure 3.21 shows the absorption spectra of various constituents in the atmosphere for a single pass through the U.S. standard atmosphere along a 45° solar illumination path. The spectra are cascaded together in the lowest curve representing the actual atmospheric transmission. In Figure 3.22, we show the exoatmospheric spectral irradiance ($E'_{s\lambda}$) cascaded together with the atmospheric transmission spectra ($\tau(\lambda)$) to yield the spectral irradiance reaching the earth

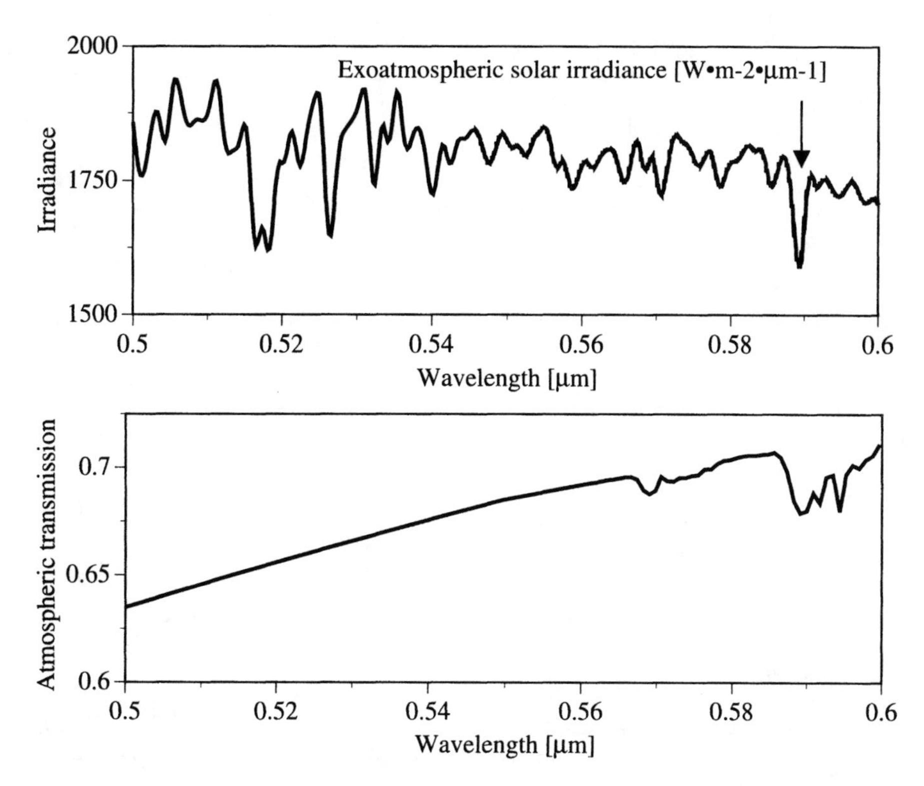

Figure 3.20 Expanded solar irradiance and transmission plots showing absorption line effects.

$E_{s\lambda}$. These spectra show how different absorbers will be important in different wavelength regions. It also shows where we might select spectral bands to avoid or to study certain atmospheric effects.

In this chapter, we have defined the basic radiometric terms and shown how radiometric concepts can be used to describe radiation propagation in the atmosphere. In the next chapter, we will forge the first few links in the image chain by using these principles to derive the governing equation for the radiance reaching the sensor.

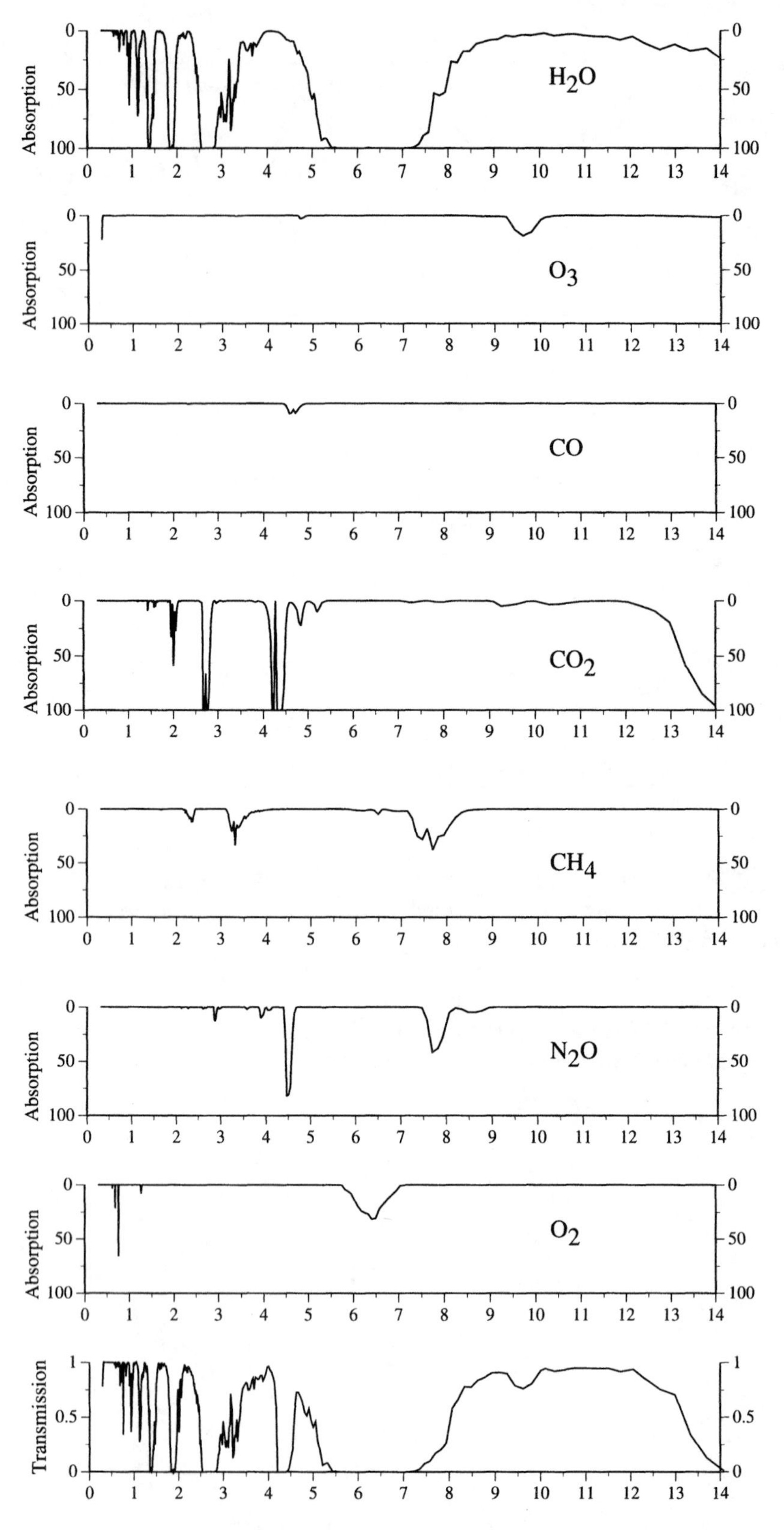

Figure 3.21 Absorption spectra of various atmospheric constituents and overall atmospheric transmission as derived from MODTRAN (cf. Berk, 1989).

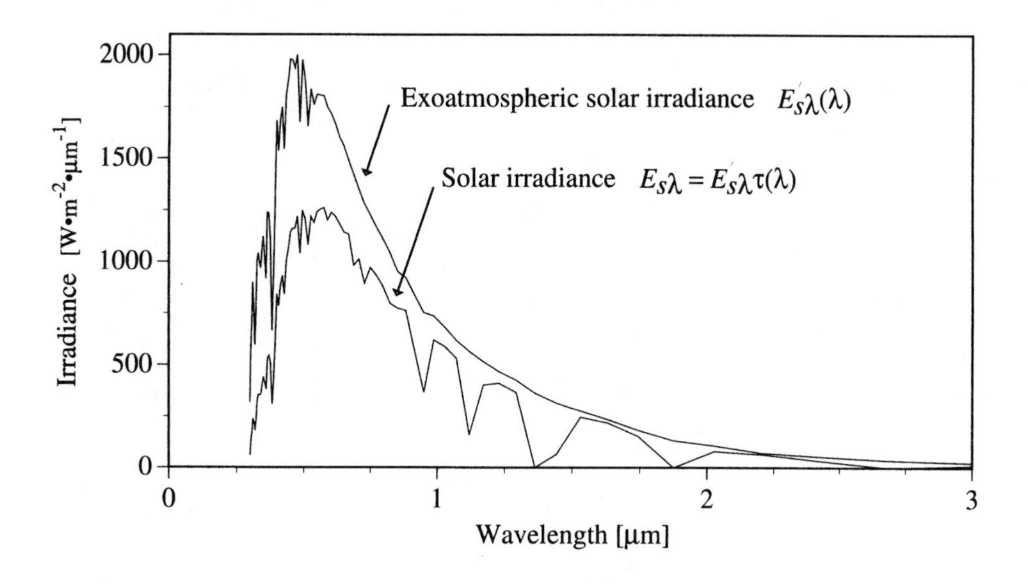

Figure 3.22 Effect of atmospheric transmission on the solar spectral irradiance reaching the earth.

3.6 REFERENCES

Berk, A., Bernstein, & L.S., Robertson, D.C. (1989). "MODTRAN: a moderate resolution model for LOWTRAN 7," GL-TR-89-0122, Spectral Sciences, Burlington, MA.

Chandrasekhar, S. (1960). *Radiative Transfer*. Dover, Minneola, NY.

Chen, H.S. (1985). *SPACE Remote Sensing Systems*. Academic, Orlando, FL.

CIE (1970). "International lighting vocabulary," CIE Publication No. 17 (E-1.1), International Commission on Illumination (CIE), Paris.

Curcio, J.A. (1961). Evaluation of atmospheric aerosol particle size distribution from scattering measurements in the visible and infrared, *Journal of the Optical Society of America*, Vol. 51, No. 5, pp. 548-551.

Einstein, A. (1905). *Annalen der Physik*. Vol. 17 (cf. English translation by A.B. Arons and M.B. Peppard, 1968), *American Journal of Physics*, Vol. 33, No. 5, pp. 367-374.

Frohlich, C. (1977). Contemporary measures of the solar constant. In *The Solar Output and Its Variation*, edited by O.R. White. Colorado Associated Univeristy Press, Boulder, CO.

Grum, F., & Becherer, R.J. (1979). *Optical Radiation Measurements:* In *Vol. 1, Radiometry*. Academic, NY.

Kaufman, Y.J. (1982). Solution of the equation of radiative transfer for remote sensing over nonuniform surface reflectivity *Journal of Geophysical Research*, Vol. 87, No. C6, pp. 4137-47.

Kneizys, F.X., Shettle, E.P., Abreu, L.W., Chetwynd, J.H., Anderson, G.P., Gallery, W.O., Selby, J.E.A., & Clough, S.A. (1988). "Users Guide to LOWTRAN7," AFGL-TR-88-0177, Environmental Research Papers, No. 1010, Air Force Geophysics Laboratory, Optical/Infrared Technology Division, Hanscom AFB, MD.

Lord Rayleigh (J.W. Stratt) (1871). *Philosophical Magazine*, Vol. 41, pp. 107-120, 274-279.

Nicodemus, Fred E., ed. (1976). "Self-study manual on optical radiation measurements: Part 1-Concepts, Chaps. 1 to 3," NBS Technical Note 910-1, U.S. Dept. of Commerce, National Bureau of Standards, U.S. Printing Office, Washington, DC.

Piech, K.R., & Schott, J.R. (1974). "Atmospheric corrections for satellite water quality studies." Proceedings of the SPIE, Vol. 57, pp. 84-89.

Planck, M. (1901). *Annalen der Physik*, Vol. 4, No. 3, 553.

Thekaekara, M.P. (1972). Evaluating the light from the sun. *Optical Spectra*, Vol. 6, No. 3, pp. 32-35.
U.S. Air Force Geophysics Laboratory (1985). *Handbook of Geophysics and the Space Environments*, 4th ed. Air Force Geophysics Laboratory, Bedford, MA.
Van de Hulst, H.C. (1981). *Light Scattering by Small Particles*. Dover, Minneola, NY.

CHAPTER

4

The Governing Equation for Radiance Reaching the Sensor

In the last chapter we introduced the basic concepts of atmospheric propagation and the fundamentals of radiometry. In this chapter we will use these principles to derive an expression for the radiance reaching a remote sensor. This governing equation must describe all the significant paths illustrated in Figures 3.1 and 3.3. To determine the relative importance of each of the terms in the governing equation, we will perform a simple sensitivity analysis (cf. Sec. 4.6) on the final equation. This governing equation represents several major links in the imaging chain which are critical to quantitative radiometric image analysis. As a result, we will derive a governing radiometric propagation equation that is quite complex. In many cases, this level of detail may not be required. We will find that simplifying assumptions can often be made without introducing significant error. At other times unexpected errors can occur because we violate some of the simplifying assumptions. For these cases, it is important to have available the more complete treatment. The approach we will use deviates somewhat from more traditional radiative transfer discussions. We try to follow the flow of photons in as intuitive a fashion as is feasible while yielding results consistent with a more rigorous traditional radiative transfer approach, (cf. Chandrasekhar 1960).

4.1 IRRADIANCE ONTO THE EARTH'S SURFACE

In developing the governing equation, we follow the flow of photons to the sensor. Initially, we will consider only solar photons with the self-emitted photons introduced in Section 4.4. In tracing the photon flux, we will first try to determine an expression for all the flux incident on a reflecting surface.

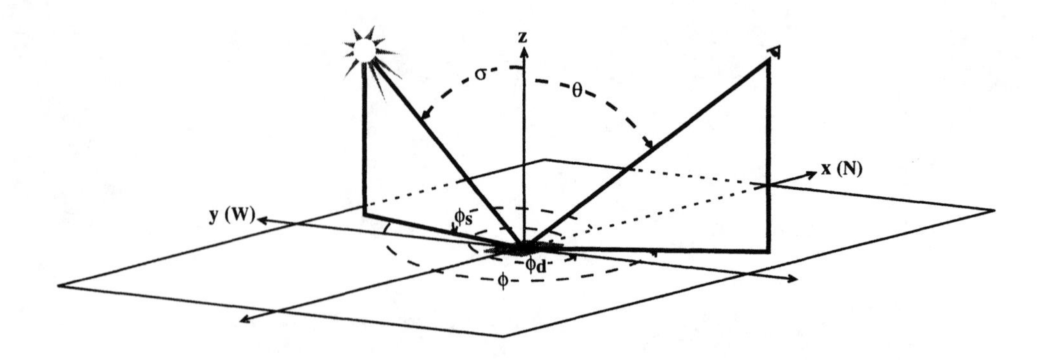

(a) World coordinate system.

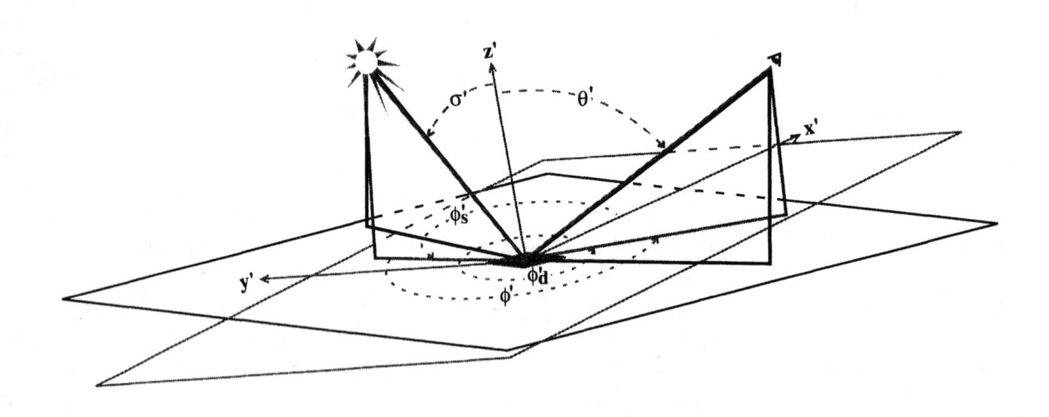

(b) Target coordinate system.

Figure 4.1 Definition of orientation angles.

4.1.1 Solar Irradiance

In general, the primary source of flux incident on a reflecting surface is the sun. In order to use the generic expressions developed in the previous section for the specific case of interest to us, we must introduce two related sets of orientation angles. These angles are depicted in Figure 4.1. The first set of angles describes the relative source-target-sensor orientation in the world coordinate system. In this system the x,y plane is defined to be tangent to the earth at the target, with the x axis due north, the y axis due west, and the z axis normal to the earth at the target. The sun location is then defined by its declination angle σ from the normal to the earth and an azimuthal angle ϕ_s defined as the angle going west from the x axis (north) to the projection of the sun onto the x,y plane. The sensor, or detector, is located through a view angle θ from the normal to the earth to the sensor and an azimuthal angle ϕ_d defined using the same measurement convention as ϕ_s. The relative azimuthal angle between the source and the detector we denote simply as ϕ.

In many cases, we need to define the source-target-sensor orientation in a target-centered coordinate system. Here we define the *x,y* plane as the plane of the target or the tangent to the target at the point of interest with the *x* axis located along the projection of the earth normal onto the plane of the target (defined to be north when the target plane and the plane of the earth are coincident). The sun location relative to the target plane is defined by the solar declination relative to the target-centered *z* axis (i.e., the normal to the target) identified as σ' and an azimuthal angle ϕ_s' measured from the target's *x* axis in the positive *y* direction to the projection of the sun onto the target plane. Similarly, the sensor location is defined by the angle from the normal to the target to the sensor (θ') and the azimuthal angle ϕ_d' to the projection of the detector onto the target plane. The relative azimuthal angle in the target plane is denoted by ϕ'.

When the target is a horizontal surface, the two coordinate systems coincide and $\theta=\theta'$, $\sigma=\sigma'$, and $\phi=\phi'$. For all other cases, a set of relative orientation angles is required to define the relationship between the coordinate systems. These angles are not required for our treatment here, so a naming convention will not be introduced. The reader should recognize that the geometric convention introduced here is arbitrary, and many naming conventions and reference coordinate systems are used in the literature.

Given the angular definitions of Figure 4.1, we can define the solar spectral irradiance onto a target as:

$$E_{s\lambda}\cos\sigma' = E'_{s\lambda}\tau_1(\lambda)\cos\sigma' \cong E'_{s\lambda}e^{-\delta(\lambda)\sec\sigma}\cos\sigma' \ [\mathrm{wm}^{-2}\mu\mathrm{m}^{-1}] \tag{4.1}$$

where $E_{s\lambda}'$ is the exoatmospheric irradiance [$\mathrm{wm}^{-2}\ \mu\mathrm{m}^{-1}$] onto a surface perpendicular to the incident beam [cf. Eq. (3.49)] and $\tau_1(\lambda)$ is the atmospheric transmission along the sun-target path, which can be written as:

$$\tau_1(\lambda) = e^{-\delta_1(\lambda)} \approx e^{-\delta(\lambda)\sec\sigma} \tag{4.2}$$

where $\delta_1(\lambda)$ is the optical depth along the sun-target path, which for single scattering and ignoring refractive index changes can be approximated as:

$$\delta_1 = \delta(\lambda)\sec\sigma \tag{4.3}$$

where $\delta(\lambda)$ is defined to be the optical depth vertically through the earth's atmosphere and $\sec\sigma = 1/\cos\sigma$. This approximation will be used in the next section and is reasonably valid for clear atmospheres and σ ranging from 0 to 60°.

It is important to realize how much the solar irradiance can change due to the relative orientation of the target to the sun even for surfaces with relatively shallow slopes (cf. Fig. 4.2). The solar irradiance will also change with atmospheric makeup (e.g., number and type of aerosols) and with elevation, since from inspection of Eq. (3.91) it is clear that the total optical depth will decrease as the elevation increases. This is particularly important because the atmosphere is denser near the earth's surface. The lowest layers make large contributions to the optical depth, that are eliminated for targets at higher elevations.

4.1.2 Downwelled Radiance (Skylight)

In addition to the direct solar irradiance term, we also recognized in Figure 3.1 that solar photons would be incident on the target due to scattering from the atmosphere. We refer to this downward scattering as *downwelled radiance*, or, in the visible, as *skylight*. In order to characterize the flux onto the target due to downwelled radiance, let's first consider the contribution from a small volume in the atmosphere, as shown in Figure 4.3. The spectral irradiance onto the face of a volume element located perpendicular to the incident flux is

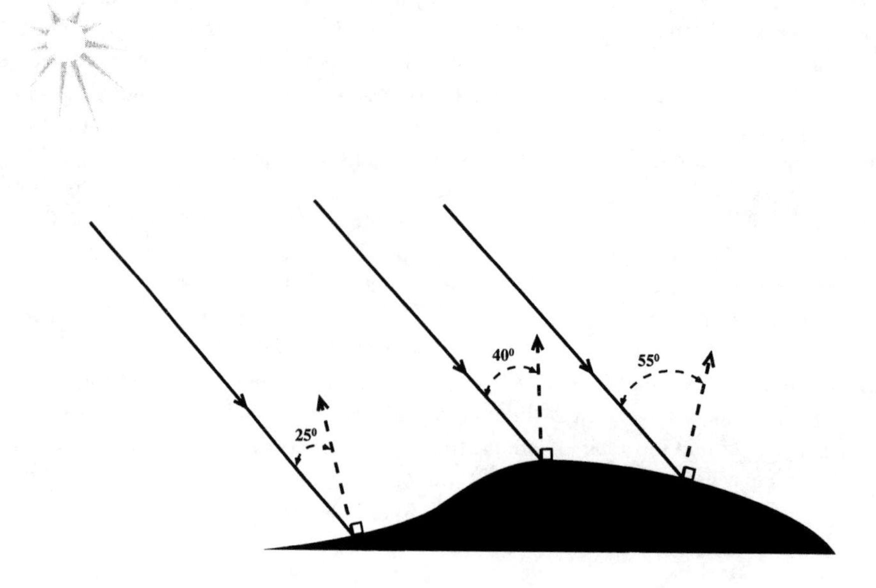

Figure 4.2 Effects of target slope on solar irradiance.

$$E_{v\lambda} = E'_{s\lambda}\tau_{L1}(\lambda)[wm^{-2}\mu m^{-1}] \tag{4.4}$$

where $\tau_{L1}(\lambda)$ is the transmission along the first leg of the *L*-shaped path from the sun to the scattering volume element to the target. If we take a unit area on the face of the volume perpendicular to the propagating ray and consider the element of spectral radiant intensity scattered toward the target in traversing a unit length along the beam (i.e., from the unit volume), we have:

$$dI_{v\lambda}[\mathrm{wsr}^{-1}\mu\mathrm{m}^{-1}] = E_{v\lambda}\beta_{\mathrm{sca}}(\lambda,\theta_v)dV[\mathrm{wm}^{-2}\mu\mathrm{m}^{-1}\mathrm{m}^{-1}\mathrm{sr}^{-1}\mathrm{m}^3] \tag{4.5}$$

where $\beta_{\mathrm{sca}}(\lambda,\theta_v)$ is the spectrally dependent angular scattering coefficient for the composite atmosphere, θ_v is the angle between the incident ray and the ray to the target, and dV is the volume element. Adapting Eqs. (3.42) and (3.49) for this case, we have the element of spectral irradiance onto the target from the volume element

$$\begin{aligned} dE_{d\lambda}[\mathrm{wm}^{-2}\mu\mathrm{m}^{-1}] &= \frac{dI_{v\lambda}\tau_{L2}(\lambda)\cos\sigma_v}{r_v^2} \\ &= \frac{E_{v\lambda}\beta_{\mathrm{sca}}(\lambda,\theta_v)\tau_{L2}(\lambda)\cos\sigma_v dV}{r_v^2} \end{aligned} \tag{4.6}$$

or, on substituting Eq. (4.4);

$$dE_{d\lambda} = \frac{E'_{s\lambda}\tau_{L1}(\lambda)\beta_{\mathrm{sca}}(\lambda,\theta_v)\tau_{L2}(\lambda)\cos\sigma_v dV}{r_v^2} \tag{4.7}$$

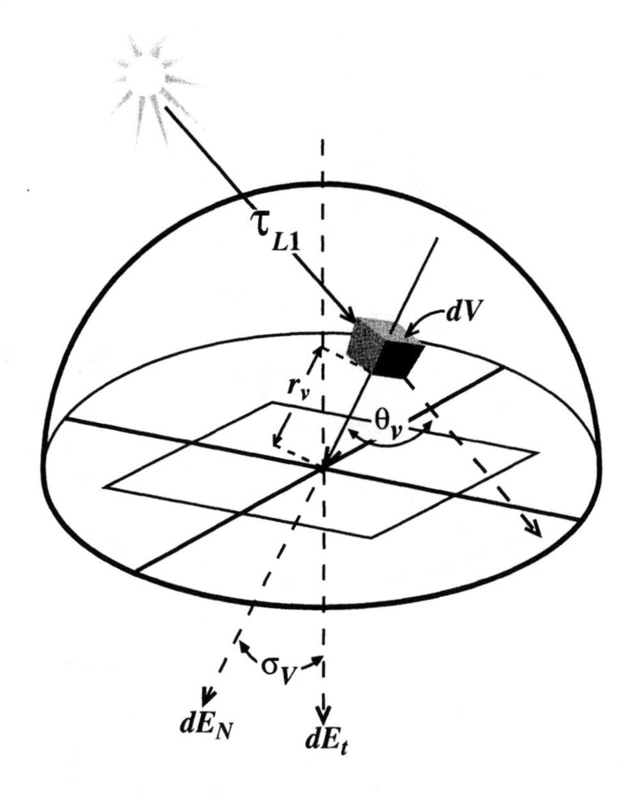

Figure 4.3 Contribution to downwelled radiance (or irradiance) from a unit volume.

where $\tau_{L2}(\lambda)$ is the transmission along the second leg of the L-shaped sun-volume-target path, σ_v is the angle between the normal to the target and the ray from the volume, and r_v[m] is the path length along the volume-target path. The total downwelled spectral irradiance $E_{d\lambda}$ could then be computed by integrating over the hemisphere above the target, yielding

$$E_{d\lambda} = \int \frac{E'_{s\lambda} \tau_{L1}(\lambda) \tau_{L2}(\lambda) \beta_{\text{sca}}(\lambda, \theta_v) \cos \sigma_v dV}{r_v^2} \tag{4.8}$$

The limits of integration would extend to space in all directions above the target, and all the parameters on the right-hand side of Eq. (4.8) except $E_{s\lambda}'$ are dependent on location (i.e., their dependence on dV must be accounted for to perform the integration). This approach is the most straightforward for seeing the dependence of the downwelled radiance (or irradiance) on the contribution from the volume scattering element. However, a numerical integration of Eq. (4.8) is somewhat laborious to set up.

A slightly more tractable approach to finding the downwelled radiance from the sky dome is to consider the directional downwelled radiance reaching the target from a particular direction and then integrating over all directions. This approach is particularly attractive because at times we are interested in the downwelled radiance from a particular direction. This approach is also attractive because it roughly parallels the numerical integration used in the MODTRAN/LOWTRAN-style computer codes that we will address in Chapter 6 (cf. Kneizys, 1988). If we consider a scattering volume element dV[m^3] subtended by the element of solid angle $d\Omega$ as illustrated in Figure 4.4, we can write the relationship

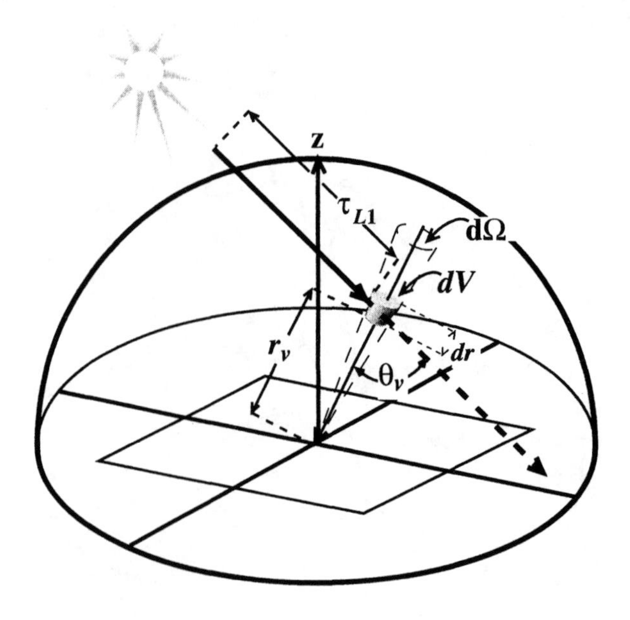

Figure 4.4 Procedure for calculation of the angular downwelled radiance.

$$dV = dAdr = d\Omega r_v^2 dr \tag{4.9}$$

where dA is the area element subtended by the solid angle element $d\Omega$ at a radial distance r_v from the target and dr is the element of radial distance. By combining Eqs. (4.4) and (4.5), we can write the element of spectral radiant intensity [wsr^{-1}μm^{-1}] scattered toward the target from a volume element as:

$$dI_{v\lambda} = E'_{s\lambda}\tau_{L1}(\lambda)\beta_{\text{sca}}(\lambda,\theta_v)dV \tag{4.10}$$

Substituting Eq. (4.9) for dV we have

$$dI_{v\lambda} = E'_{s\lambda}\tau_{L1}(\lambda)\beta_{\text{sca}}(\lambda,\theta_v)\cdot d\Omega r_v^2 dr \tag{4.11}$$

Using Eq. (3.42), the element of spectral irradiance [wm^{-2}μm^{-1}] onto a surface perpendicular to the scattered ray at the target is then given by

$$dE_{n\lambda} = \frac{E'_{s\lambda}\tau_{L1}(\lambda)\tau_{L2}(\lambda)\beta_{\text{sca}}(\lambda,\theta_v)d\Omega r_v^2 dr}{r_v^2} \tag{4.12}$$

The element of downwelled spectral radiance [wm^{-2}sr^{-1}μm^{-1}] along the scattered ray at the target's surface can then be expressed from the definition of radiance as:

$$dL_{d\lambda}(\sigma,\phi) = \frac{E'_{s\lambda}\tau_{L1}(\lambda)\tau_{L2}(\lambda)\beta_{\text{sca}}(\lambda,\theta_v)d\Omega dr}{d\Omega} \tag{4.13}$$

where we have introduced the generic direction angles σ and ϕ to describe the radiance reaching the target from the direction described by the zenith angle σ and the azimuthal angle

ϕ. The total downwelled spectral radiance reaching the target from the σ,ϕ direction can be expressed by integrating over r from zero to the top of the atmosphere according to

$$L_{d\lambda}(\sigma,\phi) = E'_{s\lambda} \int \tau_{L1}(\lambda)\tau_{L2}(\lambda)\beta_{sca}(\lambda,\theta_v)dr \tag{4.14}$$

The L-path transmission values (τ_{L1} and τ_{L2}) can be computed for each r value by numerical integration of the optical depth if the number densities and extinction cross sections are known along the L path. Finally, if the angular scattering function is known along r in the σ,ϕ direction, then Eq. (4.14) can be solved numerically.

The total downwelled spectral irradiance can be computed from the angular downwelled radiance by integration over the hemisphere above the target according to

$$\begin{aligned} E_{d\lambda} &= \int L_{d\lambda}(\sigma,\phi)\cos\sigma d\Omega \\ &= \int_{\phi=0}^{2\pi} \int_{\sigma=0}^{\frac{\pi}{2}} L_{d\lambda}(\sigma,\phi)\cos\sigma\sin\sigma d\sigma d\phi \end{aligned} \tag{4.15}$$

In the analysis of downwelled radiance we have presented thus far, we have implicitly assumed that the target is horizontal and unobstructed such that the entire hemisphere above the target is sky. This is a situation that generally only occurs in textbooks. In reality, if the target has any slope or if there are adjacent objects obstructing the sky dome, the downwelled radiance onto the target will be reduced (cf. Fig. 4.5). The most rigorous way to compute the effect from adjacent objects is to simply change the limits of integration of Eq. (4.15) so that the integral only includes that portion of the hemisphere which is sky. In most cases, it is more convenient to compute separately, or estimate, the fraction of the hemisphere above the target which is sky (F). This can most easily be accomplished by simply integrating the solid angle element ($d\Omega$) over the solid angle obstructed by background and dividing by the solid angle of the hemisphere (2π). This yields the fraction of the hemisphere which is background with the remaining fraction (i.e., 1 minus the background fraction) being sky. The computational form of this can be expressed as:

$$F = 1 - \frac{\int d\Omega}{2\pi} = 1 - \frac{\int\int \sin\sigma d\sigma d\phi}{2\pi} \tag{4.16}$$

Two simple examples of the use of this expression are to compute the shape factor (F) for the case of an adjacent building and for a sloped roof as illustrated in Figure 4.5. Piech and Walker (1971) solved for the case of an adjacent building to yield an approximation, for the point p in Figure 4.5a, of the form

$$F = 1 - \frac{\int_{\phi=0}^{\phi} \int_{\sigma=\frac{\pi}{2}-\sigma_F}^{\frac{\pi}{2}} \sin\sigma d\sigma d\phi}{2\pi} = 1 - \frac{\phi\cos\left(\frac{\pi}{2} - \sigma_F\right)}{2\pi} \tag{4.17}$$

This would produce the intuitively correct result if p were at the base of the building

$$F \cong 1 - \frac{\pi\cos\left(\frac{\pi}{2} - \frac{\pi}{2}\right)}{2\pi} = 0.5 \tag{4.18}$$

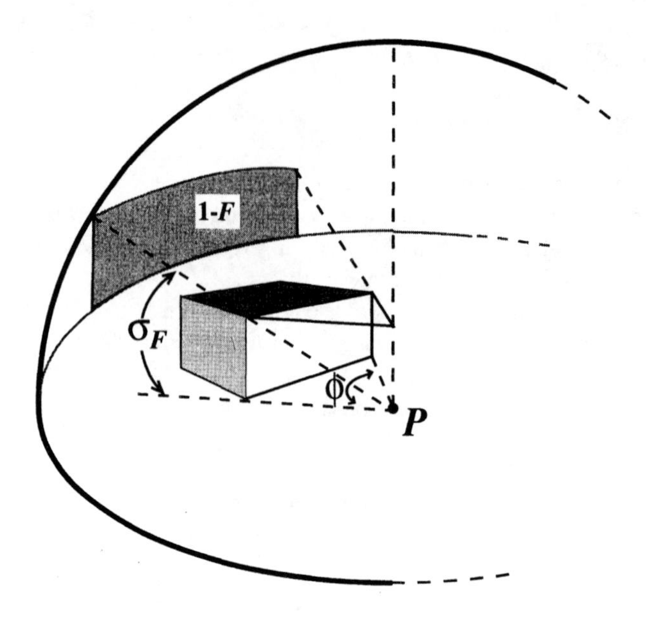

(a) The building blocks the shaded portion of the sky from irradiating the point p.

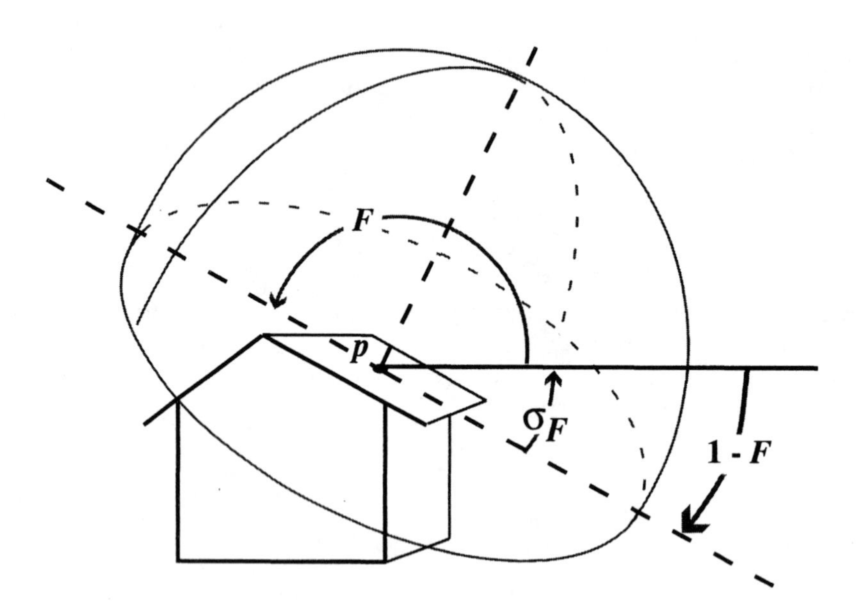

(b) The slope of the roof reduces the amount of sky seen by the point p.

Figure 4.5 Shape factor F of the exposed sky.

i.e., half the sky is visible.

For the sloped roof case, F can be approximated from

$$F \cong 1 - \frac{\int_{\phi=0}^{\pi} \int_{\sigma=\frac{\pi}{2}-\sigma_F}^{\frac{\pi}{2}} \sin \sigma d\sigma d\phi}{2\pi} = 1 - \frac{1}{2}\cos\left(\frac{\pi}{2} - \sigma_F\right) \quad (4.19)$$

The downwelled irradiance from the sky onto the target can then be expressed as:

$$E_{d\lambda\text{sky}} = \int\int L_{d\lambda}(\sigma, \phi) \cos \sigma \sin \sigma d\sigma d\phi \quad (4.20)$$

where the limits on the integrals are set to encompass the fraction of the hemisphere above the target which is sky. This can be approximated by

$$E_{d\lambda\text{sky}} \approx F E_{d\lambda} = F \int_{\phi=0}^{2\pi} \int_{\sigma=0}^{\frac{\pi}{2}} L_{d\lambda}(\sigma, \phi) \cos \sigma \sin \sigma d\sigma d\phi \quad (4.21)$$

where $E_{d\lambda}$ is the total downwelled spectral irradiance from the sky in the absence of background effects, and $E_{d\lambda\text{sky}}$ is the spectral downwelled irradiance from the fraction of the hemisphere above the target which is sky. Clearly, Eq. (4.21) will be a better approximation when the radiance from the sky is approximately a constant. Figure 4.6 illustrates that the downwelled radiance from the sky on a clear day is a function of both σ and ϕ in the reflective region and at least of σ in the thermal region. In general, under clear conditions it is a well-behaved, slowly varying function, so Eq. (4.21) should not introduce large errors. However, when scattered clouds are present or when very detailed analysis is required, a rigorous analysis of Eq. (4.20) with the proper limits of integration may be required.

4.1.3 Reflected Background Radiance

The fraction of the hemisphere above the target which is obscured by background objects (1-F) also produces some photon flux onto the target. If we let the spectral radiance reflected from a background toward the target be $L_{b\lambda}(\sigma,\phi)$ [$wm^{-2}sr^{-1}\mu m^{-1}$], where the angles σ and ϕ denote the direction from which the background radiance comes, then the effects of background radiance are analogous to the downwelled sky radiance. The irradiance onto the target from the portion of the hemisphere above the object which is background (i.e., non-sky) can be expressed as:

$$E_{b\lambda} = \int\int L_{b\lambda}(\sigma, \phi) \cos \sigma \sin \sigma d\sigma d\phi \approx (1 - F) L_{b\lambda\text{avg}} \pi \quad (4.22)$$

where we have assumed that the radiance from the background is approximately constant with a mean value of $L_{b\lambda\text{avg}}$, and the limits of integration are set to encompass the fraction of the hemisphere above the target that is background. Where the constant background assumption introduces too large an error, the full integral form of Eq. (4.22) can be used. We have not yet described a functional expression for $L_{b\lambda}$ and will defer that until we have introduced reflected radiance in the next section (4.2.2).

In summary, we have identified three sources of irradiance onto the target which yield an expression for spectral irradiance onto the target of

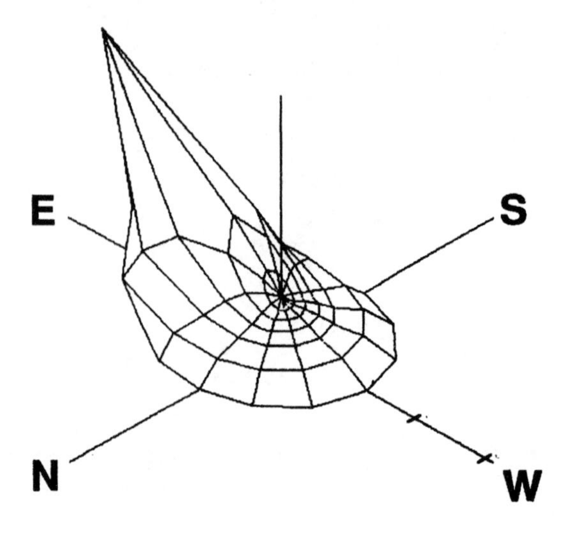

(a) Plot of downwelled radiance from the morning sky in the visible region of the spectrum.

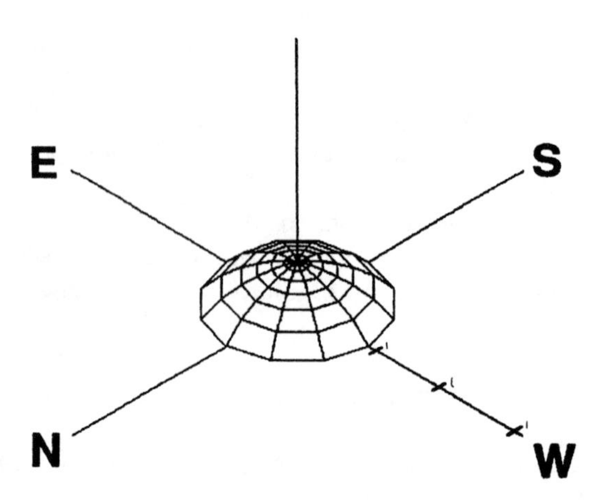

(b) Plot of downwelled radiance from the morning sky in the LWIR region of the spectrum.

Figure 4.6 Angular variation in downwelled radiance.

$$E_{\tau\lambda} = E'_{s\lambda} \cos\sigma' \tau_1(\lambda) + FE_{d\lambda} + (1-F)L_{b\lambda\text{avg}}\pi \tag{4.23}$$

4.2 REFLECTED SOLAR IRRADIANCE AND BIDIRECTIONAL REFLECTANCE

In order to convert the irradiance onto the target into radiance towards the detector, we need to consider the reflectance properties of materials. In general the reflectance properties are a function of wavelength, illumination angle, and viewing angle. We need to develop a means to express the full impact of these dependencies on the reflected radiance and to develop a simpler expression for cases when full angular reflectance data are not available.

4.2.1 Ways to Characterize Reflectance

In Chapter 3, we introduced what we will now define to be the total spectral reflectance as:

$$r(\lambda) = \frac{M_\lambda}{E_\lambda} \tag{4.24}$$

This expression for reflectance, while perfectly valid, fails to provide us with any information about the directional distribution of the reflected flux. Our experience tells us that the directional characteristics of reflectance vary considerably from mirrorlike surfaces (specular) to surfaces that appear to have little or no directional character to their reflectance (Lambertian or diffuse). In Figure 4.7, we depict the angular reflectance characteristics of several idealized surfaces, as well as a more complex object. A perfectly specular object behaves like a mirror with flux only reflected into the direction exactly opposite to the incident ray. A nearly specular object will appear to have most of the reflected energy concentrated in a cone about the specular ray. A perfectly diffuse surface appears to have the same amount of reflectance in all directions, while a nearly diffuse object will generally appear brighter in the specular direction. A less-idealized surface may appear brighter in the specular and backscatter directions and darker when viewed at grazing angles. Recall from Chapter 3 that visual brightness or apparent reflectance is directly proportional to radiance, so that the vectors in Figure 4.7 can be treated as the magnitude of the radiance in each direction. The surface that encloses the vectors can be thought of as a probability distribution function for the radiance in any direction.More formally (cf. Fig. 4.8) we can define the bidirectional reflectance to be the ratio of the radiance scattered into the direction described by the orientation angles θ_o and ϕ_o to the irradiance from the θ_i, ϕ_i direction, i.e.,

$$r_{\mathrm{BRDF}} = \frac{L(\theta_o, \phi_o)}{E(\theta_i, \phi_i)} [sr^{-1}] \tag{4.25}$$

The bidirectional reflectance distribution function (BRDF) describes these bi-directional reflectance values for all combinations of input-output angles. The BRDF values will also change as a function of wavelength, so a complete characterization would include the wavelength-dependent BRDF, using the spectral values for irradiance and radiance in Eq. (4.25). The BRDF is actually a scattering function analogous to the angular scattering coefficient, $\beta_{\mathrm{sca}}(\lambda,\theta)$ introduced in Chapter 3 to describe atmospheric scattering.

BRDF's can be measured in the laboratory (cf. Feng et al., 1993) or in the field (cf. Deering 1988). However, because of the large numbers of angles needed to fully characterize all possible combinations of source and sensor orientations, it is a cumbersome process. As a result, BRDF's are available for only a relatively restricted number of materials and land-cover types, and robust BRDF sample sets (i.e., BRDF values for different samples of the same land cover) are quite scarce. Consequently, much of the analytical work in remote sensing (cf. Chap. 6) must rely on approaches that do not require a full knowledge of BRDF values.

It is often more convenient to describe directional reflectance in a unitless form. This is accomplished by introducing the bidirectional reflectance factor (BDRF). This is the ratio of the radiance reflected into a particular direction to the radiance that would be reflected into the same direction by a perfect Lambertian radiator illuminated in an identical fashion. The perfect Lambertian radiator is defined to have a total reflectivity of unity, and because it is

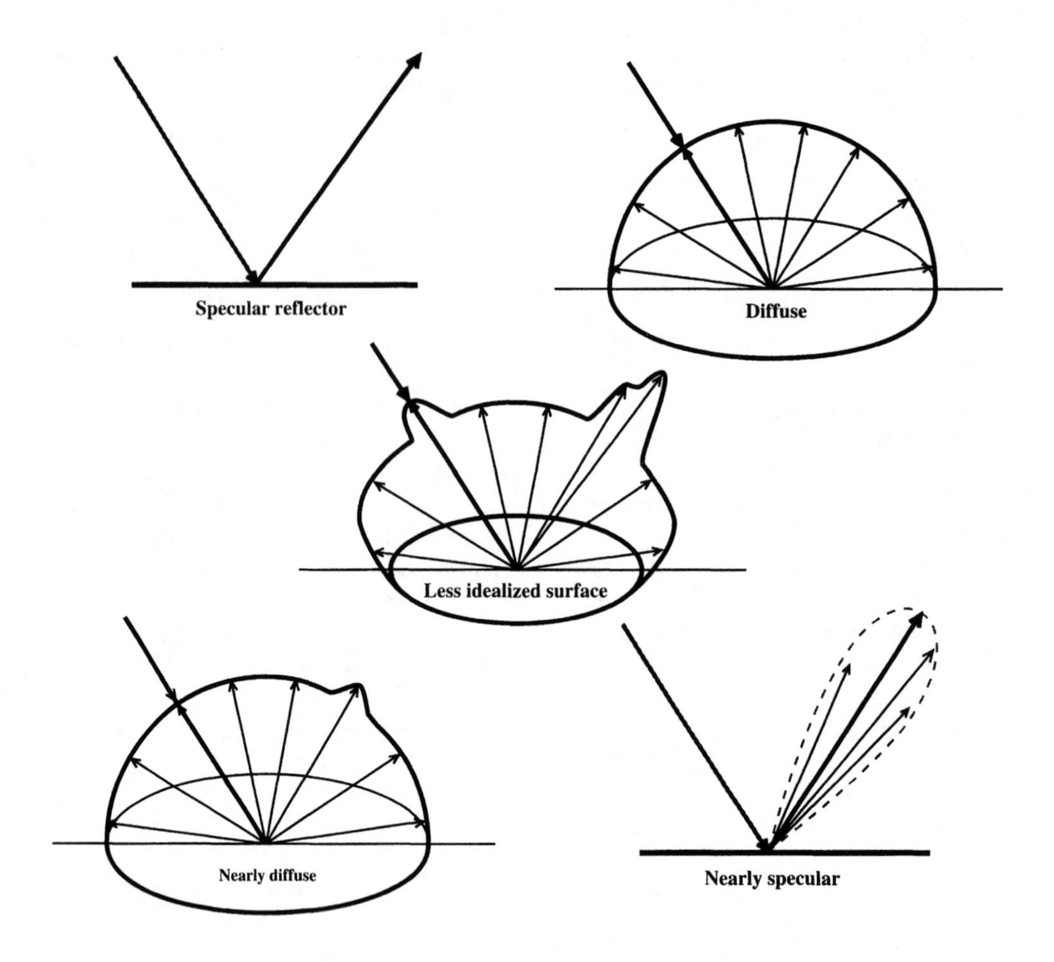

Figure 4.7 Reflectance characteristics of idealized surfaces.

Lambertian (cf. Chap. 3), it will have the same radiance in all directions. Feng et al. (1993) show how the reflectance factor (r_{BDRF}) is related to the bidirectional reflectance (r_{BRDF}) through a simple factor of π steradians (that magic π again), i.e.,

$$r_{BRDF}[sr^{-1}] = \frac{r_{BDRF}}{\pi[sr]} \tag{4.26}$$

In general we will omit the BDRF subscript and use reflectance factors for the remainder of our discussion.

In addition to the bidirectional reflectance terms, two other terms describing reflectivity are commonly used for remote sensing purposes. The first is the directional hemispheric reflectance. This is the ratio of the exitance from a target to the irradiance onto the target from a particular direction. Conversely, according to reciprocity, it is the ratio of the radiance into a particular direction to the radiance (uniform from all angles) irradiating a target.

The final reflectance term of interest is the *diffuse reflectance*. This is best described in reference to the instrumentation commonly used to measure it (cf. Fig. 4.9). The instrument is set up so that all the flux from the sample is collected by an integrating sphere, except that into a narrow cone about the specular direction. This is compared to a reading from a "perfect" Lambertian reflector measured in the same manner. The diffuse reflectance can then be

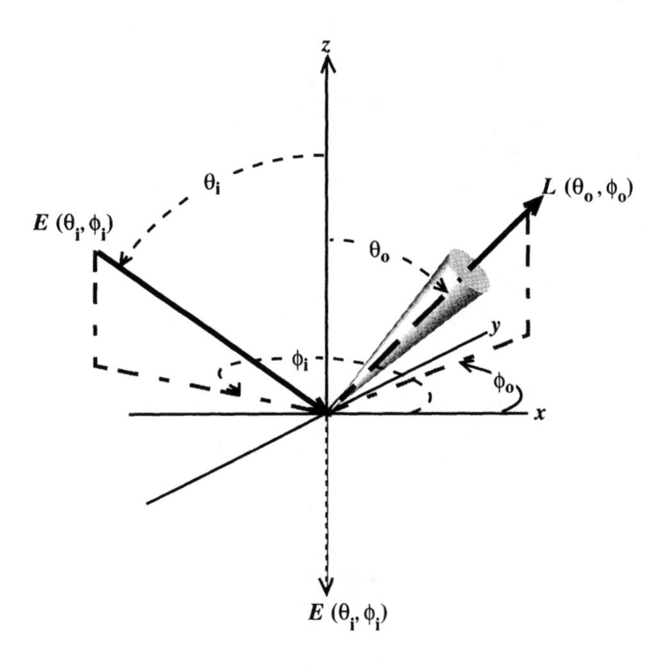

Figure 4.8 Bidirectional reflectance concept.

defined as the hemispheric reflectance with the specular component removed. If the light trap is not used, we measure the total hemispheric reflectance. The ratio of the diffuse reflectance (r_d) to the total reflectance (r_{tot}) provides a measure of the diffuseness (d) of the sample, i.e.,

$$d = \frac{r_d}{r_{tot}} \tag{4.27}$$

A Lambertian reflector would be perfectly diffuse with a diffuseness of one. The more specular (less diffuse) a sample becomes, the lower its diffuseness, with a mirror having a value of zero. The specularity is often defined as one minus the diffuseness.

4.2.2 Reflected Solar Radiance

Using the reflectance terminology we just introduced, we can develop an expression for the spectral radiance reflected toward the sensor associated with each of the irradiance terms in Eq. (4.23).

Using reflectance factor concepts, the spectral radiance $L_{sr\lambda}$ into the sensor direction (θ',ϕ_d') due to irradiance from the solar direction (σ',ϕ_s') can be expressed as:

$$L_{sr\lambda}(\theta,\phi_d) = E_{s\lambda}' \tau_1(\lambda)\cos\sigma' \frac{r(\sigma',\phi_s',\theta',\phi_d',\lambda)}{\pi} [\text{wm}^{-2}\text{sr}^{-1}\mu\text{m}^{-1}] \tag{4.28}$$

The reflected downwelled radiance L_{dr} is more difficult to express in the general case, so we will first consider the more restrictive case where we can assume that the target is approximately diffuse (Lambertian). In this case, we draw on the derivation of Eq. (3.66) for radiance from a Lambertian reflector to yield

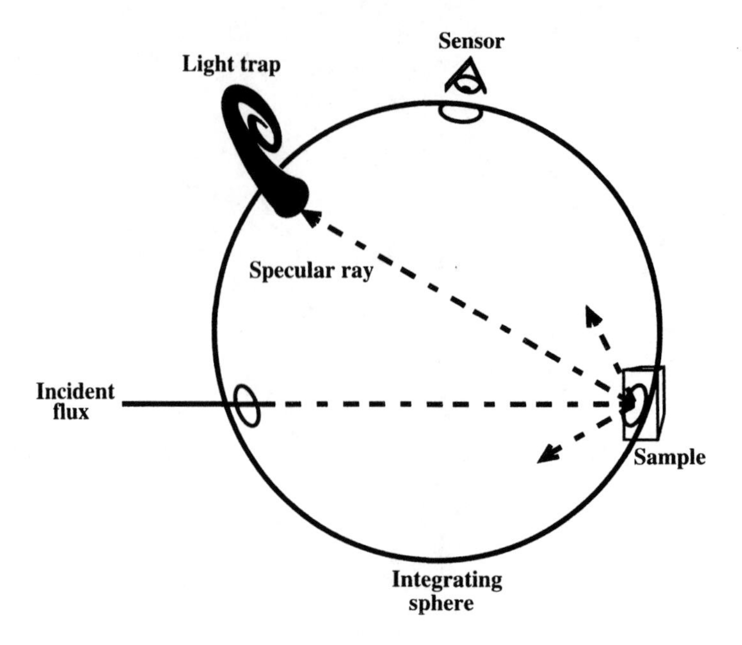

Figure 4.9 Schematic concept for measuring total and diffuse reflectance.

$$L_{dr\lambda} = E_{d\lambda} \frac{r_d(\lambda)}{\pi} \tag{4.29}$$

where r_d is the reflectance of the target if it is assumed diffuse (constant with angle). For a target with shape factor effects, this would be

$$L_{dr\lambda} = FE_{d\lambda} \frac{r_d(\lambda)}{\pi} \tag{4.30}$$

For cases where the reflectance cannot be assumed Lambertian, the reflected downwelled radiance can be expressed as:

$$L_{dr\lambda}(\theta', \phi_d') = \int_{\phi=0}^{2\pi} \int_{\sigma=0}^{\frac{\pi}{2}} L_{d\lambda}(\sigma, \phi) \cos\sigma \frac{r(\sigma', \phi', \theta', \phi_d', \lambda)}{\pi} \sin\sigma d\sigma d\phi \tag{4.31}$$

For a target with shape factor effects, the limits of integration of Eq. (4.31) would be appropriately adjusted (cf. Fig. 4.5 and related discussion).

For the reflected background radiance (L_{br}), the procedure is similar to the reflected downwelled radiance if the target can be assumed to be approximately Lambertian, i.e.,

$$L_{br\lambda} = (1 - F) L_{b\lambda\text{avg}} r_d(\lambda) \tag{4.32}$$

When the target is not Lambertian we have

$$L_{br\lambda}(\theta',\phi_d') = \int_{1-F} L_{b\lambda}(\sigma,\phi)\cos\sigma\, r_{\mathrm{BRDF}}(\lambda)d\Omega$$

or

$$= \int_{1-F}\int L_{b\lambda}(\sigma,\phi)\cos\sigma \frac{r(\sigma',\phi',\theta',\phi_d',\lambda)}{\pi}\sin\sigma d\sigma d\phi \tag{4.33}$$

where the limits of integration are over any background objects in the hemisphere above the target.

The reflected spectral radiance $L_{r\lambda}(\theta',\phi_d')$ toward the sensor can then be expressed as:

$$L_{r\lambda}(\theta',\phi'_d) = E_{s\lambda}' \cos\sigma' \tau_1(\lambda)\frac{r(\lambda)}{\pi} + \int_F L_{d\lambda}(\sigma,\phi)\cos\sigma\frac{r(\lambda)}{\pi}d\Omega$$
$$+ \int_{1-F} L_{b\lambda}(\sigma,\phi)\cos\sigma\frac{r(\lambda)}{\pi}d\Omega \tag{4.34}$$

where the functional dependence of the reflection factors $r(\lambda)$ on angle have been suppressed for more compact presentation, i.e., $r(\lambda) \equiv r_{\mathrm{BDRF}}(\lambda)$.

When the downwelled radiance or the reflectance can be assumed to be reasonably diffuse, a good approximation to Eq. (4.34) can be achieved by

$$L_{r\lambda}(\theta',\phi_d') = E_{s\lambda}' \cos\sigma' \tau_1(\lambda)\frac{r(\lambda)}{\pi}$$
$$+FE_{d\lambda}\frac{r_d(\lambda)}{\pi} + (1-F)L_{b\lambda\mathrm{avg}}r_d(\lambda) \tag{4.35}$$

Note that we still use the reflectance factor in the solar irradiance term. This term will vary a great deal with sun-target-sensor geometry, unless the target is quite Lambertian (diffuse).

In Section 4.1, we introduced the reflected background radiance without fully defining it. This can be done now, recognizing that the radiance reflected from the background onto the target results from the same factors that cause reflected radiance from the target, i.e.,

$$L_{b\lambda}(\sigma,\phi) = E_{s\lambda}' \cos\sigma_b' \tau_1(\lambda)\frac{r_b(\lambda)}{\pi} + F_b E_{d\lambda}\frac{r_{bd}(\lambda)}{\pi}$$
$$+(1-F_b)L_{ba\lambda}r_{bd}(\lambda) \tag{4.36}$$

where r_b is the reflectance factor for the background, r_{bd} is the diffuse reflectance of the background, σ_b' is the angle from the normal to the background to the sun, F_b is the shape factor for the background, and L_{ba} is the mean radiance onto the background from the hemisphere above the background which is non-sky (i.e., the background's background). L_{ba} is generally assumed to be the radiance from a horizontal Lambertian reflector with the mean reflectance (albedo) of the earth for the scene being studied. In most cases, it is not necessary to solve Eq. (4.33) for each background point in the hemisphere above the target. In general, only negligible errors will be introduced by solving for the reflected background radiance at a central point in each background object and applying that value over the entire object. (In most cases, r_b can be set equal to r_{bd}, and F_b set equal to F, with no significant error introduced.)

4.3 SOLAR RADIANCE REACHING THE SENSOR

After all this effort, we've finally got the photon flux turned around and headed toward the sensor; hopefully carrying the information we are interested in. However, the reflected radiance (L_r) headed toward the sensor is attenuated by absorption and scattering along the path. The transmission losses along the target-sensor path (τ_2) can be described in the same fashion as the losses along the sun-target path (τ_1). The optical depth can be computed in a piecewise fashion and summed to yield the overall optical depth according to Eq. (3.95). This transmission loss results in a reduction in contrast at the sensor and an overall loss in signal level. A competing effect is introduced through scattering by the atmosphere into the target sensor path. This upwelled radiance will increase the overall radiance reaching the sensor in an additive fashion. The competition between these two interrelated effects results in radiance values at the sensor which may be larger or smaller than the surface-leaving radiance values. In this section, we will consider how these interactions affect the radiance reaching the sensor.

4.3.1 Solar Scattered Upwelled Radiance (Path Radiance)

The upwelled radiance is characterized in a manner analogous to the directional downwelled radiance calculations illustrated in Figure 4.4. In this case we are concerned about the radiance from the sun scattered upward into the sensor's line of site along the sensor-target path. By analogy to the downwelled radiance case, this can be expressed as:

$$L_{u\lambda}(\theta,\phi) = E'_{s\lambda} \int \tau_{L1}(\lambda)\tau_{L2}(\lambda)\beta_{\text{sca}}(\lambda,\theta_v)dr \tag{4.37}$$

where the terms are interpreted as illustrated in Figure 4.10, the integral is along the ray from the sensor to the ground, and the angular scattering coefficient is considered to vary along that path.

The magnitude of the upwelled radiance is seen from Eq. (4.37) to result from a gain due to the angular scattering function and loss due to the transmission τ_{L2} from the scattering volume to the sensor. As a result, it is often difficult to predict whether a change in the atmospheric makeup will increase or decrease the path radiance. In general, if the number of scattering centers increases, path radiance will increase. However, these same scattering centers will scatter the flux scattered toward the sensor out of the target-sensor path, reducing path radiance. Most of the variation in scattering centers occurs in the lower atmosphere where scattering acts mostly as an energy source. The upper atmosphere, where extinction is important, is less variable. Thus, the net effect of increasing scattering centers in the atmosphere is usually an increase in path radiance.

Analysis of Eq. (4.37) also indicates that the path radiance must monotonically increase with path length along a particular line of sight. The variation in upwelled radiance with view angle is more complicated, as illustrated in Figure 4.11. The scattering phase function plays a very important role in controlling how upwelled radiance varies with view angle. This is coupled to the increase in path length with view angle from nadir, which in general will increase the path radiance. The overall effect is that path radiance will tend to increase with view angle. However, the minimum path radiance will not necessarily occur at nadir.

4.3.2 Cumulative Solar Effects

If we combine the effects due to the transmission along the target sensor path τ_2 with the

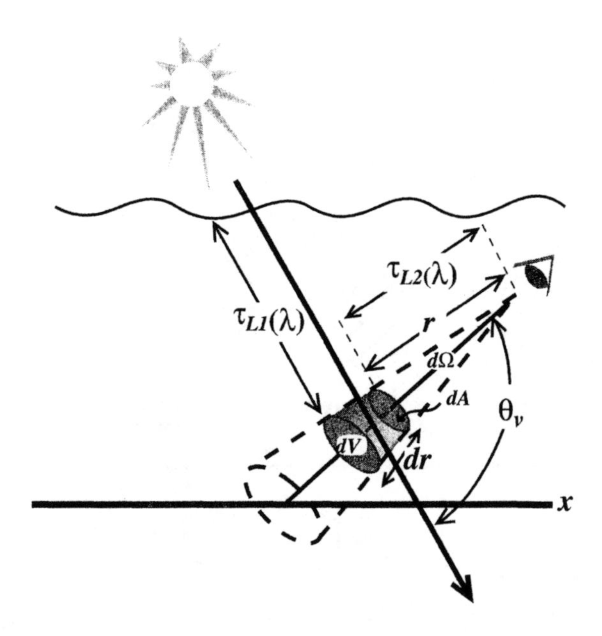

Figure 4.10 Procedure for calculation of upwelled radiance.

upwelled radiance L_u, we have an overall expression for the radiance reaching the sensor due to solar photons of

$$L_{s\lambda}[\mathrm{wm^{-2}sr^{-1}\mu m^{-1}}] = L_{r\lambda}\tau_2(\lambda) + L_{u\lambda}(\theta,\phi)$$
$$= \left[E'_{s\lambda}\cos\sigma'\,\tau_1(\lambda)\frac{r(\lambda)}{\pi} + FE_{d\lambda}\frac{r_d(\lambda)}{\pi}\right.$$
$$\left. +(1-F)L_{b\lambda\mathrm{avg}}r_d(\lambda)\right]\tau_2(\lambda) + L_{u\lambda} \tag{4.38}$$

where the angular dependencies are not explicitly indicated for clarity of presentation. At times we will find it convenient to express the transmission to the sensor as a simple function of view angle. For relatively thin atmospheres (predominantly single scattering) and restricted view angles ($\theta < 60$), the functional dependency of optical depth on view angle is nearly directly proportional to the increase in path length, so we can write the transmission to the sensor as:

$$\tau_2(\lambda) \approx e^{-\delta'(\lambda)\sec\theta} \tag{4.39}$$

where δ' is the optical depth along the vertical path to the sensor altitude. Note that for a sensor in space $\delta = \delta'$. When this approximation is used, Eq. (4.38) can be represented as:

$$L_{s\lambda} = \left[E'_{s\lambda}\cos\sigma' e^{-\delta\sec\sigma}\frac{r(\lambda)}{\pi} + FE_{d\lambda}\frac{r_d(\lambda)}{\pi}\right.$$
$$\left. +(1-F)L_{b\lambda\mathrm{avg}}r_d(\lambda)\right]e^{-\delta'\sec\theta} + L_{u\lambda} \tag{4.40}$$

For a fully exposed horizontal surface this reduces to

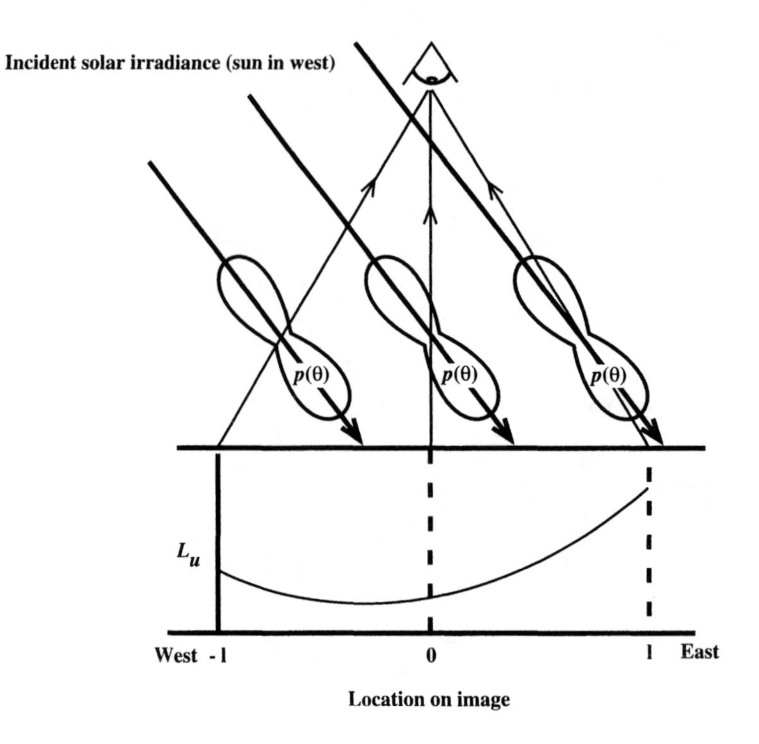

Figure 4.11 Variation in path radiance with view angle. A Rayleigh phase function is shown for reference; however, the upwelled radiance plot is for an actual atmosphere in the visible spectral region. This is the atmospheric upwelled radiance that would be seen in each line of data when flying a north-south line with a line scanner.

$$L_{s\lambda} = \left[E'_{s\lambda} \cos\sigma' \tau_1(\lambda) \frac{r(\lambda)}{\pi} + E_{d\lambda} \frac{r_d(\lambda)}{\pi} \right] \tau_2(\lambda) + L_{u\lambda} \tag{4.41}$$

The relative impact of path radiance on radiance at the sensor is shown in Figure 4.12. This plot shows radiance leaving the ground versus radiance reaching the sensor for several atmospheres with different optical depths. It is clear that the combined effects of path radiance and transmission loss will increase the radiance from low reflecting objects and often decrease the radiance from more highly reflecting targets. Furthermore, if we compare the case of the less dense to more dense atmosphere, we see that for low reflectance targets, the radiance increases with increasing turbidity (decreasing visibility and optical depth) in the atmosphere, with just the opposite occurring for high-reflectance objects (i.e., radiance will decrease with increasing turbidity). These effects are discussed in more detail and as a function of wavelength by Turner et al. (1975). They point out that while the scattering effects are reduced as you go from the blue toward the near infrared, the same basic effects occur.

4.4 THERMAL RADIANCE REACHING THE SENSOR

In Chapter 3, we identified a number of energy paths that self-emitted photons could traverse to produce radiance at the sensor (cf. Fig. 3.3). In this section, we will develop radiometric expressions to describe the radiance associated with those paths and combine them with the

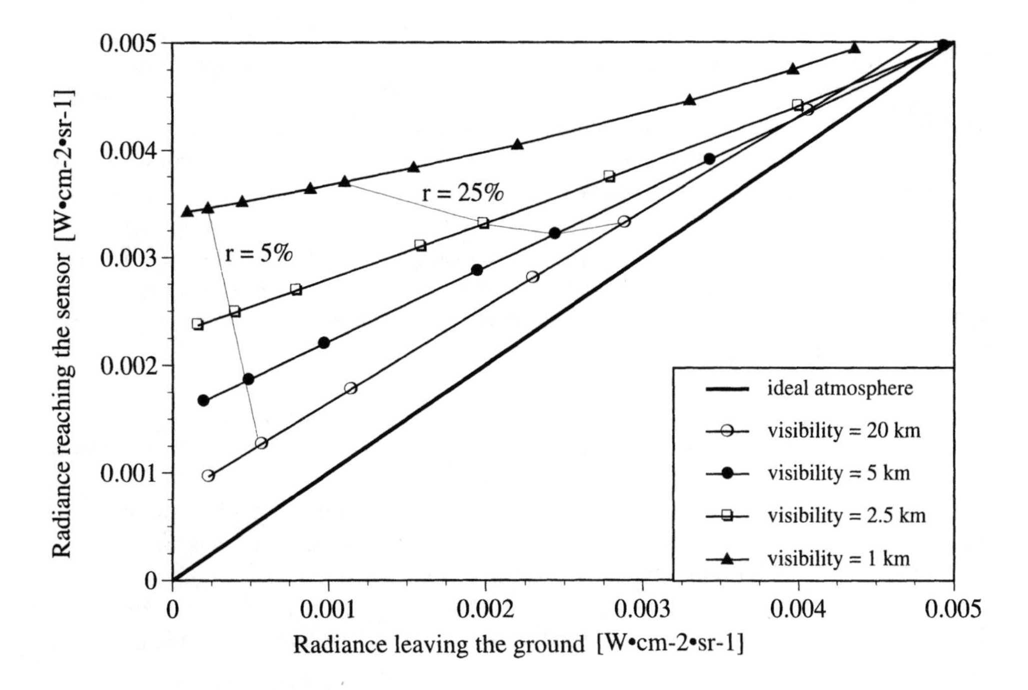

Figure 4. 12 Effects of transmission and path radiance on radiance reaching the sensor. (Data derived from the LOWTRAN atmospheric propagation model run in the visible spectral region.)

expression developed in the previous section for the solar terms.

4.4.1 Self-Emission

The energy path that is most often of interest when sensing in the thermal infrared is associated with the emission from an object due to its temperature (path *D* in Figure 3.3). This is the only path that carries information about the object's temperature. The radiance headed toward the sensor due to the target's temperature will be a function of the Planck equation modified by the wavelength-dependent emissivity of the target. This radiance will be attenuated by the transmission along the target sensor path. Combining these effects with the solar terms yields

$$L_\lambda = E'_{s\lambda} \cos\sigma' \frac{r(\lambda)}{\pi} \tau_2(\lambda)\tau_1(\lambda) + E_{d\lambda} \frac{r_d(\lambda)}{\pi} \tau_2(\lambda)$$
$$+\varepsilon(\lambda) L_{T\lambda} \tau_2(\lambda) + L_{u\lambda} \qquad (4.42)$$

where $\varepsilon(\lambda)$ is the wavelength-dependent emissivity—for non-Lambertian objects ε will also be a function of view angle (θ') and of azimuthal angle (ϕ_d')for azimuthally varying surfaces (e.g., corrugated surfaces)—and $L_{T\lambda}$ is the spectral radiance [wm^{-2}sr$^{-1}\mu$m^{-1}] for a blackbody at temperature *T* as described by the Planck equation, Eq. (3.68).

4.4.2 Thermal Emission from the Sky and Background Reflected to the Sensor

Recall from Kirchoff's law that an opaque object with emissivity ε will have reflectivity $r = 1-\varepsilon$. We need to consider reflection by the target of irradiance from the surround due to the temperature of the surround. The atmosphere has some finite temperature and will act as a source of energy to be reflected to the sensor (type E photons in Fig. 3.3). In order to determine how much radiance is reflected toward the sensor from the sky, we need to determine the contribution from (i.e., the irradiance from) each location in the sky. This is most readily conceptualized by considering an atmosphere made up of many homogeneous layers, as illustrated in Figure 4.13. If we let there be N layers between the ground and the top of the atmosphere, then the contribution to the downwelled radiance from the ith layer from the (σ,ϕ) direction can be expressed as:

$$\Delta L_{d\varepsilon}(\sigma,\phi) = L_{Ti}(1-\Delta\tau_i)\prod_{j=1}^{i-1}\Delta\tau_j = L_{Ti}(\tau_i - \tau_{i+1}) \tag{4.43}$$

where L_{Ti} is the radiance due to the temperature (T_i) of the ith layer, $\Delta\tau_i$ is the transmission through the ith layer along the beam defined by the direction angles (σ,ϕ), σ is the angle from the earth normal to the beam, ϕ is the azimuthal angle about the normal in the plane of the earth, the product on j accounts for the transmission loss along the beam from the bottom of the ith layer to the ground, and τ_i is the transmission along the beam from the bottom of the ith layer to the ground.

To make intuitive sense out of Eq. (4.43) we need to recall that conservation of energy requires that for a homogeneous layer i

$$\Delta\tau_i + \Delta r_i + \Delta\alpha_i = 1 \tag{4.44}$$

where Δr_i and $\Delta\alpha_i$ are the reflectance and absorptance of the ith layer along the beam. Since scattering is negligible at most wavelengths of interest for thermal infrared sensing (rigorously speaking, $\Delta\tau_i$ should be defined as transmission loss due solely to absorption), Δr_i is zero, leaving

$$\Delta\tau_i + \Delta\alpha_i = 1 \tag{4.45}$$

or, from Kirchoff's rule,

$$\Delta\tau_i + \Delta\varepsilon_i = 1 \tag{4.46}$$

So, we see that the effective emissivity of the layer can be expressed as:

$$\Delta\varepsilon_i = (1-\Delta\tau_i) \tag{4.47}$$

Thus we can interpret the contribution to radiance from the ith layer in Eq. (4.43) as the effective emissivity of the layer $(1 - \Delta\tau_i)$ times the radiance of the layer (L_{Ti}) times the transmission from the bottom of the layer to the ground.

The difference in transmission values on the far right-hand side of Eq. (4.43) is a simplification commonly found in the LOWTRAN literature (cf. Kneizys et al., 1988). It can be derived by recognizing that

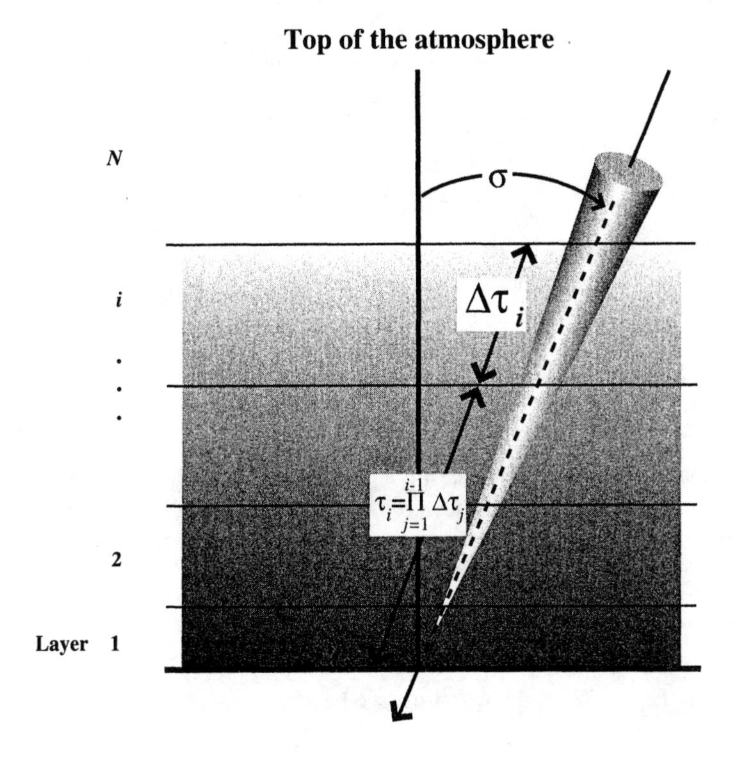

Figure 4.13 Downwelled radiance due to self-emission.

$$\prod_{j=1}^{i-1} \Delta\tau_j = \tau_i \tag{4.48}$$

and

$$\tau_{i+1} = \tau_i \Delta\tau_i \tag{4.49}$$

so that

$$\begin{aligned}(\tau_i - \tau_{i+1}) &= \prod_{j=1}^{i-1} \Delta\tau_j - \left(\prod_{j=1}^{i-1} \Delta\tau_j\right)\Delta\tau_i \\ &= (1-\Delta\tau_i)\prod_{j=1}^{i-1} \Delta\tau_j\end{aligned} \tag{4.50}$$

The total contribution to downwelled radiance from the direction defined by (σ,ϕ) can be expressed as:

$$L_{d\varepsilon}(\sigma,\phi) = \sum_{i=1}^{N} L_{Ti}(\tau_i - \tau_{i+1}) \tag{4.51}$$

where the sum is over all the layers in the atmosphere.

The element of irradiance from the σ,ϕ direction $dE_d(\sigma,\phi)$ can be expressed from the definition of radiance as:

$$dE_{d\varepsilon}(\sigma,\phi) = L_{d\varepsilon}(\sigma,\phi)\cos\sigma d\Omega \tag{4.52}$$

with the total downwelled irradiance expressed as:

$$E_{d\varepsilon} = \int_{2\pi} L_{d\varepsilon}(\sigma,\phi)\cos\sigma d\Omega \tag{4.53}$$

For a Lambertian reflector, the spectral radiance reflected to the sensor from downwelled emission from the sky would then be:

$$L_{r\varepsilon\lambda} = E_{d\varepsilon\lambda}\frac{r_d(\lambda)}{\pi} \tag{4.54}$$

where the subscript ε denotes dependence on self-emission, and the spectral dependency, which was neglected for readability, has been reintroduced. In the case of a non-Lambertian target, the element of reflected radiance can be expressed as:

$$dL_{r\varepsilon\lambda} = dE_{d\varepsilon\lambda}(\sigma,\phi)\frac{r(\lambda)}{\pi} \tag{4.55}$$

or, on substituting in Eq. (4.52) and integrating, we have

$$L_{r\varepsilon\lambda} = \int_{2\pi} L_{d\varepsilon\lambda}(\sigma,\phi)\cos\sigma\frac{r(\lambda)}{\pi}d\Omega \tag{4.56}$$

If we now introduce the case of nonhorizontal or obscured surfaces, the analysis proceeds in the same fashion as used for scattered downwelled radiance with the use of the shape factor. The radiance reflected toward the sensor due to self-emitted downwelled sky radiance and background self-emission can be expressed as:

$$\begin{aligned} L_{r\varepsilon\lambda}(\theta',\phi_d') = &\int_F L_{d\varepsilon\lambda}(\sigma,\phi)\cos\sigma\frac{r(\sigma,\phi,\theta',\phi_d',\lambda)}{\pi}\sin\sigma d\sigma d\phi \\ &+ \int_{1-F} L_{Tb\lambda}(\sigma,\phi)\cos\sigma\frac{r(\sigma,\phi,\theta',\phi_d',\lambda)}{\pi}\sin\sigma d\sigma d\phi \end{aligned} \tag{4.57}$$

where $L_{Tb}(\sigma,\phi)$ is the radiance onto the target due to the temperature of the background (T_b) coming from the (σ,ϕ) direction, and we have explicitly expressed the dependence of the reflectance factor $r(\lambda)$ on the incident and exitant angles. If the BRDF is slowly varying with angle, or the downwelled radiance and background radiance do not vary greatly, Eq. (4.57) can be expressed to a good approximation as:

$$L_{r\varepsilon\lambda} \cong FE_{d\varepsilon\lambda}\frac{r_d(\lambda)}{\pi} + (1-F)L_{b\varepsilon\lambda}r_d(\lambda) \tag{4.58}$$

where $L_{b\varepsilon\lambda}$ is the mean spectral radiance from the background due to self-emission. In most cases, background objects will all have similar temperatures that will be near earth's ambient temperature. The downwelled radiance varies with angle (cf. Fig. 4.14); however, it is a gradual function, so Eq. (4.58) is a good approximation for any target that exhibits reasonably diffuse behavior. Note in Figure 4.14(b) that under clear-sky conditions, the downwelled radi-

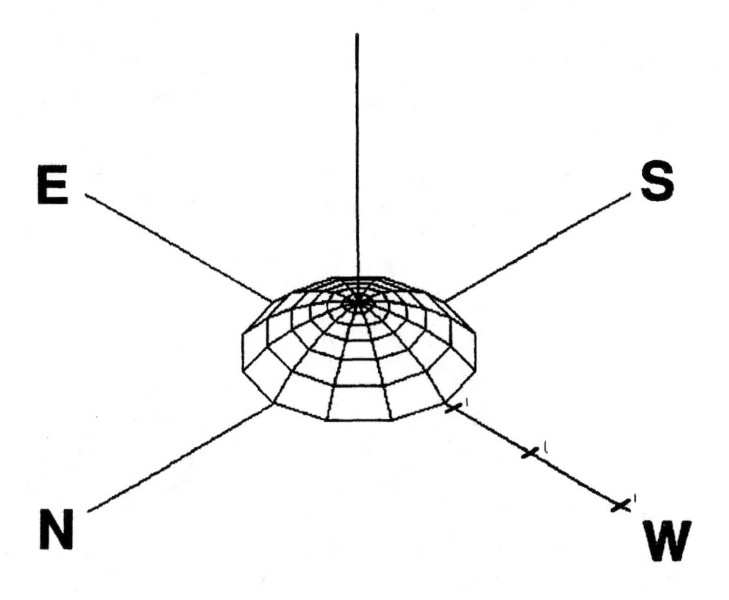

(a) Polar plot of downwelled radiance.

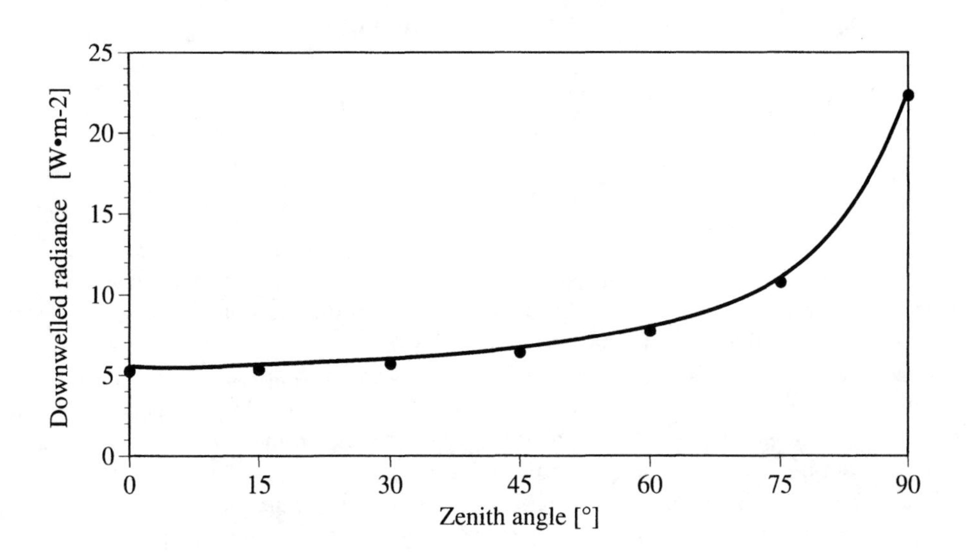

(b) Plot of downwelled radiance as a function of view angle θ for a fixed azimuth angle θ_d.

Figure 4.14 Angular variation in downwelled radiance in the 8- 14-μm region due to self-emission by the sky.

ance due to self-emission is always greater near the horizon due to the longer path through a warmer, denser atmosphere.

If we multiply Eq. (4.58) by the transmittance along the target sensor path and combine it with Eq. (4.40) and the target self-emission term, we have a new expression for the radiance reaching the sensor:

$$\begin{aligned} L_\lambda &= E'_{s\lambda} \cos\sigma' \frac{r(\lambda)}{\pi} \tau_1(\lambda)\tau_2(\lambda) + \varepsilon(\lambda) L_{T\lambda} \tau_2(\lambda) \\ &+ F\left[E_{ds\lambda} + E_{d\varepsilon\lambda}\right] \frac{r_d(\lambda)}{\pi} \tau_2(\lambda) \\ &+ (1-F)[L_{bs\lambda} + L_{b\varepsilon\lambda}] r_d(\lambda) \tau_2(\lambda) + L_{us\lambda} \end{aligned} \tag{4.59}$$

where we have introduced the subscript s on the solar reflected background radiance ($L_{bs\lambda} \equiv L_{b\lambda avg}$), the solar scattered downwelled irradiance ($E_{ds\lambda} \equiv E_{d\lambda}$), and the upwelled radiance $L_{us\lambda} = L_{u\lambda}$. Explicitly expressing the dependence on solar photons differentiates the solar terms (subscript s) from the self-emissive terms (subscript ε). Consideration of the symmetry of Eq. (4.59) or of the energy paths in Figures 3.1 and 3.3 points us to the one remaining self-emissive term yet to be considered.

4.4.3 Self-Emitted Component of Upwelled Radiance

Equation (4.59) lacks a contribution due to the self-emitted radiance from the atmosphere along the line of site. This upwelled-radiance term $L_{u\varepsilon}$ can be expressed in the same fashion as the angular contribution to the self-emitted downwelled radiance $L_{d\varepsilon}(\sigma,\phi)$, i.e.,

$$\begin{aligned} L_{u\varepsilon\lambda}(\theta,\phi) &= \sum_{i=1}^{N} L_{Ti\lambda}(\tau_i(\lambda) - \tau_{i+1}(\lambda)) \\ &= \sum_{i=1}^{N} L_{Ti\lambda}(1-\Delta\tau_i(\lambda)) \prod_{j=1}^{i-1} \Delta\tau_j(\lambda) \end{aligned} \tag{4.60}$$

where we have redefined the indexing scheme for the atmospheric layers to begin at the sensor and end with the layer just above the ground. Thus τ_i is the transmission along the sensor-target path from the sensor to the top of the ith layer.

Inspection of Eq. (4.60) indicates that for an atmosphere composed of uniform homogeneous layers, the upwelled radiance due to self-emission $L_{u\varepsilon}$ should be independent of the azimuthal view angle and symmetric about the view angle from the nadir. As view angle increases from the nadir, two competing processes take place. The increased path through each layer increases the number of emitters in the beam causing the effective emissivity of the layer $(1 - \Delta\tau_i)$ to increase. At the same time the transmission along the path decreases due to the increased path length. In general, for reasonably clear atmospheres, the incremental effects win out, and $L_{u\varepsilon}$ increases with view angle (cf. Fig. 4.15).

Adding the upwelled radiance term due to self-emission $L_{u\varepsilon}$ into our expression for the radiance at the sensor (Eq. 4.59), we have a final expression (what my students over the years have taken to calling the big equation) for the spectral radiance reaching the sensor

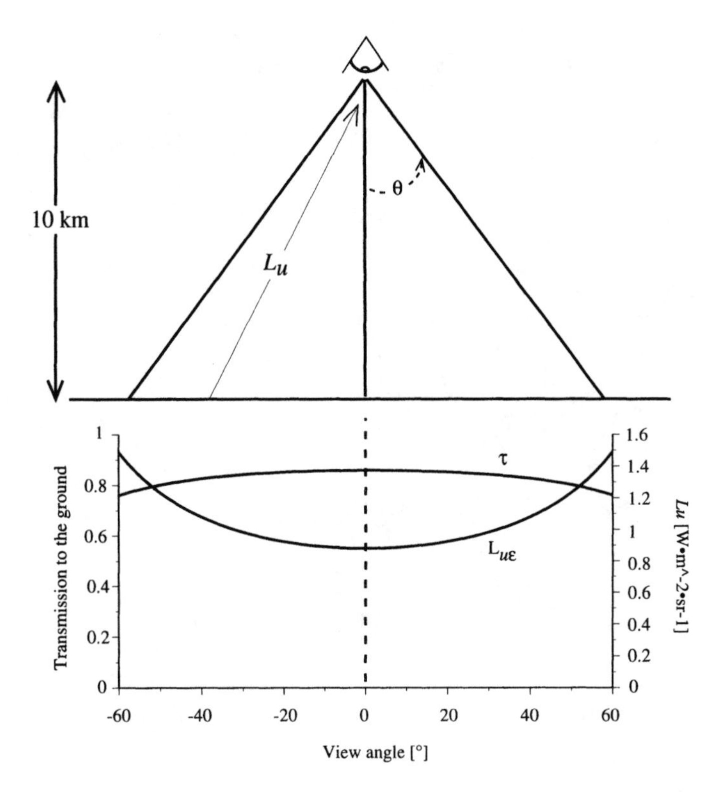

Figure 4.15 Variation in transmission and upwelled radiance with view angle due to self-emission in the 10 to 12-μm region.

$$L_\lambda = \left\{ E'_{s\lambda} \cos\sigma' \, \tau_1(\lambda) \frac{r(\lambda)}{\pi} + \varepsilon(\lambda) L_{T\lambda} + F[E_{ds\lambda} + E_{d\varepsilon\lambda}] \frac{r_d(\lambda)}{\pi} \right.$$
$$\left. + (1-F)[L_{bs\lambda} + L_{b\varepsilon\lambda}] r_d(\lambda) \right\} \tau_2(\lambda) + L_{us\lambda} + L_{u\varepsilon\lambda} \tag{4.61}$$

The relationship between each of these terms and the energy path diagrams in Figures 3.1 and 3.3 is illustrated in Figure 4.16. For clarity, we will generally use a form of Eq. (4.61) in the remainder of the book. However, if a more rigorous treatment employing full BRDF data is required, the shape factor simplifications can be replaced with the integrals over the hemisphere [cf. Eq. (4.34) and (4.57)].

4.5 INCORPORATION OF SENSOR SPECTRAL RESPONSE

The expression we have developed for spectral radiance reaching the sensor Eq. (4.61) must be cascaded with the sensor spectral response function to determine how much of the radiance is "sensed" by the system. Using the concept of effective radiance from Eq. (3.18), we can describe the effective radiance reaching the sensor as:

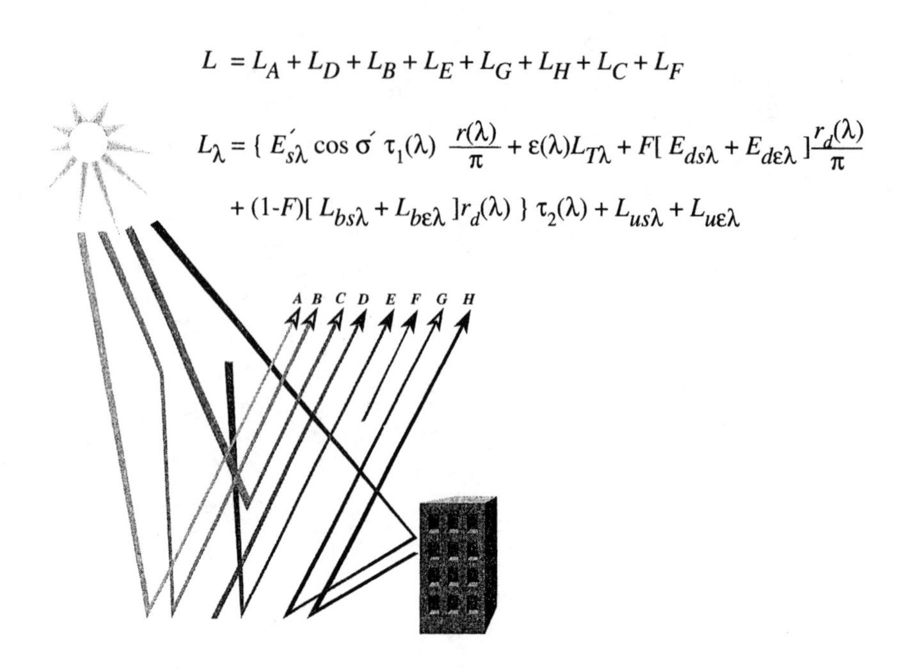

Figure 4.16 Relationship between terms in "*the big equation*" and energy paths associated with the photon flux onto the sensor.

$$L_{\text{eff}} = \int_0^\infty L_\lambda R'(\lambda) d\lambda \tag{4.62}$$

where L_λ is defined by Eq. (4.61), and $R'(\lambda)$ is the normalized spectral response function of the sensor.

In many operational situations, it is desirable to use average or effective bandpass values to simplify the calculation or representation of terms in equations such as (4.61). This can be done by approximating each term in the following fashion:

$$L_{\Delta\lambda} = \int_0^\infty \varepsilon(\lambda) L_{T\lambda} \tau_2(\lambda) R'(\lambda) d\lambda \approx \varepsilon_{\text{avg}} L_{T\text{eff}} \tau_{2\text{avg}} = \varepsilon L_T \tau_2 \tag{4.63}$$

where $L_{\Delta\lambda}$ is the effective radiance due to the target self-emission reaching the sensor, ε_{avg} is the mean emissivity across the bandpass, $\tau_{2\text{avg}}$ is the mean transmission across the bandpass, and $L_{T\text{eff}}$ is the effective radiance (due to the target's temperature) over the sensor's spectral response. For convenience, the subscripts (avg) and (eff) will be assumed, except where needed for clarity. From freshman calculus, we know that Eq. (4.63) is not true in general. It is, however, often a reasonable approximation if the wavelength-dependent terms are continuous over the bandpass and approximately constant. Figure 4.17 shows a case in the LWIR where this approximation might hold reasonably well. However, the MWIR case shown is an example where the approximation will break down. Recognizing these limitations, we will proceed to use the approximation method for clarity of presentation. The reader is cautioned that when there is substantial spectral character to the data, the wavelength-by-wavelength solution of Eq. (4.62) is required for accuracy. Using the simplified form, the

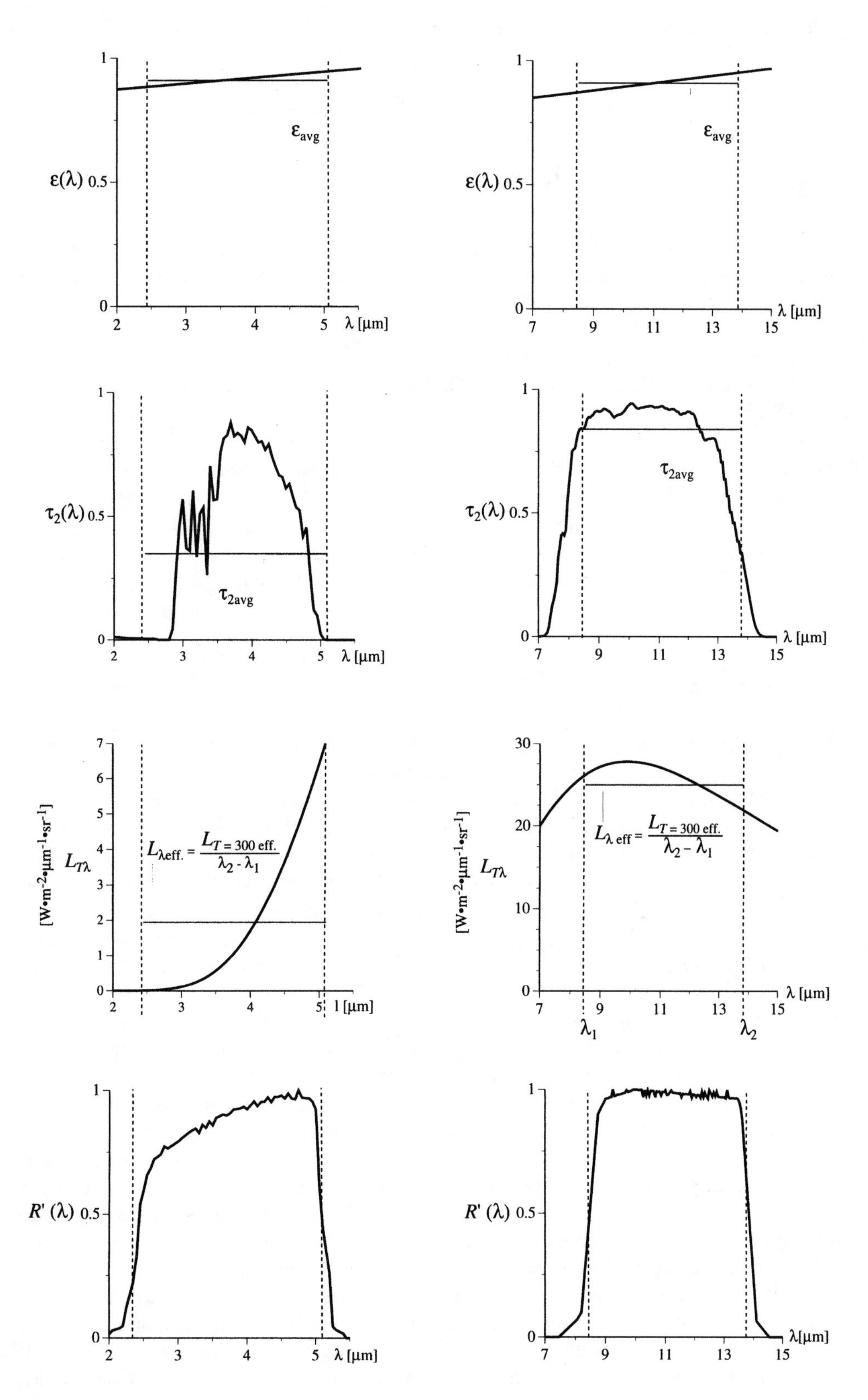

Figure 4.17 Spectral character of the parameters impacting the radiance reaching the sensor in the MWIR and LWIR spectral regions.

effective radiance in the bandpass reaching the sensor can be expressed as:

$$L = \left\{ E_s' \tau_1 \cos\sigma' \frac{r}{\pi} + \varepsilon L_T + \left(F[L_{ds} + L_{d\varepsilon}] + (1-F)[L_{bs} + L_{b\varepsilon}]\right) r_d \right\} \tau_2 + L_{us} + L_{u\varepsilon} \tag{4.64}$$

Even in this simplified form, the number of parameters needed to solve Eq. (4.64) is somewhat daunting. In the next section, we will evaluate what conditions might allow us to further simplify "the big equation."

4.6 SIMPLIFICATION OF THE BIG EQUATION AND RELATIVE MAGNITUDE ASSESSMENT

We've now come to the stage in our analysis I always loved as a physics student: where the professor drew a big line through expressions setting them equal to zero so that we no longer needed to think about them. In this section, we want to determine what the governing equation, Eq. (4.64) will look like in different portions of the spectrum and for different targets. In terms of the image chain approach, we want to determine what the governing equation for the radiometric portion of the image should be for each atmospheric transmission window.

4.6.1 Simplification

When we are trying to measure reflectance, we will seldom be interested in making measurements to better than 0.1 reflectance units (e.g., from 10.6% to 10.7% reflectance). Therefore, terms in Eq. (4.64) that result in changes significantly less than one part in several hundred will be considered negligible. Similarly, when trying to measure temperature we will seldom be interested in measuring temperature to better than 0.1 K. At 300 K, a change of 0.1 K amounts to a change in radiance of about one part in 800 in the LWIR and one part in 300 in the MWIR. Clearly the required precision will change with both the target and the application. A small change will be much more important when measuring low reflectance (e.g., water in the VIS-NIR) or low temperature (e.g., nighttime winter scenes). For any conditions that deviate substantially from those identified here, a similar analysis should be performed employing the specific parameters and precisions required. To determine the importance of the terms in "the big equation," we can perform a simple relative magnitude computation for a typical case. This involves estimating the magnitude of each parameter in Eq. (4.61), computing the value of each term by cascading in the spectral response of the sensor, (Eq. (4.62)) and summing to yield the total radiance at the sensor. If each term in turn is then set to zero and the sum of all the other terms compared to the total radiance, we can determine the relative importance of each term. If zeroing the term has negligible impact based on our required precision, we can neglect that term. Since this is not a robust assessment, we should include any terms that are close to our rejection threshold and perform a more robust test if our application varies substantially from the test presented here.

The example case we considered is presented in Table 4.1. The parameters were generated by manipulating the output from the LOWTRAN radiation propagation model using the equations developed in this chapter (cf. the discussion of LOWTRAN in Sec. 6.3.3 or Salvaggio et al., 1993). By comparing the total radiance (L) in Table 4.1 to the radiance when each term is subtracted, we can determine the relative importance of the term in each spectral interval. To simplify this assessment, we have introduced the apparent reflectance and

Table 4.1 Sensitivity Analysis of Terms in the Big Equation. (Each entry in the matrix represents the magnitude of the radiance. The entries for apparent reflectance and temperature represent the apparent reflectance or temperature if the term were set equal to zero.)

	L_A $E'_s \cos\sigma \frac{r}{\pi} \tau_1 \tau_2$	L_D $\varepsilon L_1 \tau_2$	L_B $F L_{ds} r \tau_2$	L_E $F L_{d\varepsilon} r \tau_2$	L_G $(1-F) L_{Bs} r \tau_2$	L_H $(1-F) L_{b\varepsilon} r \tau_2$	L_F $L_{u\varepsilon}$	L_C L_{us}	L_{total}
Integrated Radiance [W•cm^{-2}•sr^{-1}], $\Delta\lambda$ = 0.4 - 0.7 μm									
ε = 0.8	2.28E-03	9.10E-28	6.36E-04	6.92E-31	3.63E-05	4.55E-29	1.14E-29	1.75E-03	4.70E-03
ε = 0.9	1.14E-03	1.02E-27	3.18E-04	3.46E-31	1.81E-05	2.56E-29	1.14E-29	1.75E-03	3.22E-03
ε = 0.986	1.60E-04	1.12E-27	4.45E-05	4.48E-32	2.54E-06	3.93E-30	1.14E-29	1.75E-03	1.95E-03
Integrated Radiance [W•cm^{-2}•sr^{-1}], $\Delta\lambda$ = 3 - 5 μm									
ε = 0.8	9.51E-05	1.28E-04	1.15E-06	4.94E-06	4.98E-07	6.61E-06	3.93E-05	6.55E-07	2.77E-04
ε = 0.9	4.75E-05	1.44E-04	5.73E-07	2.47E-06	2.49E-07	3.71E-06	3.93E-05	6.55E-07	2.39E-04
ε = 0.986	6.66E-06	1.58E-04	8.03E-08	3.46E-07	3.48E-08	5.67E-07	3.93E-05	6.55E-07	2.06E-04
Integrated Radiance [W•cm^{-2}•sr^{-1}], $\Delta\lambda$ = 8 - 14 μm									
ε = 0.8	6.92E-06	5.19E-03	4.11E-08	1.86E-04	3.45E-08	2.67E-04	1.56E-03	2.39E-08	7.21E-03
ε = 0.9	3.46E-06	5.84E-03	2.06E-08	9.29E-05	1.73E-08	1.50E-04	1.56E-03	2.39E-08	7.65E-03
ε = 0.986	4.85E-07	6.39E-03	2.88E-09	1.30E-05	2.42E-09	2.29E-05	1.56E-03	2.39E-08	7.99E-03
Apparent Reflectance, $\Delta\lambda$ = 0.4 - 0.7 μm									
ε = 0.8	0.0760	0.1478	0.1278	0.1478	0.1476	0.1478	0.1478	0.0929	0.1478
ε = 0.9	0.0655	0.0913	0.1013	0.1013	0.1008	0.1013	0.1013	0.0465	0.1013
ε = 0.986	0.0564	0.0614	0.0600	0.0614	0.0613	0.0614	0.0614	0.0065	0.0614
Apparent Temperature (Kelvin), $\Delta\lambda$ = 3 - 5 μm									
ε = 0.8	282.12	277.62	291.63	291.31	291.68	291.16	288.19	291.67	291.72
ε = 0.9	283.30	267.86	288.29	288.11	288.32	287.99	284.26	288.28	288.35
ε = 0.986	284.21	253.67	284.94	284.91	284.95	284.89	280.19	284.88	284.95
Apparent Temperature (Kelvin), $\Delta\lambda$ = 8 - 14 μm									
ε = 0.8	280.58	222.35	280.64	279.13	280.64	278.46	267.15	280.64	280.64
ε = 0.9	284.03	218.31	284.06	283.34	284.06	282.90	271.07	284.06	284.06
ε = 0.986	286.64	214.09	286.64	286.55	286.64	286.48	274.12	286.64	286.64

apparent temperature concepts. These terms are often used to provide a more intuitive interpretation of radiance values in a spectral bandpass. The apparent reflectance (also referred to as the *reflectance at the top of the atmosphere*) is the reflectance a Lambertian reflector located directly in front of the sensor would have to have in the absence of any atmosphere to produce the observed radiance. It would be solved for according to:

$$r_{\text{ap}} = L_{\text{eff}} \left(\frac{E_s' \cos \sigma'}{\pi} \right)^{-1} \tag{4.65}$$

where L_{eff} is the effective radiance which we wish to express in terms of apparent reflectance.

The apparent temperature is the temperature a blackbody would have to have to produce the effective radiance observed at the sensor, i.e., it is the temperature T_{ap} which would make the following expression true.

$$L_{\text{eff}} = \int_0^\infty L_{T_{ap}\lambda} R'(\lambda) d\lambda \tag{4.66}$$

The quantitative data in Table 4.1 for the case studied verify the graphical data of Figure 3.19. In the VIS-NIR region the solar energy is so many orders of magnitude higher than the self-emitted energy that all the thermal energy paths (D, E, F, and H) are negligible. So in the reflective region, we can correctly approximate the effective radiance at the sensor as:

$$L = L_{\text{A}} + L_{\text{B}} + L_{\text{G}} + L_{\text{C}}$$
$$L = \left[E_s' \cos\sigma' \frac{r}{\pi} \tau_1(\lambda) + (FE_{ds} + (1-F)E_{bs}) \frac{r_d}{\pi} \right] \tau_2 + L_{us} \tag{4.67}$$

Furthermore, we can see that the solar irradiance term (A) and the path radiance term (C) are the largest contributors with the reflected skylight (B) and the reflected background (G) terms being significantly smaller, but not necessarily negligible. It's also clear that as the shape factor approaches 1, the reflected background radiance term will become a negligible contributor to the total radiance. This will be the case for any nearly horizontal surface that does not have taller objects or high terrain in its vicinity. It's also clear from inspection of Table 4.1 that the relative magnitude of the reflected radiance (A + B + G) and the path radiance (C) will vary considerably depending on the reflectance of the target. For dark targets (1 to 3% reflectors), the path radiance may represent 50% or more of the total radiance. In this case, a small error in computation of upwelled radiance would result in a very large error in the computed reflectance. The importance of the upwelled radiance term never vanishes as reflectance increases, but the overall radiance (and computed reflectance) becomes less sensitive to errors in the path radiance.

The case presented in Table 4.1 is for a relatively clear atmosphere. The relative importance of both the skylight term and the path radiance term will increase as the atmosphere becomes more turbid and as we shift to shorter wavelengths. Piech and Walker (1971), for example, show the spectral shape of the sunlight and skylight irradiance in the visible spectrum as plotted in Figure 4.18. They also observed that in the visible spectrum, the direct solar irradiance term could range from approximately 7 times the downwelled irradiance under clear sky conditions to nearly equal values for the two terms under very hazy conditions.

Switching to the other end of the spectrum, we see from Table 4.1 that in the LWIR region the solar photons become a vanishingly small portion of the total radiance. Even the direct solar irradiance produces too few photons to make a significant contribution to the total irradiance. As a result, in the LWIR region the effective radiance reaching the sensor can usually be expressed with acceptable error as:

$$L = L_{\text{D}} + L_{\text{E}} + L_{\text{H}} + L_{\text{F}}$$
$$L = \left[\varepsilon L_T + (FE_{d\varepsilon} + (1-F)E_{b\varepsilon}) \frac{r_d}{\pi} \right] \tau_2 + L_{u\varepsilon} \tag{4.68}$$

The target self-emission term (D) dominates this expression with significant contribution from upwelled radiance (F) (cf. Fig. 4.16). The reflected downwelled radiance (E) and reflected background radiance (H) terms are typically much smaller, though still significant contributors if measurement accuracies of tenths of a degree are desired. The relative importance of these reflected terms will decrease with increasing emissivity (decreasing reflectivity), but in general they will not be negligible until emissivity values approach 0.99. The relative importance of the downwelled radiance and the background radiance is controlled by the shape factor just as in the VIS-NIR region. For nearly horizontal unobstructed surfaces,

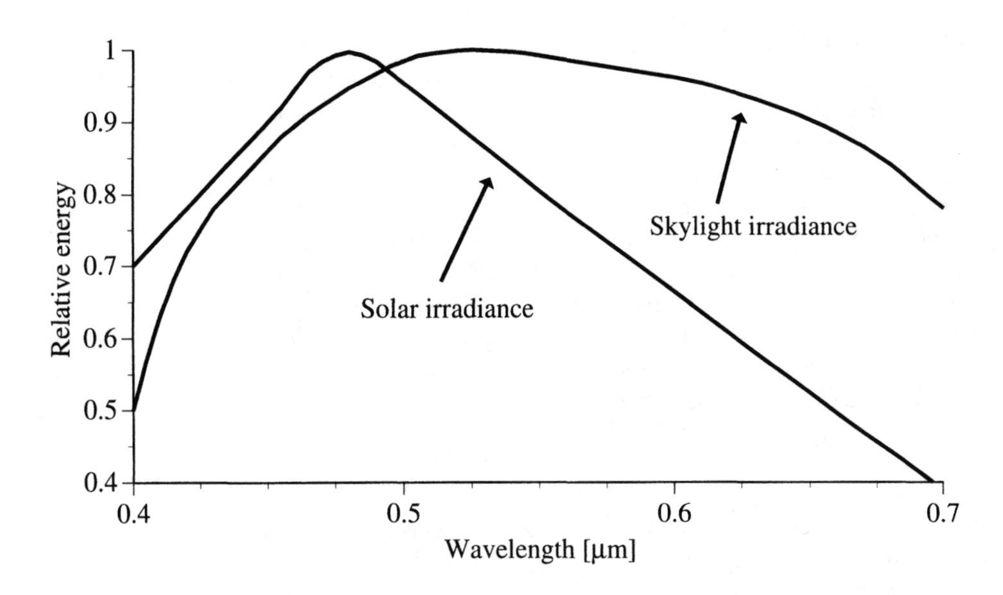

Figure 4.18 Comparison of direct solar irradiance and skylight irradiance (after Piech and Walker, 1971).

the shape factor F approaches 1.0 and the background term becomes negligible. In Chapter 6, we will use this phenomenon to select calibration targets without background effects to simplify the equations used in computing the values of atmospheric parameters.

Finally, when we consider the MWIR window region for the case treated in Table 4.1, we see the effects of the intersection of the solar irradiance and self-emission curves of Figure 3.19. The significant contribution to the photon flux from solar effects and terrestrial self-emission results in all of the terms in Eq. (4.64) being potentially significant. The solar terms will be larger under high solar elevation conditions and for cold or low emissivity targets. The thermal terms will be more important for low solar elevation angles and higher-temperature targets. However, in most cases the direct solar insolation term will be negligible only at night. In many cases for low solar elevation conditions and targets with temperatures of 290 K or higher, the reflected solar downwelled radiance (B) and solar background effects (G) may be negligible. Because of the complexity of the radiometry in this window, quantitative analysis is usually only attempted on nighttime imagery. Even with nighttime imagery, care must be taken because of the strong spectral character of several of the radiometric terms, $L_{T\lambda}$, $\tau_2(\lambda)$, in this window. The spectral integral form of the radiance equation, Eq. (4.62), must be used rather than the simplification of Eq. (4.64) if quantitative results are required.

This simple relative magnitude analysis will serve us well in determining which terms can be eliminated from our analysis. This approach can be used to obtain an intuitive sense of the relative importance of the terms in the radiance equation. It is a useful exercise whenever you begin work in a new spectral region or on targets that are substantially different from those for which you've already developed an intuitive feel. This approach does not tell us which terms or parameters will generate the dominant sources of error or how an error in one parameter will effect the radiance observed or the temperature or reflectance measured. To obtain this type of error information, a more detailed error or sensitivity analysis is required.

4.6.2 Sensitivity Analysis—Error Propagation

Before beginning this discussion, we should remind the reader of the often-neglected distinction between accuracy and precision which are both measures of error. Precision describes the repeatability of a measurement. It is often characterized by the standard deviation from the mean of many measurements. For example, if we measured the reflectance of a target 20 times and computed a mean reflectance of 0.18 with a standard deviation of 0.02, we could claim that the precision of the measurement to one standard deviation was 0.02 or two reflectance units. Accuracy, on the other hand, describes how closely an instrument or procedure can match some standardized value or what we have defined to be truth. It is often characterized by the deviation between the mean of several measured values and the true value. In the case just cited, if the true reflectance value of the sample was 0.17, we would have an accuracy associated with the measurement process of 0.01. The individual measurement error that describes how closely any individual measurement comes to truth is often taken to be the root sum square error value, i.e.,

$$S_m = (S_p^2 + S_i^2)^{1/2} \tag{4.69}$$

where S_p is the precision of the measurement, S_i is the accuracy of the measurement instrument or approach, and S_m is the total error and can be thought of as the error associated with the individual measurement (i.e., 2.2 reflectance units in our example). Note that in many cases, calibration procedures can generate unbiased errors such that the average of many readings is a very good estimate of the true value (i.e., $S_i \approx 0$). In this case, the precision of the measurement approach becomes a good estimate of the error.

In general, the error (precision, accuracy, or total) of a measurement approach is the result of errors in the procedures or values that go into that measurement. For the case where a governing equation can be used to describe a parameter of interest, a relatively simple expression can be written to describe the relationship between the errors (cf. Beers, 1957). In the simplest case, if we can define a dependent variable Y in terms of one or more independent (i.e., uncorrelated) variables Xi, i.e.,

$$Y = f(X1, X2 \cdots XN) \tag{4.70}$$

then we can express the error in Y (s_Y) as:

$$s_Y = \left[\left(\frac{\partial Y}{\partial X1} s_{X1} \right)^2 + \left(\frac{\partial Y}{\partial X2} s_{X2} \right)^2 \cdots + \left(\frac{\partial Y}{\partial XN} s_{XN} \right)^2 \right]^{1/2} \tag{4.71}$$

where S_{Xi} is the error in the individual input variables. The partials of the dependent variable (Y) with respect to the input variables describe the sensitivity of Y to small changes in X. Multiplying the partial derivative by the error on the input variable Xi generates the error in Y (s_{Yi}) associated with an error in Xi. The total error is just the square root of the sum of the squared values because independent errors tend to add in quadrature. Beers (1957) also points out that for the case where the input variables Xi's are correlated, Eq. (4.71) must be modified to reflect how an error in one input variable may be exaggerated or compensated for by the correlation with a second input variable. In this case, Beers expresses the error as:

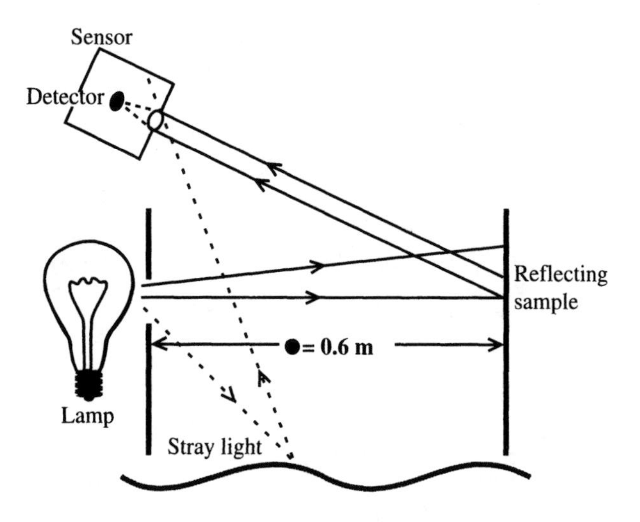

Figure 4.19 A device for measurement of reflected radiance.

$$s_Y = \left[\left(\frac{\partial Y}{\partial X1} s_{X1} \right)^2 + \left(\frac{\partial Y}{\partial X2} s_{X2} \right)^2 \cdots + \left(\frac{\partial Y}{\partial XN} s_{XN} \right)^2 + \Sigma 2 \rho_{ij} \frac{\partial Y}{\partial Xi} \frac{\partial Y}{\partial Xj} s_{Xi} s_{Xj} \right]^{1/2} \tag{4.72}$$

where ρ_{ij} is the standard correlation coefficient between variables Xi and Xj, and the sum is over all combinations of correlated variables. Note that ρ_{ij} can take on values from -1 to 1, so that the inclusion of the correlation term will reduce the error computed for negatively correlated variables.

This type of error propagation analysis not only lets us assess the overall error, but also by inspecting the individual error terms (s_{Y_i}), it tells us what parameters contribute the most to the error. This helps to prioritize error reduction efforts, since in most cases reducing the error in a small error source will have little or no impact on the total error (i.e., we first want to work on the weak links (large error sources) in the radiometric image chain).

A complete treatment of the error analysis of "the big equation" is beyond the scope of this book. However, throughout the book, we will refer to the relative importance of variables or error terms that are the result of this type of error analysis. To see how this analysis is performed, we will look at one simple case.

Take the case of a laboratory instrument that measures the radiance reflected from a sample irradiated as shown in Figure 4.19. The lamp is a calibration standard of known radiant intensity which produces an irradiance (E) onto the sample of 110 [wm^{-2}]. The expression governing the observed radiance (L) for a Lambertian reflector (r_d) can be expressed as:

$$L[wm^{-2}sr^{-1}] = E\frac{r_d}{\pi} + L_p \tag{4.73}$$

where L_p is any radiance scattered to the sensor through a process other than reflection from the target. We can rearrange Eq. (4.73) to provide the governing equation for using this device to measure reflection, i.e.,

$$r_d = \frac{(L - L_p)\pi}{E} \quad (4.74)$$

From Eq. (4.72) we can express the error in the reflectance (s_r) as:

$$\begin{aligned}(s_r) &= \left[\left(\frac{\partial r}{\partial L} s_L\right)^2 + \left(\frac{\partial r}{\partial L_p} s_{L_p}\right)^2 + \left(\frac{\partial r}{\partial E} s_E\right)^2 + \left(\frac{\partial r}{\partial \pi} s_\pi\right)^2\right]^{1/2} \\ &= \left[\left(\frac{\pi}{E} s_L\right)^2 + \left(-\frac{\pi}{E} s_{L_p}\right)^2 + \left(\frac{(L_p - L)\pi}{E^2} s_E\right)^2\right]^{1/2}\end{aligned} \quad (4.75)$$

where we have assumed s_π to be approximately zero and that all the variables are uncorrelated. For a particular measurement, we observe the radiance (L) to be 3.50 [wm^{-2}sr^{-1}], and we know that the instrument error (s_L) is ±0.17 [wm^{-2}sr^{-1}]. The scattered radiance term $L_p \pm s_{Lp}$ is observed to be 0.18 ± 0.03 [wm^{-2}sr^{-1}]. However, the lamp error is not expressed in terms of irradiance. The standards laboratory specifies the lamp assembly to have a radiant intensity ($I \pm s_I$) of 39.6 ± 1.7 [wsr^{-1}] from which we computed the irradiance onto the sample, 0.6 [m] away, to be:

$$E = \frac{I}{l^2} = \frac{39.6}{(.6)^2} = 110[\text{wm}^{-2}] \quad (4.76)$$

The error in irradiance can then be expressed as:

$$\begin{aligned}s_E &= \left[\left(\frac{\partial E}{\partial I} s_I\right)^2 + \left(\frac{\partial E}{\partial l} s_l\right)^2\right]^{1/2} \\ &= \left[\left(\frac{1}{l^2} s_I\right)^2 + \left(\frac{-2I}{l^3} s_l\right)^2\right]^{1/2}\end{aligned} \quad (4.77)$$

If we estimate the error in our distance measurement s_l to be 0.002 m, we can generate a value for the error in irradiance using Eq. (4.77):

$$\begin{aligned}s_E &= \left[\left(\frac{1}{(0.6)^2} 1.7\right)^2 + \left(\frac{-2 \cdot 39.6}{(0.6)^3} 0.002\right)^2\right]^{1/2} \\ &= [22.3 + .54]^{1/2} = 4.8[\text{wm}^{-2}]\end{aligned} \quad (4.78)$$

where we notice that the distance measurement error contributes very little to the total error.

We can now solve Eq. (4.75) for the error in reflectance. First we solve for the reflectance of the sample using Eq. (4.74), i.e.,

$$r = \frac{(3.50 - 0.18)\pi}{110} = 0.095 \quad (4.79)$$

The error in reflectance is then given by Eq. (4.75) to be

$$s_r = \left[\left(\frac{\pi[sr] \cdot 0.17[\mathrm{wm}^{-2} sr^{-1}]}{110[\mathrm{wm}^{-2}]} \right)^2 + \left(\frac{\pi \cdot 0.03}{110} \right)^2 \right.$$

$$\left. + \left(\frac{(0.18 - 3.5)\pi \cdot 4.8}{110^2} \right)^2 \right]^{1/2}$$

$$= \left[235.7 \cdot 10^{-7} + 7.3 \cdot 10^{-7} + 171.2 \cdot 10^{-7} \right]^{1/2} = 0.006 \tag{4.80}$$

We can then express the observed reflectance to be 0.095 ± 0.006. Furthermore, we see that the largest contributor to the total error is the sensor error closely followed by the error due to the lamp standard. It is important to realize that, for this type of analysis to yield reasonable results, all of the input errors must be of a common form (e.g., one standard deviation or RMS deviation from truth). In cases where a governing equation cannot be written (e.g., when iterative solutions are used), this same type of sensitivity analysis can be performed using computer simulation of the process and Monte Carlo methods (cf. Rubinstein 1981).

In this chapter, we have developed in detail a governing equation that describes the radiance reaching the sensor. We also analyzed how this equation can be simplified in certain spectral regions. Finally, we have briefly described error analysis methods that are used throughout the book to assess the errors and the importance of parameters. In the next chapter, we will analyze how the radiance reaching the sensor is converted by various sensors to recorded signal levels.

4.7 REFERENCES

Beers, Y. (1957). *Introduction to the Theory of Errors.* Addison-Wesley, Reading, MA.

Chandrasekhar, S., (1960). *Radiative Transfer.* Dover, Minneoloa, NY.

Deering, D.W. (1988). "Parabola directional field radiometer for aiding in space sensor data interpretation." Proceedings of the SPIE, pp. 924-933, Orlando, FL.

Feng, X, Schott, J.R., & Gallagher, T.W. (1993). "Comparison of methods for generation of absolute reflectance factor values for BRDF studies." *Applied Optics,* Vol. 32, No. 7.

Kneizys, F. X., Shettle, E.P., Abreu, L.W., Chetwynd, J.H., Anderson, G.P., Gallery, W.O., Selby, J.E.A., & Clough, S.A. (1988). "Users guide to LOWTRAN7," AFGL-TR-88-0177, Environmental Research Papers, No. 1010, Air Force Geophysics Laboratory, Optical/Infared Technology Division, Hanscom AFB, MD.

Piech, K. R., & Walker, J.E. (1971). Aerial color analysis of water quality. *Journal of Survey and Mapping Division, American Society of Civil Engineers,* Vol. 97, No. SU2, pp. 185-197.

Rubinstein, R.Y. (1981). *Simulation and the Monte Carlo Method,* Wiley, NY.

Salvaggio, C., Sirianni, J.D., & Schott, J.R. (1993). "Use of LOWTRAN derived atmospheric parameters in synthetic image generation models." Proceedings of the SPIE, Recent Advances in Sensors, Radiometric Calibration, and Processing of Remotely Sensed Data, Vol. 1938, Orlando, FL.

Turner, R. E., Malila, W.A., Nalepka, R.F,. & Thomson, F.D. (1975). "Influence of the atmosphere on remotely sensed data." Proceedings of the SPIE 51, 101, "Scanners and imaging systems for earth observation."

CHAPTER 5

Sensors

In Chapter 4, we derived an expression for the radiance reaching the sensor. In this chapter, one of our goals is to see how we can use imaging sensors to record and measure that radiance.

The treatment of sensors is one of the most involved and exciting aspects of remote sensing. Traditionally, the sensor has been the most complex and expensive part of the image chain. It has been the jewel designed to offset, or blind one to, any flaws in the other links. Today the sensor is still a critical and expensive component in the image chain. However, the user community is placing increasing emphasis on the processing, storage, and information extraction stages of the image chain (cf. Chaps. 6, 7, and 8). The details of sensor design or a rigorous characterization of even a few sensors are beyond the scope of this treatment. Indeed, the rapid rate of sensor evolution brought on by advances in semiconductor technology makes detailed sensor characterization obsolete almost as fast as it can be published in text form. We will, therefore, concentrate on a broad treatment of the fundamental principles of sensor technology and sensor calibration. This is intended to provide the reader with a basic knowledge of traditional sensor designs and a capability to understand more detailed studies of new systems as they appear in the current literature.

For the sake of space, we will emphasize airborne and satellite electro-optical imaging systems. The reader is referred to Chen (1985) for a more complete treatment of satellite sensing systems, including sounders and microwave systems, and to Chapter 6 of the *Manual of Remote Sensing* (cf. Colwell, 1983) for a more complete treatment of film systems. Kramer (1992) contains a comprehensive listing of the specifications for a wide range of aerospace sensing systems.

The end of this chapter links together many of the concepts in this and earlier chapters through a case study of a system design. The reader unfamiliar with the basic optical terminology referred to in this chapter should consult Hecht (1987).

5.1 CAMERAS AND FILM SYSTEMS

Much remote sensing is still, and will continue to be, conducted using photographic film camera systems. Film offers an inexpensive, high-resolution solution to many remote sensing problems. The images can be acquired with instruments ranging from the conventional 35-mm format cameras, used by many amateurs, to the large format cameras used on the space shuttle (cf. Fig. 5.1). Film also offers great geometric fidelity for mapping and mensuration at very modest costs, making it very attractive for most photogrammetric applications. Because of our interest in radiometric and computer-based image analysis, this section will address how film systems can be used as quantitative radiometric sensors and as a source of digital image data.

5.1.1 Irradiance onto the Focal Plane

As we will see in the next section, film systems are characterized in terms of the exposure (H) on the film, i.e.,

$$H[\text{joules m}^{-2}] = \int E(t)dt \cong Et \tag{5.1}$$

where t [sec] is the exposure time and E [wm^{-2}] is the irradiance onto the film. In the last chapter, we dealt extensively with the radiance reaching the front of a sensing system. In order to determine how that radiance will impact a sensor (the film in this case), we need to develop a relationship between radiance reaching the front of the sensor and irradiance at the focal plane. This is commonly referred to as the *camera equation* and is written as:

$$E = \frac{L}{G\#} \tag{5.2}$$

where the $G\#$ [sr-1] defines the throughput of the system in terms of how well it converts radiance to irradiance on the focal plane. For a simple camera, the $G\#$ is usually defined in terms of the value along the optical axis. From Chapter 3, we recall we can express the element of irradiance due to the radiance from each element of solid angle through an aperture as (cf. Fig. 5.2):

$$dE = \tau L_i d\Omega_i \cos\theta_i \tag{5.3}$$

where τ is the transmittance of the lens assembly, L_i is the radiance at location i in the aperture, $d\Omega_i$ is the solid angle subtended by an area element dA_i in the aperture, and θ_i is the angle from the optical axis to the area element dA_i. If the aperture is small compared to the focal length, then for a field of uniform radiance the aperture can be approximated as a point source with $\theta_i = 0$ and $r_i = r$. This yields:

$$E = \int_A dE_i = \int L\tau \frac{dA_i}{r_i^2} = \frac{LA\tau}{r^2} \tag{5.4}$$

For a system focused at infinity (which is the case for most remote sensing systems) r is the focal length (f) of the optical system. The irradiance can then be expressed as:

$$E = \frac{L\pi d^2 \tau}{4f^2} = \frac{L\pi\tau}{4(f\#)^2} = \frac{L}{G\#} \tag{5.5}$$

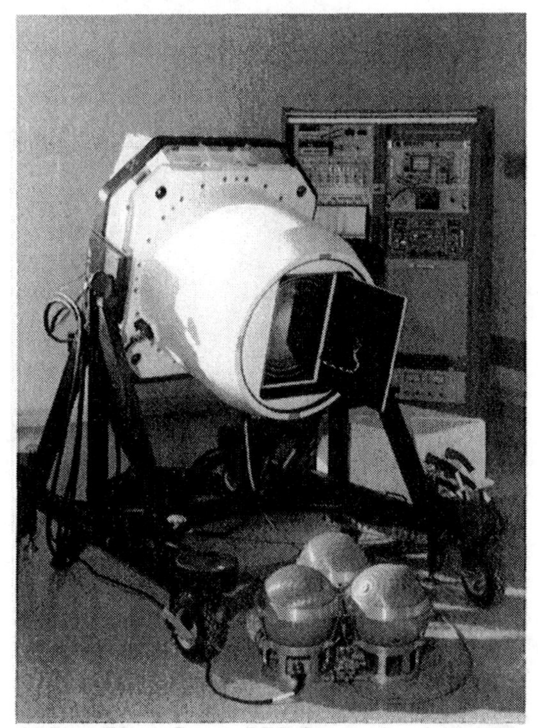

(a) Large Format Camera (LFC) flown on board the space shuttle.

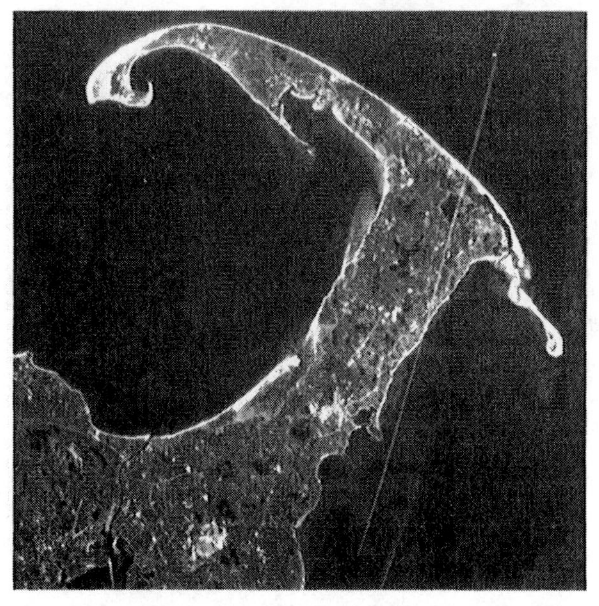

(b) Portion of an image from the LFC showing Cape Cod. Note the condensation trails and their shadows.

Figure 5.1 Examples of a specialized camera used for remote sensing. (Images courtesy of Itek Corporation.)

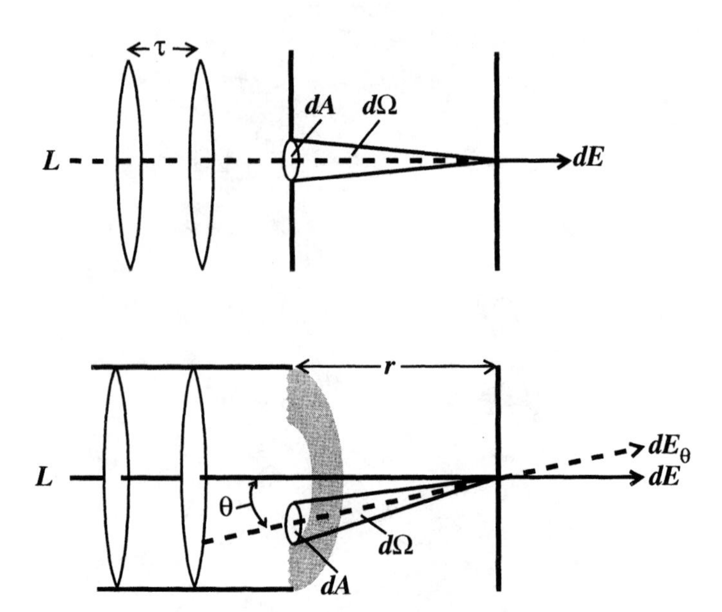

Figure 5.2 Derivation of the camera equation.

where d is the diameter of the aperture (i.e., $\pi d^2/4 = A$) and the F number ($f\#$) of the system is defined as:

$$f\# = \frac{f}{d} \tag{5.6}$$

Rearranging Eq. (5.5) we have an expression for the $G\#$ of

$$G\# = \frac{L}{E} = \frac{4(f\#)^2}{\tau\pi} \tag{5.7}$$

which is valid for large $f\#$'s. Grum and Becherer (1979) indicate that Eq. (5.7) is valid to better than 1% for $f\#$'s greater than 10, and they solve the integral of Eq. (5.3) rigorously for a simple lens to yield

$$E = \frac{\tau\pi L \frac{d^2}{4}}{\left(\frac{d^2}{4} + f^2\right)} \tag{5.8}$$

which is valid for all $f\#$'s, i.e.,

$$G\# = \frac{1 + 4(f\#)^2}{\tau\pi} \tag{5.9}$$

From our analysis of radial lens falloff in Chapter 3, i.e., Eq. (3.78), we recognize that the off-axis irradiance or exposure will be further reduced by approximately $\cos^3\theta$. In the next section, we will analyze how this exposure is recorded by the film.

5.1.2 Sensitometric Analysis

In order to describe photographic systems quantitatively, we need to introduce some terminology for numerically describing the photographic image. In general, we will be dealing with photographic transparencies (the equivalent of black and white negatives or color slides in amateur photography), and the common way to quantify them is in terms of the film density or opacity. Density is defined as

$$D = -\log_{10}\tau \tag{5.10}$$

where τ is the fraction of incident radiation transmitted through the film (transmission). Density can be defined as a function of wavelength, but it is more commonly defined over a spectral bandpass corresponding to how the data are to be presented to the viewer or instrument. For example, for color film we would have red, green, and blue density values corresponding to film transmission measurements made through filters that only transmit in the red, green, and blue spectral regions, respectively.

The density (D) of developed film is related to the exposure (H) of the film through what is called the D log$H$ curve after Hurter and Driffield (1890) who developed the method of describing film response shown in Figure 5.3. The curve shown is for one spectral emulsion layer of a positive working color transparency. The slope [gamma (γ)] of the straight-line portion is negative for film positives because the greater the exposure, the thinner (less dense) the film and, therefore, the brighter it will appear when viewed or projected in transmission mode. Note that the curve is plotted in terms of log exposure and that, at low exposures (high densities on a positive), the film becomes rapidly less responsive until changes in exposure have no impact on the film. The same effect occurs at the toe of the curve for high exposures. The approximately linear region between the shoulder and the toe of the curve is where one normally wants the image information to fall. The projection of this linear region on the log exposure axis defines the film's exposure latitude. Objects with exposure values much beyond this region will be over- or underexposed. The gamma of the film defines the contrast. Since the D max of a film is largely limited by the film type, there is an inverse relationship between contrast and exposure latitude. The D log$H$ curve is also referred to as the *characteristic curve* since it is indicative of the response characteristics of a particular film type. It is not, however, fixed for each film type. It will vary considerably between batches of film, with the physical and chemical characteristics of the film development and with the exposure time. Most films exhibit some time hysterisis (reciprocity law failure) effects such that identical exposures with significantly different exposure times will produce different density values.

The spectral sensitivity of film is controlled by the chemical composition of the film. A thorough treatment of film chemistry and analysis can be found in James (1977). The effective sensitivity of a film system is often controlled by cascading the film's intrinsic spectral sensitivity with a filter of a chosen spectral transmission. This is illustrated in Figure 5.4 where visible and IR-sensitive film is filtered with a visible-blocking (longpass) filter to achieve a film filter combination sensitive just to near infrared (NIR) wavelengths. The sensitivity of film is commonly defined as the reciprocal of the exposure needed to cause a chosen film density for the processing conditions specified. The units of sensitivity are cm^2/erg.

For color film, the film is generally composed of three absorbing layers sensitive to three different spectral regions. The density of the developed film in the red, green, and blue spectral regions is controlled by the amount of exposure in each of the three absorbing layers. This is illustrated in Figure 5.5 for the color infrared film commonly flown for vegetation studies. When flown with a minus blue (yellow) filter, the absorption layers roughly isolate green, red, and near-infrared energy. The spectral transmission of the corresponding dye layers in the developed film are also shown in Figure 5.5. From this we see that the density

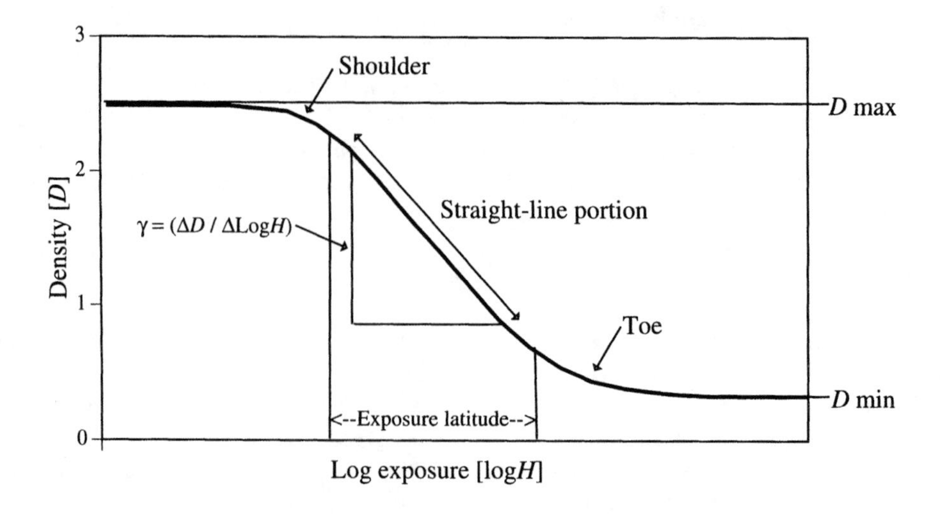

Figure 5.3 Characteristics of an H and D curve.

of the red layer (i.e., through a red filter) of the developed film depends on the NIR exposure; the density through a green filter carries red exposure information; and the blue density carries green information.

From a practical standpoint, the quantitative user of film cannot count on the chemical system remaining stable for all the variables that affect the *D* log*H* curve. Instead, the characteristic curve must be produced for each roll of film developed (ideally for each frame on long rolls). This is accomplished using a device called a *sensitometer*, which places a step wedge (or step tablet) on a portion of the film, before (header) or after (trailer) the images. The step tablet contains a sequence of known exposure steps. By plotting the density versus the log of the exposure used to create the step, the *D* log*H* curve can be produced. For color film, densities are measured with red, green, and blue filters to characterize the cyan, magenta, and yellow dye layers. In making the step wedges, care must be taken to ensure that the spectral shape of the lamp used to expose the film approximately matches the spectral shape of the flux incident on the sensor (usually the source is filtered to about 5500 to 6000 K) and that any filters to be used with the film are also used in the sensitometer. Finally, to avoid reciprocity law failure (time hysterisis effects), the exposure time used in putting the step wedge on the film should approximately match the exposure time used with the film.

After processing, the film density (for each layer if color film) is measured for each step with the same type of densitometer to be used on the image data. The plot of density versus log exposure can then be produced as shown in Figure 5.3. Film transmission and, therefore, density is dependent on the scattering properties of the film, as well as the illumination and collection optics of the densitometer (transmissometer) used (cf. Fig. 5.6). While procedures for correlating densities between instruments can be developed, it is advisable where possible to analyze the step wedge and the image data with the same instrument.

The density versus log exposure data can be fit to a polynomial of the form

$$D = b(0) + b(1)logH + b(2)(logH)^2 + b(3)(logH)^3 \tag{5.11}$$

where the constants $b(0)$, $b(1)$, etc., are determined by a least-squares regression. Alternately the density and log exposure values can be tabulated and a piecewise linear interpolation

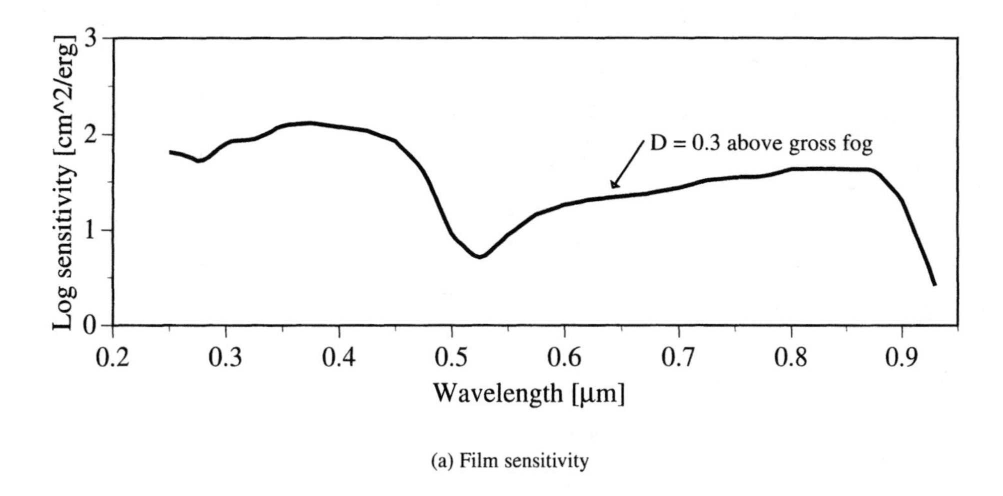

(a) Film sensitivity

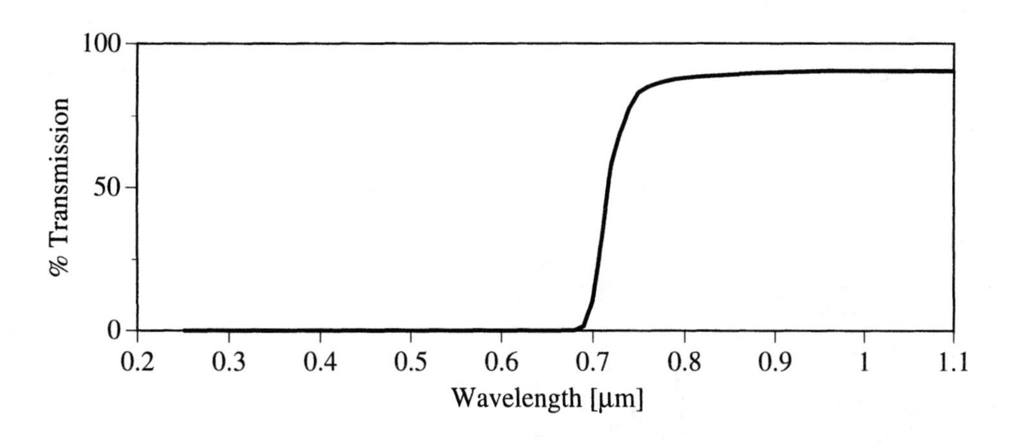

(b) Kodak Wratten #89B filter

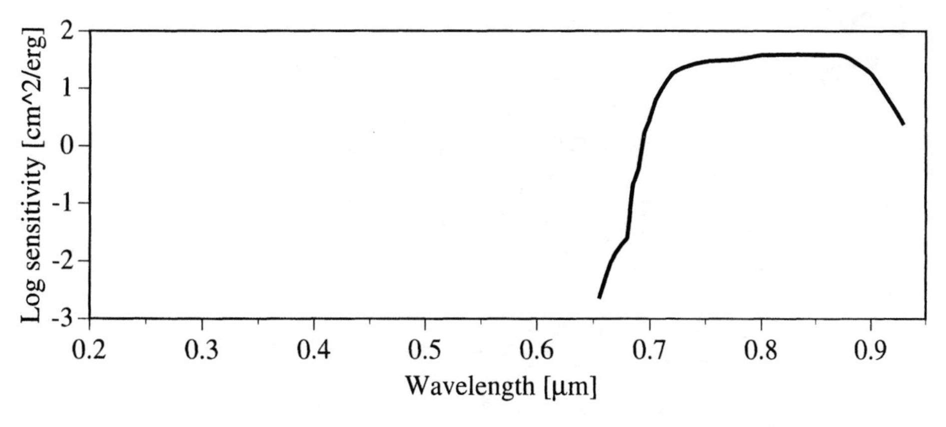

(c) Effective sensitivity of film/filter combination

Figure 5.4 Spectral sensitivity of Kodak type 2424 film and its effective sensitivity when flown with a Kodak Wratten 89B filter. (Kodak publication No. M-58.)

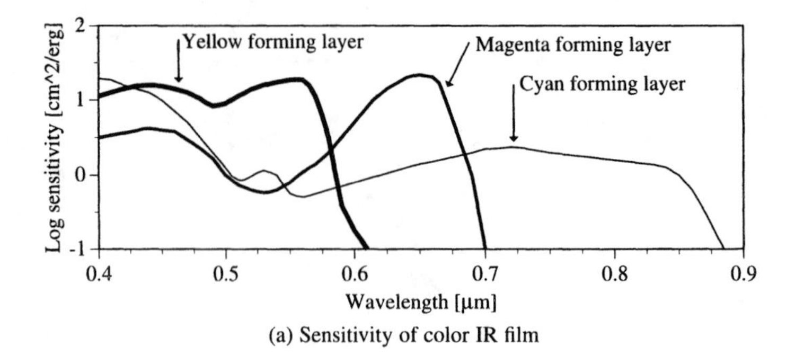

(a) Sensitivity of color IR film

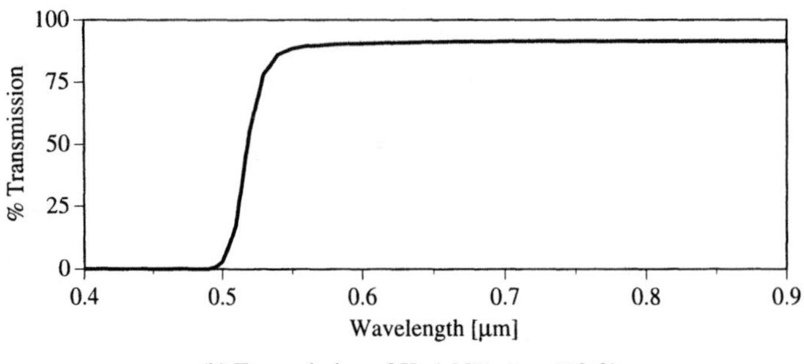

(b) Transmission of KodakWratten #12 filter

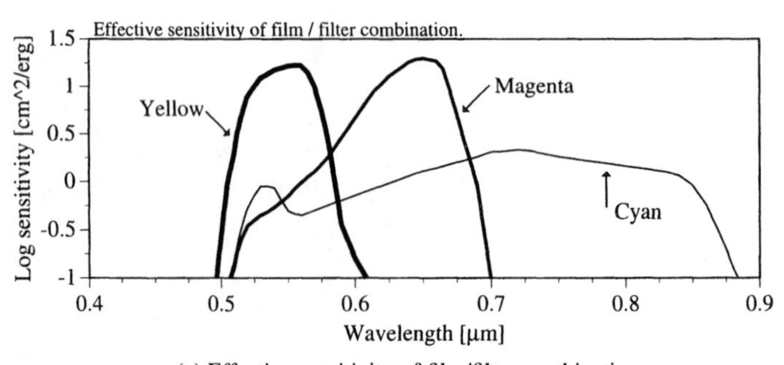

(c) Effective sensitivity of film/filter combination

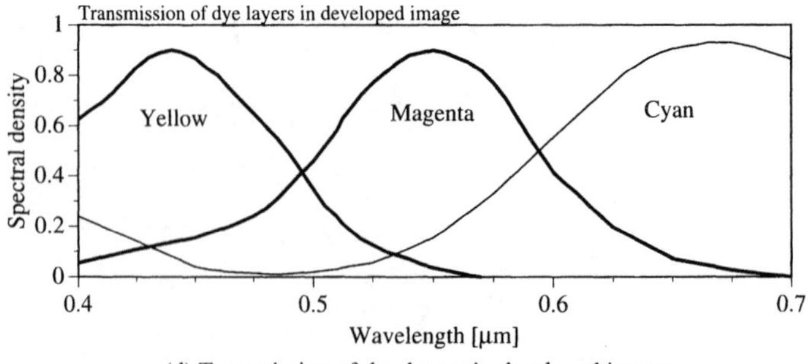

(d) Transmission of dye layers in developed image

Figure 5.5 Spectral characteristics of color infrared film. (Kodak infrared 2443 when flown with a Kodak Wratten #12 filter, Kodak publication No. M-58.)

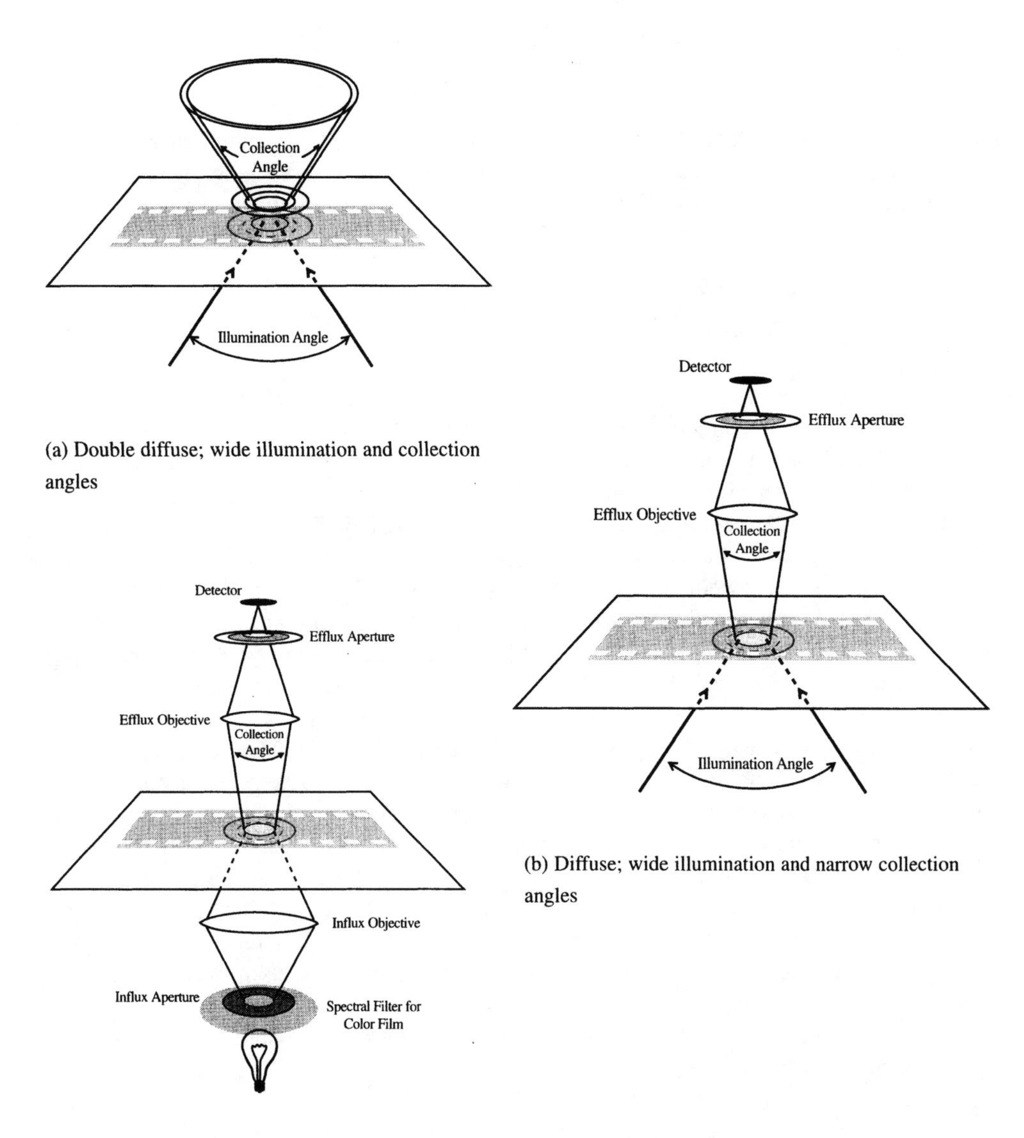

(a) Double diffuse; wide illumination and collection angles

(b) Diffuse; wide illumination and narrow collection angles

(c) Specular; narrow illumination and collection angles

Figure 5.6 Types of density readings and instrumentation.

used to convert between density and log exposure.

The radiance corresponding to any density value can be found by combining Eqs. (5.7), (5.1), and (3.78) to yield

$$L = \frac{HG\#}{t\cos^n \theta} \tag{5.12}$$

where the value of H must be measured from the D log$H$ curve for each density value. The value of n can be calculated by calibration of the performance of the lens, as discussed in Section 3.3. The view angle θ can be expressed as:

$$\theta = \tan^{-1}\left(\frac{r}{f}\right) \tag{5.13}$$

where r is the radial distance from the center of the film format to the point where the density measurement is made. The value of L is the effective radiance reaching the sensor corresponding to the value derived in Chapter 4. The normalized sensitivity is used in place of the normalized responsivity to compute the effective radiance for film systems according to

$$L = L_{\text{eff}} = \int L_\lambda \left(\frac{R_s(\lambda)}{R_{s\,\max}}\right) d\lambda \tag{5.14}$$

where $R_s(\lambda)$ is the sensitivity, $R_{s\max}$ is the maximum sensitivity, and the subscript eff is generally dropped from the effective radiance term for convenience.

Using the procedures described above, we can use a camera as a two-dimensional radiance meter. When many points on an image are to be radiometrically analyzed, or when we need to process the entire image, it is often useful to digitize the image. This involves dividing the image into a two-dimensional array of picture elements (pixels) and assigning to each pixel a digital value corresponding to the density or transmission through the film at that pixel location. Traditionally the digitizing process has been done with scanning microdensitometers. The film is scanned in a raster fashion with density values sampled at equal intervals in x and y and converted to scaled digital counts. By scanning the step wedge at the same time, a relationship between digital count and exposure can be developed for the image data and used in place of the density log exposure curve to characterize and analyze the film. With the recent advances in digital image scanning technology associated with the electronic-publishing industry, there are many low-cost image scanning systems available. Many of these can be used for quantitative remote sensing if proper care is taken to ensure the integrity of the data. (The user should test for dynamic range, spatial uniformity, geometric fidelity, and spectral separability of the film layers if color film is used.) Image digitizing services are also readily available. The Kodak photo CD is an example. With only slight adjustment to the film scanner setup controls, the photo CD instrumentation can be used even for digitizing color infrared film (cf. Fig. 5.7).

In this subsection, we have seen how we can use cameras and photographic film to produce imaging radiometers. For relatively broad spectral windows between 0.4 and 0.9 μm, film cameras can often provide a simple low-cost sensing solution. When narrow-band spectral sampling is required, film may not be fast enough. In addition, when spectral regions beyond about 0.9 μm are of interest or for unmanned satellite sensors, other approaches must be considered. This brings us to electro-optical systems, which will be the topic of the remainder of this chapter.

5.2 SIMILARITIES BETWEEN SIMPLE CAMERAS AND MORE EXOTIC ELECTRO-OPTICAL (EO) IMAGING SYSTEMS

Remote sensing film camera systems seem fairly simple to us because of their similarity to the amateur cameras we all use. In many ways, even the most exotic space-based film camera systems are just very large versions of conventional cameras. When we begin to look at electro-optical cameras (or *imagers* as they are more commonly called), many of the similarities to conventional cameras are still there, but they are less obvious. In this subsection, we will briefly describe common optical designs used in EO cameras and provide a basic breakdown of the components of EO imaging systems.

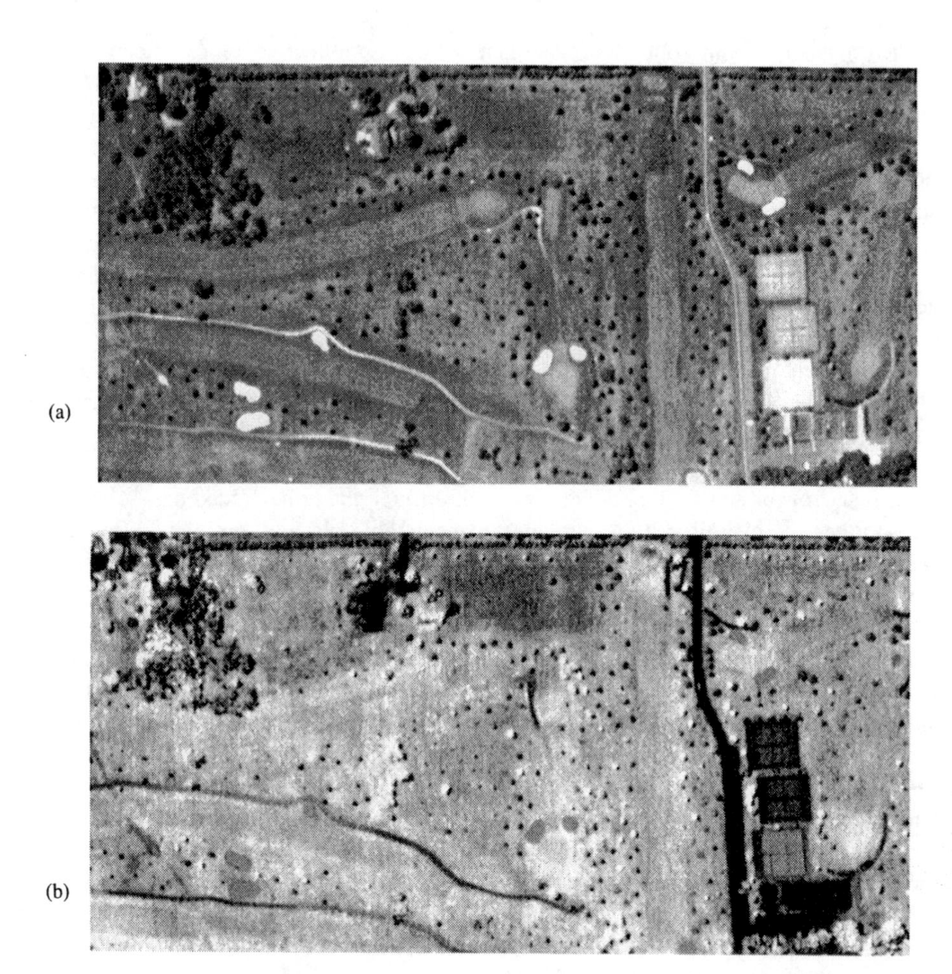

Figure 5.7 (a) Color infrared image and (b) NDVI image produced from a photo CD image scanned from 35-mm color infrared film. The NDVI image brightness increases with biomass and vegetation rigor. This image was acquired to asses turf conditions at a golf course. Note the low NDVI indicators at some of the close-cropped greens. See color plate 5.7. (Courtesy Pegasus Environmental.)

5.2.1 Optics and Irradiance at the Focal Plane

Conventional cameras, and some EO cameras, use refractive optics (e.g., glass lenses) to focus an image of the scene onto the focal plane. However, many EO systems used for remote sensing are designed to operate over a broad spectral range, which generally makes the use of refractive optics impossible or at least very expensive. Conventional optical glass, for example, does not transmit in the LWIR and the most convenient lens materials for use in the LWIR, (e.g., germanium) do not transmit in the visible wavelengths. To overcome this limitation, most EO systems use reflective optics (mirrors). This is particularly true of space-based systems because the reflective optics are much lighter than comparable refractive optics. Even for sensing in the VIS-NIR from space, reflective optics are generally used for the larger optical elements. While even an introductory treatment of optical systems is beyond the scope of this study, there are a few optical designs commonly used in EO systems which we will briefly describe. A more complete treatment of EO optical design can be found in Accetta and Shumaker (1993).

The function of reflective optics is the same as for conventional refractive optics (i.e., to focus the image of the earth onto the focal plane). Because reflective optics reflect all wavelengths the same amount, they have another advantage over refractive optics in bringing all spectral bands to focus in the same plane. A disadvantage of reflective optics is that it is difficult to use the full aperture of the system because the optical elements are opaque. Figure 5.8 shows several designs for the long focal length optical systems commonly used in remote sensing. The off-axis parabola uses only a portion of a parabolic shape to bring the image to focus alongside the primary beam where detectors or film can be introduced. This design is very long (i.e., the imaging system must be as long as the focal length of the lens), making it particularly unattractive for many space applications. It is, however, commonly used in laboratory collimators for calibration and testing of other optical systems. Test targets are placed at the focal plane and projected onto the sensor's optics simulating targets effectively located at infinity. The Newtonian design uses a simple fold to make the image accessible. This has the disadvantage of obscuring some of the incident flux, and still makes for a long optical system. Another way to make the image accessible is by placing a hole in the center of the primary mirror and folding the energy back through the hole (folded parabolic). This design shortens the overall system but causes considerable obscuration. This limitation is reduced with the Cassegrainian design that uses a parabolic primary and a hyperbolic secondary to fold the energy back through a hole in the primary. This approach has less obscuration than the simple folded configuration and a much shorter overall length to achieve the same focal length. As a result, the Cassegrain and its variants are among the most commonly used designs, particularly for satellite systems. The conventional Cassegrain design has considerable aberration. This can be largely eliminated by making both the primary and the secondary aspheric. This more complex (i.e., expensive) design is called a *Ritchey-Chretian*. A somewhat simpler variation of the Cassegrain called the *Dall-Kirkham* uses an aspheric primary and a spherical secondary. This design still has some aberration, but it is often acceptable in systems with a narrow field of view through the telescope.

Whatever the optical configuration, we will be interested in computing the irradiance onto the focal plane. For systems using Cassegrainian-type optics with a centrally obscured aperture, the *G#* of Eq. (5.9) must be corrected for the obscuration. The equation relating irradiance on the focal plane to the incident radiance can be written as:

$$E = \frac{L\tau_l \tau_o \pi d^2}{(d^2 + 4f^2)} \tag{5.15}$$

where τ_l is the transmission loss due to less-than-perfect reflection or transmission by all of the optical elements, τ_o is the transmission loss due to obscuration by the secondary, d is the diameter of the entrance aperture (usually the diameter of the primary), and f is the overall system focal length. If we assume that the spider web holding the secondary in place has negligible size (i.e., it doesn't add to the obscuration), then we can approximate the obscuration loss as:

$$\tau_o = 1 - \frac{\frac{\pi d_s^2}{4}}{\frac{\pi d^2}{4}} = 1 - \frac{d_s^2}{d^2} \tag{5.16}$$

where d_s is the diameter of the secondary. The *G#* can then be redefined to be approximately

$$G\# = \frac{1 + 4(f\#)^2}{\tau_l \tau_o \pi} \tag{5.17}$$

and used in the same way as we did for film-camera systems.

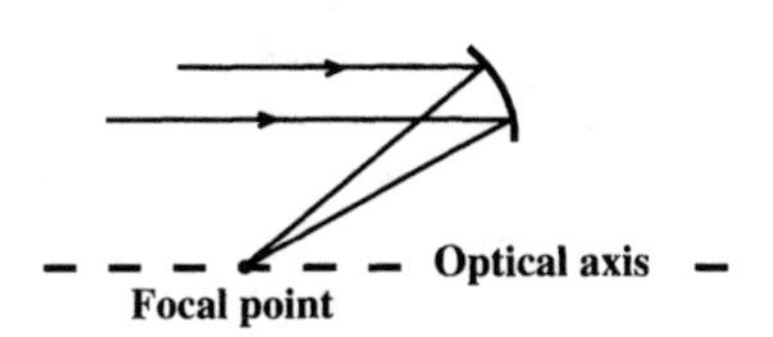

Herschel Mount (off-axis parabolic)

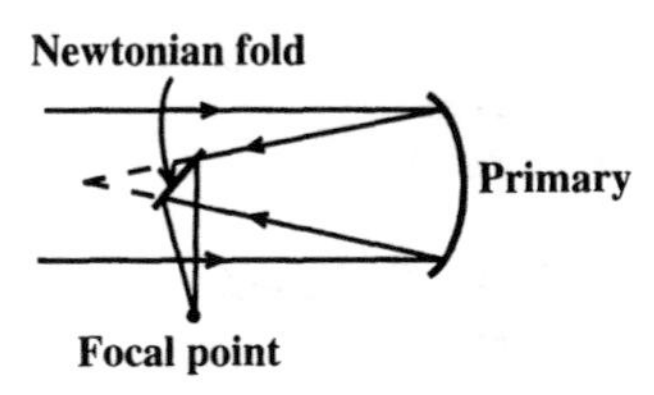

Newtonian

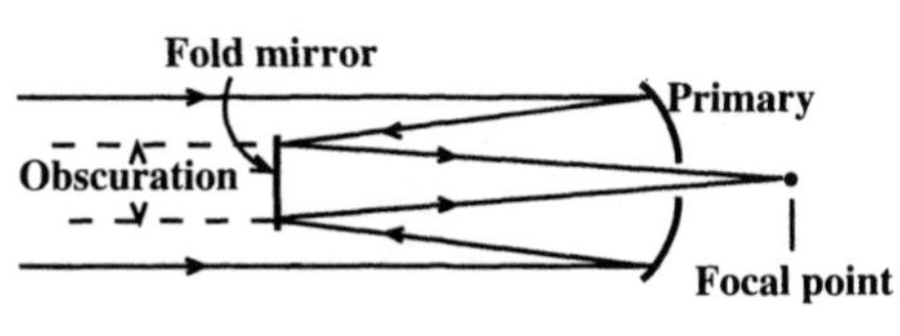

Folded parabolic

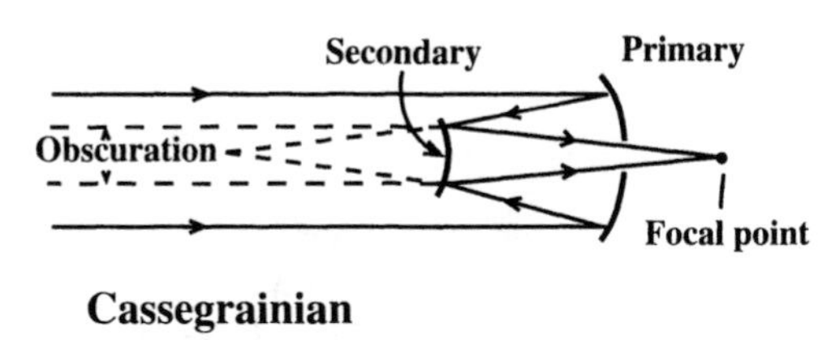

Cassegrainian

Figure 5.8 Optical designs commonly used for remote sensing. (Adapted from Wolfe and Zissis 1985.)

5.2.2 System Characterization

In many cases we will find it useful to describe EO imagers in terms of the components that make up the entire system. These components are outlined in Figure 5.9. The opto-mechanical elements are the optical elements that focus the image onto the focal plane and the mechanical elements that control what the focal plane "sees." This is the equivalent of lens, shutter, and film advance in a conventional camera system. The detectors in EO systems convert the flux incident on them into electronic signals in a fashion similar to the way film converts the flux into the latent image. The preamplifier is a critical stage in an EO system which applies a fixed magnitude, low noise gain to the very low signals from the detector. This raises the signals up to where they can be more readily processed without concern about adding significant noise. The next stage of processing, called *signal conditioning*, is done in order for the output signal to span the range needed for ease of recording, analog to digital conversion, or transmission to a ground station. Several alternative sequences are shown in Figure 5.9 for the route the conditioned signals may follow. Any one of these, or even more than one sequence, may be followed, depending on the type of system. Nearly all signals today are converted to digital form at some point in the processing. For example, a signal from 0 to 1 volt will be converted in equal steps to span from 0 to 255 (2^8) digital counts by an 8-bit system, or signals from -5 to +5 volts may be converted to digital values from 0 to 1023 (2^{10}) by a 10-bit system (cf. Chap. 7 for a more complete treatment of digitizing issues). Data recording may be analog or digital depending on the system. However, improvements in the speed, capacity, and cost of digital systems are moving most new systems to digital recording. Similarly, most data are transmitted in digital form. The reason for this is that with error correction routines, the likelihood of corruption of digital data during recording, playback, or data transmission is very low.

The digital signals are then preprocessed to reconstruct the image. This may involve correcting for sensor geometric distortion, registration of different spectral bands, noise suppression, digital signal conditioning, and radiometric corrections. This entire process from the preamplifier to the end of the digital preprocessing chain is the equivalent of the chemical processing of the latent film image to a developed image. The final digital EO image may

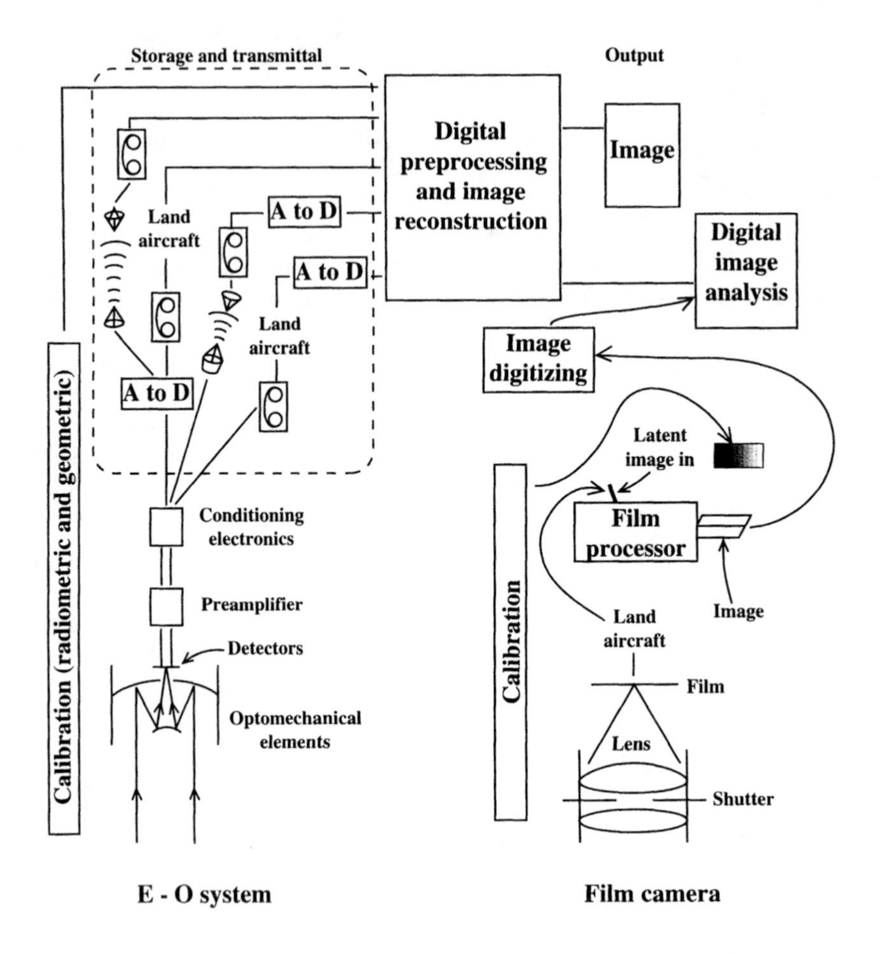

Figure 5.9 Components of an EO imaging system compared to a film camera.

be displayed on a CRT (soft copy) or written out to a film image (hard copy). The digital EO image will most likely be sent to a digital image processing system for further image analysis and information extraction. For convenience, we are treating the digital image processing system as separate from the sensor system (cf. Chap. 7) although at times the data analysis systems are packaged as part of the sensor system.

Operating in parallel with, and sometimes through, the opto-mechanical and signal processing electronics, are the calibration elements of the sensing system. These are typically sources of known radiance for radiometric calibration and orientation and timing data needed for geometric reconstruction of the data and projection into ground coordinates. For the film-camera systems, the radiometric calibration elements would be sensitometric control wedges. Global positioning and gyroscopic data would provide the X, Y, Z location of the sensor in global coordinates and the camera line-of-site angles (θ,ϕ_d, twist) for projection of the camera image to ground coordinates. In the discussions that follow, we will generally treat only the fundamental elements of the opto-mechanical systems, the detector characteristics, and relevant calibration data. For brevity's sake, the electronics and digital system characteristics are generally lumped together with only a few critical parameters addressed. A much more complete treatment of detector electronics, amplification, and signal-to-noise issues is contained in Rogatto (1993).

5.3 DETECTORS AND SENSOR PERFORMANCE SPECIFICATIONS

This section will introduce the basic terminology of detectors with an emphasis on providing some simple concepts for making rough calculations of the performance characteristics of detectors and detector noise limited systems. A more rigorous treatment, including discussion of noise types, amplifier noise, and details of various detector materials can be found in Budde (1983). Regrettably, a proper understanding of this topic requires nearly a book of its own, so we must restrict our discussion to operational terminology with the underlying semiconductor physics and noise theory left to the reader's own inquiry.

5.3.1 Detector Types

There are three fundamentally different types of EO detectors we wish to consider. They can be divided into the following categories:

A. *Thermal detectors*: These detectors absorb incident flux and undergo a temperature change. They have a high rate of change in electrical resistance with changes in temperature. As a result, when externally biased they can be made to exhibit a change in voltage across a reference resistor corresponding to changes in incident flux. (Examples—bolometer: blackened metal flake with a high rate of change of resistance with temperature; thermistor: blackened thermally sensitive semiconductor with properties similar to the bolometer.)

The next two categories are both referred to as *photon detectors* since discrete interactions of photons with electrons cause the observed electrical signals.

B. *External photo-effect detectors*: The photosensitive material used in these detectors have sufficiently low work functions that incident photons with adequate energy can free electrons from the surface of the material and produce a current in an external circuit. (Example—photomultiplier tube (PMT): electrons are emitted by the photoelectric effect at the cathode of a vacuum tube and are accelerated by an external voltage onto a metal surface. The collision of the electrons with the metal produces more free electrons, which are accelerated to a second metal surface where the process is repeated. This is continued through several stages until the multiplied electrons reach the anode and produce a current in an external circuit.)

C. *Internal photo-effect detectors*: These detectors are semiconductors in which the electrons undergo internal energy level transitions when they absorb a photon. Two types of interaction are of interest.

1. *Photoconductive detectors*: In these detectors the photon is absorbed by an electron in the valence band and excited to the conduction band of the semiconductor, where it can be observed as a change in the resistance. This can be monitored by changes in the current induced through the detector using an externally supplied voltage. (Examples—cadmium sulfide, CdS, and indium antimonide, InSb).

2. *Photovoltaic detectors*: These detectors take advantage of the internal potential difference that can develop at the junction between dissimilar materials in a semiconductor. Photons incident on this junction produce charge carriers that migrate under the internal bias and produce a voltage difference in an

external circuit. (Examples—Silicon, Si, photo diode, mercury cadmium telluride, HgCdTe).

In some cases, some of the performance characteristics of a photovoltaic detector can be enhanced by operating it in a photoconductive mode. In this case, an externally supplied reverse bias voltage is applied to the junction. Incident photons still release charge carriers; however, they travel under the influence of the induced potential in the opposite direction, producing a current that can be observed in the external circuit.

Thermal detectors have the attractive feature of being nearly uniformly sensitive over all wavelengths, with the spectral sensitivity largely governed by the method used to blacken the surface. On the other hand, because they must undergo an actual temperature change to produce a measurable signal, they tend to be somewhat slow (response times of 10^{-3} sec are typical) and, as a result, they are not used extensively in imaging devices. However, their simplicity and stability make them attractive for reference instruments, and they are widely used in radiometers for laboratory and ground truth studies. Figure 5.10 shows an example using a thermal detector in a radiometer with a thermally controlled reference cavity and a reflective chopper. The detector first "sees" the world and then the chopper. The chopper is a nearly perfect diffuse reflector, so that the detector is exposed to a reference flux level due only to the temperature of the reference cavity. The output signal is proportional to the ratio of the unchopped to the chopped signal.

Photon detectors, on the other hand, are sensitive to the amount of energy associated with each incident photon. In particular, if the amount of energy per photon ($h\nu$) does not exceed the work function for a PMT or the bandgap energy for the internal photo effect electrons, then no signal will be produced. Since the energy needed to free an electron from the conduction band (the work function) is quite large, the external photoelectric effect (PMT's) can only be used for high-energy (short-wavelength) flux. The use of PMT's is, therefore, restricted to the VIS-NIR region where their multiplicative effect makes them attractive for sensing low signal levels. However, the large size and high voltage requirements of the PMT's restrict their use to designs where only a limited number of discrete detectors are required.

As a result, the internal photo effect detectors have become the most popular class of detectors in modern imaging systems. By changing material types, the entire spectrum from 0.2 to 20 μm can be sampled using these detectors. Advances in semiconductor-based technology are providing a host of small, highly sensitive devices capable of meeting most design requirements. Indeed, as we will see in the remainder of this section, the advances in detector technology, coupled with the ability to mate the detectors to semiconductor electronic circuits, have opened the door to a revolution of new sensor designs (cf. in particular Sec. 5.5 on imaging spectrometers). To talk more specifically about the various semiconductor detectors, we need to introduce some terms for characterizing detectors.

5.3.2 Detector Figures of Merit

A number of terms have been developed over the years to describe the performance characteristics of detectors. These figures of merit are used to perform tradeoff studies between detector types, materials, and manufacturers and to evaluate what the expected performance of a system will be when a particular detector is employed. Table 5.1 lists several of the detector parameters and figures of merit we will use to characterize detectors. This is only a small sampling of the many terms EO engineers and detector manufacturers have developed to characterize specific performance features of detectors (and to confuse the innocent). The reader is referred to Rogatto (1993) for a more complete treatment.

Referring to Table 5.1, we have already introduced the wavelength-dependent respon-

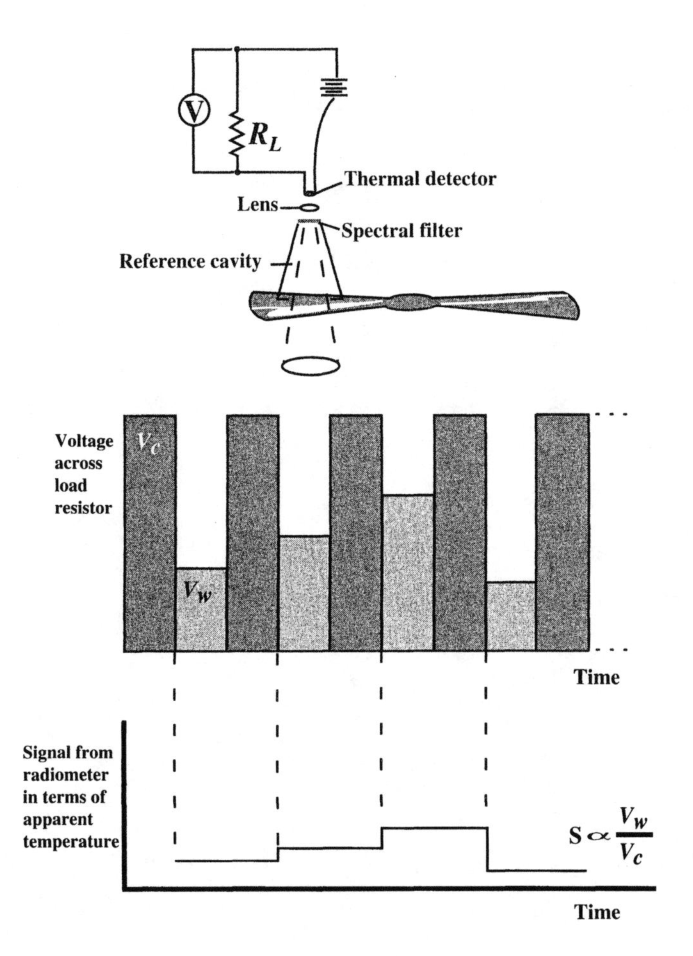

Figure 5.10 Radiometer schematic using a thermal detector.

sivity, $R(\lambda)$ to be the signal, S produced by the detector in volts or amps per unit incident flux, Φ at wavelength, λ. Similarly, recall that the total effective responsivity is the integral of the wavelength-dependent responsivity weighted by the incident spectral flux according to:

$$R = \int R(\lambda) \frac{\Phi_\lambda}{\int_0^\infty \Phi_\lambda d\lambda} d\lambda \tag{5.18}$$

making the responsivity a function of the spectral shape of the incident flux.

The signal out of the detector can then be expressed as:

$$S[\mathrm{v}] = \int R(\lambda)\Phi_\lambda d\lambda = R\Phi[\mathrm{vw}^{-1}][\mathrm{w}] \tag{5.19}$$

as long as the total flux (Φ) has the same spectral distribution as was used in computing the total effective responsivity (R), cf. Eq. (5.18). This process of weighting by the source spectral shape can be used to convert any of the spectral terms in Table 5.1 to their total effective values.

Due to thermal variation in the detector (Johnson noise), random occurrence of photon events (shot noise), variation in the thermal exchange with the surround (temperature noise),

Table 5.1 Detector Terminology for Figures of Merit

Term	Symbol	Definition	Units	Comments
Signal	S	output of Detector	[amps or volts]	
Spectral responsivity	$R(\lambda)$	$R(\lambda)=\frac{dS}{d\Phi}$	$\frac{\text{[amps or volts]}}{\text{watt}}$	
Noise	N	$\left[\frac{\sum_{i=1}^{n}(S_i - S_{avg})^2}{n}\right]^{1/2}$	[amps or volts]	RMS deviation in signal at fixed input (often zero flux)
Signal-to-Noise		S/N	[ ]	
Spectral noise equivalent power	NEP (λ)	$\frac{N}{R(\lambda)}$	[w]	Incident flux equivalent needed to yield a signal-to-noise of 1 (dark)
Spectral detectivity (λ)	$D(\lambda)$	$\frac{1}{\text{NEP}(\lambda)}$	[w^{-1}]	
D "star" or specific detectivity	$D^*(\lambda)$	$(Af)^{1/2}D(\lambda)$	[w^{-1}cm $Hz^{1/2}$]	A is detector area, f is the electronic bandwidth of the sensor
Noise equivalent temperature increment	NEΔT		K	The change in temperature of a blackbody needed to produce a S/N of unity (i.e., to produce a power level equal to the NEP)

and the random variation in signal with input frequency, there will be random variations in the signal level even when the detector is exposed to a constant flux level. These variations about the mean signal level are referred to as *noise* (N) and are usually characterized by the root mean square (RMS) variation in the instantaneous signal level (S_i) according to:

$$N_{RMS} = \left(\frac{\sum_{i=1}^{n}(S_i - S_{avg})^2}{n}\right)^{1/2} \quad [v] \tag{5.20}$$

when S_{avg} is the mean signal level and n is the number of samples. In many cases, the noise level will be a function of the signal level, so it is necessary to define *noise* relative to some flux level. A simple way to do this is to define the noise when no flux is incident on the detector (dark noise). However, for many systems, even the minimum flux levels are well above the dark level, so this noise is not indicative of what will be experienced operationally. For this reason noise levels are often specified relative to some incident flux level. For example, the noise might be defined as the RMS variation about the signal that would be generated by a flux onto the detector corresponding to the sensor viewing a 5% reflector at the top of the atmosphere for some set of solar conditions. In the thermal region, it could be the noise about the signal associated with viewing a 300K blackbody.

While noise is a measure of the quality of a signal and less noise is better, noise really only takes on meaning when viewed relative to the corresponding signal as expressed by the signal-to-noise ratio (S/N). For example, a system with a signal-to-noise ratio of 20 for signals corresponding to a 10% reflector would have approximately 10 discernible levels in signal ($2N$) between 0 and the level of a 10% reflector (i.e., steps in reflectance of less than one reflectance unit would be difficult to separate from the noise).

In many cases, it is useful to express the concept of noise in radiometric input units [w] rather than in output signal units [v]. This can be accomplished using the wavelength-dependent noise equivalent power, NEP(λ), which is the amount of incremental flux at wavelength λ required to change the signal level by an amount equal to the noise, i.e.,

$$\mathrm{NEP}(\lambda) = \frac{N}{R(\lambda)}[\mathrm{w}] \tag{5.21}$$

This term tells us that flux levels or variations in flux must be above the NEP level to have any chance of being detected. The detectivity term, $D(\lambda)$, is simply 1/NEP(λ) and has the dubious value of increasing with the quality of the detector. Of greater interest is the wavelength-dependent specific detectivity, $D^*(\lambda)$. This term makes performance comparison between detectors more valid by adjusting the detectivity by its correlation with $(Af)^{1/2}$ according to

$$D^*(\lambda) = D(\lambda)\sqrt{Af} = \frac{\sqrt{Af}}{\mathrm{NEP}(\lambda)}[w^{-1}\mathrm{cmHz}^{1/2}] \tag{5.22}$$

where A is the area of the detector and f is the signal bandwidth. This specific dependence on frequency is useful in that it reminds us that all the figures of merit are dependent on the time frequency or bandwidth of the incident signal. It is also important to realize that the response characteristics of detectors are dependent on the operating temperature of the detector. A small sampling of detector types and variation in D^* values with wavelength, temperature, and bandwidth are shown in Figures 5.11 through 5.14. Note in particular the poor temporal response of the thermistor shown in Figure 5.11 and the improved performance of the InSb detector at cryogenic temperatures.

To this point, we have been implicitly talking about discrete detectors. Increasingly, linear array and two-dimensional array detectors are being used for imaging. These arrays can be thought of as a row of discrete detectors that accumulate charge (signal) and transfer the signal to a charge coupled device (CCD). The CCD is an electronic device that can hold charge packages in discrete elements and then transfer the charge in a bucket brigade fashion through the elements and into a conventional electronic circuit at the end. It is also possible to make CCD's that are themselves photosensitive. This is particularly attractive in the VNIR region where silicon is photosensitive, since most solid-state electronics are silicon-based. Figure 5.15 shows a linear array using photo diodes as the sensing elements and a CCD as an electronic charge transfer element, as well as a linear array where the CCD is used directly as the sensor. In either case, either through a charge transfer or through the internal

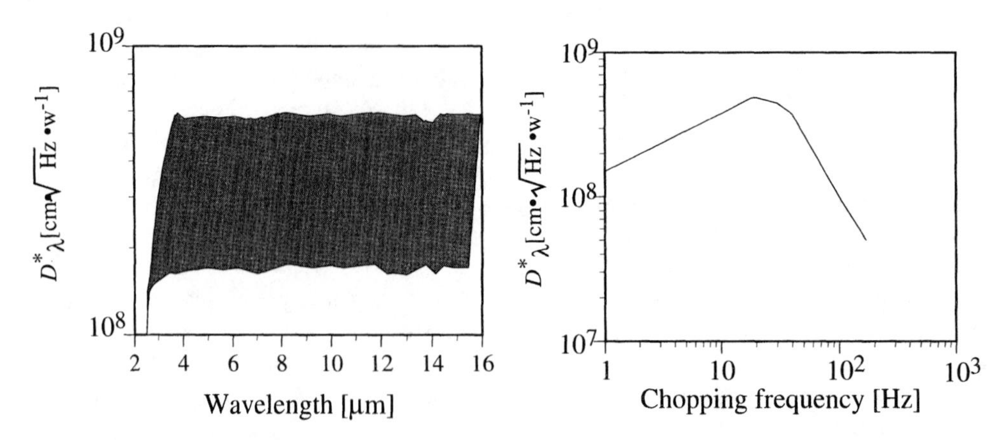

Figure 5.11 Sample performance characteristics for a thermistor at 300 K. Shaded region shows typical range. (Adapted from Wolfe Zissis, 1985.)

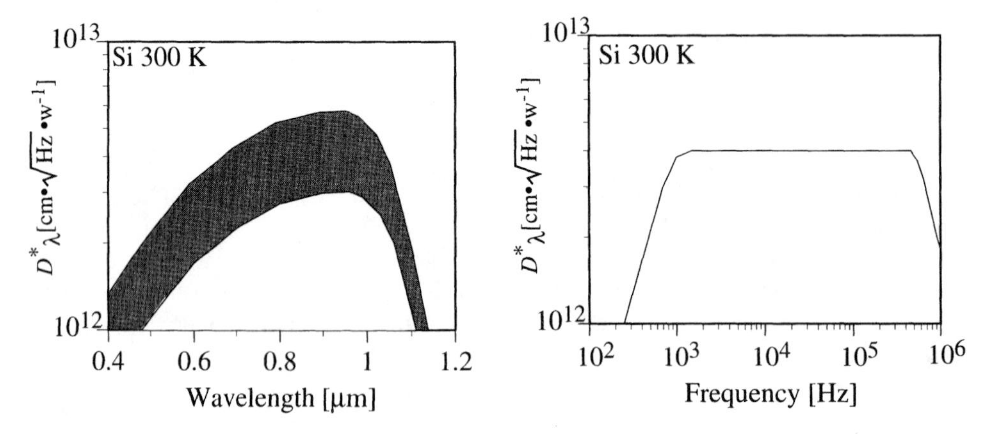

Figure 5.12 Sample performance characteristics for a silicon detector at 300 K. (Adapted from Wolfe Zissis, 1985.)

photo effect, charges are produced in a conductive layer of the material and kept in the conductive layer by an insulating layer. Discrete elements are produced by forming localized potential wells that inhibit migration of charge from where it is introduced or formed. When the signal is ready for processing, electrodes on the opposite side of the insulating layer from the conductor are set to a positive level relative to the charges, causing a migration of the charge toward the electrode. By properly sequencing the voltage pulse trains to the electrodes, the charges can be passed from element to element to the end of the CCD and into a circuit. In a two-dimensional array, the signal is usually read out a line at a time.

Array detectors offer the great advantage in imaging and spectrometry of being able to collect many lines or spectral levels simultaneously. This lets the individual detector element collect data longer, increasing the dwell time and improving the signal-to-noise ratio. These detectors have some additional noise due to inefficiencies in the charge transfer process. However, in most applications this is not sufficiently large to reduce their utility.

Dereniak and Crowe (1984) suggest that the use of figures of merit employing photon energy and electronic charge quanta are more useful for many detectors than the figures of

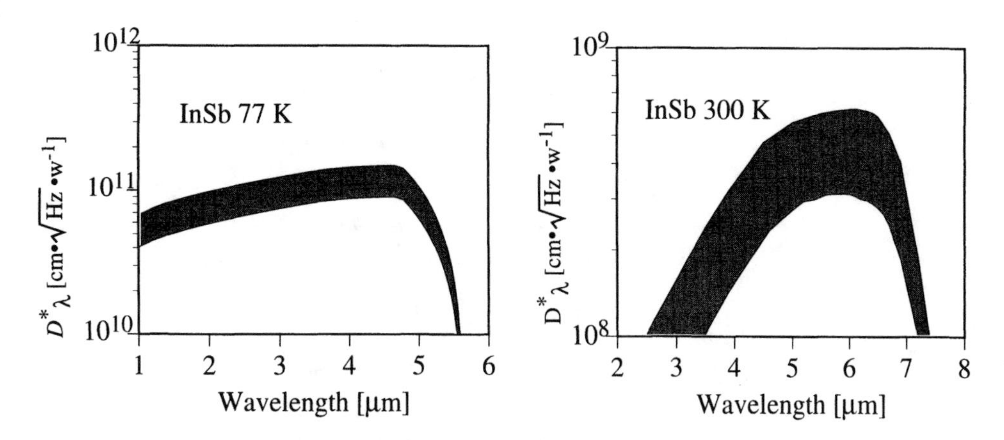

Figure 5.13 Sample performance characteristics for an InSb detector, photovoltaic mode at 77 K and photoconductive mode at 300 K. (Adapted from Wolfe Zissis, 1985.)

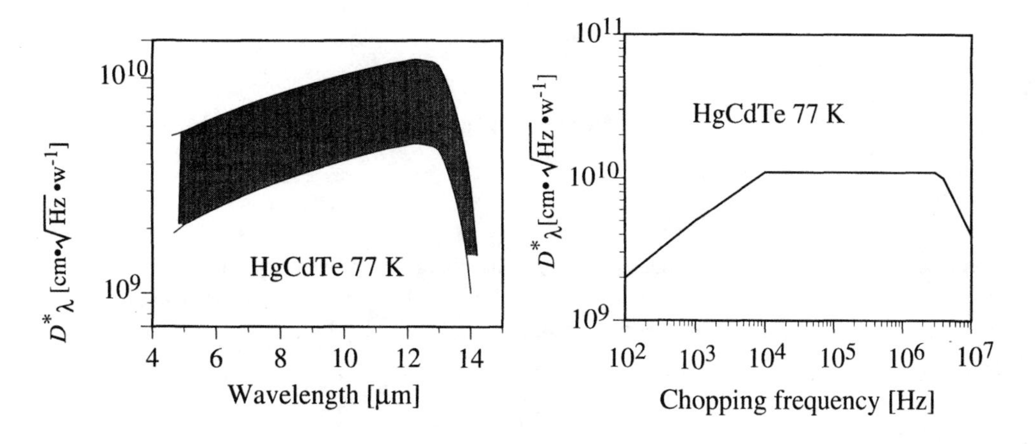

Figure 5.14 Sample performance characteristics of a HgCdTe detector operated at 77 K. (Adapted from Wolfe Zissis, 1985.)

merit discussed thus far. Many CCD arrays have their noise levels expressed in terms of numbers of noise electrons. In terms of the units we have introduced, we can express the signal (S) out of the detector-preamplifier in terms of quantum units as:

$$S[\mathrm{v}] = \Phi_p t_{\mathrm{int}} QEeCE \left[\frac{\mathrm{photons}}{\mathrm{sec}} \mathrm{sec} \frac{\mathrm{electrons}}{\mathrm{photon}} \frac{\mathrm{coul}}{\mathrm{electron}} \frac{\mathrm{volts}}{\mathrm{coul}} \right] \tag{5.23}$$

where Φ_p [photons/sec] is the incident flux expressed in terms of numbers of photons per second, t_{int}[sec] is the integration time of the detector element, QE is the quantum efficiency of the detector defined as the average number of free electrons produced per incident pho-

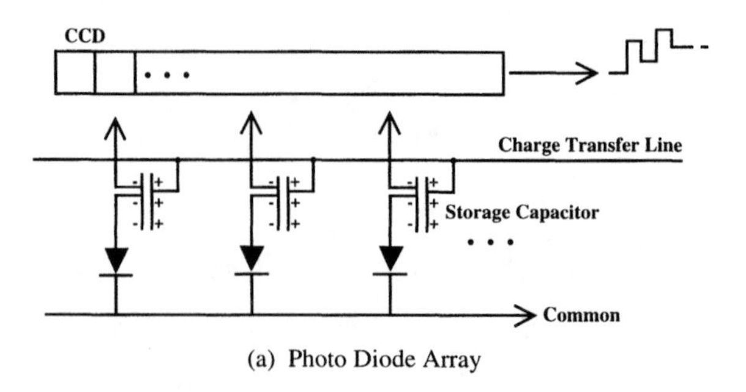

(a) Photo Diode Array

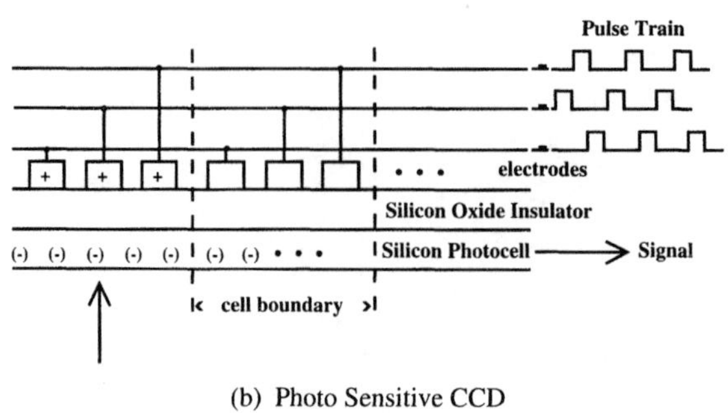

(b) Photo Sensitive CCD

Figure 5.15 Linear array concepts. (a) Diode array using CCD only for charge transfer, (b) CCD used as the photo sensitive device.

ton [electrons/photon], e [coul per electron] is the charge on an electron, and CE is the conversion efficiency of the preamplifier [volts/coul] which defines how many volts are produced per unit of charge. We can then define the photon responsivity in terms of photon flux as:

$$R_p = \frac{S}{\Phi_p} = t_{\text{int}} QEeCE \left[\frac{\text{volt sec}}{\text{photon}} \right] \tag{5.24}$$

and the noise equivalent photon power as:

$$\text{NEP}_p[\text{photons sec}^{-1}] = \frac{N}{R_p} \tag{5.25}$$

where N is the detector-preamplifier noise, including charge transfer noise, expressed in volts. The NEP_p is interpreted in the same way as NEP, except that we are now in units of photons per second required to produce a signal equal to the noise. Dereniak and Crowe (1984) point out that the responsivity expressions in Eqs. (5.24) and (5.18) are for spatially uniform signals and must be modified for spatially varying signals by the spatial frequency response (MTF) of the sensor (cf. Sec. 9.2).

5.3.3 Sensor Performance Parameters

Detectors are a critical element in a sensor's performance, and many sensors are detector limited in terms of their noise limitations (i.e., the detector is often the weakest link in the radiometric portion of the image chain in that it is the largest source of noise). However, there are many other factors that may be the weak link in the detector-electronics signal processing chain.

In some systems, the overall sensor performance may not be limited by the detector's performance but by noise in the preamplifier electronics. The preamplifier components generate most of the same types of noise as the detector, and since the incoming signal is small, the preamplifier noise can become the limiting factor in sensor performance. From the sensor user's point of view, the source of the limiting noise is not critical, only its overall effect on the system performance. It is often useful, therefore, to specify system performance characteristics (i.e., rather than detector performance characteristics) such as the noise equivalent power of the sensor. From the system NEP and knowing the optical throughput (*G#*), the noise equivalent radiance of the sensor can be computed according to

$$\mathrm{NER}(\lambda) = \frac{\mathrm{NEP}(\lambda)}{A_d}(G\#) \tag{5.26}$$

where NER is the amount of radiance or change in radiance on the front of the sensor required to produce a change in sensor output equal to the sensor's noise level, NEP is the sensor noise equivalent power (equal to the detector NEP only if the system is detector noise limited), and A_d is the area of the detector.

From the noise equivalent radiance, the noise equivalent change in reflectance (NE$\Delta\rho$) or the noise equivalent change in temperature (NEΔT) can be computed from

$$\mathrm{NE}\Delta\rho = \mathrm{NER} \cdot \frac{\Delta\rho}{\Delta L} \tag{5.27}$$

or

$$\mathrm{NE}\Delta\mathrm{T} = \mathrm{NER} \cdot \frac{\Delta T}{\Delta L} \tag{5.28}$$

where $(\Delta\rho/\Delta L)^{-1}$ represents the rate of change in radiance at the sensor corresponding to a unit change in reflectance (this is often expressed in terms of a top of the atmosphere value at set solar conditions), $(\Delta T/\Delta L)^{-1}$ represents the rate of change in radiance at the front of the sensor corresponding to a unit change in the temperature of a blackbody in front of the sensor, and all the terms on the right-hand sides of Eqs. (5.27) and (5.28) are the effective values in the sensor bandpass, cf. Eq. (5.18). These are typically the most meaningful terms for the sensor designer and user, since they provide a quick assessment of the radiometric performance in intuitive units. Recognize, however, that these values must still be adjusted for atmospheric effects to predict expected operational performance.

Two final factors need to be considered in evaluating noise-limited performance of imaging sensors. In some cases, the recorder, transmitter, or playback electronics may have worse noise specifications than the detector-preamplifier and become a limiting factor. Finally, in many systems the A to D converter that quantizes the signal may be the limiting factor. End-to-end performance assessment of the radiometric strand of the image chain is treated in greater detail in Section 9.2.

In this section, we have reviewed several of the factors that characterize a detector's or a sensor's ability to resolve small changes in the incident signal (radiance), which can be described as the sensor's radiometric resolution. The absolute magnitude of the radiance levels has not yet been considered and will be addressed in Section 5.6. For brevity's sake, we have not addressed many of the components of sensor systems as discrete entities. In the next subsections (5.4 and 5.5), we will describe several sensor systems and introduce component elements as needed.

5.4 SINGLE CHANNEL AND MULTISPECTRAL EO SENSORS

In this section, we will describe several of the critical components and design features of airborne and satellite imaging systems. With new sensors evolving at a rapid rate, we have chosen to emphasize concepts rather than the details of specific sensors (though examples of current operational sensors are included). We have somewhat arbitrarily divided sensors into those with one to ten or so spectral channels (multispectral) and those with tens to hundreds of spectral channels (hyperspectral). In general we have intermixed airborne and satellite designs, because in many ways they are more similar than dissimilar. For convenience, we will tend to treat all the sensors as though they are digital systems forming discrete pixels. This merely simplifies the terminology used in describing the systems. The reader should recognize that the digital concepts are merely the result of sampling a continuous signal that can be, and often is, recorded instead (cf. Sec. 7.1).

5.4.1 Line Scanners

Remote sensing systems are most often categorized in terms of how the image is formed (e.g., one pixel at a time, one row at a time). In many ways, the simplest imaging sensor is the line scanner (cf. Fig. 5.16). These sensors employ a spinning (scan) mirror to project the image of the detector along a line on the ground perpendicular to the aircraft or satellite ground track. By sampling the signal from the detector, the across-track image lines can be formed. During the rotation of the scan mirror, the sensor platform advances slightly, and consecutive rotations of the mirror sweep out consecutive lines on the ground, which are sampled to form the across-track lines that make up the image (cf. Fig. 5.16). The angular extent of the image across-track is referred to as the *field of view* (FOV) of the imager, and the angular extent of the individual detector element is called the *instantaneous field of view* (IFOV). The projection of the detector onto the ground is referred to as the *ground instantaneous field of view* (GIFOV) or the *ground spot of the sensor*, i.e., at nadir

$$\text{GIFOV} = H \cdot \text{IFOV} \tag{5.29}$$

where H is the flying height above ground level. The conventional line scanner design uses square detectors that are sampled along track on pixel centers and the ideal ground track advances one GIFOV per rotation to sample contiguous lines with each rotation. Thus every point on the ground is imaged (sampled) without gaps and without overlaps. Some systems employ oversampling to improve the spatial resolution of the reconstructed image (cf. Sec.9.1).

Line scanners, like most nonframing imagers we will discuss, have a unique set of geometric distortions caused by the way the image is sampled and by the motion of the sensing platform during imaging. Space-based sensor platforms are usually geometrically stabilized such that the only motion of the platform during imaging is the along-track motion of the spacecraft. Aircraft platforms are often not stabilized so that the orientation of the aircraft can change from one line to the next, or, in extreme cases, even from pixel to pixel within a line. The resulting distortions are illustrated in Figure 5.17. Aircraft aerodynamics are such that pitch and yaw (crab) are generally relatively constant errors typically removed in post-processing. Roll effects vary considerably on a line-to-line basis. Roll can be corrected either by roll stabilizing the imager or by recording the amount of roll using a gyroscope signal and advancing or delaying each line of data by the number of IFOV's of roll. Figure 5.18 shows an image before and after roll compensation was performed using signals recorded from a gyroscope. Pitch and yaw can also be corrected using data from a three-axis gyro; however, the effects on image quality typically do not warrant this degree of sophistication,

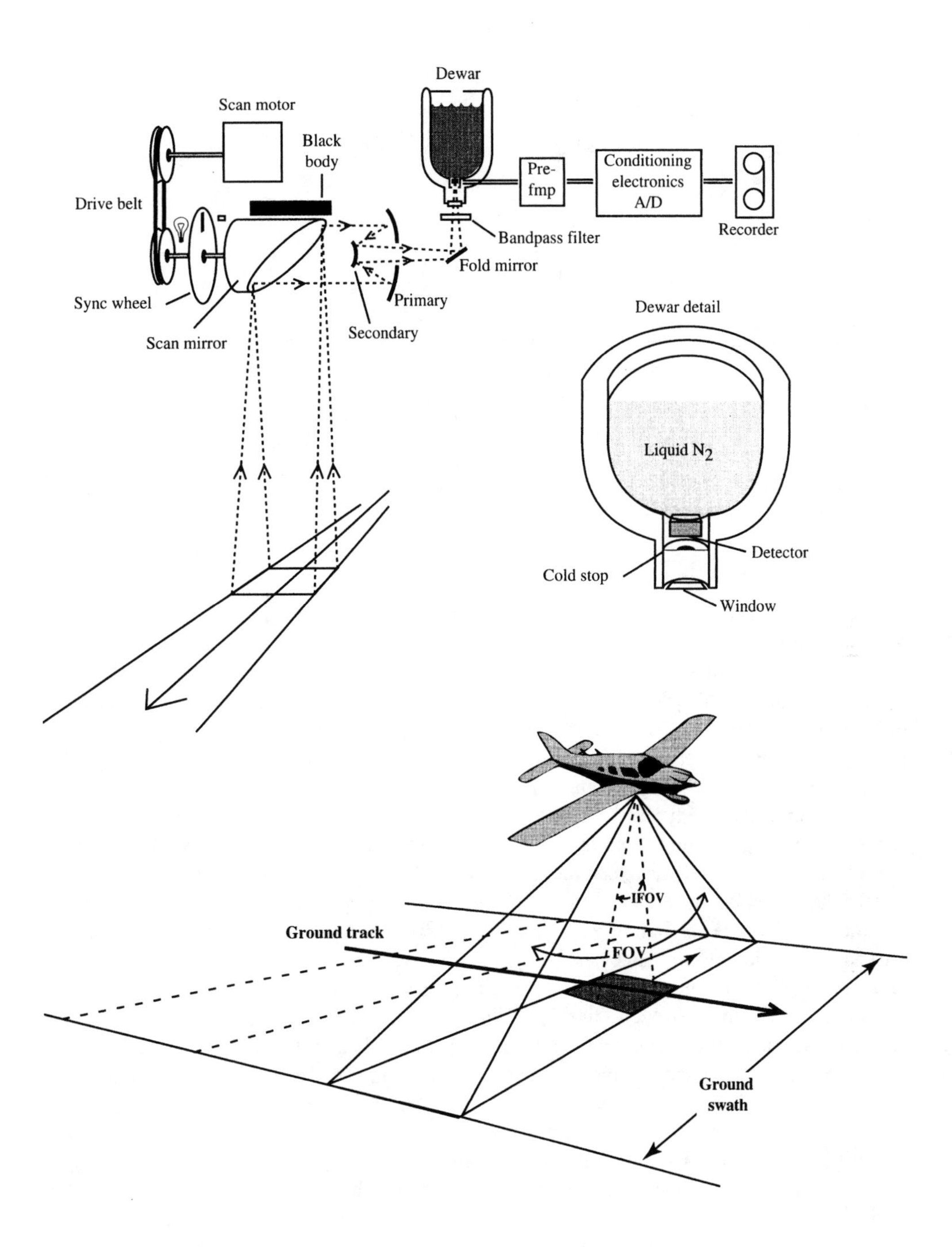

Figure 5.16 Line scanner design and collection scheme. A thermal infrared line scanner is shown with liquid nitrogen cooling of the detector and coldstop.

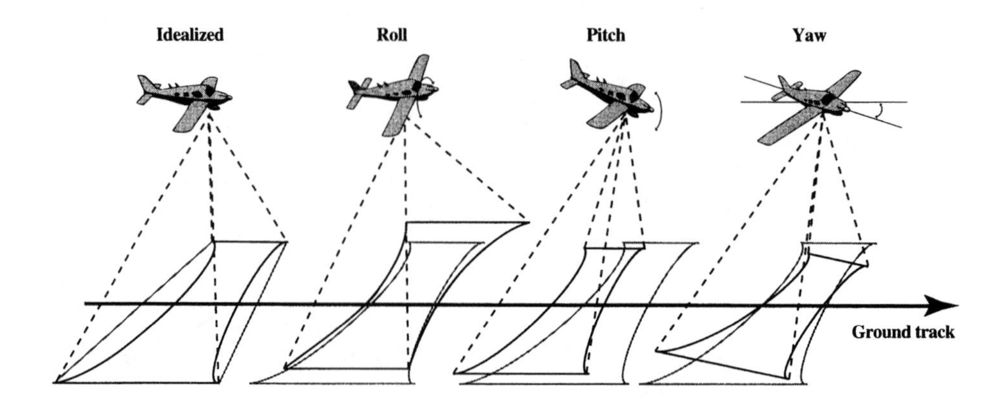

Figure 5.17 Geometric distortions due to aircraft orientation. Gray boundaries represent nominal coverage; black boundaries represent actual coverage.

and constant errors in pitch and yaw can be removed using standard resampling techniques for image or map projection (cf. Sec. 8.3.1.2).

Another type of image distortion is introduced when the scan motor speed is not matched to the aircraft speed such that the aircraft does not advance exactly one GIFOV between scan lines. The error is called V/H (velocity over altitude) error and results in longitudinal compression or stretching of the image, as shown in Figure 5.19. The final type of distortion characteristic of line scanner imagery is called *tangent distortion* and results from the fact that the data are typically sampled on equal angular steps (e.g., every IFOV), and this results in each sample representing a larger projected area on the ground as we progress off axis (cf. Fig. 5.20). The image can be thought of as the projection of the ground onto a cylinder that has the effect of foreshortening the edges of the image. This can be seen in the image in Figure 5.20 where the buildings become compressed and straight diagonal roads become curved toward the edges. This type of systematic error can be removed through geometric resampling of the image as described in Section 8.3. These corrections, however, cannot restore the resolution lost due to the larger GIFOV off axis.

Line scanners often have large fields of view (90 to 120°) providing large ground swaths. This is possible with fairly simple optics, because only the very central portion of the lenses are used for imaging with the scan mirror pointing the telescope's optical axis off nadir.

The line scanner design suffers from the disadvantage that a single detector does all the sampling, so the dwell time (the time the detector can spend gathering photons from a spot on the ground) is very short. This problem is exacerbated by the large "dead time" when the scan mirror is looking up inside itself. Most systems use some of this dead time to look at sources of known radiance inside the scanner for calibration purposes (e.g., the blackbody in Fig. 5.16). A major advantage of the line scanner besides the simplicity of its optics is the inherent registration of multispectral data in many line scanner designs (cf. Fig. 5.21). Many of these systems place the entrance aperture to a monochromator at the focal plane of the optical system. This aperture is the limiting stop in the system defining the sensor's IFOV and sample size. The monochromator disperses the data spectrally with the detector size defining the spectral bandwidth of each channel. The spectral sampling can be performed with discrete detectors or with a linear array of detectors. The signal from each detector is

(a)

(b)

Figure 5.18 Portion of a thermal infrared (8 to 14 μm) line scanner image: (a) before and (b) after lines were shifted to correct for roll distortion. (Image courtesy of the Rochester Institue of Technology, DIRS Lab.)

(a)

(b)

Figure 5.19 Thermal infrared image showing V/H distortion: (a) original image, (b) after resampling to proper V/H ratio. (Imagery courtesy of the Rochester Institute of Technology, DIRS Lab.)

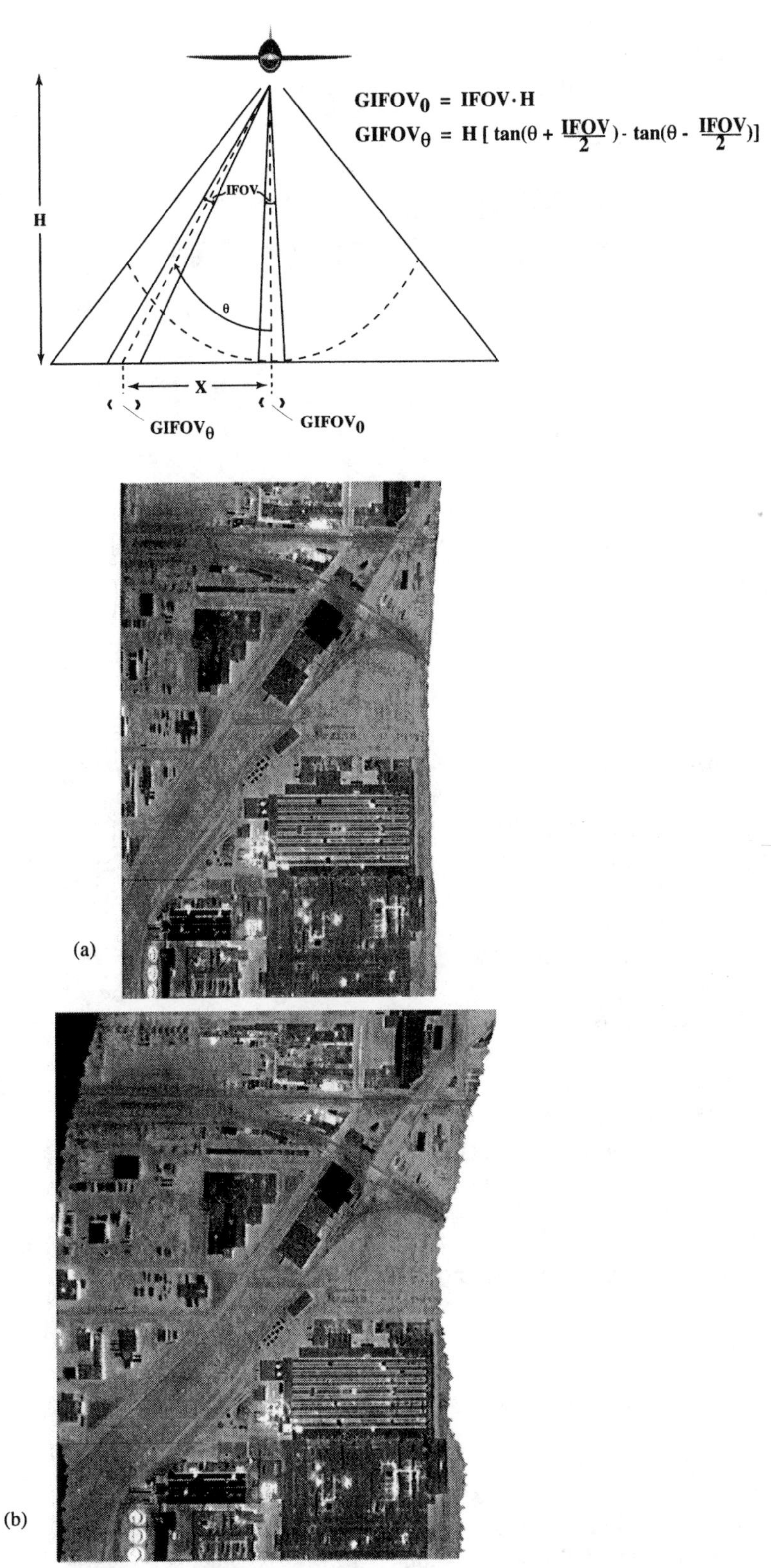

Figure 5.20 Tangent error effects. The diagonal roads in the thermal infrared line scanner image should be straight, and the tanks in the lower left roughly circular. (a) Original image, (b) image after tangent correction. (Image courtesy of the Rochester Institute of Technology, DIRS lab.)

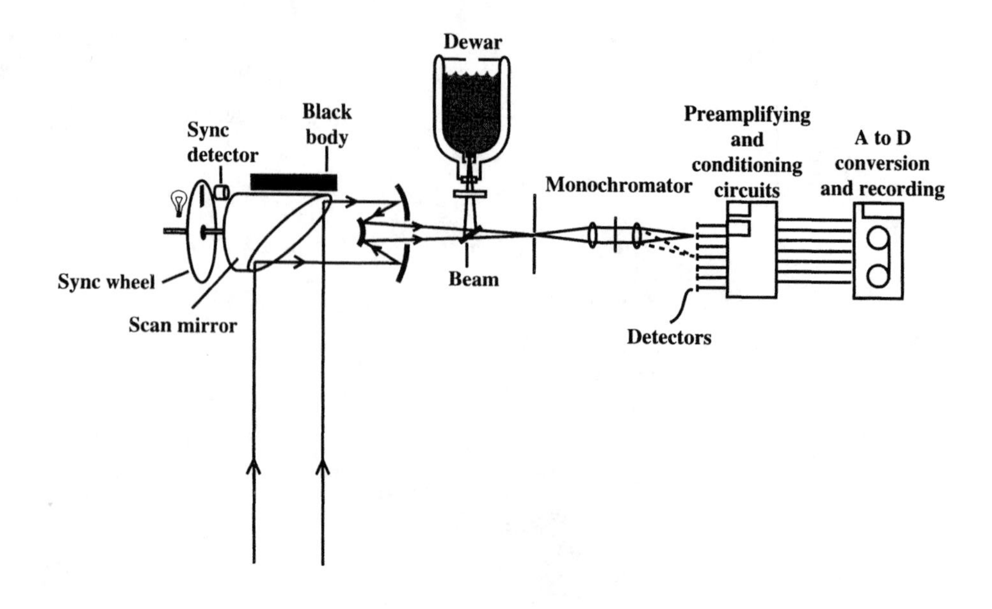

Figure 5.21 A multispectral line scanner design.

amplified and processed to become a data channel for recording or transmitting to the ground. Because the data are sampled simultaneously through a common aperture, the resulting images are inherently registered [i.e., pixel (i,j) of the green image will fall directly on pixel (i,j) of the red image]. In many multispectral scanners, the spectral range cannot be covered with a single monochromator. In the scanner design illustrated in Figure 5.21, the thermal infrared energy is folded out of the primary beam with a beam splitter and focused onto a single cryogenically cooled detector. In this design, the thermal IR detector must be shifted on its focal plane until it is sampling the same image location on the ground as the monochromator aperture. Any misalignment results in image misregistration, which must be corrected during ground processing. Line scanners commonly use a synchronization signal that lets the system know when to sample the calibration sources and when and how frequently to sample the earth (cf. Fig. 5.22).

The geosynchronous operational environmental satellite (GOES) uses an interesting variation of the traditional line scanner design called *spin scan*. Rather than use a spinning scan mirror, the entire satellite, located in geosynchronous orbit, rotates to provide the line scan effect, and the mirror oscillates to provide the line advance as shown in Figure 5.23. The GOES images provide twice-hourly coverage of the hemisphere, permitting time lapse motion sequences showing cloud dynamics.

5.4.2 Whisk-Broom and Bow-Tie Imagers

One simple way of overcoming the short dwell time of the line scanner design is to take several lines of data simultaneously. Then the time between repeat cycles of the scan mirror is reduced by the number of lines collected per sweep. This approach has been very successfully used in the multispectral scanner (MSS) instruments flown on the Landsat satellites since 1972. The scan mirror oscillates as shown in Figure 5.24, acquiring several lines of data, but only in one direction. This method of data collection is analogous to using a whisk

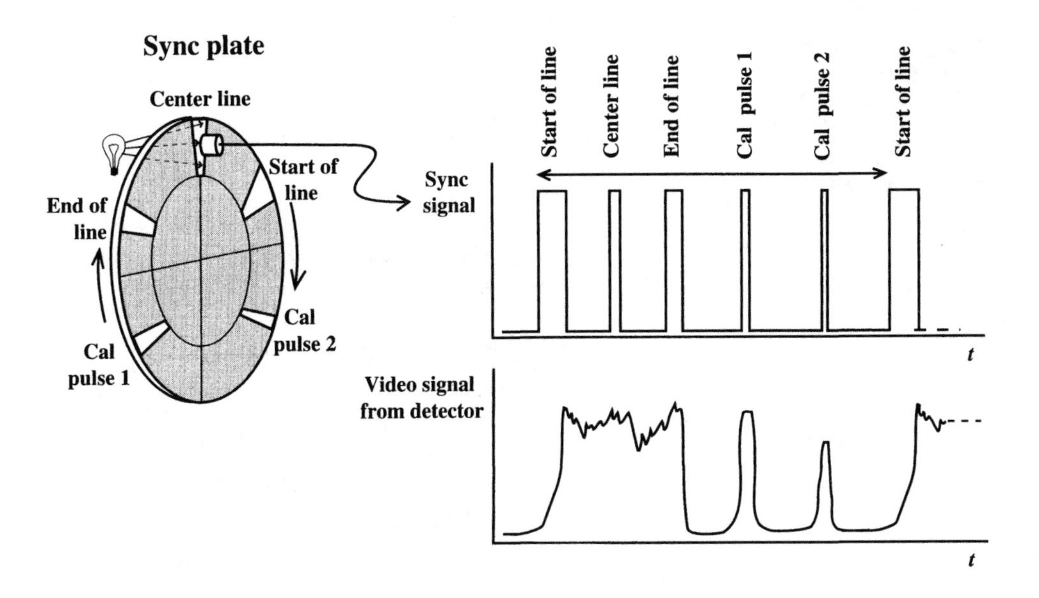

Figure 5.22 Synchronization signals.

broom to sweep dust (data) into a dust pan (data stream). In the case of the Landsat MSS, six lines of data are collected with six different detectors with each mirror sweep. The next mirror sweep is timed so that the next six lines are immediately adjacent to the last, providing continuous ground coverage. Multiple spectral bands are acquired by locating additional detectors on the focal plane in the along-scan direction. The detectors are spectrally filtered to control the wavelength sampled. As the image is swept across the focal plane by the scan mirror, each point on the ground is sampled: first in one spectral band, then moments later, by a second, etc. In reconstructing the images for spectral registration, the pixels in each spectral band must be shifted by a few pixels to properly align with the previous band. For most of the Landsat MSS instruments, four channels were collected, resulting in a total of 24 detectors. The actual detectors on the MSS are photomultiplier tubes for the green, red, and first IR channel and photo diodes for the second IR channel (cf. Table 5.2). The focal plane is sampled by light pipes (fiber-optics) which carry the signal to the detectors. This allows physically large devices (PMT's) to be effectively located in close proximity on the focal plane.

The increased dwell time of the MSS whisk broom design allows it to achieve moderate spatial (79 m GIFOV) and spectral ($\Delta\lambda \approx 0.1$ μm) resolution. The whisk broom approach still wastes approximately 1/2 the useful scan time by only taking data in one direction. The Thematic Mapper (TM), which was flown along with the MSS on Landsat 4 and 5, uses an oscillating mirror that scans in both directions (cf. Fig. 5.25). As illustrated in Figure 5.26, this approach will result in gaps and overlap regions in the ground coverage. To compensate for this, a pair of rotating parallel mirrors called the *scan line corrector* are included in the TM optical chain. These mirrors shift the image projected onto the detectors so that it is slightly ahead of the across-track location at the start of each scan and ends behind the start point. The corrector advances and retards in this fashion during both the forward and reverse scans of the primary mirror, such that the scan projections on the ground fall alongside each other (cf. Fig. 5.26). The TM takes advantage of the increased dwell time available from this bow tie correction to increase the spatial and spectral resolution over the MSS. To achieve

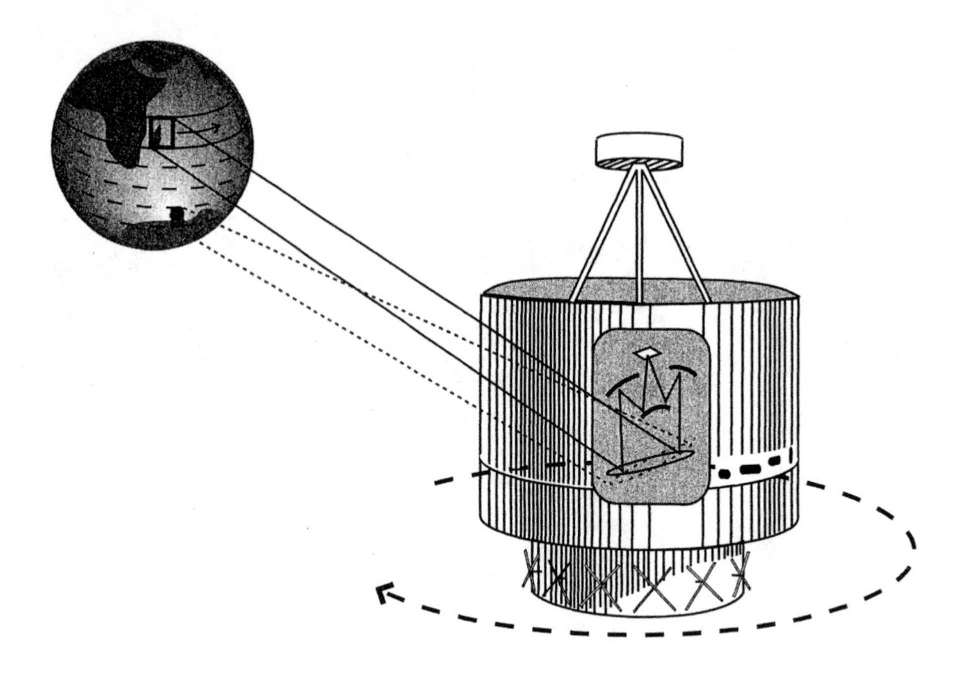

Figure 5.23 Spin scan coverage used from geosynchronous orbit (e.g., GOES).

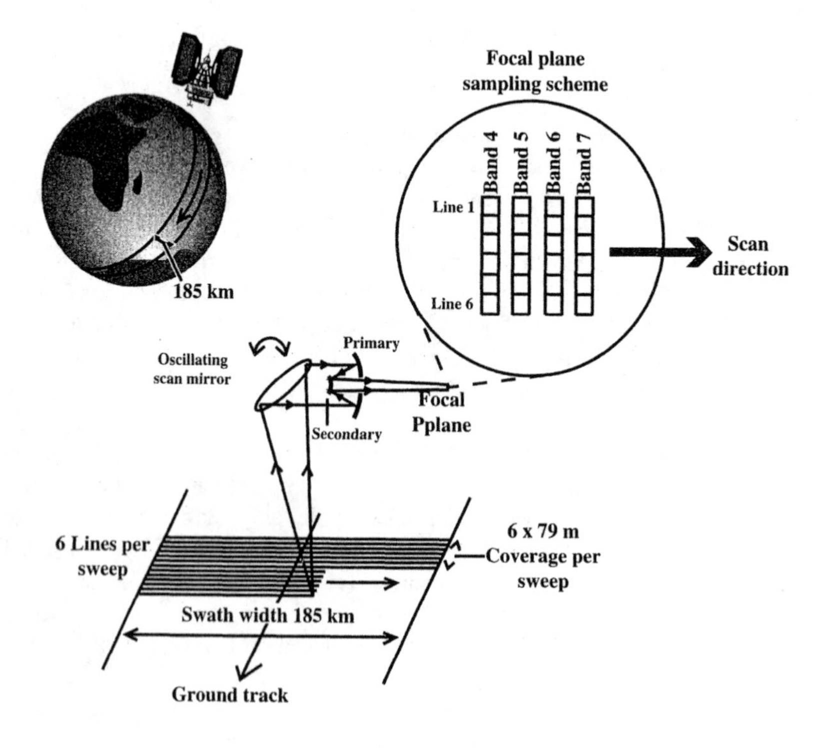

Figure 5.24 Whisk-broom design used on the Landsat MSS instrument.

Table 5.2 Characteristics of Landsat MSS and TM Sensors

Sensor	Orbits altitude	Repeat period	FOV (degrees km)	GIFOV	Nominal spectral Bands (μm)
MSS 1,2,3	Sun synchronous	18 days	11.6	79 m	# 4 0.5 - 0.6
	descending		185 km	79 m	# 5 06 - 0.7
	equatorial			79 m	# 6 0.7 - 0.8
	crossing = 9:30 a.m.			79 m	# 7 0.8 - 1.1
	913 km				
MSS 3 only				237 m	# 8 10.4 - 12.6
MSS 4&5	Sun synchronous	16 days	14.9	82 m	# 1 0.5 - 0.6
	descending		185 km	82 m	# 2 06 - 0.7
	equatorial			82 m	# 3 0.7 - 0.8
	crossing = 9:45 a.m.			82 m	# 4 0.8 - 1.1
	705 km				
TM 4&5	Sun synchronous	16 days	14.9	30 m	# 1 0.45 - 0.52
	descending		185 km	30 m	# 2 0.52 - 0.60
	equatorial			30 m	# 3 0.63 - 0.69
	crossing = 9:45 a.m.			30 m	# 4 0.76 - 0.90
	705 km			30 m	# 5 1.55 - 1.75
				120 m	# 6 10.40 - 12.5
				30 m	# 7 2.08 - 2.35

this, the ground coverage per mirror sweep is kept approximately the same, using more (16) but smaller (30-m GIFOV) detectors. In addition, the number of bands is increased to seven, as listed in Table 5.2. The TM detectors are located on the focal plane using a staggered array (cf. Fig. 5.27) to make room for the individual detectors. Relay optics are used to focus a portion of the image onto a cooled focal plane where the SWIR and LWIR detectors are located. To achieve adequate signal to noise, the GIFOV of the LWIR band is 4 times (120 m) that of the other bands.

The signals transmitted to the ground are preprocessed to account for the staggered array effects, the spectral band offsets, and offsets between the data taken in the forward and reverse sweep. In addition, both MSS and TM data are nominally corrected for the effects for the earth's rotation during the time of data collection. As the satellite travels south in its descending node, the earth's eastward rotation causes the effective ground track to drift westward. The resulting image is skewed and must be deskewed for proper ground projection. Figure 5.28 shows a full frame of Landsat data after ground processing, including the characteristic rhomboid shape indicating that it has been deskewed. A more complete treatment of the Landsat sensors can be found in Barker (1983) and Markham and Barker (1985).

The whisk-broom style sensors have largely been used only in space because of the difficulties in correcting for geometric errors in an unstable platform. They offer the advantage of increased dwell time allowing for either higher signal to noise or higher spatial and/or spectral resolution. They have the disadvantage that they require slightly larger fields of view (i.e., image quality must be maintained over the entire detector array) for the telescope, and

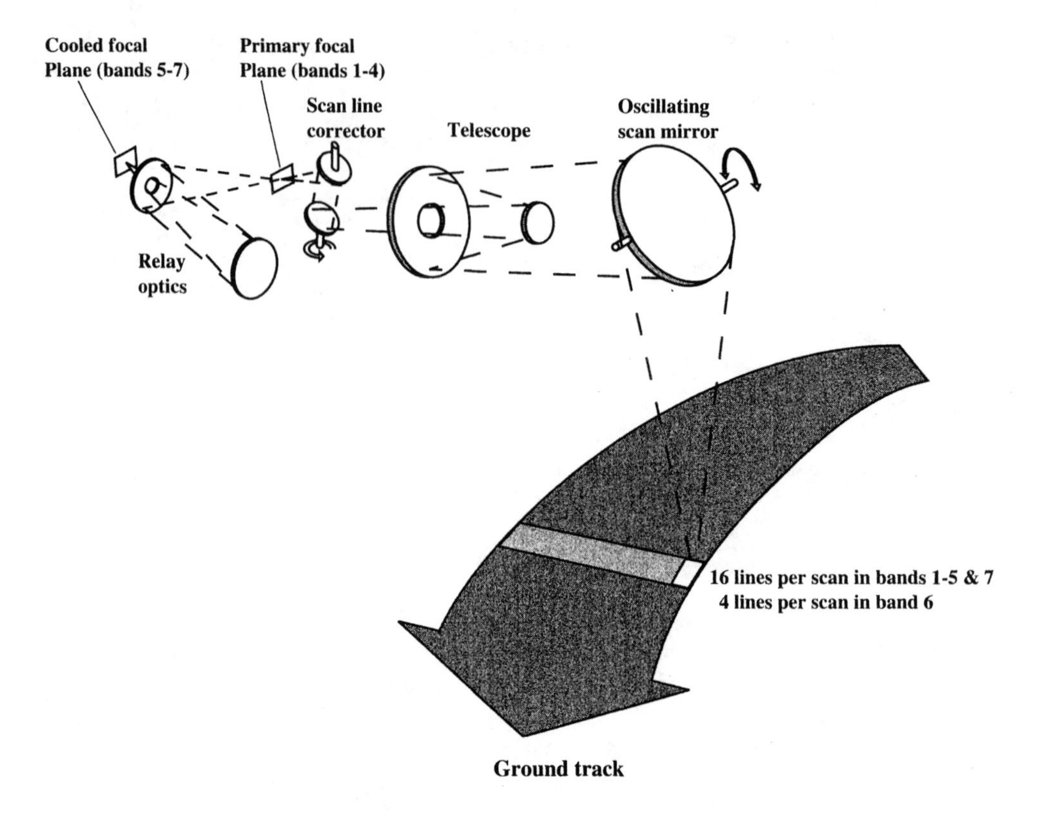

Figure 5.25 Landsat TM optical layout.

the spectral data are not inherently registered. For Landsat TM, the detectors were contained within one-quarter of a degree of the optical axis within an area approximately 1 cm on a side, so the constraint on the optics was not very severe. Wrigley et al. (1985) report that the initial band-to-band registration of the Landsat TM focal planes exhibited several tenths of a pixel misregistration. However, they go on to report that, after post-launch corrections in the ground processing, band-to-band registration met the 0.2 pixel misregistration specification and that registration is typically good to better than 0.1 pixel.

5.4.3 Push-Broom Sensors

The push-broom sensors represent a further step toward increasing the dwell time to allow system designers to make signal-to-noise or resolution tradeoffs. These sensors use linear array detectors, as illustrated in Figure 5.29, to collect entire lines of data simultaneously. Multiple spectral bands are collected, with multiple linear arrays filtered for the bands of interest. The *French Système Probatoire d'Observation de la Terre* (SPOT) instruments use a push-broom approach. With this approach, an individual detector element only needs to sample at one across-track location. This provides increased integration time for the sensor. The SPOT system uses this to achieve 10-m GIFOV's for a panchromatic band and 20-m GIFOVs for three spectral bands as listed in Table 5.3. One of the disadvantages of the push-broom approach is that very long arrays are necessary to achieve a large ground swath. For example, the SPOT ground swath is only 60 km, compared to Landsat's 185 km and

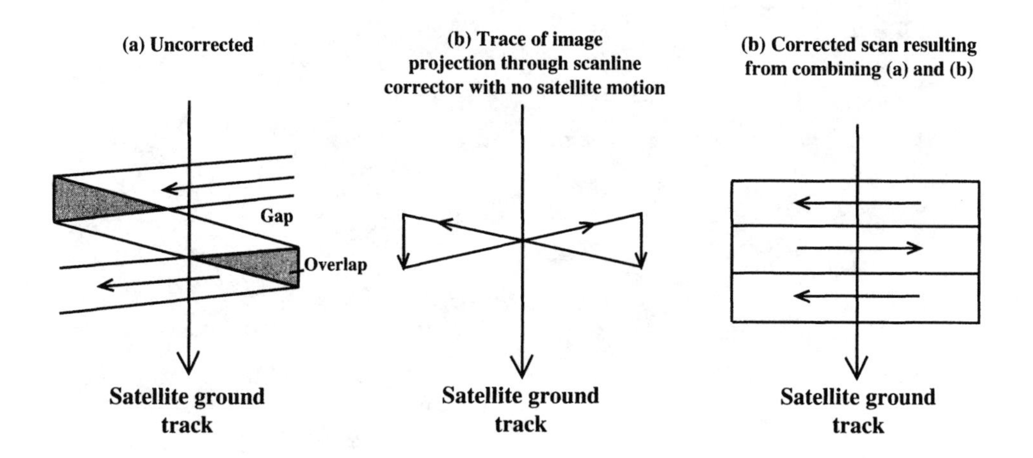

Figure 5.26 Bow-tie scan line correction used with Landsat TM.

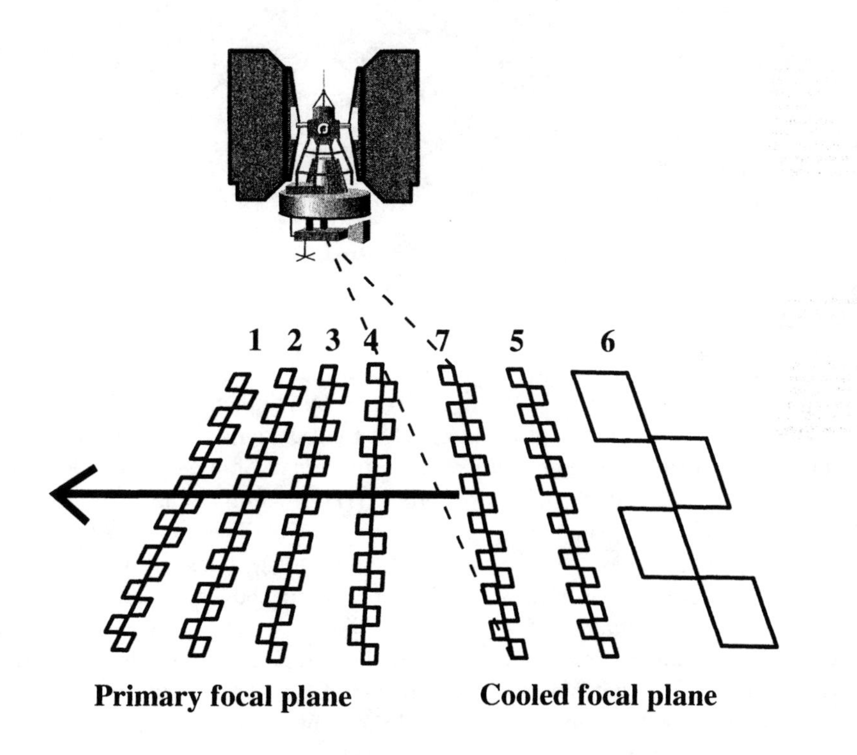

Figure 5.27 Projection of Landsat TM detectors onto ground.

Figure 5.28 Landsat TM full-scene image of Eastern Lake Ontario and the Finger Lakes region of New York State.

AVHRR's 2400 km. The SPOT system overcomes this limitation by using an across-track pointing mirror to allow it to look up to 27° to the left or right of the nominal ground track. In addition to increasing the probability of acquiring targets of interest, this allows for stereoscopic coverage using data from multiple orbits. The push-broom approach has the distinct advantage in satellite operation of having no movable parts. This increases the expected lifetime and may reduce power requirements. The detector technology, however, is far more sophisticated than the line scanner or whisk-broom approaches. Long arrays, even using the staggered-array approach illustrated in Figure 5.29, can be difficult to manufacture for some materials (e.g., HgCdTe). In addition, for focal planes that require cooling, it can be very difficult to design and provide power for cooling large focal planes. Push-broom systems also have the disadvantage of requiring wide field-of-view optics (at least in one dimension) because the entire sensor field of view is imaged at once. Despite these limitations, advances in detector technology are pushing more and more space-based systems toward push-broom solutions because of the advantages of no moving parts and the long dwell times that provide for higher spatial or spectral resolution. The spectral bands still need to be registered by shifting to correct for the offset between the arrays; however, the lack of scan motion generally improves the inherent geometric fidelity of push-broom systems. The wider field of view

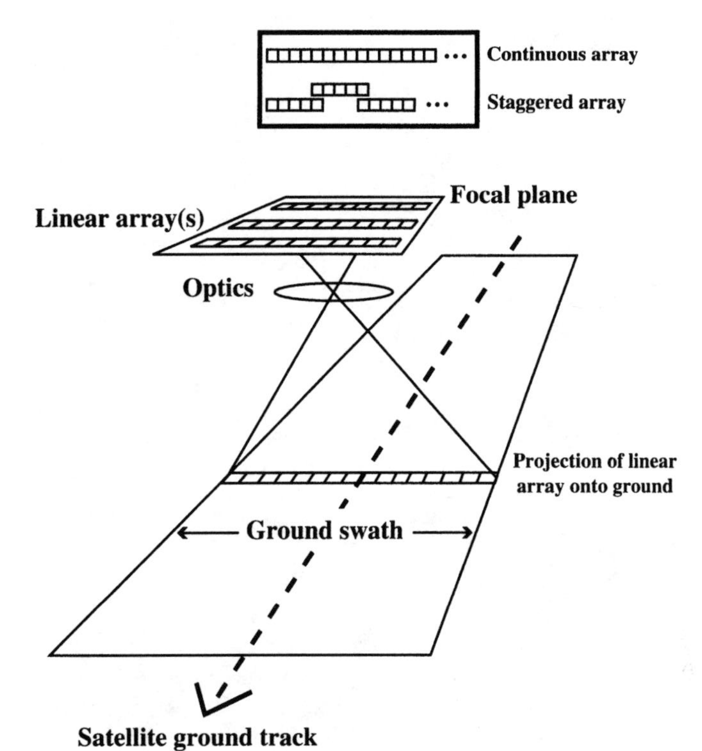

Figure 5.29 Push-broom sensor operation.

Table 5.3 Characteristics of SPOT Instrument

Instrument	Orbits altitude	Repeat cycle	FOV	GIFOV	Spectral bands (μm)
SPOT	Sun synchronous	26 days	4.13 °	10 m	0.51 - 0.73
	descending		60 km	20 m	0.50 - 0.59
	equatorial			20 m	0.61 - 0.68
	crossing ≈ 10:30 a.m.			20 m	0.72 - 0.89
	832 km				

normally required for airborne sensors and the high cost of linear arrays at longer wavelengths has limited the use of push-broom systems primarily to space systems. However, this is not a fundamental limitation, and with decreases in detector costs, more airborne push-broom systems can be expected (cf. Fig. 5.30).

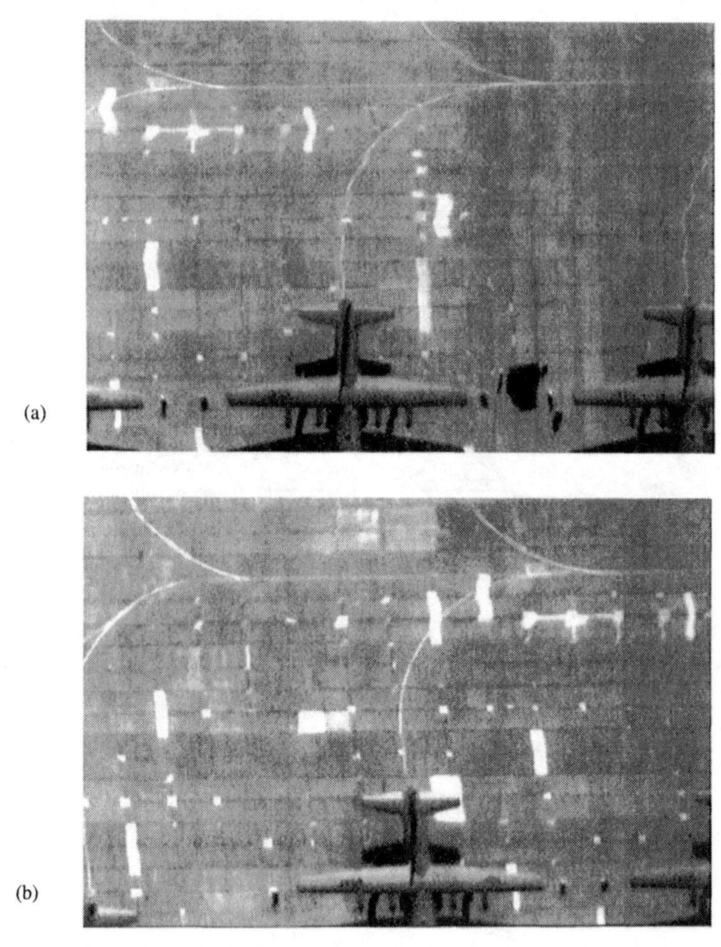

Figure 5.30 Visible images acquired with an airborne push-broom scanner. Note some residual roll effects in the lines painted on the runway in image (a). (Image courtesy Eastman Kodak Company.)

5.4.4 Framing (2-D) Arrays

Another EO imaging approach is to use a two-dimensional (2-D) array of sensors and either a mechanical or an electronic shutter to control the integration (dwell) time. This is essentially the way a camera works with the 2-D array replacing the film. Since the forward motion of the sensor will blur the image, exposure times with this type of sensor must be restricted to less than the time it takes to move the sensor one GIFOV. As a result, there is no inherent gain from a signal-to-noise standpoint over a push-broom system. The major advantage of this approach is in terms of the geometric fidelity. Since the entire image is acquired simultaneously, distortions due to within-frame sensor motion are essentially eliminated. For this reason, 2-D array sensors are very attractive for aircraft use where platform stability is a serious concern. Two-dimensional detector arrays are still difficult to manufacture at many wavelengths and large arrays (greater than 2K x 2K), even in the silicon region, become quite expensive. Another limitation of 2-D arrays is that it is difficult to obtain multispectral data. Figure 5.31 shows two approaches that are used to overcome this limitation. The sampling approach shown in Figure 5.31(a) is essentially the method used in solid-state color video cameras. A filter mask is layered on top of the array such that different detector

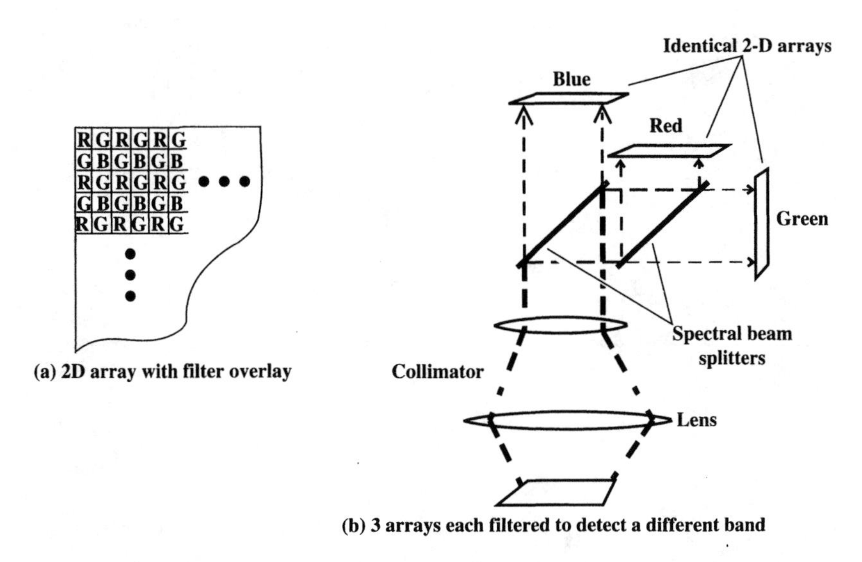

Figure 5.31 Multispectral framing sensor designs. (a) A 2-D array with flter overlay that is interpolated to form a multispectral image. (b) Multiple arrays used with beam splitters to simultaneously image in three spectral bands.

element (pixels) are sensitive to different spectral bands. Each spectral image is then produced by interpolating the values for the missing pixels (e.g., replacing those sampled in a different band), often taking advantage of the spectral correlation between bands in the interpolation process. This method reduces the spatial resolution by dividing the array among the spectral bands. An alternative approach shown in Figure 5.31(b) is to use multiple arrays. The incident image is collimated and divided by dichroic beam splitters into three components (e.g., green, red, and NIR) that are each imaged onto a separate 2-D array. The arrays must be very carefully aligned relative to each other to ensure proper registration when the multispectral images are combined (misregistration can be taken out in digital preprocessing; cf. Sec. 8.3, but this can be costly, and the resampling results in image degradation). Clearly, the beam splitting looses a considerable amount of energy and the alignment must be very precise. In general, this approach is only used when a few (usually three) bands are required for low- to-modest resolution systems (usually video quality with a few hundred by a few hundred element detector arrays). In many cases, these 2-D arrays are used to generate video compatible images at 30 frames per second. The output signals are processed to look like conventional NTSC video for ease of storage on commercial videocassette recorders (VCR's). Figure 5.32 shows examples of aerial images obtained with 2-D-array sensors.

While framing systems are extensively used for aircraft applications to avoid geometric distortion problems, they have also been used in space. The Return Beam Vidicons (RBV's) flown on Landsat-2 and 3 used a 2-D photosensitive array that was exposed and then raster-scanned with an electron beam in a fashion similar to conventional video. The RBV's on Landsat-3 had approximately 40-m resolution and used a reseau grid to help insure geometric fidelity. In the Landsat-3 era, this represented the highest-quality data available for large-area cartographic mapping from space. Figure 5.33 shows an RBV image and the corresponding portion of a Landsat MSS scene showing the improved image quality that, coupled with the geometric fidelity, increased the value of the RBV products for mapping purposes.

(a)

(b)

Figure 5.32 Examples of EO images acquired using 2-D arrays. (a) Visible image of test target acquired with a 2048 x 2048 array. (Image courtesy of Recon/Optical.) (b) MWIR image acquired with a platinum silicide array (note the cold sky in the background). (Image courtesy of Eastman Kodak Company.)

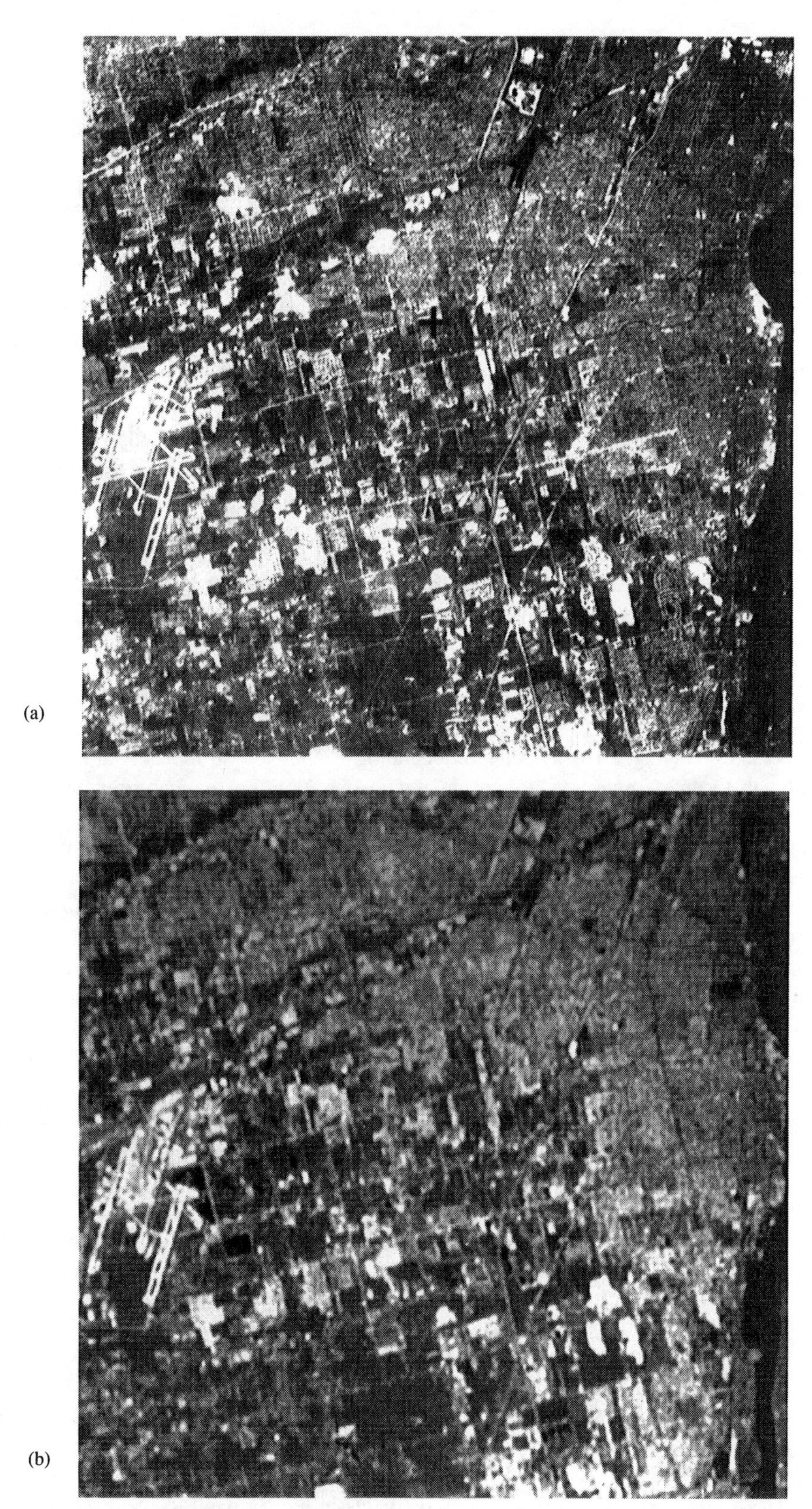

Figure 5.33 (a) Landsat-3 RBV image of Detroit and (b) a portion of a Landsat MSS image of the same area showing the increased potential of the RBV for cartographic applications. (Image courtesy of the USGS EROS Data Center.)

5.5 IMAGING SPECTROMETERS

Advances in one- and two-dimensional array detector technology, coupled with advances in digital data processing and storage, have opened the door for a new generation of sensors called *imaging spectrometers*. These sensors are designed to collect data on the spectral, as well as the spatial, characteristics of the imaged scene. Their use is motivated by the expectation that finer spectral sampling will convey more information about the makeup or condition of certain scene elements. Imaging spectrometry brings to the forefront the fundamental tradeoffs in all sensor designs between spatial, spectral, and radiometric resolution. The limited numbers of photons from a scene can only be divided into a restricted number of information bins. We can think of these bins as spatial samples (i.e., number of pixels), spectral samples (i.e., number of spectral bands), and radiometric samples (i.e., number of gray levels). In most cases, for a fixed technological approach, it takes more photons to increase the resolution (i.e., number of bins) in any category, and the only way to obtain more photons is to steal them from one of the other bins. For example, to increase the spectral resolution we must sacrifice signal-to-noise or spatial resolution. At present, these tradeoffs are largely made in an ad hoc fashion and depend on the scenario under study and the analytical algorithms being used. Hopefully, as more imaging spectrometer data become available, a practical approach to these tradeoffs can be developed that can help system designers and image analysts decide what spatial, spectral, and radiometric specifications are required for a specific analysis. In the meantime, system designs are attempting to push the limits in all three areas with a major emphasis on spectral resolution while we explore the potential of this relatively new dimension to remote sensing.

The design approaches for spectrometers are largely just extensions of traditional designs that take advantage of array detector technology. The most straightforward and most used approach to date employs a line scanner with one or more monochromators whose entrance aperture(s) is located at the primary focal plane. The dispersed spectra are then sampled with linear array detectors, as illustrated in Figure 5.34. This is essentially the same design as the multispectral scanner shown in Figure 5.21 using an array spectrometer. NASA's advanced visible and infrared imaging spectrometer (AVIRIS) is a good example of this approach. In order to cover the spectral region from 0.4 to 2.5 μm in channels with approximately 0.01 μm spectral bandwidth, it uses four spectrometers, one with a silicon array and three with InSb arrays (cf. Vane et al., 1993). Figure 5.35 is a sample image from AVIRIS. The 2-D surface image was made by assigning the red, green, and blue brightnesses to three of the 224 spectral channels. The edge pixels have their spectra from 0.4 to 2.5 μm, color-coded to provide the depth axis of the "color cube." Note that the black bands in the spectra are the atmospheric absorption bands around 1.4 μm and 1.9 μm. On a soft copy computer display, the location of the edges of the color cube can be interactively adjusted to let an analyst search spatially for spectral patterns. This is a useful quick-look tool. However, more detailed analysis is typically required using advanced computer analysis methods, since the human being is quickly overwhelmed by the volume of data associated with spectrometric images (cf. Sec. 7.4).

The AVIRIS sensor's primary function is to serve as an airborne test bed for satellite-based imaging spectrometers such as the moderate-resolution imaging spectrometer (MODIS), which is scheduled to fly on NASA's earth observation system (EOS). The MODIS approach uses a line scanner design with approximately 36 spectral channels in bands from the visible through the LWIR, with GIFOV's of 0.25 to 1 km depending on the channel. However, rather than use a monochromator for spectral dispersion, the detectors are located on the focal plane and spectrally filtered in a design reminiscent of the Thematic Mapper. The resulting raw images must then be pixel-shifted for band-to-band registration. These line scanner designs for imaging spectrometry have all the strengths and weaknesses of conventional multispectral scanners discussed in Section 5.4. In particular, the short dwell

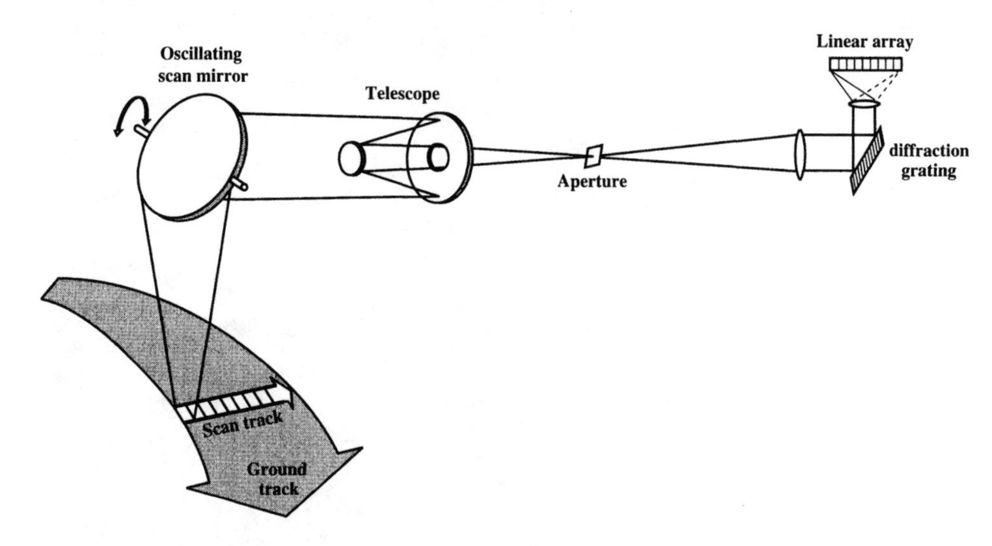

Figure 5.34 Imaging spectrometer line scanner.

time, coupled with the narrow spectral bands, means that the signal-to-noise or the spatial resolution (or both) must be sacrificed.

An alternative approach uses a push-broom design, as shown in Figure 5.36. In this design an entire line of data is imaged onto the focal plane and sampled simultaneously by a slit that is the entrance aperture to a monochromator. The beam is spectrally dispersed perpendicular to the long axis of the slit and imaged onto a 2-D array. One dimension provides the spatial sampling in a traditional push-broom fashion the other dimension provides the spectral sampling. NASA's Airborne Imaging Spectrometer (AIS) used this push-broom 2-D-array approach employing multiple monochromators to cover a spectral range from 0.4 to 2.5 μm (cf. Goetz et al. 1985). Two-dimensional-array detector technology is a limiting factor with this approach at the present time. The push-broom axis needs to be quite large for many applications, and large 2-D arrays are still difficult to manufacture (especially in the IR bands)—for example, AIS-produced images that were only 64 pixels wide. As 2-D IR arrays become more available, the push-broom approach may become more widely used to take advantage of increased dwell time.

A third possible approach to imaging spectrometry uses 2-D framing arrays and variable filters. The filters are electronically addressed, such that their spectral bandpass changes between each frame. Each frame is a full 2-D image in a different spectral region. The forward advance of the sensor, coupled with roll and other sensor motion, means that consecutive frames don't overlay and must be spatially registered during initial ground processing. This approach is further limited since the number of spectral bands must repeat before the sensor advances out of the overlap region (cf. Fig. 5.37). This approach is attractive because the individual frames have high geometric integrity and the data can be acquired with a relatively simple modification to existing 2-D array imagers. However, the amount of data processing required for image reconstruction can be prohibitive. The filters used in this approach can be discrete filters that are mechanically changed by advancing a filter wheel between frames or a single filter whose spectral transmission can be changed (e.g., tunable liquid crystal filters as described by Gunning, 1981).

Figure 5.35 AVIRIS image cube of Moffet Field California. Sensor has 224 channels, from 0.4 μm to 2.5 μm, each with a spectral bandwidth of approximately 10 nm. See color plate 5.31. (Image courtesy of NASA JPL.)

5.6 LUMINESCENCE SENSORS

As we continue to look at sensors with narrower and narrower spectral lines, a new source of photons can become significant in certain spectral bands. The source of this flux is stimulated luminescence. This occurs when a high-energy photon causes an electron in an absorbing molecule to transition from a stable level to a higher-energy state. The electron can then return to the lower-energy level by giving up the transition energy through collision (thermal energy), by emission of a photon equal to the transition energy (i.e., a photon at the same wavelength as the absorbed energy), or by a step process where the decay takes place through two or more energy levels emitting lower-energy longer-wavelength photons at each step. This last process, known as *luminescence*, is quite common. However, the flux levels involved are usually quite small and are masked by the reflected flux levels.

Luminescence can be measured in the laboratory with a luminescence spectrometer, as shown in Figure 5.38. The sample chamber can either contain a cuvette for liquid samples (most standards are dyes in varying concentrations) or solid samples. The emission spectrometer is generally located at right angles to the beam from the excitation spectrometer. Solid samples, are oriented such that any energy from a specular bounce does not enter the emission spectrometer. This reduces stray-light problems. A sample can then be characterized in terms of its emission and/or excitation spectra. The emission spectra is obtained by setting the excitation monochromator at a fixed wavelength (e.g., a laser illumination line that

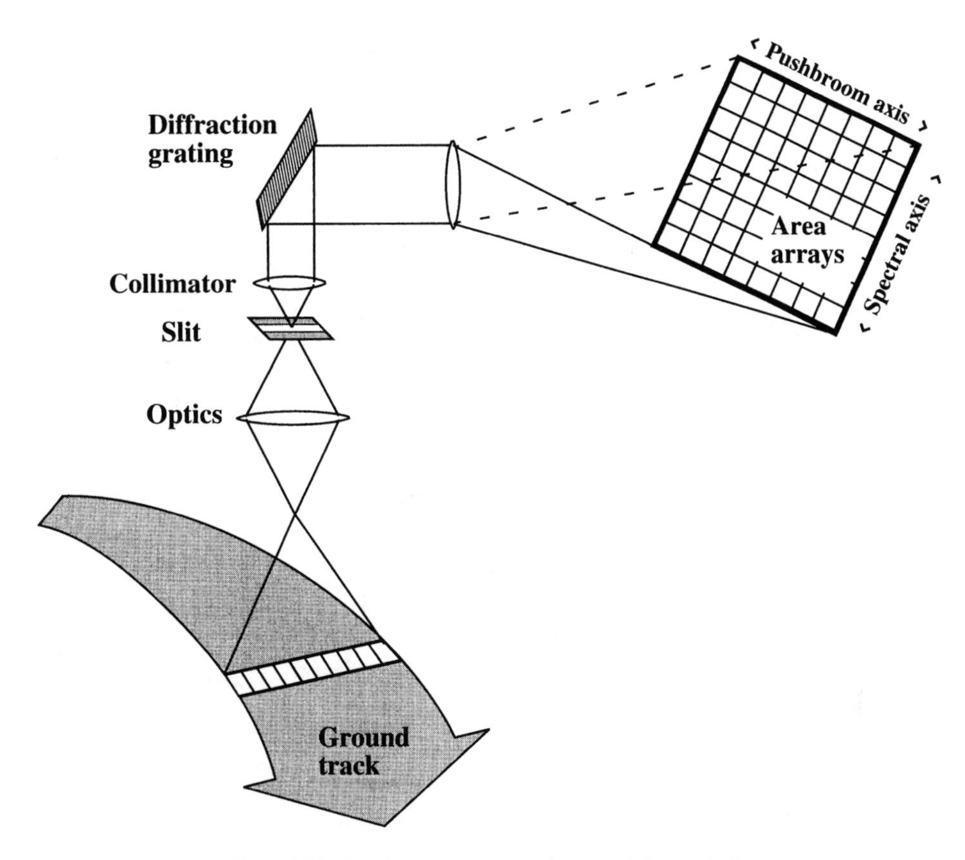

Figure 5.36 Imaging spectrometer using a push-broom design.

would be used to study laser-induced luminescence) and then scanning the emission monochromator. An excitation spectra is produced by fixing the wavelength of the emission spectrometer (e.g., at a particular line to be sensed by an imaging system) and then scanning the excitation monochromator. Sample spectra are shown in Figure 5.39 for rhodamine WT dye, which is commonly used as a luminescence standard. Watson (1981) suggests that luminescence levels can be expressed in terms of the concentration of a standard material required to produce the observed level (e.g., equivalent ppb of rhodamine WT). Luminescence can be studied with an active source such as a laser used to illuminate the target. The induced luminescence can then be observed using any of the imaging sensors we have described. To reduce the laser power levels, scanning systems are often used so that only a small region around the GIFOV of the sensor needs to be illuminated. The sensor can either collect energy in a single band at wavelengths longer than the laser or at several wavelengths to take advantage of variations in the emission spectra from different materials. Depending on the emission characteristics of the material being studied, sensing may be more effective under low-ambient illumination conditions.

An alternative approach described by Hemphill et al. (1969) uses passive sensing systems by taking advantage of the Fraunhofer absorption lines in the solar spectrum described in Section 3.5. By sensing in very narrow spectral bands centered on the Fraunhofer lines, the solar-stimulated luminescence can be detected above the reflected background flux. This can be accomplished by measuring the ratio of the irradiance in an absorption line to that in the adjacent continuum and comparing this to the ratio of the exitance in a Fraunhofer line to

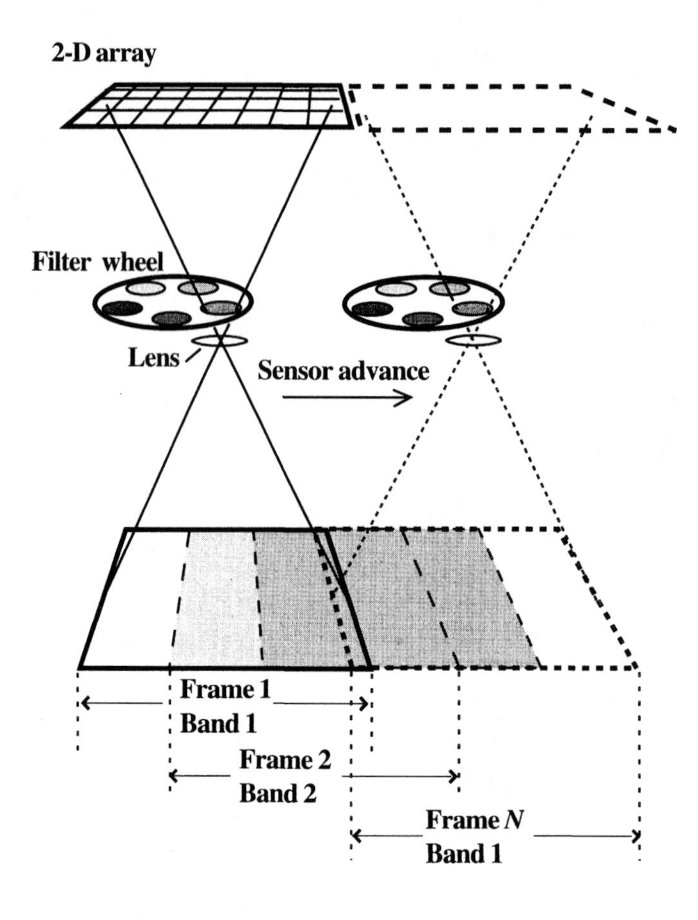

Figure 5.37 Image spectrometer using a 2-D framing array and variable filters.

that from the adjacent continuum, as illustrated in Figure 5.40. We then have the solar irradiance ratio (R_E) given by

$$R_E = \frac{E_b}{E_a} \tag{5.30}$$

and the exitance ratio given by

$$R_M = \frac{E_b r + M_\eta}{E_a r + M_\eta} = \frac{M_b}{M_a} \tag{5.31}$$

where M_η is the exitance due to stimulated luminescence, and the reflectance r is assumed constant over the narrow spectral interval a-b. Clearly, in the absence of luminescence R_E equals R_M and the difference between R_E and R_M is a measure of luminescence. We can formally define a luminescence coefficient that is a material property as:

$$\eta = \frac{M_\eta}{E_a} \tag{5.32}$$

In terms of the measurable values of irradiance and exitance, the luminescence coeffi-

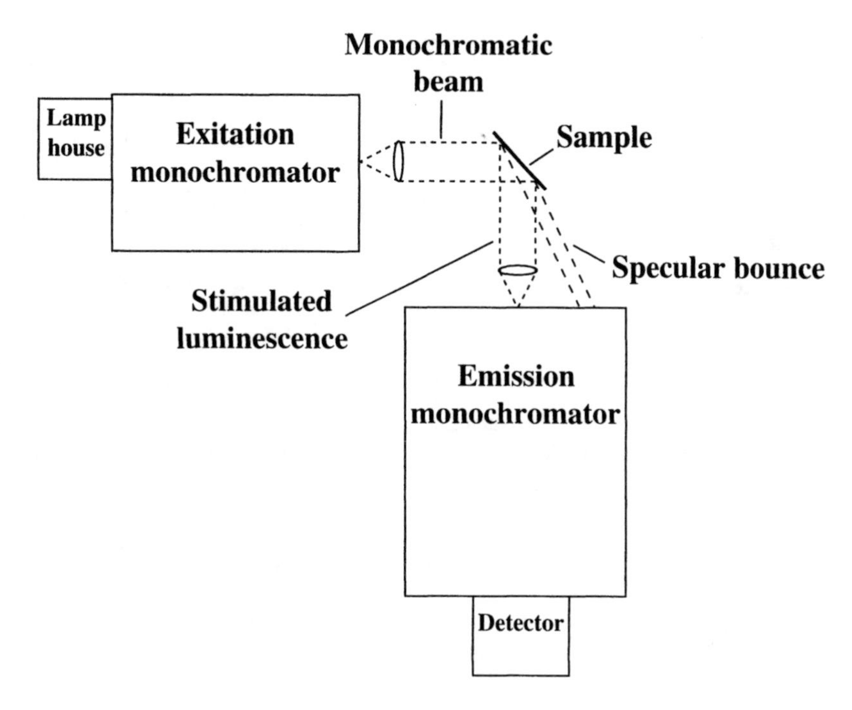

Figure 5.38 Luminescence spectrometer.

cient can be found from

$$E_b r = M_b - \eta E_a \tag{5.33}$$

$$E_a r = M_a - \eta E_a \tag{5.34}$$

and eliminating r and rearranging yields

$$\eta = \frac{E_b M_a - E_a M_b}{E_a E_b - E_a^2} \tag{5.35}$$

Watson and Theisen (1981) describe an imaging instrument called a *Fraunhofer line discriminator* (FLD) designed to produce images of solar-induced fluorescence levels in the Fraunhofer lines. It consists of a 1° field of view radiometer coupled to a line scanner configuration, as shown in Figure 5.41. The radiometer has a pair of coupled chopper blades that let it alternately look at the solar irradiance and then at the exitance from the ground. The flux is then sent through a beam splitter to a pair of Fabry-Perot etalon/filters and then to a detector to sample either the Fraunhofer line (e.g., the sodium D_2 line at 589 nm) or the adjacent continuum. Thermal control of the etalon/filter combination allows sampling in a very narrow (≈0.1 nm) band centered on the absorption line. The second etalon/filter combination is adjusted to sample just far enough away from the line center to avoid the absorption wings. Between the two fields of view (up and down) and the two spectral channels, the FLD instrument is able to acquire estimates of E_a, E_b, M_a, and M_b for use in Eqs. (5.30), (5.31, and (5.35). For simplicity, we are assuming the FLD instrument is flown low enough to minimize any atmospheric effects. These instruments have typically been fairly low resolution due to their narrow spectral band width and are flown at low altitude and low speed for improved

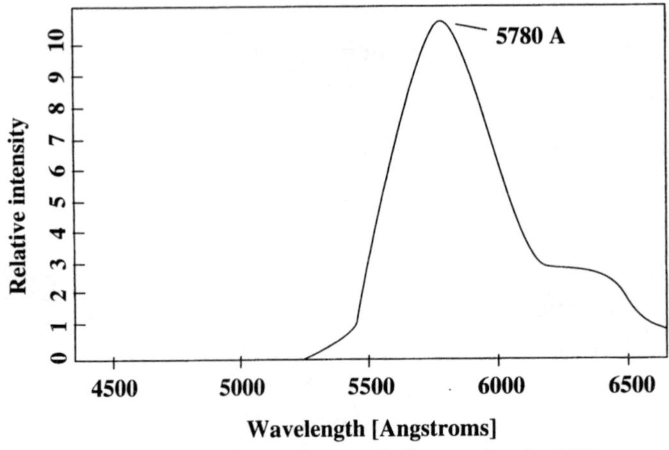

(a) Emission spectrum for an excitation wavelength of 554 nm.

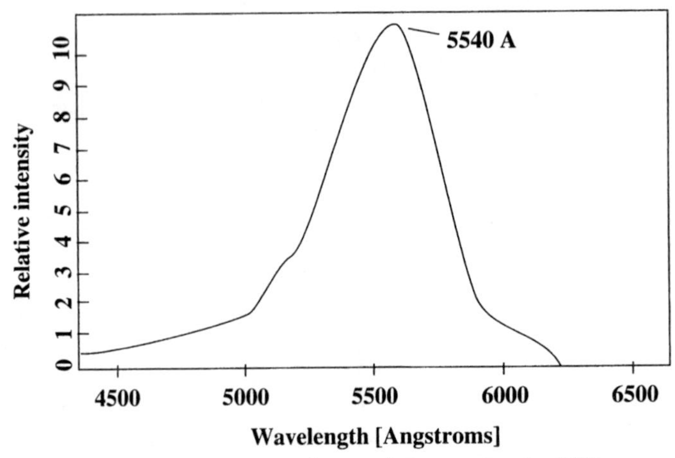

(b) Excitation spectrum for an emission wavelength of 578 nm.

Figure 5.39 Luminescence spectra of rhodamine WT. (Adapted from Hemphill et al., 1969.)

GIFOV and dwell time, respectively.

A major advantage of luminescence imagery is that the luminescence signatures are uncorrelated with reflectance or thermal signatures. As a result, they represent a new dimension in the material identification space that offers exciting opportunities for improved image analysis when used synergistically with traditional methods. Most early work has focused on single line instruments, but it is possible to build multispectral luminescence instruments for improved discrimination of luminescence signatures.

5.7 CALIBRATION ISSUES

In this section, we will be primarily concerned with ways to ensure the internal radiometric calibration of EO sensors. However, we will briefly touch on two other calibration issues. The first is geometric calibration, which can be thought of in two reference frames: How well do we know the relative orientation of one pixel to another (internal orientation)? How well

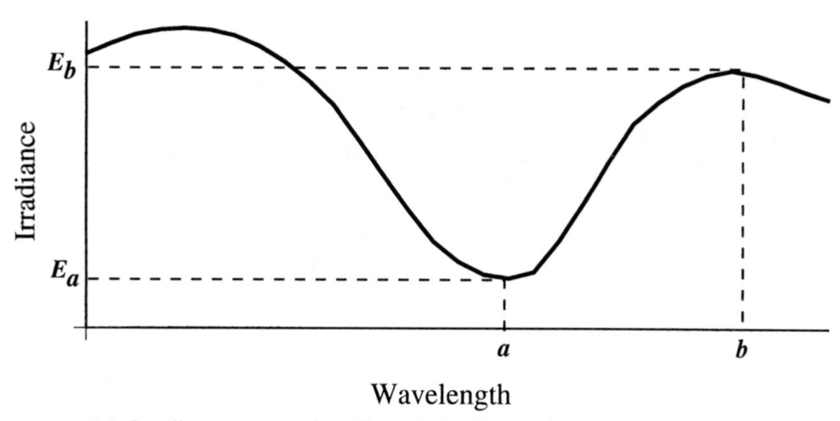

(a) Irradiance spectra in a Fraunhofer line and sampling points.

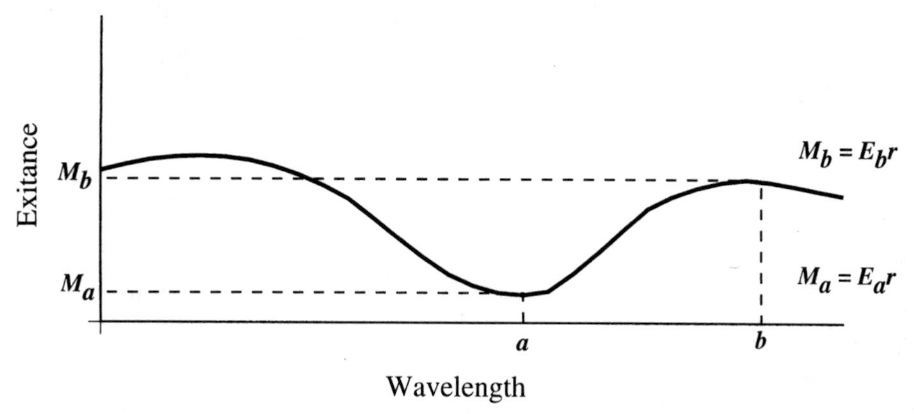

(b) Exitance spectra from the ground in the absence of luminescence.

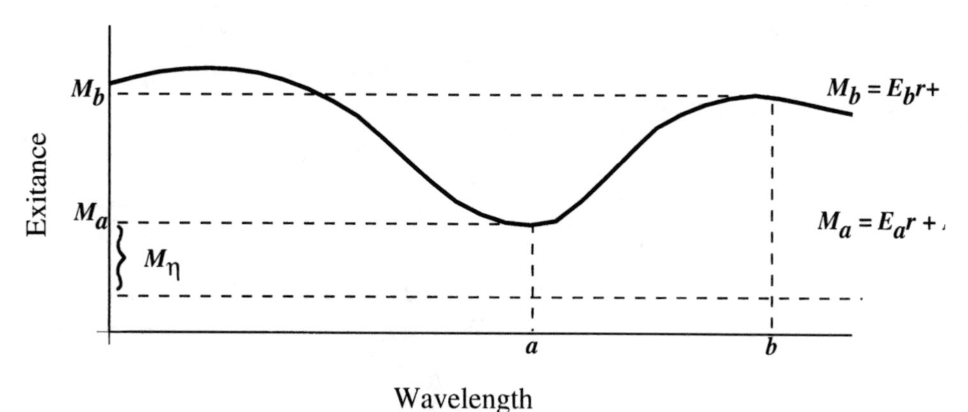

(c) Exitance spectra from the ground when luminescence occurs.

Figure 5.40 Examples of sampling points at a and b near a Fraunhofer line for (a) irradiance spectra, (b) exitance spectra in the absence of luminescence, and (c) with luminescence.

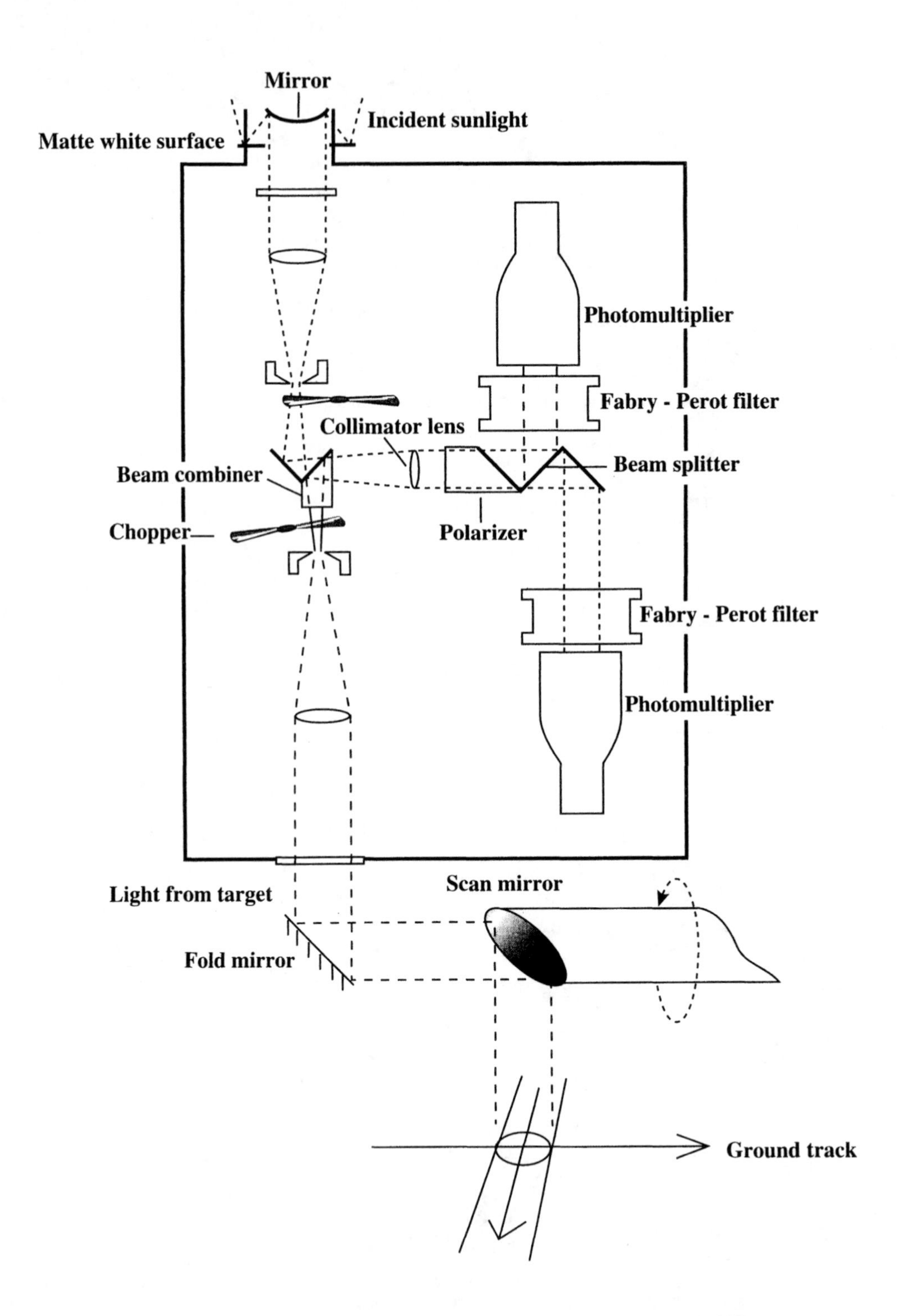

Figure 5.41 Schematic of an FLD instrument. (Adapted from Hemphill et al., 1981.)

can we project the optical axis onto the ground (external orientation)? As we mentioned in the previous section, the inherent internal geometric fidelity is best with a framing system, worse with a push broom, and poorest with a scanner. Recognize that any individual scanner may have very high geometric integrity. The designer just had to work harder to provide it. The overall geometric calibration (external orientation) is the main focus of photogrammetry (cf. Sec. 2.2) and will not be addressed here.

The second calibration topic we will mention briefly is spectral registration. In many multispectral and hyperspectral systems, the integrity of the spectral signature can be compromised due to misregistration of the various spectral bands. When this occurs, the spectral data we associate with one ground spot is actually some mixture of that spot and the surrounding pixels (i.e., we have a blurred spectral signature). This is essentially a geometry and sampling problem very closely related to the internal geometry problem discussed above. This problem is minimized when the spectral data are all sampled simultaneously through a common limiting aperture as illustrated in Figure 5.21. It is worse when pixel shifting occurs such as in whisk-broom and push-broom systems, and worse still when entire images must be resampled for registration (e.g., framing systems). As with geometric integrity, the reader is cautioned that these rules of thumb apply to the generic designs and that specific sensor performance may deviate from them if substantial effort has gone into ensuring spectral integrity.

The primary calibration issue of concern to us in this section is how to provide internal calibration to relate the observed signal (typically in digital counts) to the effective radiance reaching the sensor. This is normally done using on-board calibration standards when high radiometric integrity is required. As an alternative, laboratory preflight calibration and periodic postlaunch calibration assessment programs can be used if no calibration drift is expected or if accuracy specifications are reduced. Both methods require filling the collection aperture with known radiance levels in the laboratory and observing the output from the sensor. So we will begin by considering laboratory radiance sources.

A common procedure for laboratory calibration in the reflective portion of the spectrum is to fill the entire entrance aperture of the sensor with known radiance levels from an integrating sphere. Figure 5.42(a) shows the integrating sphere concept, and Figure 5.42(b) shows a large sphere that NASA has used to calibrate sensors. To produce the uniform radiance field, the integrating sphere must be large compared to the aperture. For large spheres to achieve radiance levels similar to what the sensor will see operationally, a number of sources are distributed inside the sphere outside of the sensor's direct line of sight. Multiple bounces from the approximately Lambertian highly reflective walls produce a nearly uniform radiance field over an area larger than the sensor's entrance aperture. To change the radiance level, sources are turned off or on. This preserves the spectral distribution of the flux. The spectral radiance from the sphere is carefully measured and uniformity verified before use (cf. Barker et al., 1985a).

The sensor output, e.g., digital count (DC), as a function of effective incident radiance (L) can then be evaluated over the expected dynamic range of the sensor. Linearity can be verified and the functional relationship between output signal and radiance defined. For a sensor with linear response, this would take on the form

$$\mathrm{DC} = gL + b \tag{5.36}$$

where g is the sensor gain [counts $\mathrm{w}^{-1}\mathrm{m}^2\mathrm{sr}$] and b is the sensor bias [counts]. The gain (g) and bias (b) are found by linear regression of pairs of DC and L values for various lamp settings in the integrating sphere. This process is repeated for each spectral band and each sensor setting (e.g., each gain setting for a multiple gain sensor).

In order to compute the effective radiance L in Eq. (5.36), the sensor's relative spectral response in each channel is observed using a monochromator and lamp standards [cf. Eq.

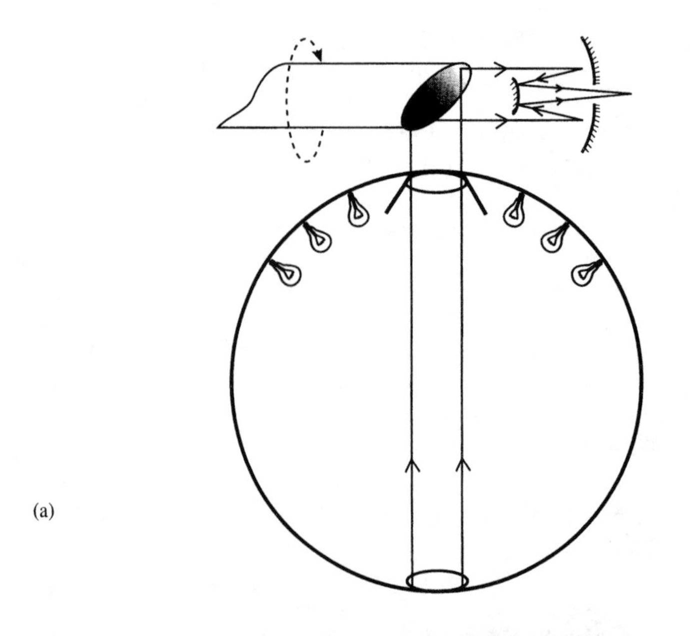

(a)

(b)

Figure 5.42 Integrating spheres used for sensor calibration: (a) sphere design, (b) sphere used in calibration of the AVIRIS Sensor. (Image courtesy of NASA Jet Propulsion Labratory.)

(3.18)]. Note that for this approach to be effective, the spectral distribution of the calibration sphere must simulate the spectral shape of the expected radiance from the earth over the spectral channel of interest (cf. Fig. 3.7).

The calibration procedures in the self-emissive region of the EM spectrum is essentially the same as for the reflective region, except that the source is changed. The most straightforward source to use is a blackbody with a uniformly radiating surface larger than the entrance aperture of the sensor. The radiance from the blackbody is changed by changing its temperature. With the use of thermistors imbedded in the blackbody for monitoring the temperature and as feedback controls in the temperature control circuits, very precise temperatures (and therefore radiance values) can be produced. If the entrance aperture is so large that maintaining thermal uniformity becomes difficult, the radiance field can be produced using a collimator with an exit aperture slightly larger than the entrance aperture of the sensor (cf. Fig. 5.43). The radiance from the collimator can be measured using laboratory instruments or computed if the throughput (*G#*) of the collimator is well known. Collimators are also used in this fashion in the reflective region of the spectrum. With a monochromator's exit aperture as the source, collimators can be used to introduce monochromatic energy for relative spectral response calibration. In general collimators are avoided, where possible, in absolute calibration, since the calibration of the collimator can introduce an additional element of uncertainty (error) in the sensor calibration. The calibration of the thermal channels proceeds in the same fashion as the reflective channels using effective radiance (L) and digital count (DC) values associated with different blackbody temperatures to define the response function, i.e.,

$$\mathrm{DC} = f(L) \tag{5.37}$$

where f is the function relating DC to L, which can often be approximated as a linear function of the form shown in Eq. (5.36). The stability of this type of calibration is very much a function of the temperature of the sensor elements (particularly the optics) and is subject to considerable drift as the sensor temperature varies.

For systems whose response is assumed stable over time, the laboratory calibration may complete the calibration process, and the resultant calibration values (g and b) are used for analysis of image data. Generally some form of validation test is performed to ensure the stability of the calibration over time. For aircraft sensors, the system is often recalibrated after each flight to ensure that pre- and postflight calibrations agree. For satellite systems, the test process can be quite a bit more difficult. Rao and Chen (1993) describe how the stability or instability of the NOAA AVHRR sensors have been monitored using uniform desert regions as stable targets after correction for viewing geometry effects. They point out a major limitation of using only preflight calibration values. Any change in sensor performance will result in a change in the calibration values, of which the user may be unaware. For example, the visible and near IR AVHRR sensors were shown to have a steady decay in gain of several percent per year. As a result, the user of sensors calibrated in this fashion must be very careful of trusting cited calibration values, unless the current validity of those values has been carefully evaluated.

To reduce the uncertainties in radiometric calibration values, many sensors employ some type of on-board calibrator. The most straightforward of these are full-aperture sources located ahead of all of the optical elements in the sensor. This approach is very commonly used in line scanners, as shown in Figure 5.16. By filling the entire aperture and being ahead of all of the sensor's optical elements, this provides essentially a full calibration of the entire sensor that reproduces in flight the laboratory calibration. Most thermal infrared systems, because of their inherent temporal instability, use two on-board blackbodies set to different temperatures (radiance levels). Each time the calibration sources are observed (e.g., every line in a line scanner), a two-point linear calibration can be performed and the sensor gain (g)

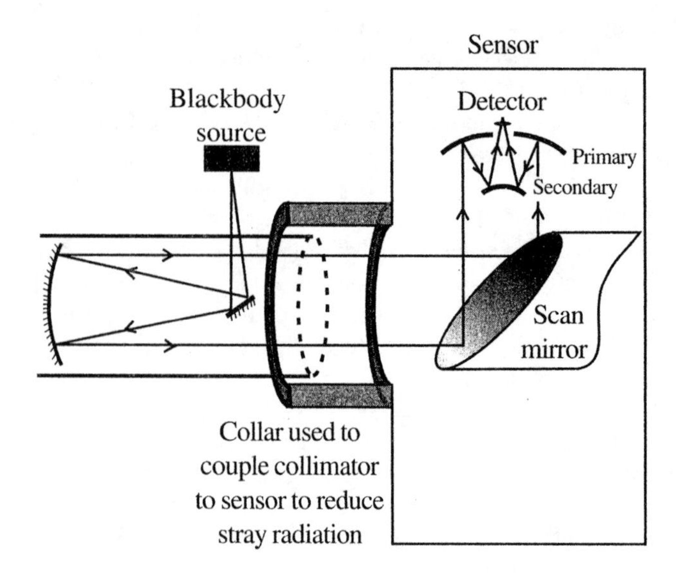

Figure 5.43 Radiometric calibration of a sensor using a calibration source and a collimator.

and bias (b) terms updated. Spectral channels with greater inherent stability may be calibrated less frequently. For example, the radiance from a calibration source in a visible channel may be changed every 100 lines. If five radiance levels are used in the calibrator, it will take 500 lines of data to develop a calibration update.

Having full-aperture calibrators can be very costly from a weight and volume standpoint as the sensor aperture increases. As a result, many alternative methods are used. For example, in the Thematic Mapper, a calibration "wand" is "waived" in the optical path while the scan mirror is reversing direction (cf. Barker et al., 1985b). The calibrator is introduced after the telescope when the beam diameter has been considerably reduced. Using fiber optics from three lamps, it irradiates the detectors with controllable irradiance levels as it is swept across the beam. The temperature of a surface on the calibrator provides one thermal IR calibration point, and a mirror on the calibrator reflects the energy from a small blackbody onto the IR detectors to provide a second calibration level. Thus, the thermal band is fully updated each line.

In all cases where on-board calibrators are used, the calibration sources must themselves be calibrated. This is particularly true when something other than a full-aperture source of known radiance is used (e.g., the TM calibrator). The on-board calibrators are usually calibrated by first carefully calibrating the sensor using laboratory sources as described above. Then, for those sensor conditions, the digital-count-to-radiance relationship is well-known, and any digital count can be converted to a known radiance level. If no changes are made to the sensor configuration except to let it "see" the on-board calibration source(s), then the digital count observed can be converted to the calibration radiance level that will be associated with that source condition. This process is repeated until all source settings (e.g., all calibration lamp levels) are calibrated to known radiance levels. This transfers the calibration of the laboratory radiance standards to the on-board calibration sources. It is important to recognize that the radiance from the standards is not necessarily derived by this process. Instead, we derive the value for what the radiance reaching the front of the optical system would have to be to produce the same flux on the detector as is being produced by the calibration source.

On-board calibration sources would appear to be the solution to sensor drift, instability, and sensor and electronic-aging problems. Because the sensor's calibration is continually updated, we should always be able to convert the observed signal to the effective radiance reaching the sensor. Regrettably, there are several obstacles that may continue to limit this process for certain sensors. The first limitation was introduced in describing the Thematic Mapper's calibration approach. If the calibrator is not ahead of all of the sensor optical elements (including windows), then it cannot account for changes that may occur in the temperature or surface characteristics of the forward surfaces. For space-based systems like the TM, the fouling of the mirror surfaces is expected to be a very slow process having only very minor long-term impact on the sensor. Furthermore, the temperature of the fore optics can be monitored or controlled and slight corrections introduced into the calibration equation to attempt to adjust for changes. The net result is that designs that do not have the sources ahead of all optical elements can be calibrated. However, the calibration can be in error since changes in the system can occur that are not corrected for in the calibration process. As a result, it is important to perform periodic calibration checks on this type of system. For systems that do not use scan mirrors (e.g., push-broom and framing systems), it can be very difficult to view calibration sources on a regular basis. The calibration of these sensors is further complicated by the increased physical size of the focal plane and the increased angular field that must be exposed to a uniform radiance field. If the calibrator flux is not uniform, detector-to-detector calibration error will be introduced which affects not only the radiometric calibration but also the appearance of the images by introducing striping and shading artifacts. One way to calibrate array systems is to periodically turn the pointing mirror, if there is one (e.g., SPOT), so that the sensor looks internally at calibration sources. Alternately, the entire sensor can be turned to look at calibration sources (i.e., deep space or the moon). Both these maneuvers are risky with satellite systems since they place the sensor in a nonimage collection mode where it could become "stuck." Consequently, this type of calibration is usually performed much less frequently than is commonly the case with scan mirror systems. In summary, as Norwood and Lansing (1983) point out, radiometric calibration difficulties increase as we go from scanning, to push broom, to framing sensors.

An additional difficulty plagues all satellite calibration systems (and even some aircraft sensors if flight schedules preclude periodic calibration maintenance). This difficulty arises from decay in the calibrator itself. Calibration relies on the sources producing the same flux levels over time. On satellites, this may be over periods of many years. Over time, the sources, the calibration control electronics, or the calibration monitoring sensors can change. Thus it is necessary to periodically calibrate the calibrator. One way to do this is with a redundant calibrator that is not used operationally. If it is activated on a periodic basis, its prelaunch radiance can be compared to the radiance predicted by the operational calibrators. Any discrepancies can then be used to update the calibration of the calibrator. This redundant approach can be very expensive, and with full aperture calibration it may be nearly impossible to accomplish. In addition, in some cases it just raises the question of whether the backup calibrator or the operational units are in error.

Teillet et al. (1990) suggest that one sensor system can be used to check the calibration on a second. For sensors that have nearly coincident acquisitions, the spectral radiance at the top of the atmospheres along a common line of sight should be the same. By correcting for differences in spectral response functions, some level of cross-calibration can be achieved.

Nearly all the problems described above can be overcome by periodic laboratory calibration updates. For satellite systems, this can be a bit difficult unless we use the world as our laboratory. Thome et al. (1993) describe how this has been done in the reflective region using large, nearly uniform, reflectance surfaces (White Sands, New Mexico) observed through a very clear, dry, well-characterized atmosphere. Under these conditions, they estimate that by measuring the reflectance and modeling the atmosphere, the radiance at the top of the atmosphere can be predicted to approximately 3.5%. When these measurements are made coincident with a satellite overpass, they provide a single well-known calibration point. The white sands represent a high radiance level where DC and radiance are known. If the sensor sees essentially zero flux (e.g., in the back scan) and is biased such that the digital count associated with this flux is observed, then a second point is available to provide a linear calibration update for the system.

Schott (1993) describes how a similar approach can be used to calibrate the thermal channels on satellite systems. A calibrated thermal sensor on an aircraft is used to underfly a satellite sensor with a similar spectral response. The airborne sensor is used to sense the radiance near the top of the atmosphere (e.g., 7 km), and then atmospheric propagation models are used to predict the slight changes in the radiance that the satellite would see at the very top of the atmosphere. By observing several targets over a range of radiance values, a digital-count-to-radiance calibration update for the satellite can be developed. He indicates that, when care is taken in target selection and the use of radiometric correction techniques, the error in top-of-the-atmosphere radiance values using this technique can be reduced to better than 0.9 K. This result is for the LWIR spectral region, with the error expressed in apparent temperature.

Before leaving this subject, we should point out an overarching problem associated with sensor calibration. This is the fundamental difficulty associated with making accurate measurements of radiometric quantities. Grum and Becherer (1979) point out that, unlike other physical quantities where measurements to small fractions of a percent are common, measurements of radiant energy to the 1% level is extremely difficult, and in many cases errors of up to 10% are considered adequate. A great deal of care and experience is required to obtain measurements approaching the 1% level, even in the laboratory, and approaching similar levels on airborne or space-based systems is extremely difficult. We should point out that in the thermal region where temperature measurements can be used to assist the process, absolute levels of better than 1% are readily achieved. However, in this region applications requirements push accuracy needs below the 1% level (cf. Chap. 6). In summary, sensor calibration is still difficult to achieve with high accuracy. As the remote sensing community increasingly tries to make its results repeatable and quantitative, absolute calibration becomes more and more important. Particularly as we move to an era of trying to make long-term environmental studies using remote sensing, it will be critical that we have highly calibrated sensor data. Chapter 6 will address in depth the overall issues associated with the use of calibrated sensor data.

5.8 SENSOR CASE STUDY

In this section, we will perform some simple calculations of sensor performance to attempt to tie together the concepts introduced in this chapter. To make a tractable example, we will not consider a full sensor design but a sensor upgrade such that the degrees of freedom are reduced. Furthermore, to make the example more interesting, we will take an actual problem for our case study. The sensor in question is a thermal infrared line scanner using a design similar to that shown in Figure 5.16. The original system performance specs are listed in Table 5.4. Ground spot size (scene element) will be a critical issue in any upgrade considerations, so we first need to convert system specs to a meaningful ground sample size. To compute the GIFOV of the system, we need to find the IFOV and know the flying height. The IFOV can be expressed as:

$$\text{IFOV} = \frac{l_o}{f} = \frac{l_o}{F\#d} = \frac{0.75[\text{mm}]}{4 \cdot 76.2[\text{mm}]} = 2.5 \cdot 10^{-3}[\text{radians}] \tag{5.38}$$

Table 5.4 Specification of the Original IR Line Scanner Used in the Case Study

Optics	Dall-Kirkham-Cassegrain
Aperture	d = 3 inches
F number	$F = 4$
Mirror scan rate	$S_o = 110$ Hz
8 - 14 μm filter	Filter factor = 0.7
IR Tran Window	Filter factor = 0.9
Detector specifications, - HgCdTe	
•Field of view as defined by cold shield	20°
•Length of side	l = 0.75 mm
•Specific detectivity	$D_\lambda^* = 3.31 \cdot 10^{10}$ cm Hz$^{1/2}$w^{-1}
•Responsivity	R = 1306 v/w
System noise expressed as apparent temperature	N_o = 0.25-0.35 K (gain dependent)
Total field of view	FOV = 120°
Aircraft speed range	V = 100 to 180 mph
Derived values	
Focal length	$f = F\#\cdot d = 4\cdot 76.2$ mm = 304.8 mm
Instantaneous field of view	IFOV = l/f = 0.75 mm/305.5 mm = $2.5\cdot 10^{-3}$=2.5 milliradians
Bandwidth	$\Delta f = 0.14\cdot 10^6$ Hz
Throughput	G# = 40.4sr^{-1}
NEΔT (optics and detector)	$NE\Delta T_0 = 0.09$ K
Noise equivalent power	$NEP_0 = 8.5\cdot 10^{-10}$ w

where l_0 is the side dimension of the square detector, f is the focal length, d is the diameter of the entrance aperture (3 inches), and $F\#$ is defined to be f/d. This 2.5-milliradian system was usually flown at 1000 ft, AGL (H) to yield a GIFOV of

$$\text{GIFOV} = H \cdot \text{IFOV} = 1000[\text{ft}] \cdot 2.5 \cdot 10^{-3} = 2.5[\text{ft}] \tag{5.39}$$

At the time in question, a primary use of the line scanner was to study heat loss from residential structures using data from nighttime flyovers during the winter. A combination of engineering studies and pressure from competitors had defined a requirement for a scanner with a GIFOV at nadir of 1 ft, a system noise equivalent temperature difference of less than 0.3 K, and a ground swath of at least 2000 ft. From an engineering standpoint, the easiest solution was to buy a commercially available system that met the specifications. The management response was: *that solution exceeded the value the company placed on the upgraded system by more than an order of magnitude, and a more affordable solution was required.* So with very limited resources, our task was to identify and implement an upgrade to the system to meet the improved specifications. We already met the noise and FOV specs, but we needed to improve on GIFOV. The easiest solution was to fly lower. However, to meet the GIFOV spec, we would have had to recommend flying the system at 400 ft. This is below safety limits for operational collections and would not provide an adequate ground swath. As part of this consideration, it was determined that the current operational flying height of 1000 ft was also a practical minimum. This led to a computation that the system IFOV must be

$$\mathrm{IFOV} = \frac{\mathrm{GIFOV}}{H} = \frac{1\ [\mathrm{ft}]}{1000\ [\mathrm{ft}]} = 1 \cdot 10^{-3} [\mathrm{radians}] \tag{5.40}$$

or less. A 1-milliradian system would have a detector size of

$$l_n = \mathrm{IFOV} \cdot f = 1 \cdot 10^{-3} \cdot 304.8 [\mathrm{mm}] = 0.3 [\mathrm{mm}] \tag{5.41}$$

where l_n is the side dimension of the detector, and the subscript n will be used as necessary to designate the new system and o the original system. Detectors of this size were readily available in HgCdTe, and this still appeared to be the most appropriate detector material. Improvements in fabrication technology allowed the vendors to spec an effective D^* value for a new detector of $3.3 \cdot 10^{10}$ [$\mathrm{w}^{-1}\mathrm{cm\ Hz}^{1/2}$] and an effective responsivity of 6300 [vw^{-1}]. They also agreed that (for a small fee) they would hand-select to try to exceed these values for a single detector. After considering a variety of alternatives, we decided to evaluate whether the system could be improved sufficiently by using a smaller detector. As we will see, this is not as simple a solution as it may at first sound.

Knowing that the flux on the detector would decrease in proportion to the area (6.25 times) and that we could only count on a 4 times increase in responsivity and possibly no increase in specific detectivity, a more detailed assessment to determine if the current system was detector noise limited seemed in order. This would help us determine if we needed to consider only detector specifications or if improvement in the preamplifier (preamp) noise specifications should also be considered. We also need to determine if the scan rate is sufficient to allow full coverage with the smaller GIFOV. The minimum aircraft speed (V) is 100 [mph]. The ground advance between lines (X_o) is:

$$X_o = Vt_o = \mathrm{V}/s_o = \frac{100\ [\mathrm{mph}]}{110\ [\mathrm{Hz}]} = 1.33\ [\mathrm{ft}] \tag{5.42}$$

where t_o is the time per revolution of the scan mirror and s_o is the scan rate expressed in revolutions per second. At the original scan rate and minimum air speed, the improved scanner will have a gap between scan lines of 0.33 ft when flown at 1000 ft. To have adjacent scan lines, the scan mirror would have to spin at

$$s_n = \frac{V}{X_n} = \frac{100\ [\mathrm{mph}]}{1\ [\mathrm{ft}]} = 147\ [\mathrm{revolutions\ sec}^{-1}] \tag{5.43}$$

This increase in scan mirror rate would require a change in the drive motor and would reduce the dwell time, further increasing our already substantial concerns about meeting the noise specifications. For the present, we will plan on making this modification, but consider it a tradeoff option.

While we are considering scan speeds, we should also analyze the frequency response required by the original and the upgraded system since we will need this for the noise calculations. First we compute the time per sample, or the dwell time, as the time it takes the mirror to sweep out one IFOV, i.e.,

$$t_{so} = t_o\left(\frac{\#\,\text{IFOV}}{\text{rev}}\right)^{-1} = \frac{1}{s_o}\left(\frac{\#\,\text{IFOV}}{\text{rev}}\right)^{-1}$$

$$= \frac{1}{110}[\text{sec}]\left(\frac{2\pi\,[\text{radians}]}{2.5\cdot 10^{-3}\,[\text{radians}]}\right)^{-1} = 3.6\cdot 10^{-6}[\text{sec}] \tag{5.44}$$

The frequency bandwidth (f_o) is computed as:

$$f_o = \frac{1}{2t_{so}} = \frac{1}{2\cdot 3.6\cdot 10^{-6}[\text{sec}]} = 0.14\cdot 10^{6}[\text{Hz}] \tag{5.45}$$

For the proposed 1-milliradian system with the increased scan rate, the dwell time (t_{sp}) would be

$$t_{sp} = t_p\left(\frac{\#\,\text{IFOV}}{\text{rev}}\right)^{-1} = \frac{1}{s_n}\left(\frac{\#\,\text{IFOV}}{\text{rev}}\right)^{-1}$$

$$= \left(\frac{1[\text{sec}]}{147}\right)^{-1}\left(\frac{2\pi\,[\text{radians}]}{1\cdot 10^{-3}[\text{radians}]}\right)^{-1} = 1.1\cdot 10^{-6}[\text{sec}] \tag{5.46}$$

where t_p is the time per revolution at the increased speed, and the subscript p is used to indicate proposed values.

The resulting frequency bandwidth would be

$$f_p = \frac{1}{2t_{sp}} = \frac{1}{2\cdot 1.1\cdot 10^{-6}[\text{sec}]} = 0.45\cdot 10^{6}[\text{Hz}] \tag{5.47}$$

If the scan rate is not increased, we would have a dwell time t_{sn} of

$$t_{sn} = t_0\left(\frac{\#\,\text{IFOV}}{\text{rev}}\right)^{-1} = \frac{1}{s_o}\left(\frac{\#\,\text{IFOV}}{\text{rev}}\right)^{-1}$$

$$= \frac{1}{110[\text{Hz}]}\left(\frac{2\pi[\text{radians}]}{1\cdot 10^{-3}[\text{radians}]}\right)^{-1} = 1.45\cdot 10^{-6}[\text{sec}] \tag{5.48}$$

with a corresponding frequency bandwidth f_n of

$$f_n = \frac{1}{2t_{sn}} = 0.35\cdot 10^{6}[\text{Hz}] \tag{5.49}$$

As seen in Figure 5.14, the mercury-cadmium telluride detectors can pass these signals without a problem. By trying to pass a 450-kilohertz signal (cf. Eq. (5.47)) through the scanner recorder system, it was found that the preamplifier was the only element that would not pass this signal. A new preamp was planned which would roll off at the required detector bandwidth, so this was not a problem.

We needed to know the NEΔT of the detector relative to the noise performance of the sensor. To begin, we solve for the NEP of the original system as:

$$\text{NEP}_o = \frac{1}{D} = \frac{1}{D*}\sqrt{A_o}\sqrt{f_o} = \frac{0.075[\text{cm}]\sqrt{0.14\cdot 10^{6}[\text{Hz}]}}{3.3\cdot 10^{10}[\text{cmHz}^{1/2}\text{w}^{-1}]}$$

$$= 8.5\cdot 10^{-10}[\text{w}] \tag{5.50}$$

where A_o is the area of the original detector. The NEΔT of the original system can then be expressed as:

$$\text{NE}\Delta\text{T}_o = \frac{\text{NEP}_o}{\Delta\Phi / \Delta T} \tag{5.51}$$

where $\Delta\Phi/\Delta T$ is the change in flux on the detector associated with a unit change in temperature. To solve for this value, we first define the change in radiance on the sensor per unit change in temperature to be

$$\frac{\Delta L}{\Delta T} = \int_8^{14} L_{\lambda 301} d\lambda - \int_8^{14} L_{\lambda 300} d\lambda = 8.41 \cdot 10^{-5} [\text{wcm}^{-2} sr^{-1} \text{K}^{-1}] \tag{5.52}$$

where we are implicitly assuming all the specifications are for a 300-K target.

Using the $G\#$ concept from Eq. (5.7), we can describe the change in irradiance on the detector per unit change in temperature as:

$$\frac{\Delta E}{\Delta T} = \frac{\Delta L}{\Delta T} \frac{1}{G\#} \tag{5.53}$$

The change in flux per unit change in temperature for the original sensor can then be expressed as:

$$\frac{\Delta\Phi}{\Delta T} = \frac{\Delta E}{\Delta T} A_o = \frac{\Delta L}{\Delta T} \frac{1}{G\#} A_o \tag{5.54}$$

The $G\#$ for the original system can be approximated by:

$$G\#_o \cong \frac{4F\#^2}{\tau_k \tau_o \pi} = \frac{4 \cdot 4^2}{0.8(0.9 \cdot 0.7)\pi} = 40.4[\text{sr}^{-1}] \tag{5.55}$$

where τ_k is the transmission loss due to obscuration by the secondary optic in the Cassegrain telescope (0.8 for this system), τ_o is transmission loss due to the bandpass filter (0.7), and the window on the dewar (0.9) in the original system.

Inserting this result into Eq. (5.54) yields

$$\begin{aligned}\frac{\Delta\Phi_o}{\Delta T} &= 8.41 \cdot 10^{-5} [\text{wcm}^{-2} \text{sr}^{-1} \text{K}^{-1}] \cdot \frac{1}{40.4[\text{sr}^{-1}]} \cdot (0.075[\text{cm}])^2 \\ &= 1.17 \cdot 10^{-8} [\text{wK}^{-1}]\end{aligned} \tag{5.56}$$

which combined with Eq. (5.51) yields

$$\text{NE}\Delta\text{T}_o = \frac{\text{NEP}_o}{\frac{\Delta\Phi_o}{\Delta T}} = \frac{8.5 \cdot 10^{-10} [\text{w}]}{1.17 \cdot 10^{-8} [\text{wK}^{-1}]} \cong 0.1[\text{K}] \tag{5.57}$$

Since the observed system noise N_o is 0.25 to 0.35 [K] and depends on the system gain, we conclude that the system is not quite detector noise limited. If we assume that the bulk of the remaining system noise is from the preamplifier, we can solve for the preamp noise as:

$$N_{po} = (N_o^2 - \text{NE}\Delta\text{T}_o^2)^{1/2} = \left((0.25 \text{ to } 0.35)^2[\text{K}] - (0.1\text{K})^2\right)^{1/2}$$
$$= 0.23 - 0.33[\text{K}] \tag{5.58}$$

In any new system, we must recognize that in using a similar preamp design with components with comparable specifications, we must anticipate similar noise levels for similar signal levels. Because the preamp noise increases with gain, we must expect that if we have lower signal levels, we will have even greater noise associated with the preamp.

We can perform a quick assessment of the relative signal levels expected from the new detector (S_p) compared to the original (S_o) as follows:

$$\frac{S_p}{S_o} = \frac{A_n}{A_o}\frac{R_p}{R_o}\frac{t_{sp}}{t_{so}}\frac{\tau_n}{\tau_o} \tag{5.59}$$

since we know that the signal out of a sensor will be proportional to area, responsivity, dwell time, and the transmission of the optics. In our case, knowing that we need more signal, we plan to improve the filter factor by having the bandpass filter incorporated into the dewar window for an overall transmission of 0.8. The expected signal ratio using the vendor's estimated responsivity would be:

$$\frac{S_p}{S_o} = \frac{(0.03[\text{cm}])^2}{(0.075[\text{cm}])^2}\frac{6300[\text{vw}^{-1}]}{1306[\text{vw}^{-1}]}\frac{1.1 \cdot 10^{-6}[\text{sec}]}{3.6 \cdot 10^{-6}[\text{sec}]}\frac{(0.8)}{(0.9 \cdot 0.7)} = 0.3 \tag{5.60}$$

This means that we would require nearly 3 times as much gain from the preamp to achieve the same signal levels. This places us at the high end of the preamp noise range.

At this point, it was determined to try to maintain as much signal level as possible so that the scan mirror motor speed would not be changed. This decision was augmented by concerns that scan mirror deformation at higher speed might degrade image quality (cf. Feng et al., 1994). Based on this decision, we can compute the expected NEΔT of the new detector using the vendor's predicted performance for D^*. The expected noise equivalent power would be:

$$\text{NEP}_p = \frac{\sqrt{A_n}\sqrt{f_n}}{D^*_p} = \frac{.03[\text{cm}]\sqrt{0.35 \cdot 10^6[\text{Hz}]}}{3.3 \cdot 10^{10}[\text{cmHz}^{1/2}\text{w}^{-1}]} = 5.38 \cdot 10^{-10}[\text{w}] \tag{5.61}$$

and the expected noise equivalent change in temperature due to the detector specs would be:

$$\text{NE}\Delta\text{T}_p = \frac{\text{NEP}_p}{\Delta\Phi / \Delta T} = \frac{\text{NEP}_p G\#_n}{\Delta L / \Delta T \cdot A_n}$$
$$= \frac{5.38 \cdot 10^{-10}[\text{w}]32[\text{sr}^{-1}]}{8.41 \cdot 10^{-5}[\text{wcm}^{-2}\text{sr}^{-1}\text{K}^{-1}](0.03[\text{cm}])^2} = 0.2[\text{K}] \tag{5.62}$$

where the $G\#$ of the new system is

$$G\#_n = \frac{4F\#^2}{\tau_K \tau_o \pi} = \frac{4 \cdot 4^2}{0.8 \cdot 0.8\pi} = 32[\text{sr}^{-1}] \tag{5.63}$$

If we predict a preamp noise (N_{pp}) of 0.3 [K] (due to low signal and high gain), we would have a system noise expressed in apparent temperature of

$$N_p = \left[(\mathrm{NE\Delta T}_p)^2 + (N_{pp})^2\right]^{1/2}$$
$$= \left[(0.2[\mathrm{K}])^2 + (0.3[\mathrm{K}])^2\right]^{1/2} = 0.4[\mathrm{K}] \tag{5.64}$$

This expected performance is slightly worse than our specification. In addition, we have been doing simplified calculations with mean, or in some cases peak, values for our figures of merit, rather than spectrally integrated expected values over the bandpass. This will tend to exaggerate our performance figures. Nevertheless, we decided to proceed based on the detector vendor's efforts to exceed specifications and our plans to improve the preamp to reduce noise. Finally, we held out as a rather costly fallback option, the possibility of using three or four detectors and a time delay and integration (TDI) approach. This would involve three detectors arranged on the focal plane in the along-scan direction in much the same way the Landsat MSS multispectral detectors are arranged (cf. Fig. 5.24). In this case, rather than collecting multispectral images of the scene, we acquire multiple copies of the same scene that could be pixel-shifted (time-delayed) to be in registration and then averaged (integration) to form an image with reduced noise. The noise in the resulting image will be reduced by the square root of the number of detector elements. In our case with three detectors, this should yield.

$$N_p(\mathrm{TDI}) = \frac{N_p}{\sqrt{3}} = \frac{0.4[\mathrm{K}]}{\sqrt{3}} = 0.23[\mathrm{K}] \tag{5.65}$$

which would be an acceptable noise figure for this system. The relatively large cost of this option was due to two incremental detectors, preamplifiers and recorder input electronics, as well as the cost of having the detectors mounted in the dewar. While not large in absolute dollars, it would nearly triple the cost of this supposedly low-budget upgrade.

So, a much younger version of your author held his breath and ordered the single detector, hoping to keep the costs and his fledgling career on track. Knowing we would not be detector noise–limited, we asked the vendor to look in particular for a high-responsivity detector and to push the $D*$ value up if possible. The resultant detector specifications were a specific detectivity (D_n*) of $9.6 \cdot 10^{10}[\mathrm{cmHz}^{1/2}\mathrm{w}^{-1}]$ and a responsivity (R_n) of $11.9 \cdot 10^3[\mathrm{v/w}]$.

These numbers made the single detector option continue to look feasible. The $\mathrm{NE\Delta T}_n$ of the new detector should be:

$$\mathrm{NE\Delta T}_n = \frac{\mathrm{NEP}_n}{\frac{\Delta\Phi}{\Delta T}} = \frac{\mathrm{NEP}_n G\#}{\Delta L / \Delta T \cdot A_n}$$
$$= \frac{1.85 \cdot 10^{-10}[\mathrm{w}][32\mathrm{sr}^{-1}]}{8.41 \cdot 10^{-5}[\mathrm{wcm}^{-2}\mathrm{sr}^{-1}\mathrm{K}^{-1}](0.03[\mathrm{cm}])^2} = 0.1\mathrm{K} \tag{5.66}$$

where

$$\mathrm{NEP}_n = \frac{\sqrt{A_n}\sqrt{f_n}}{D*_n} = \frac{0.03[\mathrm{cm}]\sqrt{0.35 \cdot 10^6[\mathrm{Hz}]}}{9.6 \cdot 10^{10}[\mathrm{cmHz}^{1/2}\mathrm{w}^{-1}]} = 1.85 \cdot 10^{-10}[\mathrm{w}] \tag{5.67}$$

This indicates that we should see similar noise levels from the new detector as from the original.

The relative signals from the new detector compared to the original should be

(a)  (b) 

(c) 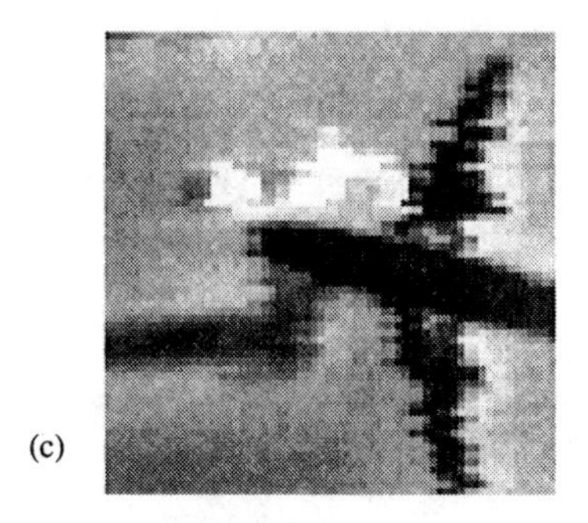 (d) 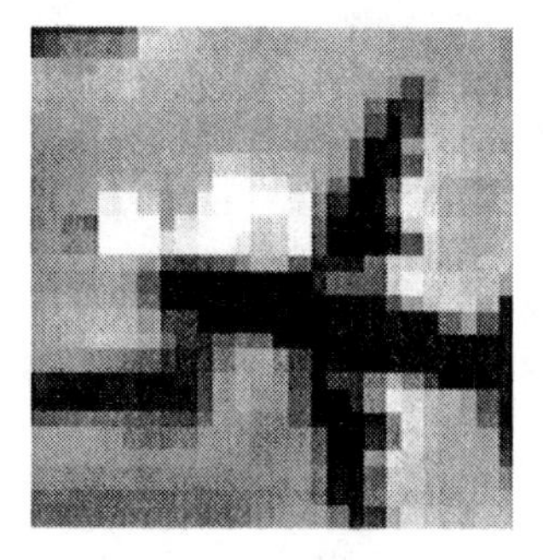

Figure 5.44 Portion of an LWIR image acquired with an infrared line scanner (a) after sensor upgrade, (b) simulation of how the scene would have been imaged by the original system. Enlargements (c) and (d) show a portion of image (a) and (b), respectively. Note that the line-to-line variations have not been conpensated for in this image. (Image courtesy of RIT's Digital Imaging and Remote Sensing Laboratory.)

$$\frac{S_n}{S_o} = \frac{A_n}{A_o}\frac{R_n}{R_o}\frac{t_{sn}}{t_{so}}\frac{\tau_n}{\tau_o}$$
$$= \frac{(0.03[\text{cm}])^2}{(0.075[\text{cm}])^2}\frac{11.9\cdot 10^3[\text{v / w}]}{1306[\text{v / w}]}\frac{1.45\cdot 10^{-6}[\text{sec}]}{3.6\cdot 10^{-6}[\text{sec}]}\frac{0.8}{(0.9)(0.7)} = 0.75 \quad (5.68)$$

indicating that gain levels from the new preamp will only need to be slightly higher than the old. The actual observed noise level for the entire system after assembly was approximately 0.3 K, just meeting our design specifications.

Figure 5.44 shows a portion of an image acquired with the upgraded system, along with a simulated image of how the same scene would have appeared if imaged by the original system. Note that in addition to the esthetic differences, small details would not have been observed in the original scene. For example, the shape/structure of the hose/cable running from the support vehicle to the airplane is difficult to see in the 2.5-milliradian image.

5.9 REFERENCES

Accetta, J.S., & Shumaker, D.L., eds. (1993). *The Infrared and Electro-Optical Systems Handbook,* SPIE Optical Engineering Press, Bellingham, WA.

Barker, J.L., Abrams, R.B., Ball, D.L., & Leung, K.C. (1985a). "Radiometric calibration and processing procedures for reflective bands on Landsat-4 Protoflight Thematic Mapper." Proceeding of the Landsat-4 Science Characterization Early Results Symposium, NASA, Goddard, 1983, NASA Conference Publication 2355, pp. 47-86.

Barker, J.L., Ball, D.L., Leung, K.C., & Walker, J.A. (1985b). "Prelaunch absolute radiometric calibration of the reflective bands on the Landsat-4 Protoflight Thematic Mapper." Proceedings of the Landsat-4 Science Characterization Early Results Symposium, NASA Goddard, 1983, NASA Conference Publication 2355, pp. 277-372.

Barker, J.L., ed. (1985). "Proceedings of the Landsat-4 Science Characterization Early Results Symposium," Feb. 1983, NASA Goddard, NASA Conference Publication 2355, NASA, Goddard.

Budde, W., (1983). *Optical Radiation Measurements: Vol. 4, Physical Detectors of Optical Radiation.* Academic, NY

Chen, H.S. (1985). *Space Remote Sensing Systems. An Introduction.* Academic, Orlando, FL.

Colwell, R.N., ed. (1983). *Manual of Remote Sensing,* 2d ed., American Society of Photogrammetry and Remote Sensing, Bethesda, MD.

Dereniak, E.L., & Crowe, D.G. (1984). *Optical Radiation Detectors.* Wiley, NY.

Feng, X., Schott, J.R., & Gallagher, T.W. (1994). Modeling the performance of a high-speed scan mirror for an airborne line scanner. *Optical Engineering*, Vol. 33, No. 4, pp. 1214-22.

Goetz, A.F.H., Vane, G., Solomon, J.E., & Rock, B.N. (1985) Imaging spectrometry for earth remote sensing. *Science*, Vol. 228, pp. 1147-53.

Grum, F., & Becherer, R.J. (1979). *Optical Radiation Measurements: Vol. 1 Radiometry.* Academic, NY.

Gunning, W.J. (1981). Electro-optically tuned spectral filters: a review. *Optical Engineering*, Vol. 20, No. 6, pp. 837-845.

Hecht, E. (1987). *Optics.* 2d. ed., Addison-Wesley, Reading, MA.

Hemphill, W.R., Stoertz, G.E., & Markle, D.A., (1969). "Remote sensing of luminescent materials." Proceedings of the 6th International Symposium of Remote Sensing of Environment, University of Michigan, Ann Arbor, MI.

Hurter, F., & V.C., Driffield (1890). Photo chemical investigations and a new method of determination of the sensitivity of photographic plates. *Journal of the Society of Chemical Industrials* No. 9, pp. 455-469.

James, T.H. (1977). *Theory of the Photographic Process*, 4th ed., Macmillan, NY.

Kramer, H.J. (1992). *Earth Observation Remote Sensing: Survey of Missions and Sensors.* Springer-Verlag, Berlin.

Markham, B.L., & Barker, J.L., eds. (1985). Special Landsat image data quality assessment (LIDQA) *Photogrammetric Engineering and Remote Sensing*, Vol. 51, No. 9.

Rao, C.R.N., & Chen, J. (1993). Calibration of the visible and near infrared channels of the advanced very high resolution radiometer (AVHRR) after launch. Proceedings of the SPIE, Vol. 1938, pp. 56-66.

Rogatto, W.D., ed. (1993). "Electro-optical components." In Vol. 3: *The Infrared and EO System Handbook*, J.S. Accetta, & D.L. Shumaker eds. SPIE Optical Engineering Press, Bellingham, WA.

Schott, J.R. (1993). "Thermal infrared calibration of satellite sensors." Proceedings of the Workshop on Atmospheric Correction of Landsat Imagery, Geodynamics, Torrance, CA.

Slater, P.N., Biggar, S.F., Holm, R.G., Jackson, R.D., Mao, Y., Moran, M.S., Palmer, J., & Yuan, B. (1987). Reflectance and radiance-based methods for in-flight absolute calibration of multispectral sensors. *Remote Sensing of Environment*, Vol. 22, pp. 11-37.

Teillet, P.M., Slater, P.N., Ding, Y., Santer, R.P., Jackson, R.D., & Moran, M.S. (1990). Three methods for absolute calibration of the NOAA AVHRR sensors in-flight. *Remote Sensing of Environment*, Vol. 31, pp. 105-120.

Thome, K.J., Biggar, S.F., & Slater, P.N. (1993). Recent absolute radiometric calibration of Landsat-5 TM and its application to the atmospheric correction of aster in the solar reflective region. Proceedings of the Workshop on Atmospheric Correction of Landsat Imagery, Published by Geodynamics Corporation, Torrance, CA.

Vane, G., Green, R.O., Chrien, T.G., Enmark, H.T., Hansen, E.G., & Porter, W.M. (1993). The airborne visible/infrared imaging spectrometer (AVIRIS). *Remote Sensing of Environment*, Vol. 44, pp. 127-143.

Watson, R.D. (1981). Quantification of luminescence intensity in terms of rhodamine WT standard. Workshop on Applications of Luminescence Techniques to Earth Resources Studies, W.R. Hemphill & M. Settle, eds. LPI Tech Rpt. 81-03, Lunar and Planetary Institute, Houston, TX, pp. 19-21.

Watson, R.D., & Theisen, A.F., (1981). Electronic and optical modification of the engineering model FLD and the evaluation of peripheral equipment. Workshop on Applications of Luminescence Techniques to Earth Resources Studies, W.R. Hemphill & M. Settle, eds. LPI Tech Rpt. 81-03, Lunar and Planetary Institute, Houston, TX, pp. 15-18.

Wolfe, W.L. & Zissis, G.J., eds. (rev. 1985). *The Infrared Handbook.* Environmental Research Institute of Michigan, Ann Arbor, MI.

Wrigley, R.C., Hlavka, C.A., Card, D.H., & Buis, J.S. (1985). Evaluation of Thematic Mapper inter-band registration and noise characteristics. *Photogrammetric Engineering and Remote Sensing*, Vol. 51, No. 9, pp. 1417-25.

CHAPTER

6

Atmospheric Calibration: Solutions to the Governing Equation

In Chapter 4, we developed the governing equation Eq. (4.61) for the radiance reaching a remote sensing system, "the big equation." In the last chapter, we discussed these remote sensing systems and how they could be calibrated using laboratory or on-board calibrators. As a result of this calibration process, we will assume that when required we can convert the remotely sensed signal (i.e., digital count, voltage, or film density) to radiance reaching the front of the sensor. This means that for calibrated sensors, we will assume that we can solve for the radiance reaching the sensor at any image location. This yields the radiance value on the left-hand side of the big equation. In this chapter, we will address procedures for solving for the remaining variables in the big equation so that we can associate ground temperatures or reflectance values with each point (pixel) in an image. This process of atmospheric calibration is one of the most difficult tasks facing the remote sensing community. At present, no single approach has proved sufficiently simple, accurate, and robust enough to be widely accepted and operationally used. As a result, a number of methods for atmospheric calibration exist which are useful for a particular type of problem or accuracy level. In this chapter, we will look at a range of calibration approaches aimed at covering most remote sensing situations. These approaches are sometimes referred to in the literature as atmospheric correction techniques. Because our overall goal is to have a "calibration" system that relates image digital counts to surface parameter (e.g., reflectance or temperature) we will refer to this process as atmospheric calibration.

The reader should keep in mind the fundamental assumption behind the value of atmospheric calibration, i.e., that we have identified a relationship between the quantitative value of one or more remotely sensed variables (e.g., temperature and/or spectral reflectance) and one or more parameters of interest (e.g., vegetation rigor). While the functional form of this relationship is very application-specific and, therefore, beyond the scope of this effort, we will cite a few examples here to provide motivation for this process. Gordon (1983) presents a relationship between the reflectance of the ocean and water quality parameters. Piech et al. (1978) cite a correlation between the reflectance of water in the visible region and several water quality variables (e.g., chlorophyll, suspended solids, and gelbstoffe), and used this relationship to map chlorophyll variations in fresh water using calibrated reflectance values. Piech and Walker (1974) show relationships between soil texture and moisture and reflectance values, and suggest how moisture and texture signatures can be separated using broad-band spectral reflectance values. The geologic community has long used laboratory reflectance spectra to assist in the characterization of rock samples, and Marsh and Lyon (1980) show the correlation between surface data and satellite-derived reflectance values. Numerous studies have shown relationships between reflectance and various vegetation parameters, including chlorophyll concentration (cf. Thomas and Gausman, 1977), water content (cf. Thomas et al., 1966) and Leaf Area Index (LAI) (cf. Suits, 1972). In order to fully utilize these relationships, it is first necessary to remove the effects of the atmosphere.

Similarly, Schott and Wilkinson (1982) discuss how absolute temperature measurements can be related to heat loss from buildings. Heilman and Moore (1980) discuss how temperature is related to soil moisture, and numerous studies have shown relationships between water temperature and water circulation (cf. Legeckis, 1978). This small sampling of studies shows that laboratory and field reflectance (and/or temperature) values have been correlated with a variety of parameters of interest to applications scientists. The development and use of these relationships is a major undertaking, with many application-specific factors confusing these relationships. As a result, we will not address the development of these application-specific relationships here. Instead, we want to stress that one of the goals of the remote sensing community is to be able to make quantitative measurements of reflectance and temperature so that these measurements can be directly used by the applications scientists developing laboratory and field correlations between these variables and application specific parameters (e.g., vegetation condition, soil moisture, etc.). In order to quantitatively measure reflectance or temperature, we must first account for the effect of the atmosphere on the observed radiance.

In this chapter, we will consider atmospheric calibration methods employing field measurements (ground truth) at the time of overflight, methods employing measurements from the images themselves (in-scene methods), and methods using radiation propagation models. We will also consider some calibration methods used to normalize atmospheric effects between multiple images of the same site from different days (multidate normalization methods) and image-processing methods designed to reduce the apparent effects of the atmosphere on the output image.

6.1 TRADITIONAL APPROACH: CORRELATION WITH GROUND-BASED MEASUREMENTS

A labor-intensive but often effective way to calibrate a remote sensor is to acquire ground truth at the time of image acquisition. Ground truth can take on many forms, but its purpose is to improve the analyst's ability to extract information from the remotely sensed images. In some cases, this means taking measurements so that the atmospheric parameters in the governing equation are known. It can also involve measurements to define the relationship between the remotely sensed variable (e.g., reflectance) and the parameter under study (e.g., stress in a forest canopy). Or, in many cases, ground truth can involve extensive measurements to directly characterize, usually quantitatively, the parameter(s) under study (e.g., soil moisture).

One common way to calibrate an image is to skip the atmospheric calibration step and attempt to develop directly a functional relationship between the remotely sensed values and the parameter under study using observed ground truth values for the parameter. This often takes on the form

$$Y = f(\mathrm{DC}_1, \mathrm{DC}_2 \cdots \mathrm{DC}_N) \tag{6.1}$$

where Y is the parameter under study (e.g., soil moisture), $\mathrm{DC}_1 \cdots \mathrm{DC}_N$ are the digital count values in spectral bands 1 through N (e.g., red, green, and blue), and f, simply denotes that there is a functional relationship between the dependent parameter Y and the observed values ($\mathrm{DC}_1 \cdots \mathrm{DC}_N$). In many cases, no form is known or presumed for the function f, and a number of regressions are tried to minimize the residual errors in equations of the form

$$Y_i = a_0 + a_1 \mathrm{DC}_{1i} + a_2 \mathrm{DC}_{2i} + a_N \mathrm{DC}_{Ni} + e \tag{6.2}$$

$$Y_i = b_0 + b_1 \mathrm{Log}\mathrm{DC}_{1i} + b_2 \mathrm{Log}\mathrm{DC}_{2i} \cdots + b_N \mathrm{Log}\mathrm{DC}_{Ni} + e \tag{6.3}$$

$$Y_i = c_0 + c_1 \mathrm{DC}_{1i} + c_2 \frac{\mathrm{DC}_{1i}}{\mathrm{DC}_{2i}} + c_3 \mathrm{DC}_{2i} + c_4 \frac{\mathrm{DC}_{2i}}{\mathrm{DC}_{3i}} \cdots + c_{2N} \frac{\mathrm{DC}_{(N-1)i}}{\mathrm{DC}_{Ni}} + e \tag{6.4}$$

where the subscript i indicates that the data values are coming from the ith ground truth location corresponding to image location i having digital count values DC_{1i}-DC_{Ni}. The values a_0-a_N, b_0-b_N, etc., are regression coefficients arrived at by including all sets of parameters and independent digital count values, and e is the residual error of the regression. The regression coefficients are chosen to minimize the residual error (e) for the postulated function, and then the functional form (f) with the lowest mean-square error is selected. In general, a random search for a functional form for the relationship between an independent variable (Y) and digital count values should only be performed if large amounts of data are available and no physical form for the nature of the relationship can be postulated. Even then, the functional form of the relationship should be kept simple or an overconstrained nonrobust solution can easily result. Such a solution might generate a relatively small error value (e) for the ground truth data, but might generate very large errors when applied to other digital count values elsewhere in the scene. The only way to ensure any degree of robustness to solutions of this form is to have a great deal of data (compared to the degrees of freedom in the regression equation) and to make sure, within reason, that all combinations of digital count values are included in the sample data set. This is generally a prohibitively expensive solution, especially because it is only valid for the data set under study. This is because the regression coefficients will be a function of the atmospheric conditions at the time of image acquisition and will change from acquisition to acquisition. As a result, a major ground truth program, potentially involving hundreds of ground truth samples, is required each time imagery is collected.

This type of regression approach can often be pursued with more confidence and requires less ground truth if the nature of the relationship between reflectance and the parameter of interest (Y) is known. In this case, we can break the overall functional form up into subrelations of the form

$$\mathrm{DC} = f'(L) \tag{6.5}$$

$$L = f''(r) \tag{6.6}$$

$$Y = f'''(r_1, r_2 \cdots r_N) \tag{6.7}$$

or

$$Y = f(\mathrm{DC}_1, \mathrm{DC}_2 \cdots \mathrm{DC}_N) \tag{6.8}$$

In many cases f' and f'' are approximately linear, so that

$$\mathrm{DC}_1 = m_1' L_1 + b_1' \tag{6.9}$$

$$L_1 = m_1'' r_1 + b_1'' \tag{6.10}$$

where m_1' and b_1' are the linear coefficients relating the radiance in spectral band 1 (L_1) to the digital count in band 1 (DC_1) and m_1'' and b_1'' are the linear coefficients between the reflectance in band 1 (r_1) and the radiance in band 1 (L_1). Then, if from laboratory or field studies a functional form for f''' can be postulated, e.g.,

$$Y \cong a_0 + a_1 r_1 + a_2 r_2 \tag{6.11}$$

it can be shown that

$$Y_i = b_0 + b_1 \mathrm{DC}_{1i} + b_2 \mathrm{DC}_{2i} + e \tag{6.12}$$

Again, a least-squares regression can be used to find the coefficients b_0, b_1, etc.; however, in this case the function form used is defined by the form of Eq. (6.11). The robustness of the solution can be ascertained from the laboratory or field data used to derive Eq. (6.11), and a much smaller ground truth program can be used. This method has been used for a variety of applications. Johnson et al. (1981), for example, used this general approach to map chlorophyll concentrations for water quality studies. For this approach to work satisfactorily, it is necessary to have a wide range of ground truth data that corresponds to the dynamic range of the parameter being investigated. Likewise, there is an implicit assumption that the regression coefficients apply equally well over the entire image (i.e., the atmosphere is spatially constant). This general approach also requires the reflectance values to be effectively Lambertian over the sensor's field of view unless special efforts are made to account for any non-Lambertian behavior.

In order to take advantage of a broad range of functional relationships between reflectance and applications-oriented parameters, more general ground truth calibration methods are often employed. These methods are aimed at computation of the necessary atmospheric parameters to solve directly for the value of the surface reflectance in each spectral band. These reflectance values can then be used to study any number of functions of the form of Eq. (6.7) where the nature of the function is known or can be derived at a later time from laboratory or field studies. These approaches are more appropriately called *atmospheric calibration techniques* and will be treated in the thermal region in Section 6.3 and in the reflective region in Section 6.4.

We should point out that while the discussion of the ad hoc image calibration methods in this section (6.1) have emphasized the reflective bands, one need only replace reflectance with temperature as the variable of interest and the approach can be used for calibration of thermal imagery as well.

6.2 APPROACHES TO ATMOSPHERIC CALIBRATION

Many calibration techniques exist that are targeted at different spectral regions, sensor configurations, or operational constraints. We have attempted to include a sampling that will provide the user with a starting point for coping with most collection conditions. The treatment is not intended to be all inclusive, nor were the methods necessarily selected because they are clearly the most accurate or widely used. In fact, there are very few widely used or generally accepted calibration methods, in large part because so few have been objectively tested over a robust data set. This chapter makes a concerted effort to describe all of the methods in terms of a self-consistent nomenclature using standard radiometric terminology to facilitate intercomparison.

Often in the sections that follow (6.3 and 6.4), we are assuming that the image data are from a calibrated sensor. By this we mean that the recorded signal (typically in terms of digital counts, analog voltage, or film density) can be converted to the effective radiance reaching the imaging system. In several cases, the image calibration methodology combines sensor calibration for sensors where radiance and recorded signal are linearly related. As mentioned in the previous chapter, increasing care in recent years has gone into the design and implementation of sensor calibration. In addition, operators of satellite sensors have initiated postlaunch, as well as periodic checks, of ongoing sensor calibration as discussed in Section 5.6 Regrettably, these improvements in sensor calibration can only be fully utilized if the remaining unknowns associated with atmospheric and illumination effects can be adequately calibrated.

There is no clear definition of what constitutes adequate atmospheric calibration. Each user will have a different definition of what calibration level is required for his or her application. Where expected errors associated with a calibration method are clearly demonstrated, we will attempt to cite them. However, the reader should be aware that in many cases the cited numbers are often from a limited study, usually performed by the developer of the method who may have used several "ad hoc tricks" not clearly defined in the simplified treatment presented here.

Depending on one's perspective, atmospheric calibration can be thought of as either forward or reverse engineering. In the forward-engineering case, the objective is to predict the radiance reaching the sensor and compare that to the sensed radiance to evaluate the calibration of the sensor. This would be our perspective if we were interested in postlaunch or periodic evaluation of on-orbit satellite systems (cf. Sec. 5.7). The more common way to analyze the data would be to assume that the sensor provides accurate radiance values and then to work backwards (reverse engineering) to extract information about the absolute or relative reflectance or the absolute or relative temperature of scene elements. Both approaches require similar information about atmospheric conditions. However, the forward-engineering sensor calibration methods can assume that a great deal of ground truth and ancillary data are available, since they only need to be used occasionally and at well-controlled locations. The more common reverse-engineering methods would normally be used under operational conditions where ground truth and ancillary data may be limited or nonexistent. Since most users are in the position of assuming that the sensor is calibrated and need to extract information from the imagery, this chapter will emphasize the reverse-engineering approaches.

The calibration approaches can be further divided based on the type of information required. The most rigorous calibration approaches are aimed at extracting absolute surface reflectance or temperature values. However, in many cases, only relative reflectance or temperature values are needed. For example, in multispectral analysis it may be sufficient to know the reflectance ratio between two spectral bands without needing to know the reflectance in either band. At other times, it may only be necessary to know the change in reflectance over time. For example, in change detection studies, or process monitoring studies, it is often sufficient to know how much the reflectance of an object changed over time or to be able to plot relative reflectance values as a function of time. Finally, there are times when it is only necessary to reduce atmospheric effects to improve the potential for comparing data over time or within a scene. For example, the brightness values of forest pixels on opposite sides of a hill are different, and a method to normalize these illumination variations is required (i.e., neither the actual magnitudes or change in magnitude must be known). The next two sections (6.3 and 6.4) will address various methods for calibration of image data to temperature and reflectivity, respectively.

In these sections we will, for simplicity, often assume that only small errors are introduced by using effective bandpass values for radiance, transmission, etc. (cf. Sec. 4.5). The reader is cautioned to check the validity of this assumption against the calibration precision required and use more accurate spectral solutions if necessary and possible.

6.3 APPROACHES TO MEASUREMENT OF TEMPERATURE

This section will address methods for calibration of thermal images where the objective is temperature measurement. In general, when considering calibration for temperature we will assume that we are sensing in the long-wave infrared (LWIR) region or in the midwave infrared (MWIR) region at night, since this is where most quantitative thermal imaging is undertaken (cf. Sec. 4.6). The methods considered will be ground truth, in-scene methods, and atmospheric modeling approaches. In each case, one or more examples of calibration approaches will be presented. These are not intended to be comprehensive, but only an attempt to show some of the more common classes of calibration techniques and to point out the general pros and cons.

6.3.1 Ground Truth Methods (Temperature)

The ground truth methods presented here and in Section 6.4.1 should not be confused with the methods described in Section 6.1. Both approaches use ground truth data; however, the methods presented here are aimed at calculation of atmospheric calibration values (e.g., τ and L_u), whereas the methods in Section 6.1 are aimed at simple one-time ad hoc correlation solutions between digital count values and applications-oriented parameters.

The ground truth methods are the most common and straightforward approaches to calibration. Simply put, they involve measuring the temperature or radiance from several objects in the study scene and then computing the transmission and path radiance. For a blackbody, the governing equation can be approximated as:

$$L = L_T e^{-\delta' \sec\theta} + L_u = L_T \tau_2 + L_u \tag{6.13}$$

For a calibrated sensor, the observed radiance values (L) can be computed from the recorded signal levels and regressed against the radiance due to the temperature (L_T) of the measured objects. This approach was used effectively by Scarpace et al. (1974). They set up several children's wading pools that could be easily resolved and whose water temperatures could be varied and measured (cf. Fig. 6.1). For near-vertical viewing in the LWIR, the emissivity of water is quite high ($\approx$0.986), so the blackbody approximation introduces little error. The assumption is made that the calibration measurements are made on level surfaces free from background effects, so the shape factor F in Eq. (4.68) is set to 1. In addition, since their interest was in studying water temperature alongside the calibration site, any residual error due to reflected downwelled radiance would have been included as part of L_u and corrected for in the calibration, i.e.,

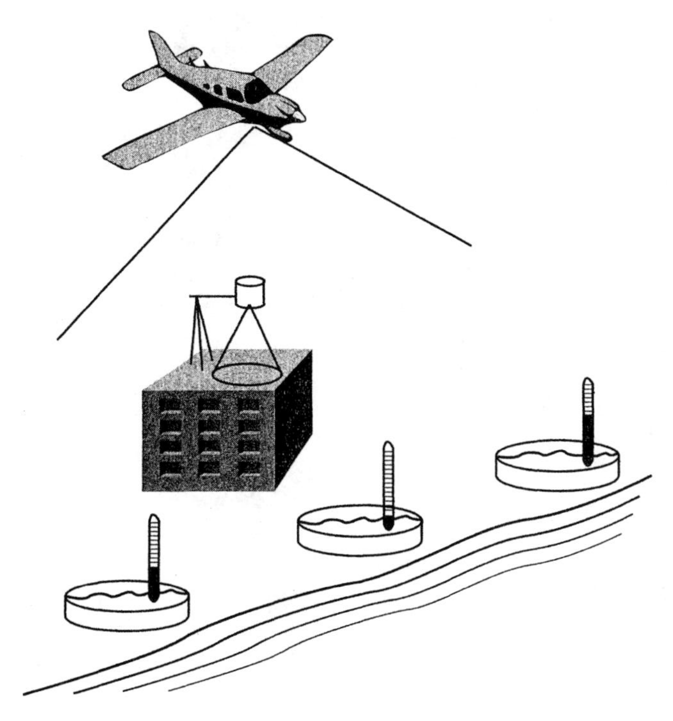

Figure 6.1 Temperature and radiance ground truth measurements can both be used to calibrate thermal infrared images.

$$L = L_T \varepsilon \tau_2 + \tau_2 r L_d + L_u = m L_T + b \qquad (6.14)$$

where $m = \tau_2 \varepsilon$ is the slope of the regression and $b = \tau_2 r L_d + L_u$ is the intercept.

In practice, the water temperature in the study area adjacent to the calibration site was computed by solving for L_T, i.e.,

$$L_T = (L - b) / m \qquad (6.15)$$

and converting the radiance to temperature through the Planck equation.

This ground truth method will work well for blackbodies and will be self-correcting for gray bodies if it is only applied to objects with the same emissivity as the calibration standards (e.g., in this case the term $\tau_2 r L_d$ is a constant for the pools and the water body under study). On the other hand, a more involved approach is necessary for surfaces having varying emissivity.

In principle, the value of εL_T for ground truth studies can be computed from contact thermistors and emissivity values (ε) determined by laboratory or field measurements, or by using normal "textbook" values (cf. Gubareff, 1960). This approach is seldom used in practice because of the difficulty associated with making accurate contact temperature measurements of solids (particularly surfaces with low thermal conductivity) and the limited emissivity database for naturally occurring backgrounds. The problem with using solids as standards is compounded by timing problems associated with taking ground truth. Because the surface temperatures can change rapidly with environmental conditions, the data must be taken concurrent with the sensor overflight for precise calibration.

To reduce the error associated with temperature and emissivity measurements, ground truth of solid surfaces is more appropriately acquired with a calibrated radiometer (ideally one with the same spectral bandpass as the overflight sensor). By taking several radiometer readings of targets with different radiance values, a simple linear regression can be performed to yield τ_2 and L_u from

$$L = \tau_2(\varepsilon L_T + rL_d) + L_u = \tau_2(L_0) + L_u \tag{6.16}$$

where L_0 is the surface-leaving radiance measured by the radiometer. Regrettably, because of the thermal instability of most surfaces, the surface radiance readings must still be concurrent with the overflight, and care must be taken to avoid background radiation from neighboring objects (including the experimenter) which would change the value of L_d (assumed constant in the above discussion).

Given τ_2 and L_u from Eq. (6.16), L_T and hence T can be computed for any scene object by solving for L_0 and rearranging to yield

$$L_T = (L_0 - rL_d) / \varepsilon \tag{6.17}$$

where ε is determined by ground truth or by using tabulated values. Ideally, L_d can be determined by a radiometer, which integrates the downwelled radiance from the sky in the spectral band of interest. Alternately, nomograms have been developed (Bell et al., 1960) which can be used to approximate L_d based on air temperature and atmospheric conditions. A more rigorous solution for L_d can be obtained by using an atmospheric propagation model such as LOWTRAN (cf. Sec. 6.3.3). In general, a fair amount of error in downwelled radiance (L_d) can be tolerated, since the term $r\tau_2 L_d$ is usually small compared to εL_T. For example, if L_d is 80% of L_T and r is 0.1, a 10% error in L_D will result in an error of less than 1% in L_T.

When quality ground truth data are used and emissivities well characterized, surface temperature errors of less than several tenths K should be achievable. However, because of a variety of logistical problems, it is often very difficult to acquire quality ground truth data. It can be particularly difficult to acquire data for low-resolution systems where the average temperature or radiance of large surfaces must be known with great accuracy. In addition, there are often occasions when, for operational reasons, it is not possible or cost effective to take ground truth. This forces us to look for alternative methods for calibrating thermal infrared images.

6.3.2 In-Scene Calibration Techniques (Temperature)

A family of atmospheric calibration techniques has emerged that take advantage of characteristics of the data in the images themselves for calibration. These approaches are known as *in-scene* methods and often rely on multiple ways of looking at the same scene. We will look at three of these methods that involve multiple images of the same scene taken at multiple altitudes, multiple view angles, or multiple wavelengths.

6.3.2.1 Profile (Multiple Altitude) Calibration Technique

This technique was developed to eliminate the problems associated with ground truth approaches. It is used on airborne systems and involves flying the imaging system at a series of altitudes above the same ground area. When analyzing the images, targets having a range of radiance values are identified and their radiance measured at each altitude from the flight altitude down to the lowest practical altitude. These radiance values are then plotted as a function of altitude and extrapolated to zero altitude, as shown in Figure 6.2. The data are

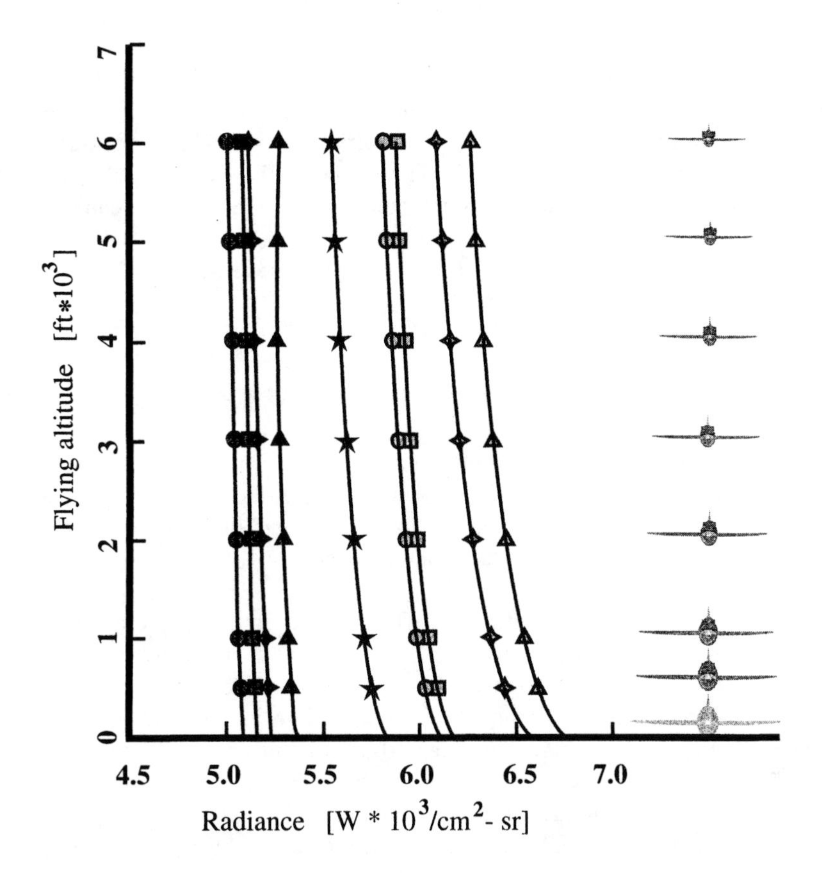

Figure 6.2 Data collected using the multiple-altitude (profile) calibration technique. Symbols represent sampled points for each target, and the line through each point is extrapolated to zero altitude.

usually collected near vertical (such that $\theta = 0$). Thus the radiance at the flight altitude can be expressed as:

$$L(h,\theta) = L(h,0) = \tau(h,0)L(0,0) + L_u(h,0) \tag{6.18a}$$

$$= mL(0,0) + b \tag{6.18b}$$

where the extrapolated radiance at the surface is $L(0,0)$, and we have expressed the functional dependence on altitude (h) and view angle (θ) for the relevant variables. If the radiance at altitude $L(h,0)$ is regressed against the radiance for that same object extrapolated to the surface value $L(0,0)$ for each of several objects, then the slope (m) is the atmospheric transmission $\tau(h,0)$ and the intercept (b) is the upwelled radiance $L_u(h,0)$. In effect, we use the data extrapolated to zero altitude as ground truth. To first order, the transmission at any angle can be approximated as [cf. Eq. (4.2)]:

$$\tau(h,\theta) \approx \tau(h,0)^{\sec(\theta)} \tag{6.19}$$

and the upwelled radiance can be approximated as

$$L_u(h,\theta) \approx L_u(h,0)\sec(\theta) \tag{6.20}$$

for clear atmospheres and small θ (less than 60°).

The downwelled radiance can be computed using the LOWTRAN model (cf. Sec. 6.3.3). However, an alternative approach is to observe a non-Lambertian surface of known angularly dependent emissivity such as water at two angles. If the reflected downwelled radiance is assumed to be approximately constant, then the following analysis can be performed

$$L(h,\theta) = \varepsilon(\theta)L_T\tau(h,\theta) + r(\theta)L_d\tau(h,\theta) + L_u(h,\theta) \tag{6.21}$$

and

$$L(h,0) = \varepsilon(0)L_T\tau(h,0) + r(0)L_d\tau(h,0) + L_u(h,0) \tag{6.22}.$$

Equating the L_T terms for the same object viewed at two angles and solving, yields an expression for L_d (cf. Schott, 1979). Experiment has shown that this method is not particularly accurate. However, as discussed in Section 6.3.1, for most surfaces the reflected portion of the signal is a small component, so that the overall analysis is not particularly sensitive to errors in L_d.

Schott (1979) reports on a blindfold test where the profile technique was used in the 8- to 14-μm region to compute water surface temperatures. The standard error using the method was 0.4 K when compared to simultaneously acquired ground truth values. This technique has also been used to acquire "ground truth" for satellite imaging systems. Schott and Schimminger (1981) utilized aircraft imagery calibrated with the profile technique to measure the radiance of large areas that had been simultaneously measured with a satellite with a 0.5-km GIFOV. The satellite data could then be accurately calibrated to approximately 1.1 K using this approach (cf. Fig. 6.3). The profile technique has the advantage of demonstrated accuracy and no requirements for external data. It has a significant disadvantage in that it cannot be utilized with satellite imaging systems except in the rare case where a simultaneous aircraft underflight is available.

6.3.2.2 Multiple View Angle Techniques

Saunders (1967), McMillan (1975), and Chedin et al. (1982), among others, have reported on methods to extract surface temperatures using some form of multiple-view angle technique. The method presented here was suggested by Byrnes and Schott (1986) and is presented mostly for its consistency with the previous methods. All the methods rely on the assumption that the radiance from a target observed at two different look angles changes only due to the difference in path length through an atmosphere comprised of homogeneous layers (cf. Fig. 6.4). If, for convenience, we take one angle to be vertical, then the radiance reaching the sensor from the same target can be expressed for viewing through an angle θ as:

$$L(h,\theta) = \tau(h,\theta)L(0,\theta) + L_u(h,\theta) \tag{6.23}$$

and for vertical viewing as:

$$L(h,0) = \tau(h,0)L(0,0) + L_u(h,0) \tag{6.24}$$

From Chapter 3, we recall that for Lambertian surfaces the surface-leaving radiance into the θ direction [$L(0,\theta)$] and the radiance normal to the surface [$L(0,0)$] are the same. Thus Eqs.

(a)

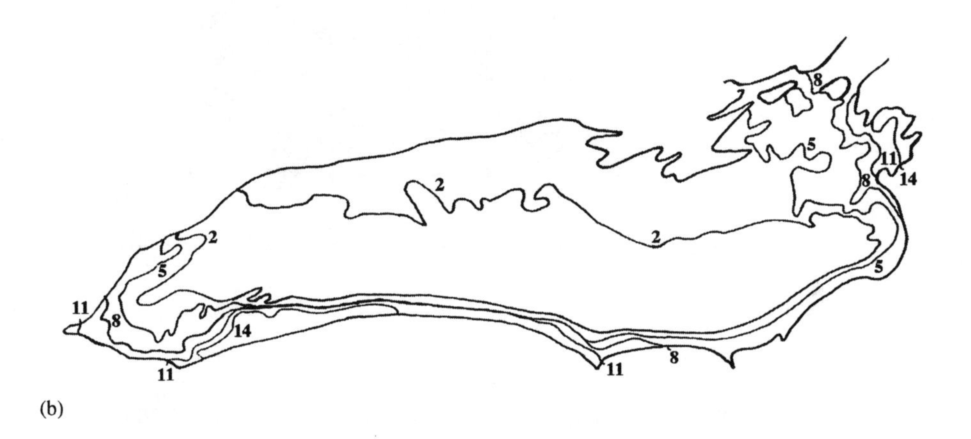

(b)

Figure 6.3 (a) HCMM image of Lake Ontario with aircraft underflight coverage shown in white boxes. (b) Thermal map produced after underflight calibration. The warm ring of near-shore water is associated with the spring thermal bar. (Adapted from Schott and Schimminger, 1981.)

(6.23) and (6.24) can be combined by eliminating the surface radiance term to yield

$$L(h,\theta) = \frac{\tau(h,\theta)}{\tau(h,0)} L(h,0) + \left[L_u(h,\theta) - \frac{\tau(h,\theta)}{\tau(h,0)} L_u(h,0) \right] \tag{6.25}$$

If several Lambertian objects having a range of radiance values are imaged both vertically and at some known angle, then a linear regression of the data will yield a slope term

$$m = \tau(h,\theta) \,/\, \tau(h,0) \tag{6.26}$$

and an intercept

$$b = L_u(h,\theta) - \frac{\tau(h,\theta)}{\tau(h,0)} L_u(h,0) \tag{6.27}$$

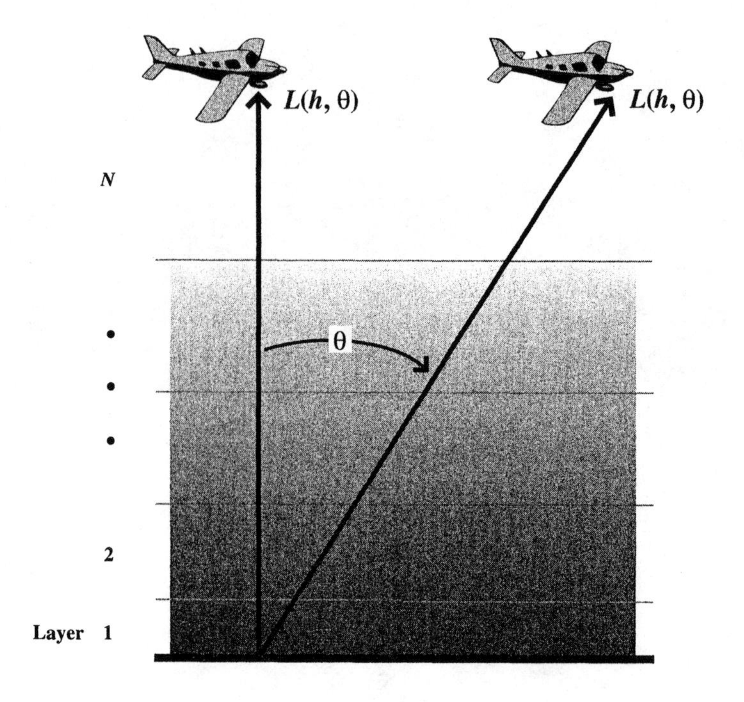

Figure 6.4 The multiple look angle calibration technique assumes vertical layering of a horizontally homogeneous atmosphere.

Using Eq. (6.19), the slope term can be analyzed to yield the transmission according to

$$\tau(h,0) = m^{1/(\sec(\theta)-1)} \tag{6.28}$$

In a similar manner using Eq. (6.20), the upwelled radiance for small angles ($\theta = 0^\circ$ to 60°) and relatively clear atmospheres can be approximated as:

$$L_u(h,0) = \frac{b}{\sec(\theta) - m} \tag{6.29}$$

Most variations of the multiple-view angle technique use clear atmosphere approximations similar to those of Eqs. (6.19) and (6.20). As a result, the use of this method is often restricted to certain altitudes, atmospheric conditions, or atmospheric types. Nevertheless, the potential value of not requiring ground truth data or low-flying aircraft is a considerable advantage. Chedin et al. (1982) and Byrnes and Schott (1986) both point to potential errors of several degrees Kelvin using simplified approaches to the multiple-view angle technique in the 8- to 14-μm bandpass. Mericsko (1993) has suggested empirically derived refinements to this approach that suggest that considerably better results can be obtained using better approximations to Eqs. (6.19) and (6.20). This approach is still often limited to aircraft use, because most currently operational space-based sensors cannot readily obtain multiple views of the same object through the same atmosphere.

Plate 1.1 Color photographic image of the earth acquired by Skylab astronauts. (Image courtesy of NASA Goddard.)

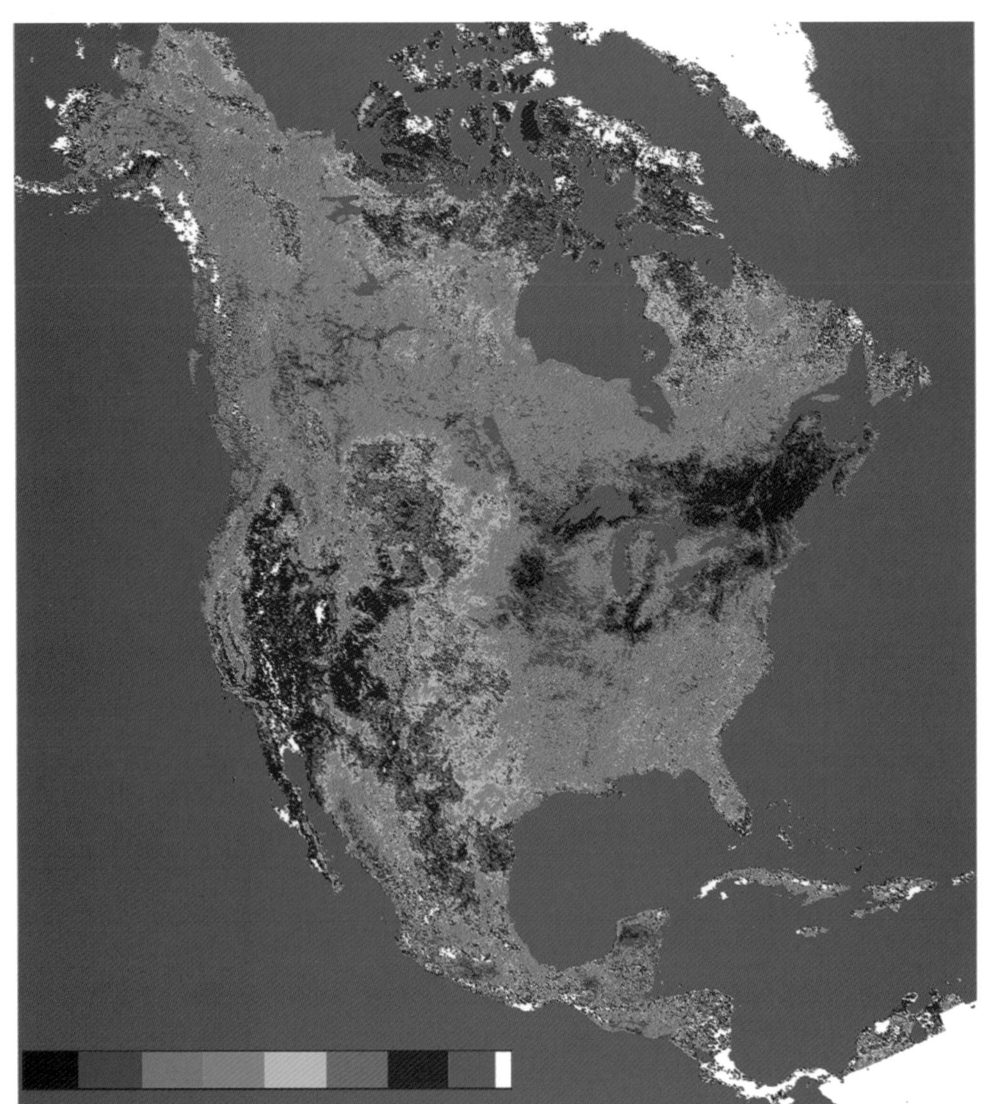

Plate 1.2 Color-coded map of the normalized difference vegetative index of North America derived from AVHRR data. The color bar shows coding of green vegetation biomass decreasing from left to right. (Image produced jointly by the Canada Center for Remote Sensing and the EROS Data Center.)

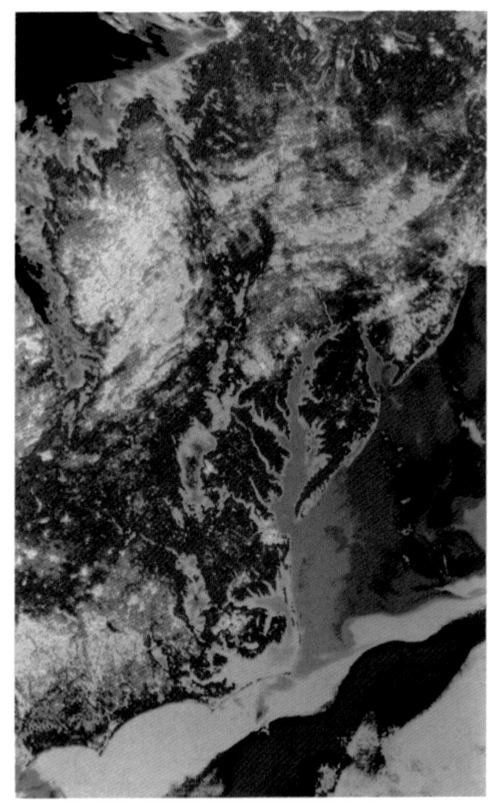

Plate 1.3 Color-coded image of Eastern U.S. produced from the thermal channel of the Heat Capacity Mapping Mission's (HCMM) radiometer. Note the warm urban areas and the warmer water in the Gulf Stream. (Image courtesy NASA Goddard.)

Plate 1.4 Color infrared aerial photograph of a forested area.

(a)

(b)

Figure 1.7 Images from portions of Landsat TM images. The region is a section of the Custer National Forest (a) before and (b) after a forest fire. (Image courtesy of EROS Data Center.)

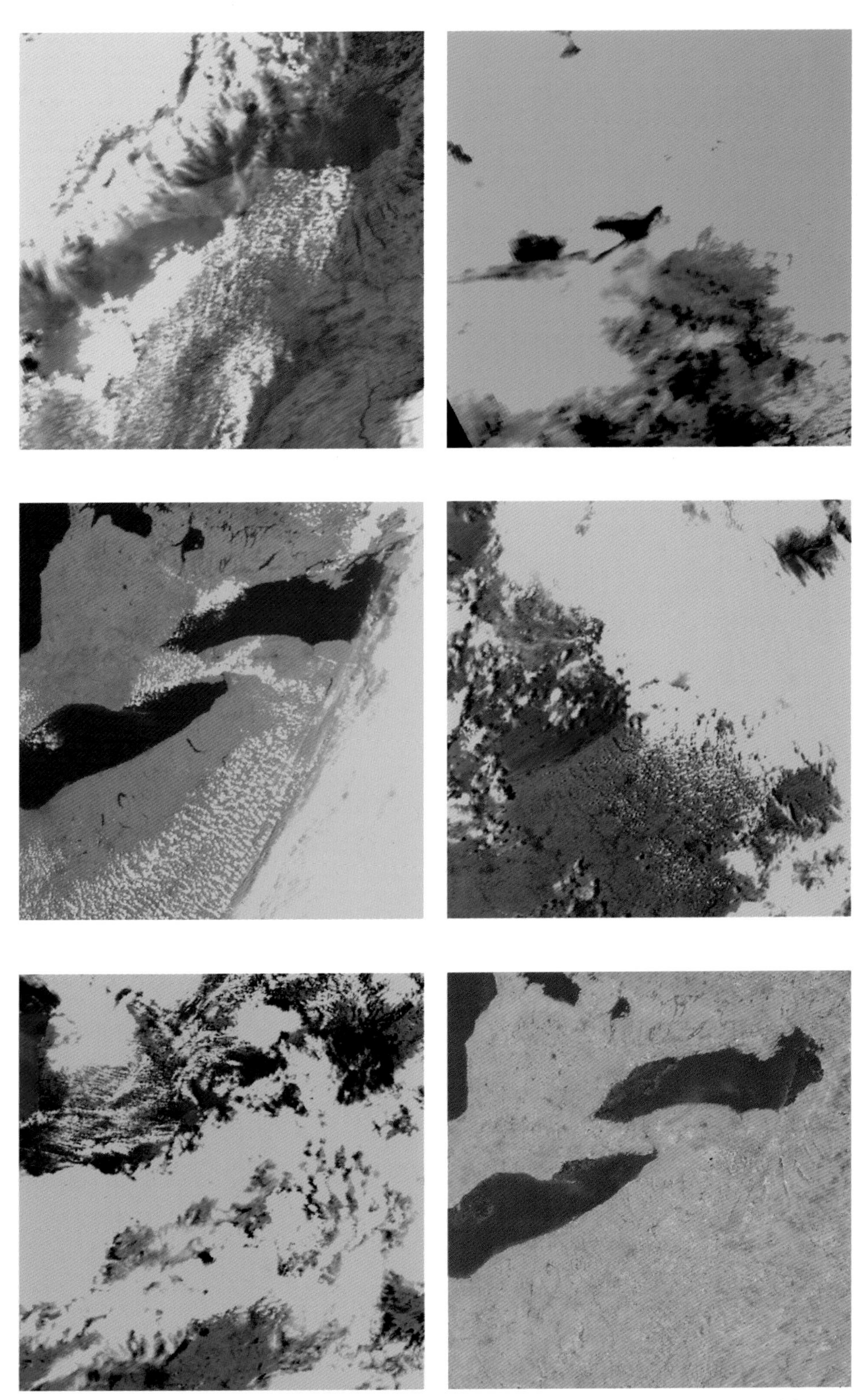

Plate 1.8 Five-day sequence of AVHRR images of the Eastern Great Lakes and a cloud-free composite made by combining the five-day sequence. In this false-color infrared representation, brighter reds are indicative of denser vegetation.

Plate 2.5 Color infrared image of camouflaged objects.

Plate 2.6 Color IR photo showing vegetation condition. A gypsy moth infestation has defoliated most of the trees in the image. "Cosmetic" spray programs along the roadways have prevented severe defoliation in these areas. (Image courtesy of RIT's DIRS laboratory.)

(a) 

(b) 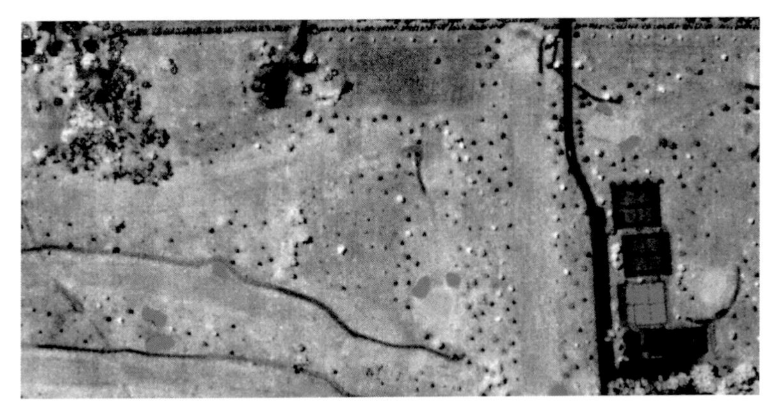

Plate 5.7 (a) Color infrared image and (b) NDVI image produced from a photo CD image scanned from 35-mm color infrared film. The NDVI image brightness increases with biomass and vegetation rigor. This image was acquired to assess turf conditions at a golf course. Note the low NDVI indicators at some of the close-cropped greens. (Courtesy Pegasus Environmental.)

Plate 5.35 AVIRIS image cube of Moffet Field California. Sensor has 224 channels, from 0.4 μm to 2.5 μm, each with a spectral bandwidth of approximately 10 nm. (Image courtesy of NASA JPL.)

1982 image

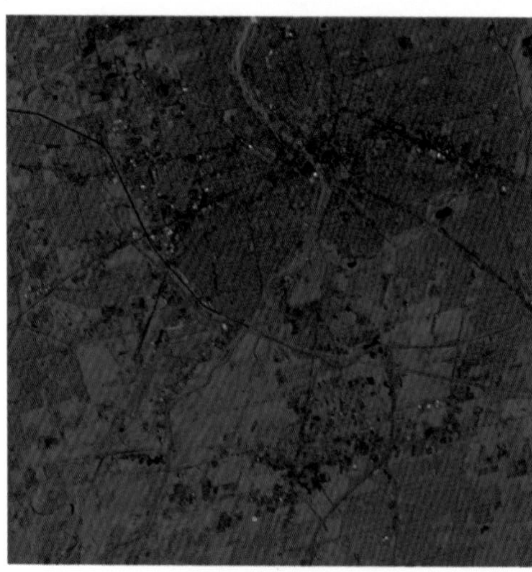

1984 image

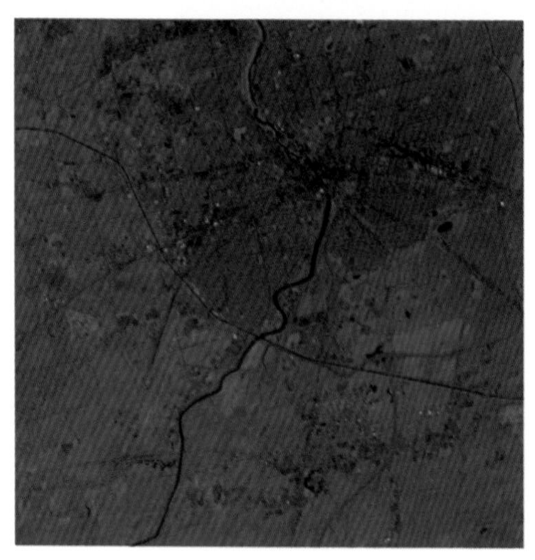

1982 image transformed to "look like" the 1984 image

Plate 6.16 Pseudo-invariant feature (PIF) transformation applied to a portion of a Landsat scene.

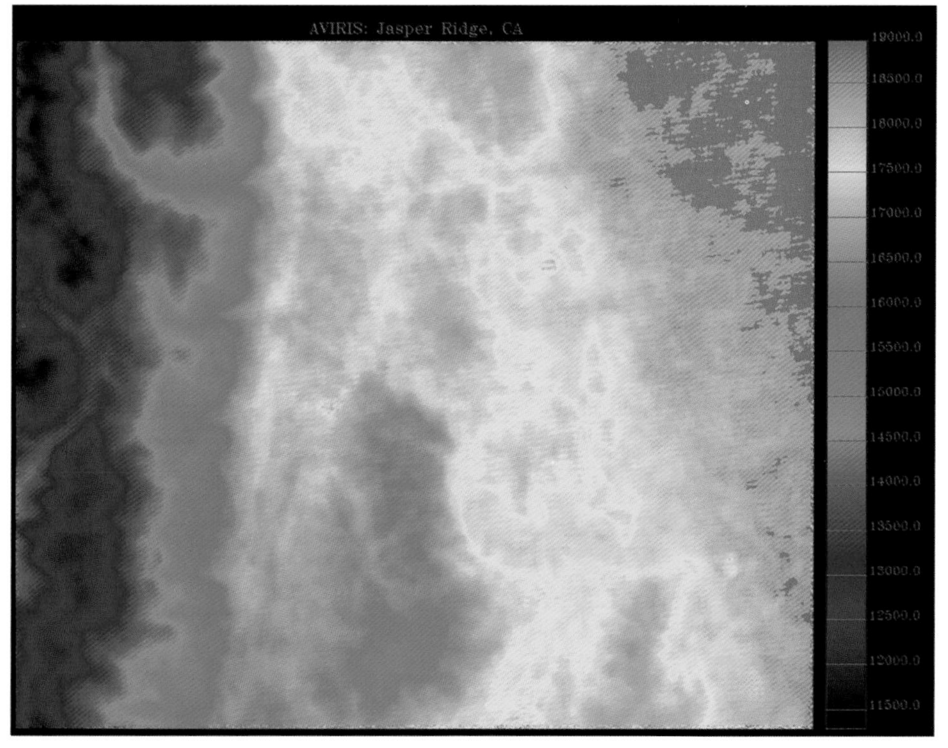

Plate 6.18 (a) AVIRIS image of Jasper Ridge, CA, and (b) a map of column water vapor content derived from the 940-μm absorption line data in the AVIRIS spectra. (Image and map courtesy of NASA JPL.)

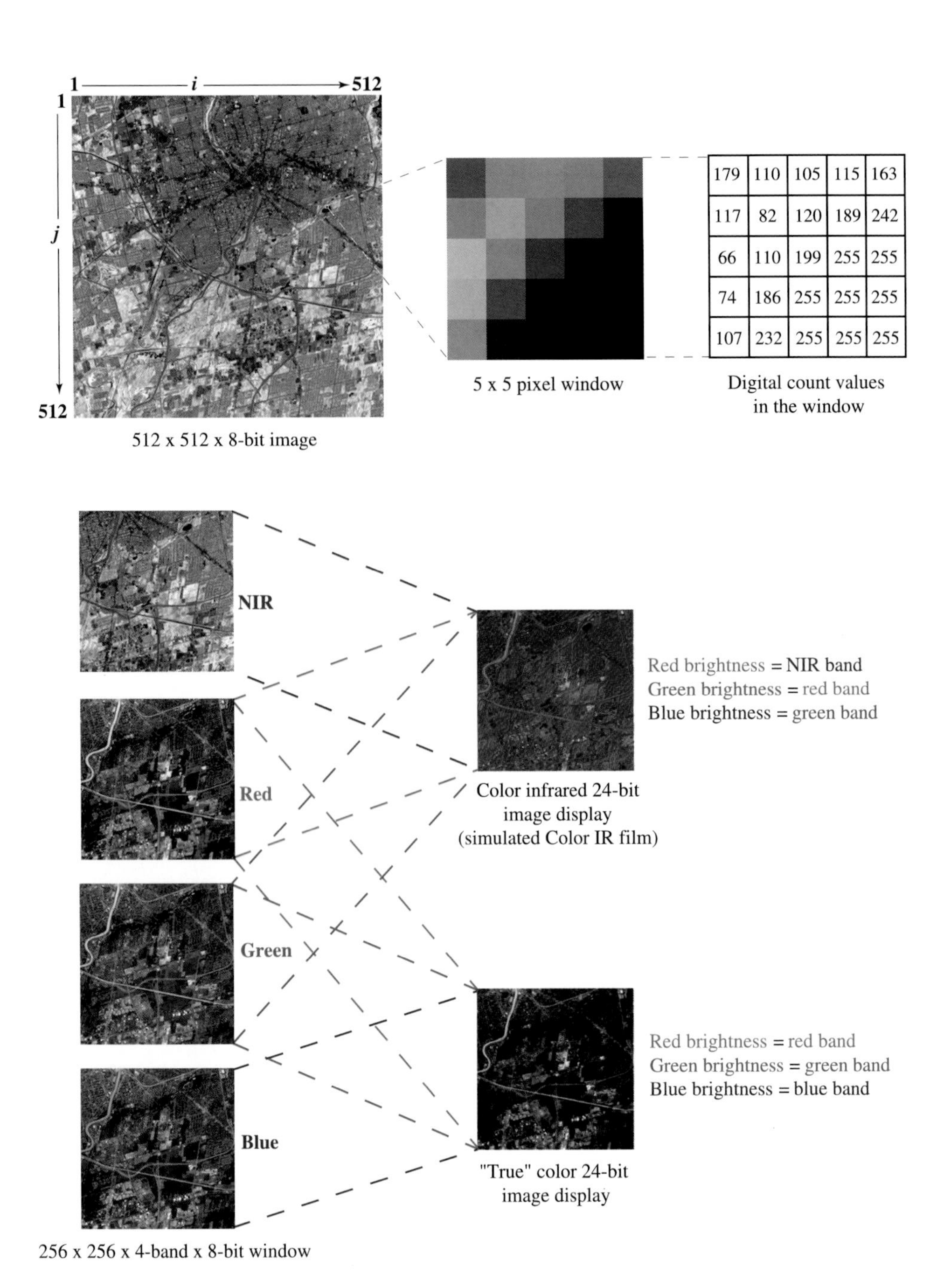

Plate 7.1 Digital image concepts.

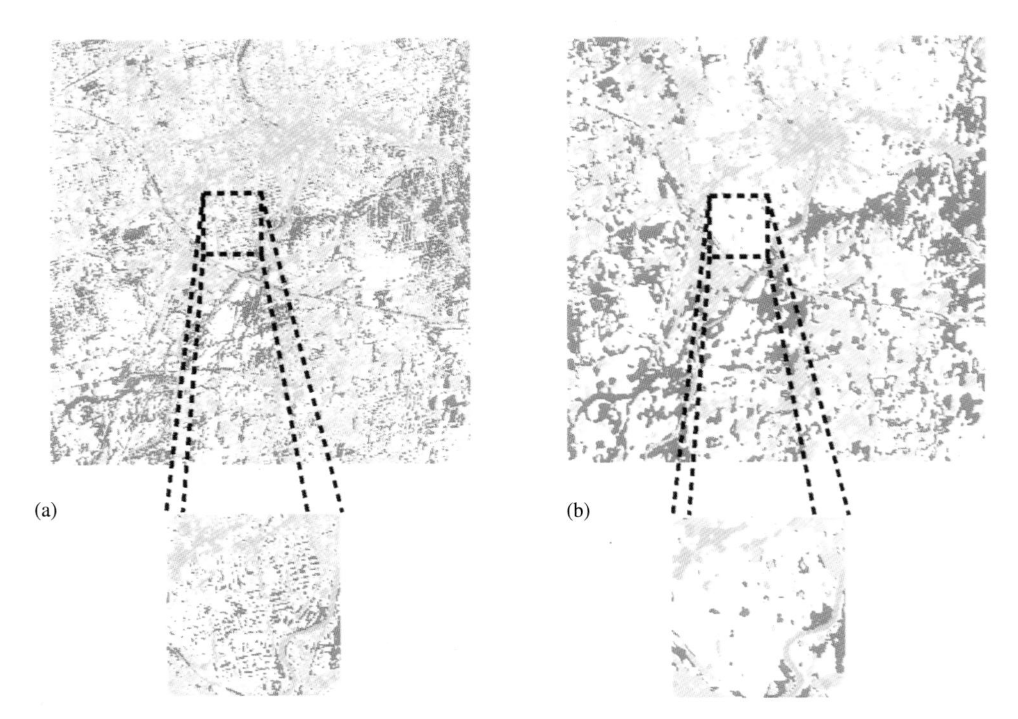

Plate 7.9 Mode filter operation and application to a class map image. (a) Original class map of a portion of a Landsat image of Rochester, New York. (b) Class map after application of a mode filter.

Plate 7.31 Example of the image shown in figure 7.10 classified using texture metrics images.

Color-coded fused image (IHS to RGB)

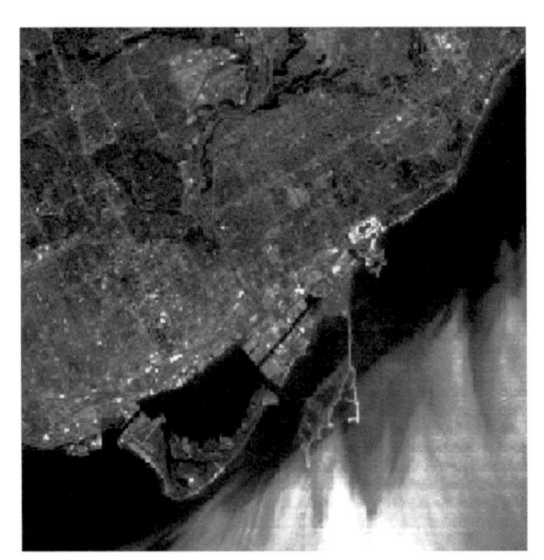

Red channel of TM (intensity)

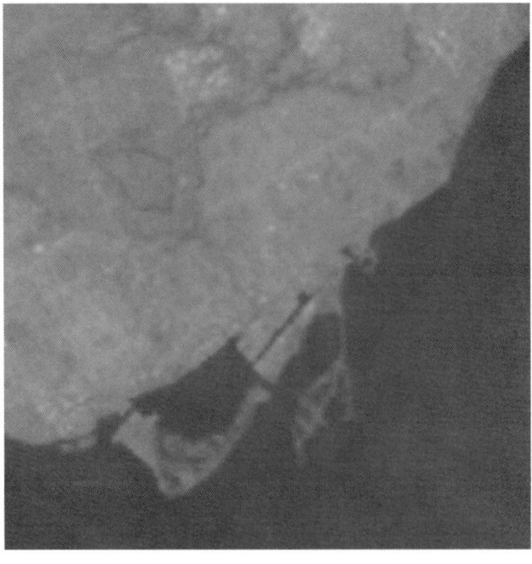

LWIR channel (hue)

Plate 8.15 Image fusion of the red channel of TM with the LWIR channel using the IHS transform.

TM false color IR

Fused false color IR

Spot panchromatic

Plate 8.16 Fusion of multispectral TM data with geometrically registered and resampled SPOT data.

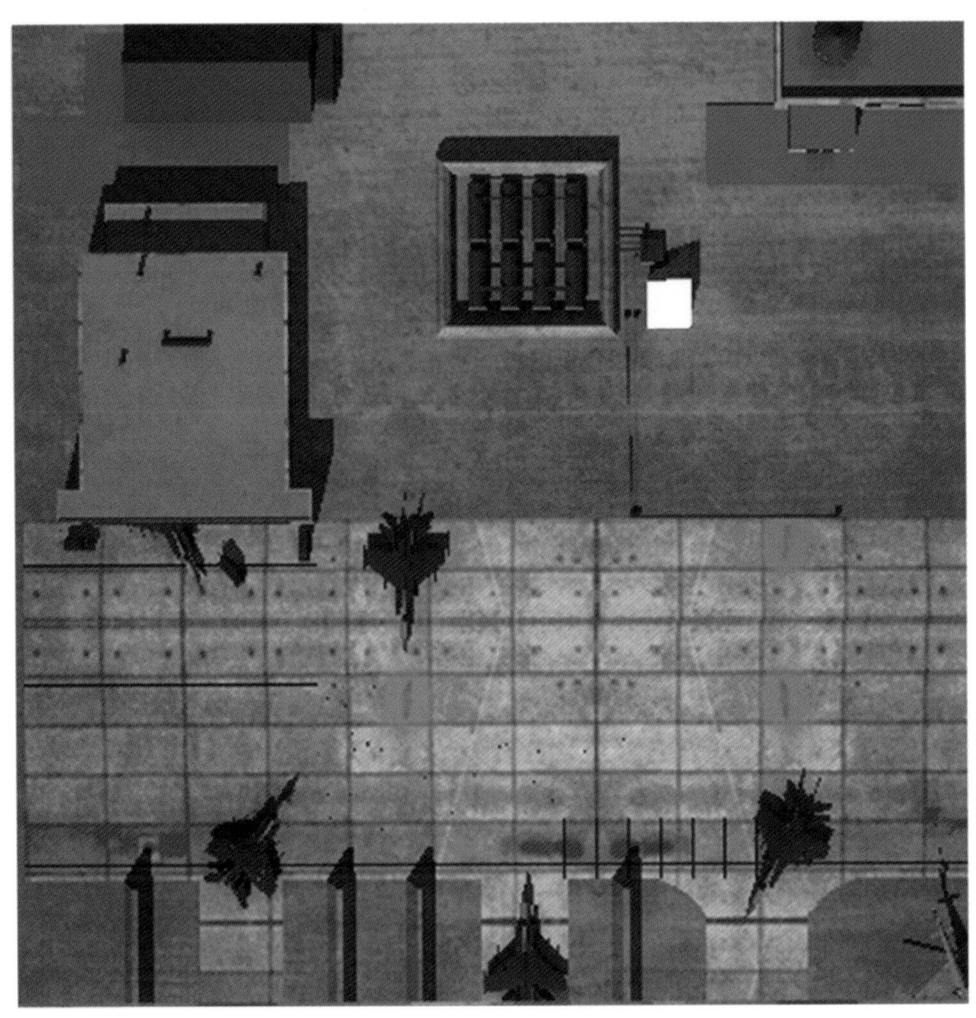

Plate 10.14 A false-color synthetic scene produced by DIRSIG showing texture effects.

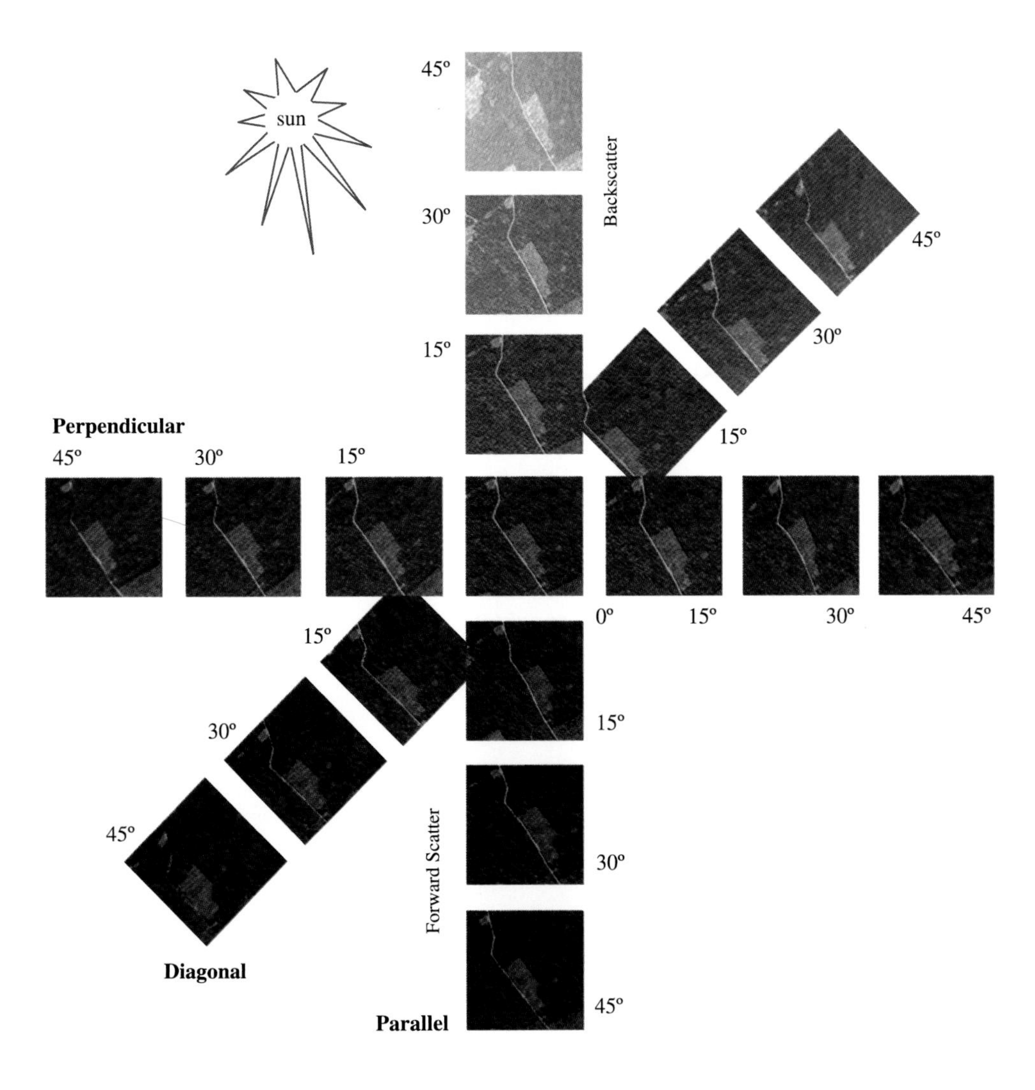

Plate 10.16 Color IR version of ASAS images from three flight lines showing combined effects of atmospheric and bidirectional reflectance variation. (Image courtesy of NASA Goddard.)

Plate 10.20 A synthetic sunset produced by DIRSIG. This effect is possible due to the extensive spectral modeling DIRSIG incorporates with the help of MODTRAN.

6.3.2.3 Multiple-Bandpass Technique

Prabhakara et al. (1974) and Anding and Walker (1975) discuss a method for computation of surface radiance based on differential absorption in two spectral bands (channels) within a broader atmospheric transmission window. This is very similar in theory to the multiple-angle technique where the differential atmospheric effects are assumed to be well-defined due to path length changes. In the multichannel approach, the basic radiative transfer equation is expressed as:

$$L(h,\theta) = L(0,\theta)\tau(h,\theta) + L_{TA}[1 - \tau(h,\theta)] \tag{6.30}$$

where $L(h,\theta)$ is the radiance at the sensor, $L(0,\theta)$ is the radiance at the earth's surface headed toward the sensor, L_{TA} is the radiance from the air column between the target and the sensor due to its mean effective temperature, and $[1-\tau(h,\theta)]$ is the effective emissivity of the air column (N.B. $L_u(h,\theta) = L_{TA}\,[1-\tau(h,\theta)]$). From Section 3.4, we recall that for atmospheres dominated by absorption effects, the transmission can be expressed as:

$$\tau(h,\theta) = e^{-C_{\text{ext}}mz} \tag{6.31}$$

where C_{ext} is the extinction crosssection [m^2], m is the number density [m^{-3}], and z is the path length [m] between the earth and the sensor.

For clear atmospheres, this can be expanded using a Taylor series and truncated to yield as a good approximation:

$$\tau(h,\theta) \approx 1 - C_{\text{ext}}mz \tag{6.32}$$

then Eq. (6.30) becomes

$$L(h,\theta) = L(0,\theta) - [L(0,\theta) - L_{TA}]C_{\text{ext}}mz \tag{6.33}$$

Expanding the Planck radiance equation about temperature and keeping only linear terms yields

$$T_i(h,\theta) \cong T(0) - [T(0) - T_A]C_{\text{ext}i}mz \tag{6.34}$$

where $T_i(h,\theta)$ is the apparent temperature at the sensor in the ith bandpass, $T(0)$ is the apparent temperature at the surface, T_A is the apparent temperature of the atmosphere corresponding to L_{TA}, and $C_{\text{ext}i}$ is the absorption cross section in the ith bandpass.

Thus, apparent temperature and the extinction cross section in the bandpass are seen to be approximately linearly related for atmospheric windows where the apparent temperature of the earth and the apparent temperature of the atmosphere (T_A) are each constant. For many adjacent regions (i.e., subwindows) within an atmospheric window, this is approximately true. Then if the extinction crosssections are known for two or more spectral bands, the apparent temperatures at the sensor in each band need only be plotted against the corresponding extinction crosssection to yield the apparent surface temperature as the intercept (cf. Fig. 6.5). When only two spectral bands are used, the apparent surface temperature can be expressed in terms of the ratio of the extinction cross sections between the bands which is often easier to obtain than an absolute value. The apparent surface temperature can then be expressed as:

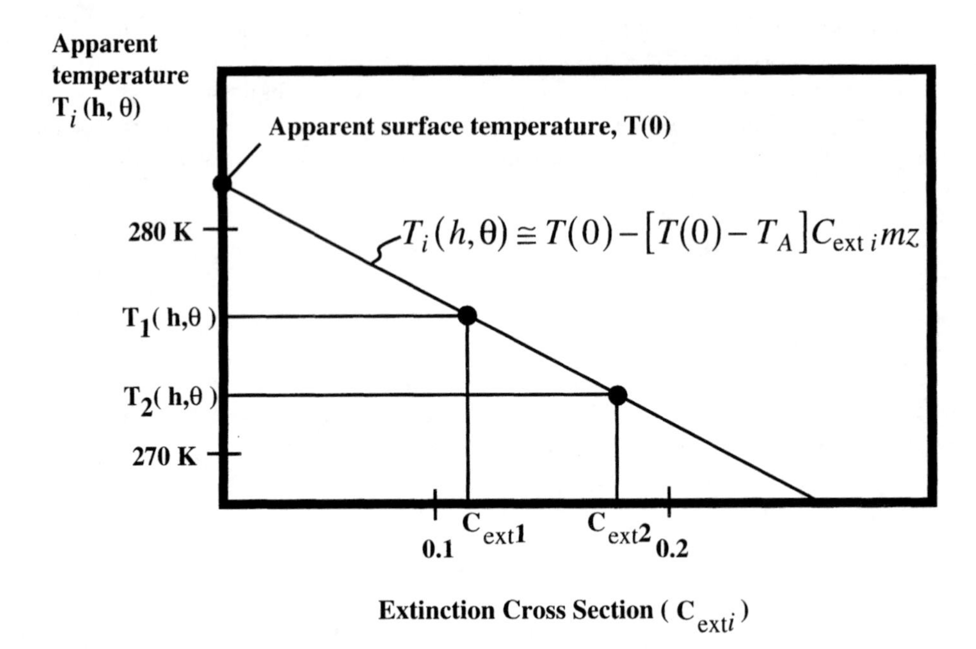

Figure 6.5 The multiple-bandpass calibration technique relies on the difference in the extinction coefficient between two usually adjacent spectral bands.

$$T(0) = \frac{T_1(h,\theta) - RT_2(h,\theta)}{1 - R} \tag{6.35}$$

where $T_1(h,\theta)$ and $T_2(h,\theta)$ are the apparent temperatures in bands 1 and 2, respectively, and $R = C_{ext1}/C_{ext2}$. The apparent temperature can be converted to the surface-leaving radiance value using the Planck equation, and corrected for emissivity and downwelled radiance as described in Section 6.3.1 for Eq. (6.17). (Note that this analysis can also be done in terms of the extinction coefficient β_{ext}, where $\beta_{exti} = mC_{exti}$.)

For this approach to be effective, the two windows need to have significantly different extinction crosssections. Anding and Walker (1975) discuss methods for computation of the extinction coefficients using radiation propagation models. They go on to indicate the results of tests of this method using multiple bands in the 8- to 14-μm window. These results indicate errors of the order of 1 K in apparent surface temperature. They also indicate that better results can be obtained using more than two bandpasses.

The advantage of this approach is that no ground truth is required, and one does not have to image from multiple angles or at multiple altitudes. It does, however, require having multiple images of the target in different bands. The NOAA TIROS N series of satellites were specifically designed for this type of calibration using multiple bands in the 10.5- to 12.5-μm window (cf. Price, 1984 and Walton et al., 1990). Because these sensors have a 1-km ground spot, it is difficult to evaluate calibration accuracy except over very large uniform temperature regions. Where this has been done for sea surface temperatures, errors of less than 1 K are reported.

6.3.3 Atmospheric Propagation Models (Temperature)

The logistics of ground truth or in-scene calibration techniques often make their use impractical or impossible. As a result, atmospheric propagation models are often an attractive solution when other calibration methods cannot be utilized. These models are attractive in that they can be applied to any image (i.e., there are no multiple look angle or multiple altitude, etc., requirements). In addition, they can be applied to existent imagery where little or no ground truth is available. As a result, they offer a very attractive solution that only involves the running of a propagation model. The disadvantage is that they can only effectively model the propagation through an atmosphere that is well characterized in terms of its temperature and the distribution of constituent elements. In general, they are designed to yield solutions, even if only rough estimates of atmospheric conditions are available. However, accurate results require a detailed knowledge of the atmosphere. (The adage "garbage in equals garbage out" is very applicable here.)

The LOWTRAN model (Kneizys et al., 1983 and 1988) and the higher spectral resolution MODTRAN model (Berk et al., 1989) is probably the most widely used and readily available of the propagation models, so we will use them here to exemplify this approach. Generally these models assume that the atmosphere is divided into a number of homogeneous layers as shown in Figure 6.6. The temperature of each layer can be determined from radiosonde data acquired at the time of the data collection or from generic profiles stored in the LOWTRAN model. Similarly, the concentration of the permanent gases and water vapor can be estimated from radiosonde air pressure and relative humidity data as a function of altitude. User-supplied data on meteorological visibility, season, and air mass type (continental, maritime, etc.) can be used to estimate aerosol numbers and size distributions. Once again, LOWTRAN provides several standard atmospheric profiles if no detailed data are available from the time of the collection.

At any wavelength the transmission through the ith layer of the atmosphere can be approximated as:

$$\tau_{ik} = e^{-m_{ik} C_k z_i} \tag{6.36}$$

where m_{ik} is the number density of the kth atmospheric constituent (e.g., CO_2) in the ith layer, C_k is the spectral extinction crosssection of that constituent, and z_i is the path length for propagation through the ith layer.

This expression is only rigorously true for discrete wavelengths and nonscattering media. In practice, LOWTRAN uses an empirical approximation to Eq. (6.36) and solves for τ_{ik} based on number densities derived from atmospheric profile data (e.g., radiosonde) and database values for the spectral extinction crosssections of the atmospheric constituents. The transmission ($\tau_{i\lambda}$) through the ith layer along a path length z_i at wavelength λ is simply the product of all the τ_{ik} values, i.e.,

$$\tau_{i\lambda} = \prod_k \tau_{ik} \tag{6.37}$$

In wavelength regions where solar reflection effects are negligible (e.g., the 8- to 14-μm window or 3- to 5-μm window at night), the LOWTRAN-style propagation equation can be approximated as (cf. Fig. 6.6):

$$L_\lambda = \varepsilon L_{T\lambda} \prod_{i=1}^{N} \tau_{i\lambda} + \sum_{i=1}^{N} \left[(1 - \tau_{i\lambda}) L_{T_i\lambda} \prod_{j=i+1}^{N} \tau_{j\lambda} \right] \tag{6.38a}$$

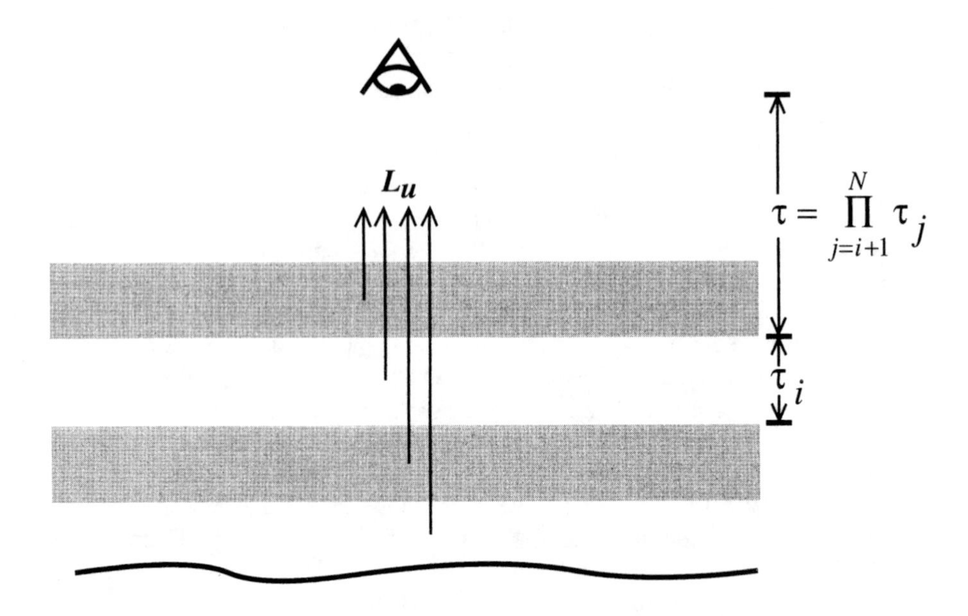

Figure 6.6 Summation of the contributions from each layer in the atmosphere to obtain the cumulative path radiance reaching the sensor.

$$= \varepsilon L_{T\lambda} \tau_\lambda(h,\theta) + L_{u\lambda}(h,\theta) \tag{6.38b}$$

where $L_{Ti\lambda}$ is the blackbody equivalent spectral radiance associated with the temperature (T_i) of the ith layer.

The effective radiance reaching the sensor is then expressed as:

$$L = \int R'(\lambda) L_\lambda d\lambda \tag{6.39}$$

where $R'(\lambda)$ is the normalized spectral response function of the sensor.

Most radiation propagation models will allow the user to input a target surface temperature T and emissivity ε, to be used in generating the surface-leaving spectral radiance ($\varepsilon L_{T\lambda}$), and solve an equation similar to Eq. (6.38a) for the spectral radiance reaching the sensor. Equation (6.39) would then be numerically integrated to yield the effective radiance reaching the sensor. To solve for the effective transmission and path radiance reaching the sensor for use in calibrating thermal images, it is often necessary to modify the radiation propagation source code or "cheat" the code into providing the required values. In our case, we require the effective bandpass transmission $\tau(h,\theta)$, upwelled radiance $L_u(h,\theta)$ and downwelled radiance L_d. The $\tau(h,\theta)$ and $L_u(h,\theta)$ terms can be found by inserting several target temperatures T for blackbodies that cover the range of temperatures in the scene and generating the corresponding effective radiance values at the sensor $L(h,\theta)$. (LOWTRAN allows the user to define the sensor location and view angle.) If one also solves for the effective bandpass radiance from these same targets on the ground, $L(0)$, through numerical solution of Eq. (6.39), then a simple linear regression of $L(0)$ with $L(h,\theta)$ can be performed according to:

$$L(h,\theta) = mL(0) + b \tag{6.40}$$

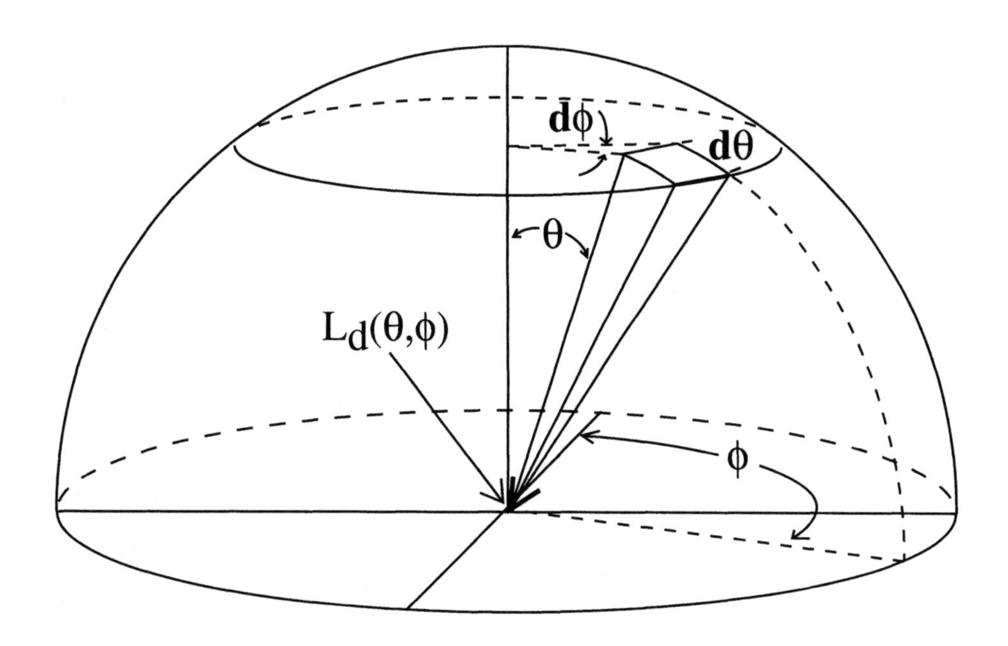

Figure 6.7 Illustration of eq (6.41) used for computation of the total downwelled radiance from angular values supplied by multiple LOWTRAN runs.

where m = $\tau(h,\theta)$ is the effective bandpass transmission and $b = L_u$ is the effective upwelled radiance in the bandpass. Note that in general, the effective transmission $\tau(h,\theta)$ will *not* equal the LOWTRAN-derived average transmission. Also, the LOWTRAN propagation model Eq. (6.38a) does not account for reflected downwelled radiance. However, the model will allow a sensor on the ground to look to space, allowing a computation of the directional downwelled radiance $L_d(\theta,\phi)$ for a declination angle θ from the zenith and an azimuthal angle ϕ. Repeated runs of LOWTRAN looking at a range of angles will allow generation of sufficient data to compute the integrated downwelled radiance L_d by numerical integration of (cf. Fig. 6.7)

$$L_d = \frac{E_d}{\pi} = \int \pi^{-1} L_d(\theta,\phi)\cos\theta d\Omega \tag{6.41a}$$

Substituting $d\Omega = \sin\theta\, d\theta d\phi$ yields:

$$L_d = \int_{\phi=0}^{2\pi} \int_{\theta=0}^{\frac{\pi}{2}} \pi^{-1} L_d(\theta,\phi)\cos\theta \sin\theta d\theta d\phi \tag{6.41b}$$

and for $L_d(\theta, \phi)$ independent of ϕ (i.e., no azimuthal variations in the sky),

$$L_d = 2\pi \int_{\theta=0}^{\frac{\pi}{2}} \pi^{-1} L_d(\theta,\phi)\cos\theta \sin\theta d\theta \tag{6.41c}$$

$$= 2 \int_{\theta=0}^{\frac{\pi}{2}} L_d(\theta, \phi) \cos\theta \sin\theta d\theta \tag{6.41d}$$

where E_d is the irradiance from the sky [w/m^2] onto the earth's surface, $d\Omega$ is the element of solid angle [sr^{-1}], and the integral is over the hemisphere above the target at the earth's surface. In essence, LOWTRAN can be made to perform a numerical approximation to the downwelled radiance computations described in Section 4.1.2.

Thus, the standard radiation propagation models can be manipulated to yield the required inputs to a governing equation of the form

$$L(h,\theta) = [\varepsilon L_T + (1-\varepsilon)L_d]\tau(h,\theta) + L_u(h,\theta) \tag{6.42}$$

where $L(h, \theta)$ is derived from image data, and the only unknowns are ε and L_T. If the user can estimate emissivity, then L_T and, hence, T can be found.

The most serious limitation to this approach is that the accuracy of the solution is very dependent on the quality of the input data on atmospheric conditions. Thus, if a radiosonde is launched at the study site at the time of the image data acquisition, the results should be very good (Schott 1993 suggests that errors of less than 1.0 K can be expected in the 8- 14-μm window). Where this is not possible, and standard atmospheres are used or radiosonde data are interpolated across space and time, substantial errors can occur (greater than 7 K). As a result, when concurrent and coincident radiosonde data are not available, methods to improve atmospheric interpolation or to augment radiosonde data must be employed. Byrnes and Schott (1986) describe simple techniques for adjusting the lower atmospheric profiles as a function of time using surface temperature, humidity, and visibility data (cf. Fig. 6.8). In addition, much work is now concentrating on the use of satellite-based multispectral infrared and microwave sounders to extract temperature and humidity data as a function of altitude (cf. Hanel et al., 1992). While these methods are not yet as precise as radiosonde data, they may be a valuable tool in assisting us to interpolate sparse radiosonde data over space and time.

As pointed out earlier, the LOWTRAN-style approach also has the advantage of relative ease of use, so it can supplement other atmospheric calibration techniques by generating any of the variables they do not. For example, the multiple-bandpass techniques do not generate a value of L_d, and this can be obtained from LOWTRAN.

6.3.4 Emissivity

All of the approaches presented above have a common problem in that they cannot separate temperature from emissivity effects. In order to separate out what component of the surface-leaving radiance is due to temperature and what is due to emissivity, we need to estimate the emissivity. This can be quite difficult because only a limited amount of data are available on the emissivity of surfaces, particularly irregular surfaces as they are commonly found in imagery (e.g., road surfaces, grass, gravel, etc.). Where data are available (cf. Gubareff, 1960), they are often not spectral; they do not cover the correct spectral bandpass, or only total hemispheric data are available. Becker et al. (1985) have described an instrument for measuring the spectral and directional emissivity of surfaces, and Schott (1986) describes a pair of instruments that can be used in the field to measure both normal hemispheric and angular hemispheric emissivity values. More recently, Salisbury and D'Arian (1992) describe an instrument and measurements of the spectral emissivity of a variety of natural surfaces. Knowledge of emissivity is very important, since an error in emissivity of one unit

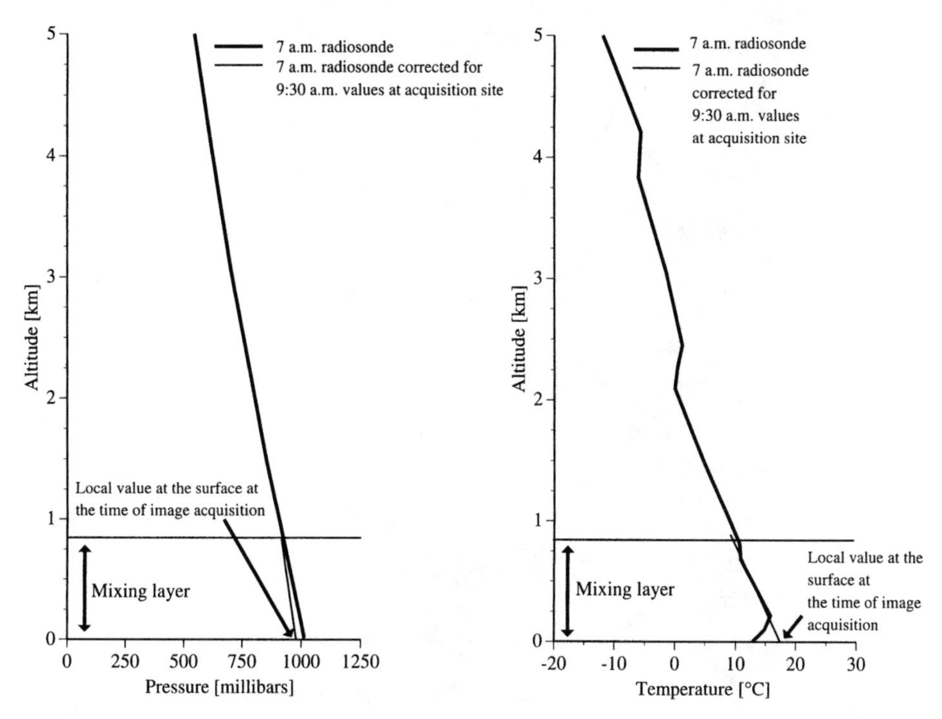

Figure 6.8 A simple method of correcting radiosonde data for local conditions. The radiosonde data were collected in Buffalo, New York at 7:00 a.m. The data were corrected for an acquisition over Rochester, New York (100 km away) using surface data acquired at the time of the overflight (9:30 a.m.). The correction involved a simple straight-line extrapolation from the top of the near-surface mixing layer to the surface.

(in the 8- 14-μm window) can commonly result in temperature errors of approximately 1 K. The importance of emissivity and emissivity error is downplayed in most of the quantitative results sited in the literature. This is because water is commonly used as the test target (particularly for satellite studies). The emissivity of water for near-normal viewing is wellknown and very high. This minimizes any error in temperature that would result due to emissivity effects and underestimates errors that will occur for objects with lower or less well-known emissivities. Lowe (1978) demonstrated an interesting approach for dealing with this problem. Airborne images were acquired with two sensors. An active system imaged the ground where it was illuminated with a CO_2 laser operating at 10.6 μm ,and passive thermal images were collected in the 10- to 12.6-μm region (cf. Fig. 6.9). The brightness in the active images was assumed to vary primarily as a function of emissivity (reflectivity), thus producing a linear emissivity map that could be calibrated from two or more known targets. These emissivities could then be used with the passive images to solve for temperature under the assumption that the objects in the scene behaved as gray bodies over the 10- to 12.6-μm region. Lowe (1978) points out that this may be a poor assumption for the silicates due to strong Restrahlen effects from quartz in this spectral band.

6.3.5 Summary of Thermal Calibration Methods

In summary, numerous approaches have been identified for measurement of surface temperature. Most of them have been shown under certain circumstances to yield accurate temper-

(a)

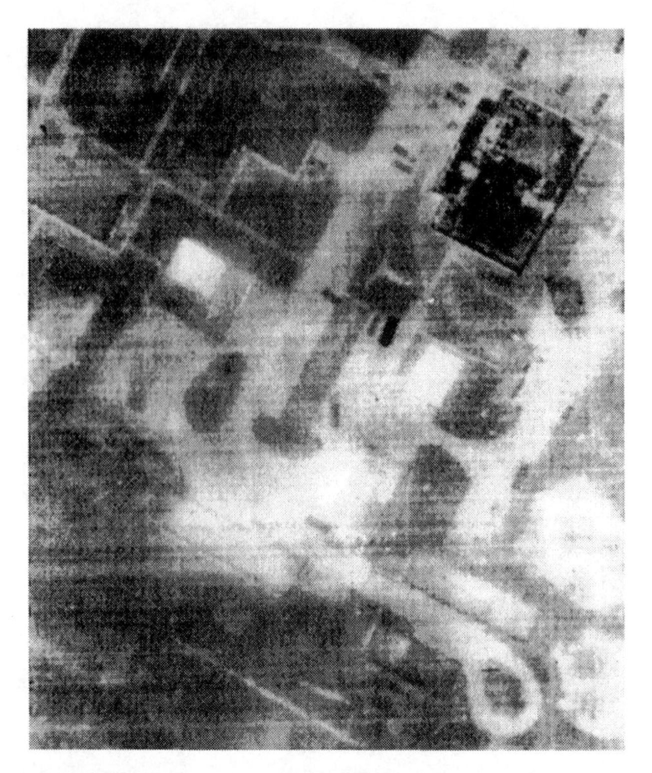

(b)

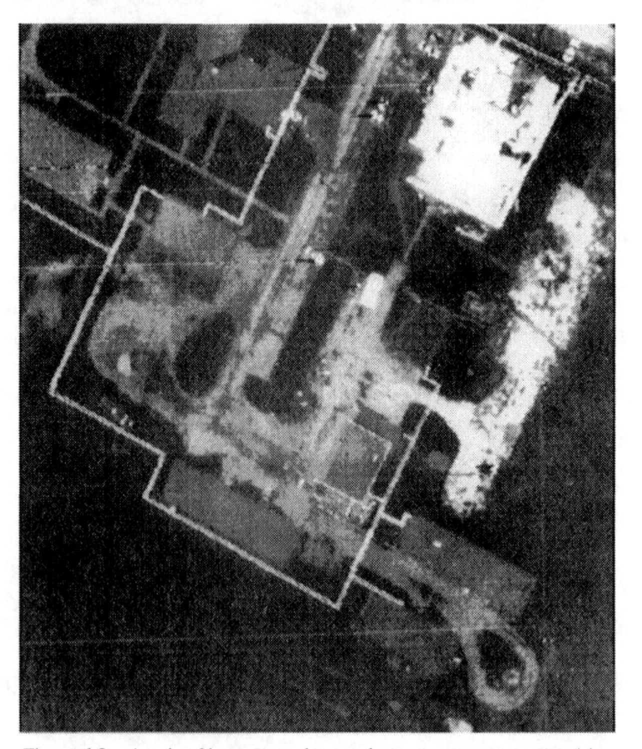

Figure 6.9 A pair of images used to produce a temperature map: (a) a passive thermal infrared image in the 10 to 12.6-μm band: (b) a reflective image obtained by illuminating the same scene as in (a) with an active CO2 laser at 10.6 um. [Images courtesy of D. Lowe (ERIM).]

ature measurements to better than 1 K. However, very few have undergone rigorous tests of robustness under operational conditions. Furthermore, some of the most accurate and best proven, such as the multiple-altitude or split-band approaches, require specific platform or sensor characteristics that can be prohibitive. We are thus often left with a situation where adequate calibration approaches still do not exist, or where the possible approaches have not been tested to the point where the expected results are known. In short, the old adage of caveat emptor applies, and the user should become quite familiar with a calibration technique in known situations before trusting the results.

The atmospheric calibration techniques discussed in this section are generally not applicable in regions of the EM spectrum, where reflective solar flux dominates the observed radiance. There are, however, a parallel set of calibration methods applicable in the solar reflective region, as we will see in the next section.

6.4. APPROACHES TO MEASUREMENT OF REFLECTIVITY

In the VIS-NIR and SWIR, the scientific questions have revolved around how to measure reflectance or, barring that, how to measure the relative reflectance between spectral bands. In this section, we will parallel the treatment used for calibration of images for temperature measurement by looking at ground truth, in-scene, and atmospheric propagation models for calibration of image data to reflectance. The underlying premise is that the type or condition (e.g., vegetation health, water quality, soil moisture) of the targets of interest are a function of reflectance. Therefore, if the image data can be calibrated to reflectance, they can be directly compared to laboratory and field measurements of reflectance to yield improved classification and condition assessment.

6.4.1 Ground Truth Methods (Reflectance)

The calibration of aerial and satellite systems using ground truth has generally relied on the use either of control panels or ad hoc control surfaces of known reflectance. The control panel approach is attractive because the range of reflectance values can be controlled by the fabrication process, and care can be taken to ensure that the samples are approximately Lambertian to minimize any errors that could be introduced by sensor view angle effects. The use of panels is the most-attractive calibration method if quality ground truth panels three or more times the ground instantaneous field of view of the sensor can be imaged at the time of data collection (cf. Fig. 6.10). The calibration is then a simple matter of regressing observed radiance values against known reflectance values in each band according to

$$L = (E_s \pi^{-1} + L_d) \tau_2 r_d + L_u \tag{6.43a}$$

$$L = m r_d + b \tag{6.43b}$$

where $E_s \cong E_s' \cos \sigma' \tau_1$, $\tau_2 \cong e^{-\delta' \sec\theta}$, $m = E_s \pi^{-1} \tau_2 + L_d \tau_2$ is the slope of the regression, and $b = L_u$ is the intercept. We will assume for most of this section that calibration sites are chosen such that the shape factor (F) is unity. In addition, we assume that τ_1, τ_2, and L_u are constant over the scene. When this is not the case (e.g., sensors with a wide field of view or a large area of coverage), corrections must be made for changes in the atmosphere over the scene. In many cases, this involves recalibrating the scene at several locations.

The reflectance of any Lambertian object or the apparent Lambertian reflectance of any object in the scene can then be computed by rearranging Eq. (6.43b). In fact if the sensor's

Figure 6.10 Ground control panels used for atmospheric calibration. (a) Control panels, (b) Image of control panels.

radiance is a linear function of digital count, the sensor need not be radiometrically calibrated to use this approach. The panel digital count values can be regressed with the reflectance values according to

$$\mathrm{DC} = gL + b_o \tag{6.44a}$$

$$= gmr_d + gb + b_o \tag{6.44b}$$

$$= m' r_d + b' \tag{6.44c}$$

where g and b_o are the unknown sensor gain and bias for the bandpass, and m' and b' are the slope and intercept of the regression of image DC values with known panel reflectance values.

In cases where logistics preclude the use of ground truth panels, the same effect can be achieved by measuring the reflectivity of several Lambertian objects in the scene with a range of reflectance values and again regressing against the observed radiance or digital count. Care must be taken in the selection and measurement of these ad hoc control surfaces to ensure that they are sufficiently diffuse over all relevant angles, and that they are uniform enough that the field sample will be characteristic of the image sample. This can be particularly difficult with satellite images where the ground sample distance may be tens or hundreds of meters. Figure 6.11 shows an image of the "reflectomobile" used by Slater et al. in characterizing the reflectance of large surfaces for satellite sensor calibration as described in Section 5.7 (cf. Slater et al., 1987).

It is important to note the assumptions implicit in the use of these standard ground truth methods. First, the calibration targets are assumed flat and level, with no neighboring obscuration; second, they are assumed homogeneous; and third, Lambertian. If these conditions are met and the calibration target reflectance values are wellknown, then the solution that

Figure 6.11 The reflectomobile used by Slater et al., 1987, in calibration of satellite sensors. The radiometers are calibrated in the field with reflectance standards and then driven over a study site to generate mean reflectance over a large, approximately uniform surface. (Courtesy of Phil Slater, University of Arizona.)

results should be very accurate. However, they will only be completely valid for targets with similar characteristics. Sloped surfaces and non-Lambertian surfaces will have errors that will be a function of how significantly they deviate from the calibration assumptions.

The slope effects can easily be corrected for Lambertian surfaces if the relative amount of skylight to total illumination (l) is known. In this case, it can be shown that the radiance from any angled surface is

$$L_i = \left[(m - lm)\cos\sigma_i' / \cos\sigma' + lm\right] r_d + b \tag{6.45}$$

where m and b are the slope and intercept from Eq. (6.43b), l is the ratio of skylight to total illumination in the bandpass $L_d/(E_s\pi^{-1}+L_d)$, and σ'_i is the angle from the normal of the ith target to the sun (recall that σ' is the angle from the normal of the calibration targets to the sun). Rearranging Eq. (6.45) yields the diffuse reflectance for a target at any orientation. The skylight to total illumination ratio in the bandpass can be determined with a radiometer located in the scene which records the irradiance onto a horizontal surface. This yields the total irradiance. By simply casting a shadow on the irradiance sensor with a disk whose shadow just blocks the sun, the skylight irradiance can be determined and the ratio computed (cf. Fig. 6.12). Note that the instruments used need only be linear and have no dark level (bias). No careful calibration is required. Since the skylight to total illumination ratio (l) can vary significantly with time and sky conditions, care must be taken to make these measurements coincident with the image acquisition and in the spectral band (s) of interest. The value of l can also be determined using in-scene methods as described in Section 6.4.2.

For non-Lambertian surfaces, solving for reflectance values requires assumptions regarding the shape of the bidirectional reflectance function. That treatment is beyond our

Figure 6.12 Eppley total irradiance pyroheliometer (a) and skylight pyroheliometer with sun band (b). These instruments measure the total (a) and skylight irradiance (b) onto a horizontal surface. The devices shown are for broad-band measurement. To obtain bandpass values, similar instruments with appropriate filters are used.

scope , so we will assume that we will be reporting the apparent diffuse reflectance of all samples (i.e., reflectance factors as described in Sec. 4.2.1).

6.4.2 In-Scene Methods (Reflectance)

In many cases, the logistics associated with acquiring good ground truth are impractical or impossible to meet. As a result, many methods of in-scene calibration have been developed.

One of the most interesting approaches takes advantage of the difference in radiance levels observed at shadow edges. Piech and Walker (1974) have shown how this radiance difference can be used to compute the upwelled radiance term (L_u) and the skylight to total illumination factor (l). The radiance in the bandpass observed just outside a shadow edge (L_s) and just inside a shadow (L_{sh}) cast on the same diffuse material can be expressed as:

$$L_s = \left[E_s \pi^{-1} \tau_2 + F L_d \tau_2\right] r_d + L_u \tag{6.46}$$

$$L_{sh} = F L_d \tau_2 r_d + L_u \tag{6.47}$$

where F is the fraction of the hemisphere above the target area that is sky [i.e., (1-F) is the fraction obscured by neighboring objects as shown in Fig. 6.13]. Combining Eqs. (6.46) and (6.47) by substitution of r_d and rearranging yields

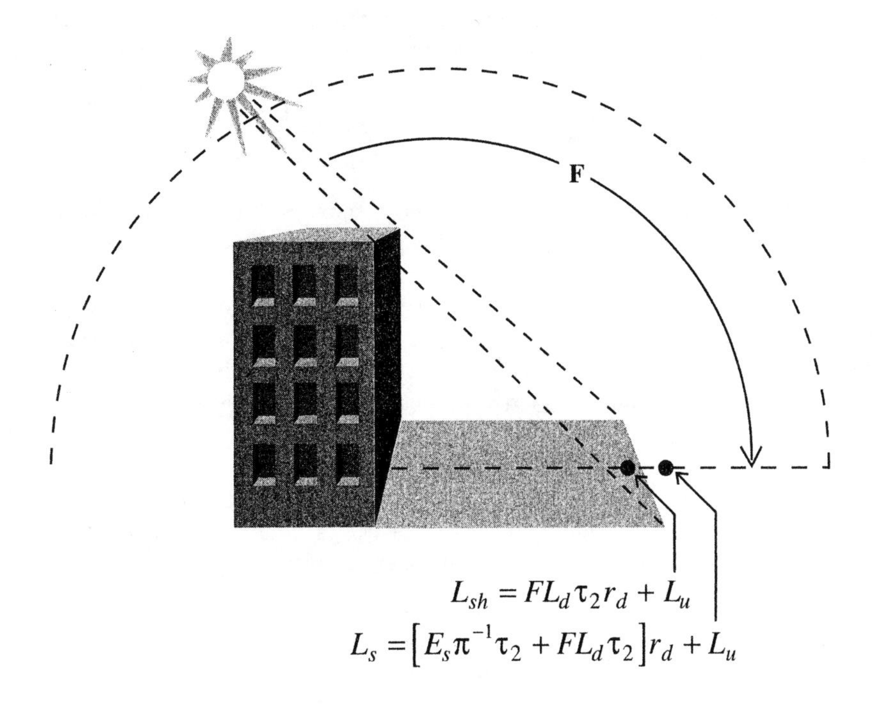

Figure 6.13 The radiance from the same surface just beyond a shadow edge L_s and just inside the shadow L_{sh}. The shape factor F is the fraction of the hemisphere above the shadow edge illuminated by skylight.

$$L_s = \frac{E_s\pi^{-1} + FL_d}{FL_d}L_{sh} - \frac{E_s\pi^{-1} + FL_d}{FL_d}L_u + L_u \tag{6.48a}$$

$$L_s = mL_{sh} + b \tag{6.48b}$$

where $m = (E_s\pi^{-1}+FL_d)/FL_d$ and $b = (1\text{-}m)L_u$.

For narrow fields of view, L_u can be assumed constant, and if objects of similar shape are selected, then F is also approximately constant. Therefore, m and b in Eq. (6.48b) are approximately constant and can be solved for by a linear regression of the radiance just in the shadow (L_{sh}) versus the radiance just in the sun (L_s) for several targets of varying reflectance. L_u can then be found from the intercept (b) and slope term (m) to be:

$$L_u = \frac{b}{1-m} \tag{6.49}$$

Rearranging the intercept term yields an expression for the skylight to total illumination (l) to be:

$$l = \frac{1}{1+F(m-1)} \tag{6.50}$$

where F can be computed from geometry for similarly shaped shadow-casting objects used in the regression (e.g., buildings).

Piech et al. (1978) suggest that, given the upwelled radiance (L_u) from Eq. (6.49), the total incident radiance term can be solved for using a statistical estimate of the mean observed radiance for a class of objects whose mean reflectance in the bandpass can be well estimated. In their studies, they commonly chose concrete. The necessary equations take on the form:

$$\left[E_s\pi^{-1}\tau_2 + L_d\tau_2\right] = \frac{L_{avg} - L_u}{r_{davg}} \tag{6.51}$$

where L_{avg} is the mean radiance observed for many samples of the standard material (concrete) sampled in the scene, and r_{davg} is the mean reflectance based on numerous laboratory or field measurements of the standard material (concrete).

Piech (1980) found that in applying this "scene color standard" approach to color aerial photography that reflectance errors were typically less than 1 unit below 10 reflectance units, and less than approximately 10% of the reflectance value above 10 reflectance units. Regrettably, this method cannot be used with imagery where the resolution is insufficient to show sharp shadow edges (i.e., most satellite data). As a result, alternative approaches to solve for the upwelled radiance are required.

One of the simplest methods called the *dark object subtraction* or *histogram minimum method* (HMM) is to set the minimum scene radiance to be the upwelled-radiance on the assumption that it represents the radiance from a scene element with near zero reflectivity (cf. Chavez, 1975 and Switzer et al., 1981), i.e.,

$$L = (E_s\pi^{-1} + L_d)\tau_2 r_d + L_u \cong L_u \text{ for } \mathrm{r_d} \approx 0 \tag{6.52}$$

This method works reasonably well in spectral bands where near-zero reflectors are available in close proximity to the study region, i.e., within a region where the atmosphere may be assumed constant, and when applied to brighter objects where any errors are relatively small. For example, in the near infrared, clear water has very low reflectance, and most studies are aimed at vegetation whose reflectance is quite large. On the other hand, this method often breaks down in the visible region where minimum scene reflectances may still be a few percent, which is also the reflectance of water and vegetation in this region. This limitation can be overcome if one considers multispectral data where the upwelled radiance (L_u) can be found with confidence in one spectral band (e.g., the NIR using clear water). Then an estimate for L_u in other bands can be found if we assume that the reflectance in the two bands are approximately correlated with zero bias such that

$$r_2 = Cr_1 + e \tag{6.53}$$

where r_1 and r_2 are the reflectance values in band 1 and band 2, C is approximately a constant, and e is the error due to the lack of perfect correlation between r_1 and r_2. If we then write the expression for the radiance observed from level objects in spectral bands 1 and 2 as:

$$L_1 = (E_{s1}\pi^{-1} + L_{d1})\tau_{21} r_{d1} + L_{u1} = m_1 r_{d1} + L_{u1} \tag{6.54a}$$

$$L_2 = (E_{s2}\pi^{-1} + L_{d2})\tau_{22} r_{d2} + L_{u2} = m_2 r_{d2} + L_{u2} \tag{6.54b}$$

and combine using Eq. (6.53), we have

$$L_2 \cong \frac{m_2 C}{m_1}(L_1 - L_{u1}) + L_{u2} + e \tag{6.55}$$

If we let band 1 be the NIR band where the upwelled radiance is known using the dark-object method, then for many objects [preferably level and of materials where we believe Eq. (6.53) is valid], we can regress the radiance in band 2 (e.g., any visible band) against the radiance in band 1 after dark-object subtraction (i.e., $L_1 - L_{u1}$). The regression minimizes the error term (e), such that the intercept is approximately the upwelled radiance in band 2 (L_{u2}.) This process can be repeated for each band of interest.

To complete the full calibration, one must still solve for m_1 and m_2 using a method such as the scene color standard approach described above. However, many investigations only require the variation in the relative spectral reflectance ratio. This value can be expressed as:

$$\frac{L_1 - L_{u1}}{L_2 - L_{u2}} = \frac{m_1 r_{d1}}{m_2 r_{d2}} \propto \frac{r_{d1}}{r_{d2}} \tag{6.56}$$

where m_1/m_2 can be assumed constant (e.g., in level terrain). Although not rigorously valid in sloped terrain, Eq. (6.56) is approximately valid when the skylight term is small compared to the sunlight term (i.e., the cos σ' effects—which vary from pixel to pixel—approximately cancel), i.e.,

$$\frac{L_1 - L_{u1}}{L_2 - L_{u2}} = \frac{E'_{s1}\pi^{-1}\tau_{11}\tau_{21}\cos\sigma' + \tau_{21}L_{d1}}{E'_{s2}\pi^{-1}\tau_{12}\tau_{22}\cos\sigma' + \tau_{22}L_{d2}} \frac{r_{d1}}{r_{d2}} \propto \frac{r_{d1}}{r_{d2}} \tag{6.57}$$

Equation (6.57) has been used in geologic studies to minimize pixel-to-pixel terrain effects and increase variations due to material changes as characterized by reflectance ratios. Crippen (1988) discusses the importance of correcting for the additive (path radiance) effects when using band ratios in regions with topographic relief. This is illustrated in Figure 6.14 in which the simple band ratio image is dominated by topographic effects, though reversed in contrast from the original images. After subtracting for additive effects, the topographic effects are greatly reduced, and a mine and tailings show up distinctly as bright (high ratio) objects.

We should point out that most of these in-scene methods do not require absolute sensor calibration if the sensor output (e.g., digital count) is a simple linear function of incident radiance. In these cases, the relationships of Eq. (6.44) can be used. Instead of using radiance values, digital count values can be used directly in the in-scene methods presented here. For example, Eq. (6.55) reduces to:

$$\mathrm{DC}_2 = \frac{g_2 m_2 C}{g_1 m_1}(\mathrm{DC}_1 - \mathrm{DC}_{u1}) + \mathrm{DC}_{u2} + e \tag{6.58}$$

where $\mathrm{DC}_{ui} = g_i L_{ui} + b_{oi}$ is the digital count in the ith band that would be produced by the upwelled radiance in the ith band (L_{ui}), and the magnitude of the error term is modified by the sensor gain. DC_{u1} can be approximated in the NIR band using the minimum digital count or dark-object method. The regression of DC_2 values versus (DC_1 - DC_{u1}) values produces DC_{u2} as the intercept. Substitution of digital count values in the LHS of Eq. (6.57) results in a solution for the RHS that is similar to Eq. (6.57):

$$\frac{\mathrm{DC}_1 - \mathrm{DC}_{u1}}{\mathrm{DC}_2 - \mathrm{DC}_{u2}} = \frac{g_1}{g_2} \frac{(E'_{s1}\pi^{-1}\tau_{11}\tau_{21}\cos\sigma' + \tau_{21}L_{d1})}{(E'_{s2}\pi^{-1}\tau_{12}\tau_{22}\cos\sigma' + \tau_{22}L_{d2})} \frac{r_{d1}}{r_{d2}} \propto \frac{r_{d1}}{r_{d2}} \tag{6.59}$$

A similar analysis can often be performed using digital count values with the sensor calibration incorporated into the atmospheric calibration. In many cases, combining the calibrations can reduce the overall errors in the calibration process by simultaneously minimizing errors.

With all of these methods, residual errors in low-reflectance targets are still of concern both when full calibration to reflectance is required and when using reflectance ratios. The

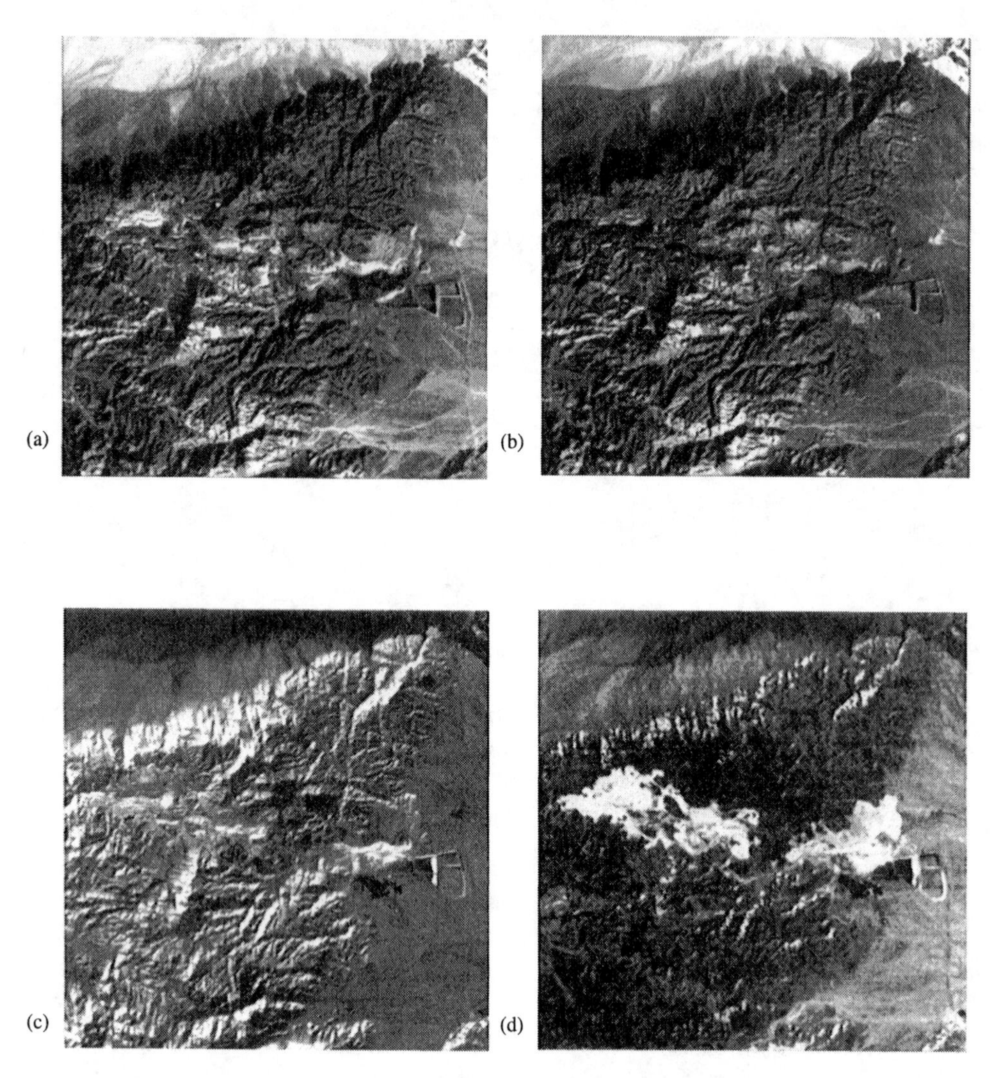

Figure 6.14 Images showing how illumination effects can be reduced in ratio images if additive effects due to the sensor bias and path radiance are removed. (a) Landsat TM band 1 subscene of the Eagle Mountains Iron Mine, (b) band 4, (c) band 1/ band 4 (note reversal of topographic effects,) (d) band 1/ band 4 after subtraction for additive effects (note how the topographic effects are reduced and the high-ratio signature of the mine and tailings becomes apparent.) (Images courtesy of R. E. Crippen NASA/JPL.)

need for a reflectance standard in the scene is also often a problem in regions without man-made features or where the resolution precludes isolating unique materials (e.g., pure pixels of concrete). As a result, atmospheric propagation models must sometimes be considered.

6.4.3 Atmospheric Propagation Models (Reflectance)

The LOWTRAN model described in Section 6.3.3 can also be used for computing the terms in the reflective radiance equation. The LOWTRAN-7 model deals with multiple scattering effects, and when properly manipulated, this type of model will yield estimates for all of the atmospheric terms in the governing equation. Most atmospheric propagation models use numerical solutions to radiance equations similar to those introduced in Sections 4.1 to 4.3 coupled with tables of data on the spectral properties of the atmospheric constituents. The main limitation with LOWTRAN and other similar atmospheric propagation models in the wavelength regions where solar scattering is important is the difficulty in adequately characterizing the makeup of the atmosphere (particularly aerosols). As a result, the angular scattering coefficients associated with the atmospheric layers are often poorly characterized.

Slater et al. (1987) have shown how atmospheric propagation models can be used to characterize the atmosphere when supplemented with ground-based sun photometer readings. The sun photometers are used over time to measure the optical depth of the atmosphere using the Langley plot method (cf. Slater, 1980). The total atmospheric optical depth δ' is computed as minus the slope of the logarithmic relation between the observed irradiance (E_{obs}) and the solar declination angle (σ), i.e.,

$$\ln E_{obs} = \ln(E_s' e^{-\delta \sec\sigma}) = \ln E_s' - \delta \sec \sigma \tag{6.60}$$

where E_{obs} is the irradiance recorded by a sensor aimed directly at the sun with a field of view just large enough to fully encompass the solar disc. To simplify the treatment, the total optical depth can be treated as:

$$\delta = \delta_g + \delta_w + \delta_a \tag{6.61}$$

where δ_g is the optical depth associated with the permanent gasses, δ_w is the optical depth associated with water vapor, and δ_a is the optical depth associated with aerosols. If δ_g and δ_w can be well estimated by atmospheric propagation models where good radiosonde data are available, then aerosol optical depth (δ_a) can be found from Eq. (6.61) using model derived values for δ_g and δ_w and the experimental value of the overall optical depth (δ). The modeled values of aerosol concentration can then be adjusted such that the model predicts the observed value (by working at several wavelengths significant improvements in the characterization of the aerosols is possible). The "calibrated" model can then be used to compute the necessary terms in the governing equation. Slater et al. (1987) have shown that this type of approach can work quite well, particularly in low-turbidity atmospheres. Ongoing limitations are associated with the need for good radiosonde data, surface irradiance measurements, and stable atmospheres, all of which can be prohibitive. It is particularly important to realize that if one does not use some method of enhancing the calibration of the atmospheric models (such as the Langley plot method), significant errors can occur, particularly at shorter wavelengths where scattering is important. This is because scattering is such a significant component of the transmission, illumination, and path radiance terms, and the radiosonde data, upon which these models rely, is insufficient to predict the magnitude and shape of the aerosol scattering function.

The radiation propagation models are very attractive, because they will yield a complete solution for all of the parameters in the governing equation at all wavelengths. In addition, these models include corrections for many of the variations in the atmospheric parameters due to changes in elevation, slant range, and view angle. Most of the other methods assume that these values are constant over the study area. However, the errors from the models in spectral regions where scattering is important can be very large if the models are not somehow adjusted for the magnitude and often the shape (phase function) of the scattering coefficient. We should point out that this is not a fundamental limitation of the models, but of our knowledge of the input parameters the models require. As methods evolve to calibrate the inputs to these models, they will undoubtedly play an increasing role in atmospheric calibration. At present, essentially the only way to properly calibrate the models is with extensive ground truth, e.g., field radiometry measurements that are impractical for most operational uses.

Because of the difficulties in achieving a full calibration to reflectance, many users have opted for various types of relative calibration techniques. Also, in many cases, our knowledge of the relationship between reflectance and application-specific parameters is still poorly characterized and is often only expressed in relative terms (e.g., spectral reflectance ratios or changes in reflectance over time). Thus, in many applications relative calibration may provide much of the data that can be readily used by our current knowledge base.

6.5 RELATIVE CALIBRATION (REFLECTANCE)

Of all the various relative calibration or normalization approaches, we will concentrate on two general categories. The first is the use of ratios or relative spectral information between multiple spectral bands of the same scene. The second involves the use of normalization techniques to reduce atmospheric variations between multiple images of the same area acquired at different times.

6.5.1 Spectral Ratio techniques

As introduced earlier (Sec. 6.4.2), in certain cases where the absolute reflectance is not required, the relative reflectance between spectral bands can provide significant information. Piech et al. (1978) point out how the blue-to-green reflectance ratio can be very useful in the study of water quality, and many investigators have used near infrared to red reflectance ratios to characterize vegetation condition. Piech et al. (1978) describe how reflectance ratios can be estimated given that one knows the upwelled radiance (L_u) in each spectral band (e.g., by using the shadow method or the minimum radiance method). The radiance in spectral bands 1 and 2 can then be expressed as

$$L_1 = (E_{s1}\pi^{-1}\tau_{21} + L_{d1}\tau_{21})r_{d1} + L_{u1} = \alpha_1 r_{d1} + L_{u1} \tag{6.62}$$

$$L_2 = (E_{s2}\pi^{-1}\tau_{22} + L_{d2}\tau_{22})r_{d2} + L_{u2} = \alpha_2 r_{d2} + L_{u2} \tag{6.63}$$

where L_1 and L_2 are the observed radiance values in bands 1 and 2 for an object with diffuse reflectivity values r_{d1} and r_{d2}. These equations can be combined to yield

$$L_1 = \frac{\alpha_1}{\alpha_2}\frac{r_{d1}}{r_{d2}}L_2 - \frac{\alpha_1}{\alpha_2}\frac{r_{d1}}{r_{d2}}L_{u2} + L_{u1} \tag{6.64}$$

Let's assume a class of objects can be identified whose spectral reflectance ratios approximate a constant of known value. For example, Piech et al. (1978) observed that the class of manmade "gray" objects (e.g., concrete, asphalt, gravel roofs) exhibited this behavior. They generated average reflectance ratio values for the visible and near-infrared reflectance regions based on many laboratory and field samples. For such a class of objects, Eq. (6.64) becomes a linear equation of the form

$$L_1 = \frac{\alpha_1}{\alpha_2}k_{12}L_2 - \frac{\alpha_1}{\alpha_2}k_{12}L_{u2} + L_{u1} \tag{6.65}$$

$$L_1 = mL_2 + b \tag{6.66}$$

where m and b are the slope and intercept of a linear regression of the radiance observed in band 2 against the radiance observed in band 1 for samples from the class of objects with constant reflectance ratio $k_{12} \cong r_{d1}/r_{d2}$. The reflectance ratio for any unknown sample can then be expressed as:

$$\frac{r_{d1}}{r_{d2}} = \frac{L_1 - L_{u1}}{L_2 - L_{u2}} \frac{k_{12}}{m} \tag{6.67}$$

The precision of this method was observed to be approximately 10% of the measured ratio when the upwelled radiance values (L_u) were computed using the shadow method.

Many investigators have found that by simply ratioing raw digital counts, they can crudely approximate reflectance ratios with significant, but often tolerable errors. The commonly used normalized difference vegetation index (NDVI), suggested by Rouse et al. (1973), is a slightly more sophisticated version of this approach, which attempts to reduce atmospheric and illumination effects by using a differencing and ratioing method applied to observed digital count values, i.e.,

$$\text{NDVI} = \frac{\text{DC}_{\text{IR}} - \text{DC}_{\text{R}}}{\text{DC}_{\text{IR}} + \text{DC}_{\text{R}}} \tag{6.68}$$

where DC_{IR} and DC_{R} are respectively the digital count values in the near infrared and red spectral bands. Figure 1.2 illustrates how several AVHRR images can be combined to generate an NDVI image over very large regions. By producing sequences of NDVI images over time, large-scale changes in biomass can be monitored.

6.5.2 Scene-to-Scene Normalization

In cases where only the relative change between two images is of interest, or where we only want to improve the appearance of an image, radiometric normalization techniques can be used.

The simplest of these is a method called *histogram matching* or *histogram specification* (cf. Gonzales and Wintz, 1987). The method involves passing one image through a look-up table that attempts to adjust the histogram to match some specified histogram (cf. Sec. 7.1.1). Schott et al., (1985) utilized this approach to visually enhance satellite images by removing the apparent atmospheric effects. This was accomplished by matching the histogram of each spectral band to the corresponding histogram of a very low-altitude image of similar scene content. Since the low-altitude scene had almost no atmospheric effects, the resulting satellite image "appears" to have the atmospheric effects removed. This is a nonquantitative approach aimed solely at improving the visual appearance.

This same approach has been used to normalize images of the same scene taken at different times. The process removes the dominant brightness variations between the scenes caused by illumination and sensor response variations. The resultant scenes "appear" as if they were taken under the same imaging conditions (i.e., same sensor response, same atmospheric effects, and same illumination levels). However, because the histogram matching technique is a simple numerical fit, it can mask real changes between the images. For example, given a scene where a major portion of the image area was water which was turbid in the first scene and clear in the second, the histogram matching approach would try to force the two scenes to look the same. As a result, after modification the second scene would not only look like the first in terms of the imaging conditions but also what was clear water would appear turbid, thus masking rather than enhancing real differences between the scenes. Because the transforms are based on whole-image statistics, this masking effect will be minimized when the differences between the scenes only represent a small number of pixels. Small-scale effects like new buildings or roads will be clearly differentiated. Large-scale subtle changes, e.g., water quality effects, vegetation condition, etc., will tend to be masked by the process.

Building on the histogram matching approach, several authors have suggested a more quantitative scene-to-scene normalization built on radiation propagation principles and designed to eliminate the undesirable effects of the global histogram matching approach. To derive the normalization transform, the radiance on each day is expressed as [cf. Eq. (6.62)]:

$$L_i = \alpha_i r_{di} + L_{ui} \tag{6.69}$$

The digital count on each day can then be expressed as:

$$\mathrm{DC}_i = g_i L_i + b_{oi} \tag{6.70}$$

$$= g_i \alpha_i r_{di} + g_i L_{ui} + b_{oi} \tag{6.71}$$

if the sensor's response is linear with radiance. For each band, the relationship between the digital count on day 1 (DC_1) and day 2 (DC_2) for pixels with the same reflectance can be written as:

$$\mathrm{DC}_1 = \frac{g_1\alpha_1}{g_2\alpha_2}\mathrm{DC}_2 - \frac{g_1\alpha_1}{g_2\alpha_2}(g_2 L_{u2} + b_{o2}) + g_1 L_{u1} + b_{o1} \tag{6.72a}$$

$$= m\mathrm{DC}_2 + b \tag{6.72b}$$

where m and b are the coefficients that would transform a digital count on day 2 to have the same digital count on day 1 for pixels that have the same reflectance on each day (cf. Fig. 6.15). That is, this transform would make the day-2 image appear quantitatively as though it were taken under the same sensor response (gain and bias), illumination, and atmospheric conditions (α and L_u) as the day-1 image. As a result, after the transform any differences between the images would be real since pixels with the same reflectance on the 2 days would have the same digital count.

The normalization process reduces to solving for the linear transform coefficients [m and b in Eq. (6.72b)] for each spectral band of interest. The simplest solution (cf. Hall et al., 1991) is to identify two objects, or classes of objects, whose reflectance is assumed constant between the two images. To reduce errors, a dark object and a bright object are chosen, and the digital counts (or mean digital counts from a sample) for the objects are acquired from each image. These "dark object – bright object" samples yield a two-point solution to the simple linear equation relating the digital counts on the 2 days [Eq. (6.72b)]. This method assumes that the objects chosen (or their mean if sampling is done) will have the same reflectance on both days. This method can be difficult to apply in scenes where unique objects with invariant reflectances are difficult to identify. Even where invariant objects exist, due to the small sample size, a slight error in one term can lead to large normalization errors (e.g., if the dark object is water and the water on day 2 is more turbid than on day 1). A regression method used by Jensen (1983) overcomes this sample limitation by spatially registering the two scenes and then performing a simple linear regression of day-1 digital

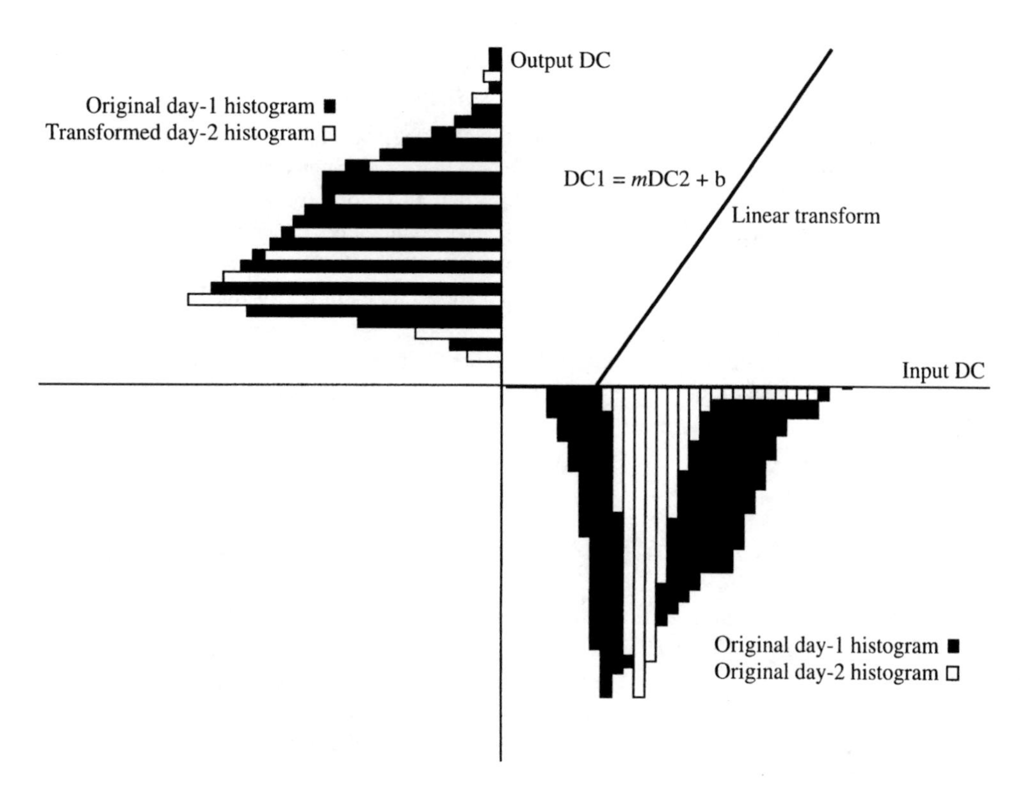

Figure 6.15 Linear lookup table utilized to transform a day-2 image to appear as if imaged under the same conditions as the day-1 image.

counts against day-2 digital counts to find the coefficients (m and b) in Eq. (6.72b). This method implicitly assumes that the average reflectance values on day 1 and day 2 are the same and that the large sample size and least-squares method will reduce the errors due to those pixels whose reflectance has changed. This can be a problem when changes occur over a large portion of an image, since these real changes will tend to be normalized out by the regression process. Problems can also be introduced if significant changes occur between the two dates in a feature class that represents sufficient pixels to perturb the regression result.

Yuan and Elvidge (1993) have suggested a method for interactively selecting a subset of the total scene for input to the regression calculation. This method is designed to reduce effects due to class differences between the scenes, including effects such as clouds, snow, and shadows. However, subtle changes in vegetation or water covering large portions of the scene will still tend to be masked by this enhanced regression method. Schott et al. (1988) proposed a method that uses the invariant reflectance concept employed by the "dark object – bright object" method but employs large sample sizes (like the regression method) to reduce errors. It is premised on the user's ability to isolate a class of pixels called *pseudo-invariant features* (PIF) whose reflectance distribution is assumed to remain nearly constant over time (e.g., manmade surfaces such as roads and rooftops). The values of m and b in Eq. (6.72b) can then be expressed as:

$$m = \frac{\sigma_1}{\sigma_2} \tag{6.73}$$

$$b = \mathrm{DC}_{1\mathrm{avg}} - \frac{\sigma_1}{\sigma_2}\mathrm{DC}_{2\mathrm{avg}} \tag{6.74}$$

where σ_1 and σ_2 are the standard deviations of the class of invariant pixels segmented on day 1 and day 2, respectively, and $\mathrm{DC}_{1\mathrm{avg}}$ and $\mathrm{DC}_{2\mathrm{avg}}$ are the means of these same classes. Salvaggio (1993) indicates that changes of more than about one reflectance unit should be above the noise level of this process in Landsat TM scenes where a good sampling of PIF features is possible. This method is limited, however, to scenes where a reasonable number of invariant features can be isolated. Figure 6.16 shows an example of the PIF transformation applied to a portion of a Landsat TM scene.

In addition to the simple ratios and NDVI methods presented in this section, a number of image processing techniques have been employed to reduce the effect of the atmosphere on image analysis. Among the most common of these is the principle component transform described in Section 7.3.

6.6 ATMOSPHERIC CALIBRATION USING IMAGING SPECTROMETER DATA

A large proportion of the new work in the field of atmospheric calibration is focused on the potential of new sensors. In particular, hyperspectral imaging (introduced in Sec. 5.5), in addition to providing a new way to sample information about surfaces, also provides much more information about the atmosphere. The narrow bandpass data allow us to sample the absorption lines of several of the relevant atmospheric constituents, and thereby back out their effects on the overall scene. These sensors offer the potential for directly measuring the detailed condition of the atmosphere using overhead imagery. For example, Gao and Goetz (1990) show how water vapor values can be spatially mapped by analyzing airborne imaging spectrometer data.

This method uses the water vapor absorption lines at 0.94 or 1.14 μm, as well as the region on either side of the absorption lines as observed by the radiance recorded by an imaging spectrometer. If the scattered terms (L_u and L_d) are assumed to be small, the product of target reflectance times total path transmission can be expressed as a radiance ratio ($R_{\Delta\lambda}$), i.e.,

$$R_{\Delta\lambda} = \frac{L_{\Delta\lambda}}{E'_{s\Delta\lambda}\pi^{-1}\cos\sigma} \cong \tau_{1\Delta\lambda}\,\tau_{2\Delta\lambda}\,r_{d\Delta\lambda} \tag{6.75}$$

where $L_{\Delta\lambda}$ is the radiance in a narrow bandpass $\Delta\lambda$ observed with a calibrated imaging spectrometer, $E'_{s\Delta\lambda}\pi^{-1}\cos\sigma$ is the radiance in the bandpass for a 100% reflector at the top of the atmosphere (this can be obtained from tabulated solar spectral data, and the earth/sun distance), $\tau_{1\Delta\lambda}\,\tau_{2\Delta\lambda}$ represents the transmission along the total sun-target-sensor path, and $r_{d\Delta\lambda}$ is the reflectance in the narrow bandpass. From spectrometer data of a single pixel, a spectral shape for the total path transmission modified by reflectance can be represented as the radiance ratio from the spectrometer data. The reflectance is assumed to be approximately linear over a narrow spectral region, including the absorption line and its shoulders. If the transmission just outside the absorption line to either side is assumed constant, then the shape of the reflectance "line" in the region of interest can be defined using a straight-line fit between the wings of the absorption line. Using an initial estimate for the atmospheric makeup and a high spectral resolution atmospheric propagation model, an estimate of the total path transmission ($\tau_{1\Delta\lambda}\,\tau_{2\Delta\lambda}$) can be produced. This is modified by the spectral shape of the reflectance and convolved with the spectral response of the spectrometer to produce an ini-

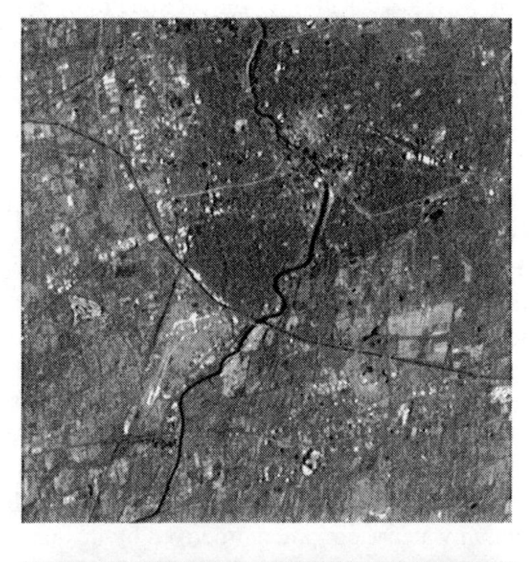

1982 image

1984 image

1982 image transformed to "look like" the 1984 image

Figure 6.16 Pseudo-invariant feature (PIF) transformation applied to a portion of a Landsat scene. See color plate 6.16.

tial estimate of the shape of the radiance ratio ($R_{\Delta\lambda}$) as calculated from the spectrometer data. The estimate is then iteratively adjusted by changing the water vapor content in the atmospheric propagation model until an adequate match to the observed spectra is obtained. This process can be repeated for every pixel in the scene. The resultant data can be corrected for path length effects and converted to the amount of water in the vertical atmospheric column over the pixel, based on the atmospheric model chosen. This simplified treatment is valid for nonvegetated regions where no liquid water is present. A more rigorous treatment is required over vegetation and where path-scattered radiance is significant (cf. Gao and Goetz, 1990).

Green (1993) describes a similar approach where the Rayleigh atmosphere and water vapor are estimated along with the reflectance of the target and, where relevant, the shape of the liquid water transmission spectra in the spectral region of interest. The leaf water absorption is important over vegetated regions, since it will change the spectral shape of the radiance in the atmospheric absorption region for water vapor. The atmospheric estimates are derived from MODTRAN, and the radiance at the sensor is repeatedly estimated for different-column water vapor amounts for each pixel until a best fit between the predicted and observed spectra is obtained (cf. Fig. 6.17). Figure 6.18 shows three bands of an AVIRIS image of Jasper Ridge and the vertical column water vapor amount derived by this method. Note in particular how the column water vapor is inversely correlated with topography due to the longer path through the atmosphere in the central portion of the image compared to the ridge along the left side of the image. Goetz (1993) points out the importance of observing this type of spatial variation in the atmosphere in terms of its impact on the apparent reflectance observed by Landsat TM sensors. He points out, for example, that in TM band 5 (the TM band most susceptible to water vapor variation), retrieved reflectance values could be in error by as much as 15% of the actual value if variations in water vapor content across a scene are not accounted for.

6.7 SUMMARY OF ATMOSPHERIC CALIBRATION ISSUES

We have presented a number of approaches to atmospheric calibration in this section. They reflect the wide range of tools being used in different spectral bands, for different sensors, and for differing user requirements. The large number of methods is probably most indicative of the fact that well-accepted operational calibration methods applicable to a broad range of sensors do not yet exist. Much ongoing work is directed at trying to develop more accurate and more robust calibration methods. Some of the most interesting work involves the use of spectrometer data to help characterize the atmosphere (cf. Sec. 6.6). Not all of this work involves imaging systems. For example, infrared microwave sounders are being used from space to retrieve atmospheric temperature profiles and total column absorber numbers of certain gases. If these sounders are flown in formation with imaging satellites, they can provide input to atmospheric propagation models. Geumlek et al. (1993) estimates that the performance specifications for atmospheric retrieval of temperature and water vapor on the current defense meteorological satellite program (DMSP) sounders may be sufficient for retrieval of apparent surface temperatures in the LWIR to within 1 K (one sigma). These types of "hardware" solutions (e.g., flying satellite sensors in formation) are very exciting for the long run, but they are not yet readily available. In the interim, many users are looking to hybrid solutions that combine aspects of the various approaches described in this chapter. A good example of this is discussed by Moran et al. (1993) where they show that the in-scene method of dark-object selection for estimating upwelled radiance when coupled with atmospheric propagation models can potentially provide an operational calibration in the reflective region. These results from a limited study under clear atmosphere conditions indicate that reflectance errors of 2 reflectance units (0.02) may be achievable with this type of hybrid method.

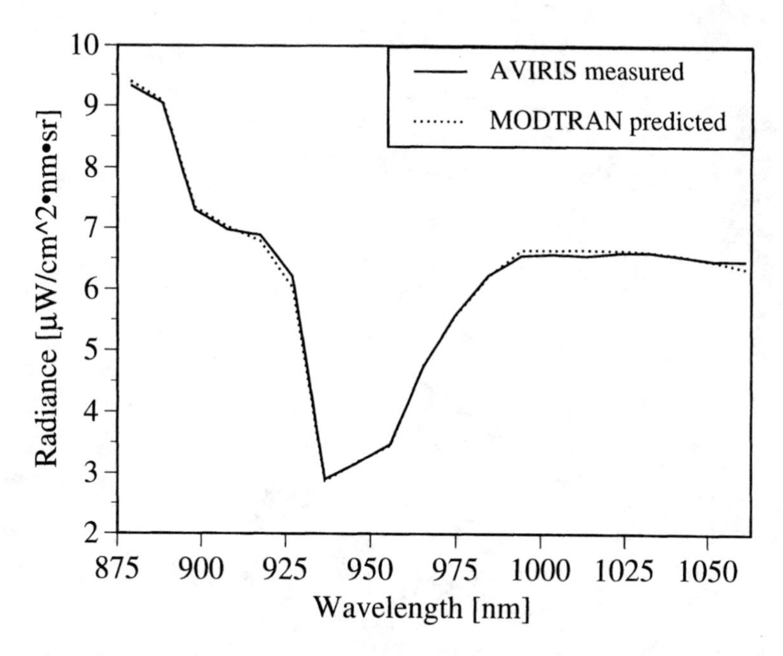

(a) Comparison of measured and predicted spectral characteristics near the 940-nm water absorbtion band.

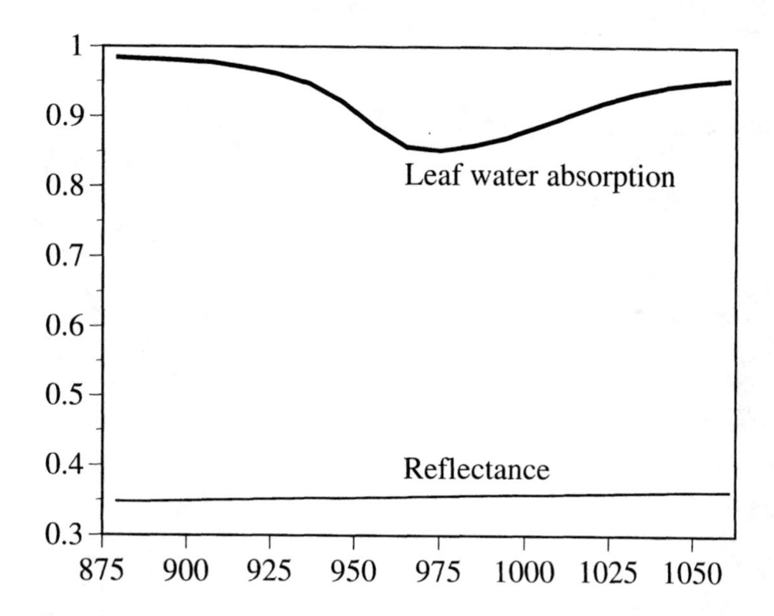

(b) Leaf water absorbtion spectra and reflectance spectra used in the estimation model to generate the predicted spectra shown in (a).

Figure 6.17 Iterative methods can be used to generate column water vapor amounts on a pixel to pixel basis by matching observed and predicted sperctral radiance values (a) when reflectance shape and target absorber spectra are included in the prediction process (b). Data courtesy of NASA JPL.

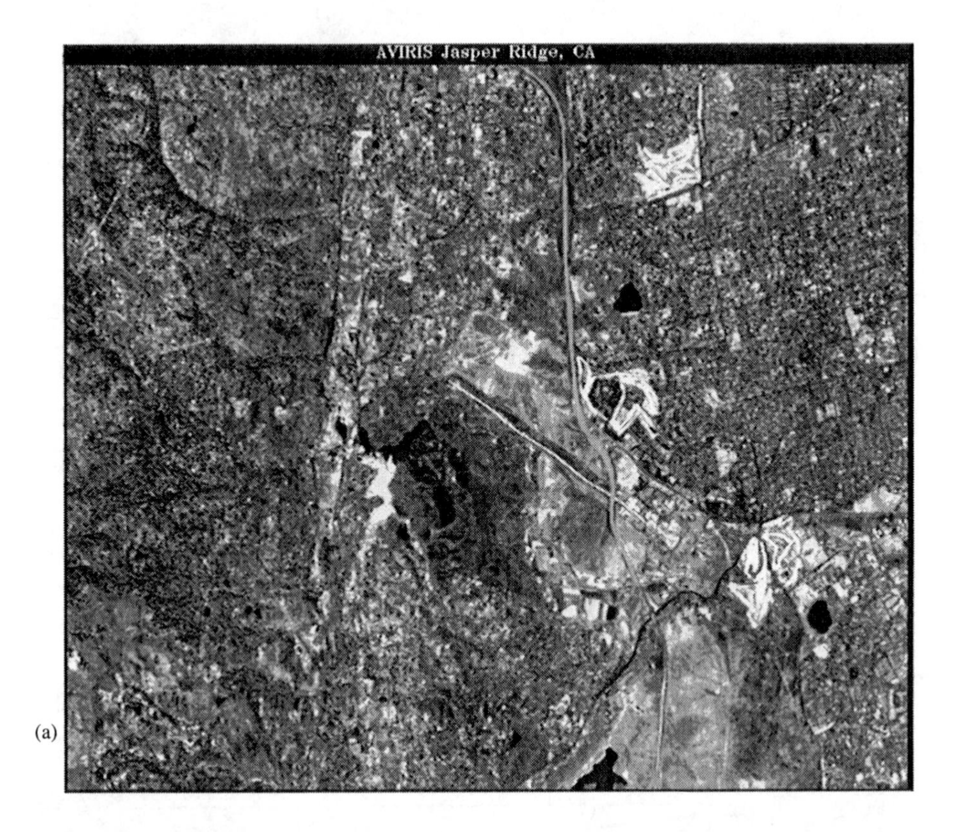

(a)

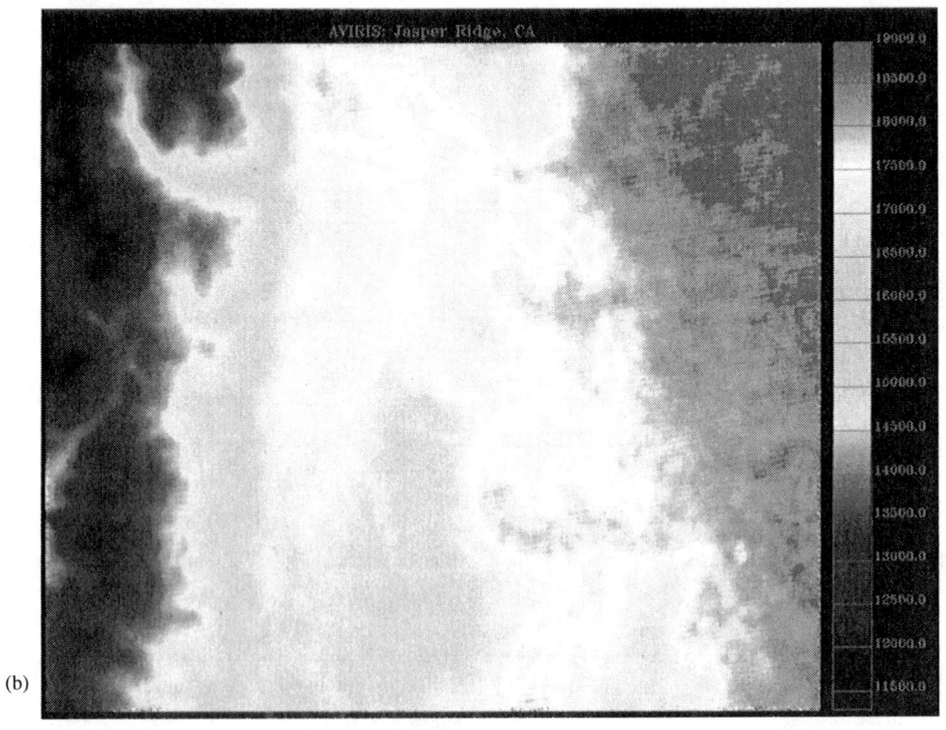

(b)

Figure 6.18 (a) AVIRIS image of Jasper Ridge, CA, and (b) a map of column water vapor content derived from the 940-μm absorption line data in the AVIRIS spectra. See color plate 6.18. (Image and map courtesy of NASA JPL.)

In summary, a number of potential methods for atmospheric calibration exist. However, in general the burden is still on the user to determine the accuracy (e.g., temperature or reflectance errors) required for the specific task at hand, and then to determine which of the possible calibration methods can meet the requirements.

6.8 REFERENCES

Anding, D. C. & Walker, J.P. (1975). "Use of Skylab EREP data in a sea surface temperature experiment." Science Applications, Ann Arbor, MI, NASA-CR-144479.

Becker, F., Ramanantsizehena, P., & Stoll, M. (1985). Angular variation of the bidirectional reflectance of bare soils in the thermal infrared band. *Appl. Opt.*, Vol. 24, No. 3.

Bell, E. F., Eisner, L., Young, J., & Oetjen, R.A. (1960). Spectral radiance of sky and terrain at wavelengths between 1 and 20 microns II, sky measurements. *J. Opt. Soc. Am..*, Vol. 50, No. 12, pp. 1313-02.

Berk, A., Bernstein, L.S., & Robertson, D.C. (1989). MODTRAN: a moderate resolution model for LOWTRAN 7. GL-TR-89-0122, Spectral Sciences.

Byrnes, A.E. & Schott, J.R. (1986). Correction of thermal imagery for atmospheric effects using aircraft measurement and atmospheric modeling techniques. *App. Opt.*, Vol. 25, No. 15.

Chavez, P.S. (1975). "Atmospheric, solar, and MTF corrections for ERTS digital imagery." Proceeding of the American Society of Photogrammetry, Falls Church, VA., p. 699.

Chedin, H., Scott, N.A., & Berroir, A. (1982). A single channel, double-viewing angle method for sea surface temperature determination from coincident METEOSAT and TIRO-N radiometric measurements. *J. of App. Meteor. Soc.*, Vol. 21, pp. 613-618.

Crippen, R.E. (1988). The dangers of underestimating the importance of data adjustments in band ratioing. *International Journal of Remote Sensing*, Vol. 9, No. 4, pp. 762-776.

Gao, B.C., & Goetz, A.F.H. (1990). Column atmospheric water vapor and vegetation liquid water retrievals from airborne imaging spectrometer data. *J. of Geophysical Res. - Atm.*, Vol. 95, pp. 3549-64.

Goetz, A.F.H. (1993). "Effects of water vapor and cirrus clouds on TM-derived apparent reflectance-based on AVIRIS experience." Proceeding of the Workshop on Atmospheric Correction of Landsat Imagery, Torrance, CA, sponored by the Defense Landsat Program Office, pp. 55-60.

Gonzalez, R.C., & Wintz, P. (1987). *Digital Image Processing*, Addison-Wesley, Reading, MA.

Gordon, H.R., Brown, D.K., Brown, J.W., Evans, O.B., Broenkow, R.H.B., & Broenkow, W.W. (1983). Phytoplankton pigment concentraitons in the Middle Atlantic Bight: comparison of ship determinations and CZCS estimates. *Appl. Opt*, Vol. 22, pp. 20-36.

Green, R.O., Conel, J.E., & Roberts, D.A. (1993). Estimation of aerosol optical depth, pressure evaluation, water vapor and calculation of apparent surface reflectance from radiance measured by the Airborne Visible/Infrared Imaging Spectrometer (AVIRIS) using a radiative transfer code. *SPIE*, Vol. 1937, *Imaging Spectrometry of the Terrestrial Environment*, pp. 2-11.

Gubareff, G.G., et al. (1960). *Thermal Infrared Radiation Properties Survey: A Review of the Literature*, 2d ed. Honeywell Research Center, Minneapolis-Honeywell Regulator Company, Minneapolis, MN.

Hall, F.G., Strebel, D.E., Nickeson, J.E., & Goetz, S.J. (1991). Radiometric rectification: toward a common radiometric response among multidate, multisensor images. *Remote Sensing of Environment*, Vol. 35, pp. 11-27.

Hanel, R.A., Conrath, B.J., Jennings, D.E., & Samuelson, R.E. (1992). *Exploration of the Solar System by Infrared Remote Sensing*. Cambridge University Press, NY.

Heilman, J.L., & Moore, D.G. (1980). Thermography for estimating near surface soil moisture under developing crop canopies. *J. Appl. Meteorology*, Vol. 9, pp. 324-328.

Jensen, J.R., ed. (1983). Urban/suburban land use analysis. In Colwell, R.N., ed. *Manual of Remote Sensing*, Vol. 2, 2d ed., pp. 1571-66.

Johnson, R.W., Bahn, G.S., & Thomas, J.P. (1982). Synoptic thermal and oceanographic parameter distributions in the New York bight apex. *Photogrammetric Engineering and Remote Sensing*, Vol. 48, No. 11, pp. 1593-98.

Kneizys, F. X., Shettle, E. P., Gallery, W.O., Chetwynd, J.H., Jr., Abreu, L.W., Selby, J.E.A., Clough, S.A., & Fenn, R.W. (1983). "Atmospheric transmittance/radiance: computer code LOWTRAN 6." Optical Physics Division, Air Force Geophysics Laboratory, Air Forces Systems Command, Hanscom AFB, MA.

Kneizys, F.X., Shettle, E.P., Abreu, L.W.. Chetwynd, J.H., Anderson, G.P., Gallery, W.O., Selby, J.E.A., & Clough, S.A. (1988). "Users guide to LOWTRAN 7," AFGL-TR-88-0177, Environmental Research Papers, No. 1010, Optical/Infrared Technology Division, Air Force Geophysics Laboratory, Hanscom AFB, MA.

Legeckis, R. (1978). A survey of worldwide sea surface temperature fronts detected by environmental satellites. *J.of Geophysical Res.* Vol. 83 (C9), pp. 4501-22.

Lowe, D.S. (1978). "Effects of emissivity on airborne observation of roof temperature." Proceedings of Thermosense 1, sponsored by the American Society of Photogrammetry.

Marsh, S.E. & Lyon, R.J.P. (1980). Quantitative relationship of near-surface spectra to Landsat radiometric data. *Remote Sensing of Environment*, Vol. 10, No. 4, pp. 241-261.

McMillan, L.M. (1975). Estimation of sea surface temperatures from two infrared window measurements with different absorptions. *J. Geophysical Res.*, Vol. 80, pp. 5113-17.

Mericsko, R. (1993). "Enhancements to atmospheric-correction techniques for multiple thermal images." Proceedings of the Workshop on Atmospheric Correction of Landsat Imagery, sponsored by Defense Landsat Office, Torrance, CA.

Moran, M.S., Jackson, R.D., Slater, P.N., & Teillet, P.M. (1993). "Simplified procedures for retrieval of surface reflectance factors from Landsat TM sensor output." Proceeding of the Workshop on Atmospheric Correction of Landsat Imagery, Torrance, CA, sponsored by the Defense Landsat Program Office, pp. 85-90.

Piech, K.R., & Walker, J.E. (1974). Interpretation of soils. *Photogrammetric Engineering and Remote Sensing*, Vol. 40, pp. 87-94.

Piech, K.R., Schott, J.R., & Stewart, K.M. (1978). The blue-to-green reflectance ratio and lake water quality. *Photogrammetric Engineering and Remote Sensing*, Vol. 44, No. 10, pp. 1303-19.

Piech, K.R. (1980). Material identification using broad band visible data. *Proceedings of the SPIE Image Processing for Missile Guidance,* Vol. 238.

Prabhakara, C., Dalu, G., & Kunde, V.G., (1974). Estimation of sea surface temperature from remote sensing in the 11 to 13 μm window region. *J. of Geophysical, Res.*, Vol. 79, No. 33.

Price, J.C. (1984). Land surface temperature measurements from the split window channels of the NOAA 7 advanced very high resolution radiometer. *J. of Geophysical Res.*, Vol. 89, No. D5, pp. 7231-37.

Rouse, J.W., Haas, R.H., Schell, J.A., & Deering, D.W. (1973). "Monitoring vegetation systems in the Great Plains with Third ERTS." ERTS Symposium, NASA No. SP-351, pp. 309-317.

Salvaggio, C. (1993). "Radiometric scene normalization utilizing statistically invariant features." Proceeding of the Workshop on Atmospheric Correction of Landsat Imagery, Torrance, CA, sponsored by the Defense Landsat Program Office, pp. 155-160.

Salisbury, J.W., & D'Arian, D.M. (1992). Emissivity of terrestrial material in the 8-14 μm atmospheric window. *Remote Sensing of Environment*, Vol. 42, pp. 83-106.

Saunders, P.M. (1967). Aerial measurement of sea surface temperature in the infrared. *J. of Geophysical Res.*, Vol. 72.

Scarpace, F.L., Madding, R.P., & Green, T. III (1974). "Scanning thermal plumes." Ninth International Symposium on Remote Sensing of Environment.

Schott, J.R. (1979). Temperature measurement of cooling water discharged from power plants. *Photogrammetric Engineering and Remote Sensing*, Vol. 45, No. 6, pp. 753-761.

Schott J.R., & Schimminger, E.W. (1981). "Data use investigations for applications explorer mission A (Heat Capacity Mapping Mission)." Calspan Report No. 6175-M-1, NASA Accession No. E81-10079.

Schott, J.R., & Wilkinson, E.P. (1982). Quantitative methods in aerial thermography. *Opt. Eng.*, Vol. 21, No. 5.

Schott, J.R., Biegel, J.D., & Volchok, W.J. (1985). "Comparison of methods for removal of atmospheric effects from remotely sensed color images." Presented at SPSE's 25th Fall Symposium – Imaging '85, Arlington, VA.

Schott, J.R. (1986). "Incorporation of angular emissivity effects in long wave infrared image models." Proceedings of the SPIE Symposium, Infrared Technology XII, Vol. 685, pp. 44-52.

Schott, J.R. Salvaggio, C., & Volchok, W.J. (1988). Radiometric scene normalization using pseudoinvariant features. *Remote Sensing of Environment,* Vol. 26, No. 1, pp. 1-16.

Schott, J.R. (1993). "Methods for estimation of and correction for atmospheric effects on remotely sensed data." Presented at SPIE's OE/Aerospace and Remote Sensing 1993, Vol. 1968, No. 51, pp. 448-482, Orlando, FL.

Slater, P.N. (1980). *Remote Sensing: Optics and Optical Systems.*, Addison-Wesley, Reading, MA.

Slater, P.N., Biggar, S. F., Holm, R. G., Jackson, R. D., Mao, Y., Moran, M. S., Palmer, J., & Yuan, B. (1987). Reflectance- and radiance-based methods for the in-flight absolute calibration of multispectral sensors. *Remote Sensing of Environment*, Vol. 22, pp. 11-37.

Suits, G.H. (1972). The calculation of the directional reflectance of a vegetative canopy. *Remote Sensing of Environment*, Vol. 2, pp. 117-125.

Switzer, P., Kowalik, W.S., & Lyon, R.J.P. (1981). Estimation of atmospheric path-radiance by the covariance matrix method. *Photogrammetric Engineering and Remote Sensing,* Vol. 47, No. 10, pp. 1469-76.

Thomas, J.R., Myers, V.I., Heilman, M.D., & Wiegand, C.O. (1966). "Factors affecting light reflectance of cotton." Proceedings 4th Symposium on Remote Sensing of Environment, Ann Arbor, MI, pp. 305-312.

Thomas, J.R. & Gausman, H.W. (1977). Leaf reflectance vs. leaf chlorophyll and carotenoid concentration for light crops. *Agronomy J.*, Vol. 69, pp. 799-802.

Walton, F.F., McClain, E.P., & Sapper, J.F. (1990). "Recent changes in satellite-based multi-channel sea surface temperature algorithms." MTS90, Marine Technology Society, Washington, DC.

Yuan, D. & Elvidge, C.D. (1993). "Application of relative radiometric rectification procedure to Landsat data from use in change detection," Proceeding of the Workshop on Atmospheric Correction of Landsat Imagery, Torrance, CA, sponsored by the Defense Landsat Program Office, pp. 162-166.

CHAPTER 7

Digital Image Processing for Image Exploitation

The field of digital image processing (DIP) has and continues to grow at a very rapid rate. This growth is spurred on by improvements in the speed and reductions in the cost of computing power, along with technological advances in the detector and A-to-D conversion fields that have placed digital images at every user's fingertips. The treatment presented here is a very small sampling aimed at providing the user with a common terminology and a limited exposure to some DIP tools frequently used in remote sensing. A more thorough general treatment of DIP can be found in Gonzalez and Woods (1992), Rosenfeld and Kak (1982), and Pratt (1991), to name only a few. Some of the DIP techniques specific to remote sensing, such as multispectral classification, are treated in more detail by Schowengerdt (1983) and Richards (1986). We will present most of the general image processing tools in Section 7.1 in the form of examples. The diversity of applications for these tools in remote sensing is almost endless, but generally represents simple logical extensions of the examples presented here adapted to a user's specific needs. The background mathematics for the algorithms is elementary algebra or in the case of Section 7.1.4, basic linear systems that are well treated elsewhere (cf. Gaskill, 1978). Consequently, we will not derive the algorithmic expressions but only present the operational form or, in some cases, the resultant images. Because multispectral image classification and processing are so fundamental to much of remote sensing, the background concepts and equations for these topics are presented in greater detail in the rest of the chapter.

For simplicity, in this chapter we will assume that all images are in digital form, meaning that they have been sampled into picture elements (pixels) and that the image can be considered to be made of an array of pixels (usually square), with the brightness of each pixel represented by a numerical value (usually an integer) described as the gray level or digital count (DC) level (cf. Fig. 7.1). A digital image is commonly characterized by the number of pixels in the x and y directions, the number of spectral bands, and the number of bits required to store the range of gray levels for each pixel in each band. For example, a digital image with 512 pixels in the x dimension and 1024 pixels in the y direction, three spectral bands,

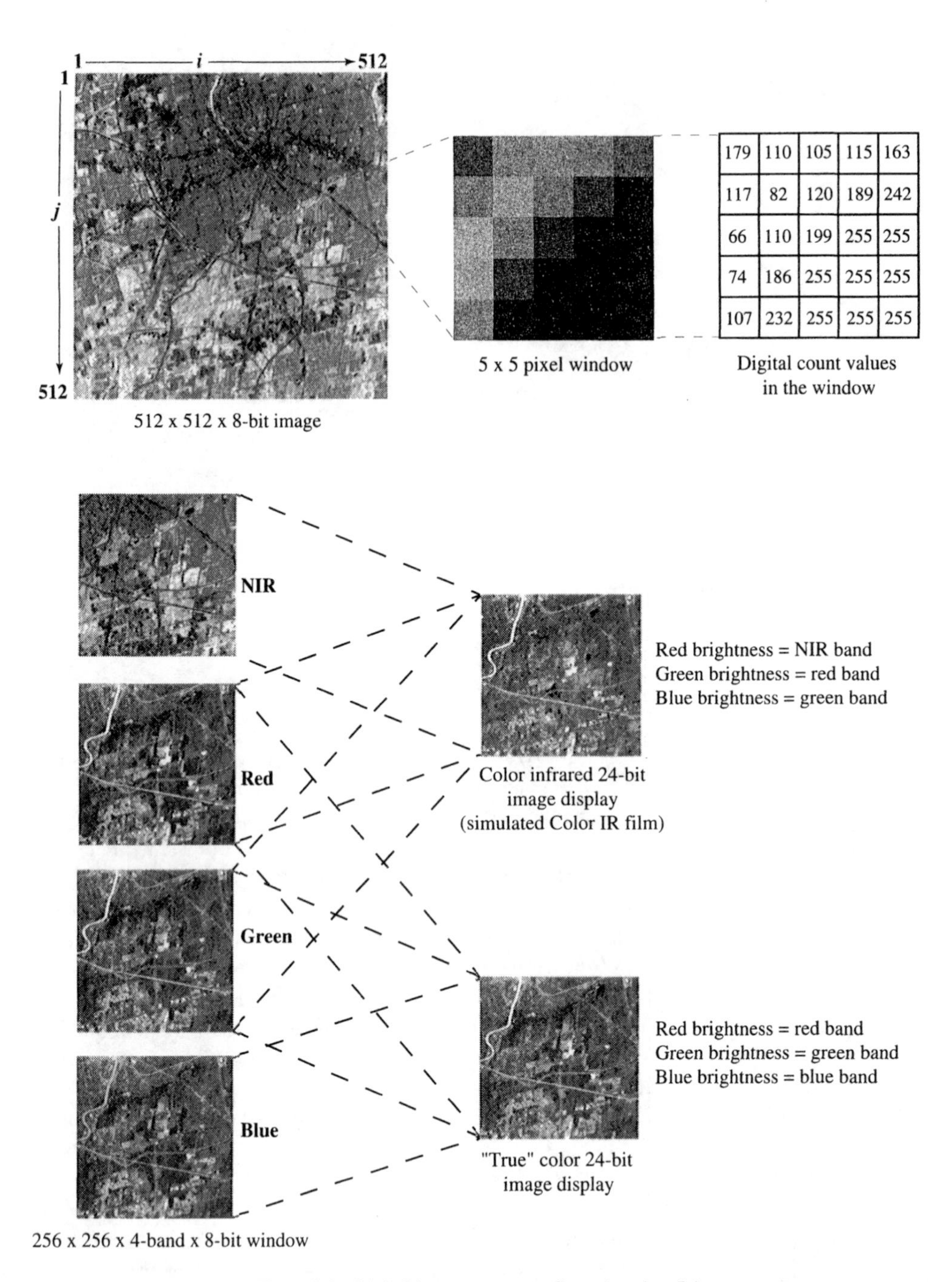

Figure 7.1 Digital image concepts. See color plate 7.1.

and 256 gray levels per spectral band, would be called a 512 by 1024 by 3-band by 8-bit image. Because many images use 8 bits (256 gray levels) of dynamic range, this is often abbreviated to 512 x 1024 x 24 bits where the division of the data into three 8-bit bands is assumed. Unless otherwise specified, we will assume that multiband images are co-registered such that the information in the (i,j)th pixel location in band 1 corresponds to the same point on the ground as the (i,j)th pixel location in bands 2, 3, etc.

In most cases, the images from EO systems are produced such that each pixel corresponds to one GIFOV of the sensor at nadir. In the case of film systems, the images can be digitized at a range of "resolutions." If the full resolution of the film image is to be preserved, sampling theory dictates that two samples be taken per just-resolvable line pair. So a film system that resolves 50 line pairs per mm in each direction will require

$$S_x = \frac{50 \text{ line pairs}}{\text{mm}} \cdot \frac{2 \text{ samples}}{\text{line pair}} = 100 \text{ samples/mm} \tag{7.1}$$

For 9-inch (nominal) color film from a mapping camera (actual format size 230 mm), the number of pixels at this resolution is:

$$N = S_x X \, S_y Y = (100 \text{ samples/mm} \cdot 230 \text{ mm})^2 = 0.53 \cdot 10^9 \text{samples} \tag{7.2}$$

where $S_x = S_y$ are the sample rates in the x and y directions, and X and Y are the film format dimensions in the x and y directions. For three 8-bit spectral bands, this represents a total number of bits of

$$V_b = 3_{\text{bands}} \frac{8 \text{ bits}}{\text{band}} \cdot N = 12.7 \cdot 10^9 \text{bits} \tag{7.3}$$

which is more commonly expressed as bytes of data storage:

$$V = \frac{V_b \text{bits}}{8 \text{ bits / byte}} = 1.6 \cdot 10^9 \text{ bytes} \tag{7.4}$$

This is an enormous amount of computer storage and reflects the high resolution and high information density of photographic film. Often film data are digitized at less than full resolution if resolution is not a critical issue in the digital image analysis task. The original film data are still used directly for obtaining spatial detail. EO images can also become very large with a 6000 x 6000 pixel, 7-band, 8-bit Landsat TM image requiring 250 megabytes of data storage. Today's general purpose workstation computers can handle storage and processing of images of this size. However, processing a full image can still be prohibitively slow. Often a portion of an image (e.g., a 512 x 512 window) or a sub-sampled version of the image (e.g., every tenth pixel in x and y) is analyzed first to define all processing steps and then the entire image analyzed using the "script" developed interactively on the image subset. Image subsets of 512 x 512 to 2048 x 2048 are most commonly used for interactive image processing because soft copy displays (computer monitors) are usually restricted to these sizes.

The speed of today's high-speed computer workstations is such that all the algorithms presented here can be run on fair-size images (e.g., 1024 x 1024) in a matter of minutes. When required, a host of specialized image processing hardware is available to expedite processing of large images or implementation of more complex algorithms.

7.1 STANDARD IMAGE PROCESSING TOOLS

This section presents some of the DIP tools commonly used to enhance or exaggerate the appearance of images to facilitate visual image analysis, as well as methods used to preprocess images to improve machine processing approaches. The machine processing approaches considered in later sections of this chapter emphasize statistical pattern recognition of spectral patterns. The whole field of spatial pattern recognition is of keen interest to many aspects of remote sensing, but is beyond the scope of this treatment. The interested reader should consider Tou and Gonzalez (1974), Rosenfeld and Kak (1982), and Fu (1974).

7.1.1 Point Processing

The simplest and most widely used class of image processing operations is called *point processing* or gray-level manipulation. It involves adjusting the DC level of pixels from one value to another without concern for the values of neighboring pixels (i.e., the operation is on each individual point as opposed to a neighborhood). The operation can be thought of as passing the image through a lookup table (LUT), which takes every gray level in the input image and assigns it a new gray level. This can most easily be seen in a graphical plot of input vs. output values as shown in Figure 7.2 and is represented mathematically as:

$$g(i,j) = \mathrm{LUT}(f(i,j)) \tag{7.5}$$

where $f(i,j)$ and $g(i,j)$ are the DC values of the input image (f) and output image (g) at each location (i,j) and LUT is the lookup table function.

Point processing operations are often viewed in terms of their impact on the histogram of the image. The histogram of an image (f) is designated HIS (f) and is simply a one-dimensional vector whose components are the number of pixels in the image having DC = 0, DC = 1 ... DC = DC_{max}. Note that if the components of the histogram vector are divided by the total number of pixels in the image (N), the normalized histogram (HIS′) becomes an estimate of the discrete gray level probability distribution for the image, i.e.,

$$\mathrm{HIS}'(f) = \frac{\mathrm{HIS}(f)}{N} \tag{7.6}$$

We recognize that the value of the nth element in the vector HIS′ is the probability (p_{DC}) of a pixel having a DC (brightness level) equal to n. We often plot the normalized histograms of the input and output images along with the LUT as depicted in Figure 7.2 to clarify the impact of the LUT on the image. In most cases, we will also plot the identity LUT for reference. The identity operation maps each input DC level back to that same DC level in the output image.

The input image in Figure 7.2 utilizes only a small portion of the eight bit dynamic range of the system. Many 8-bit systems have fixed gain and offset such that bright objects with high sun angles and dark objects at low sun angles must all be accommodated in the same 8-bit dynamic range. As a result, the quantization must be somewhat coarse, and any given image will generally occupy only a limited portion of the available dynamic range. Polar and equatorial thermal ranges must also be included over a common 8-bit range for thermal bands. The resulting images often have low contrast. By passing the image through a linear contrast enhancing LUT, as depicted in Figure 7.2, the full 8-bit dynamic range can be utilized and the apparent image contrast improved. Note that this process adds no new information or true contrast to the scene, but only makes the existent contrast more apparent to the observer. Examples of the use of lookup table operations include enhancing dark regions of an image (e.g., shadows) as shown in Figure 7.3, and isolation of certain DC levels (e.g., for isotherm mapping) as shown in Figure 7.4.

Another commonly used LUT operation is histogram equalization. This operation is designed to normalize contrast by spreading the output image data approximately equally across the available dynamic range. It is based on the cumulative distribution function (CDF) of the input image histogram, which is simply a one-dimensional vector where each element is that fraction of the pixels in the image with DC values less than or equal to the number of the element where the index starts at zero, i.e.,

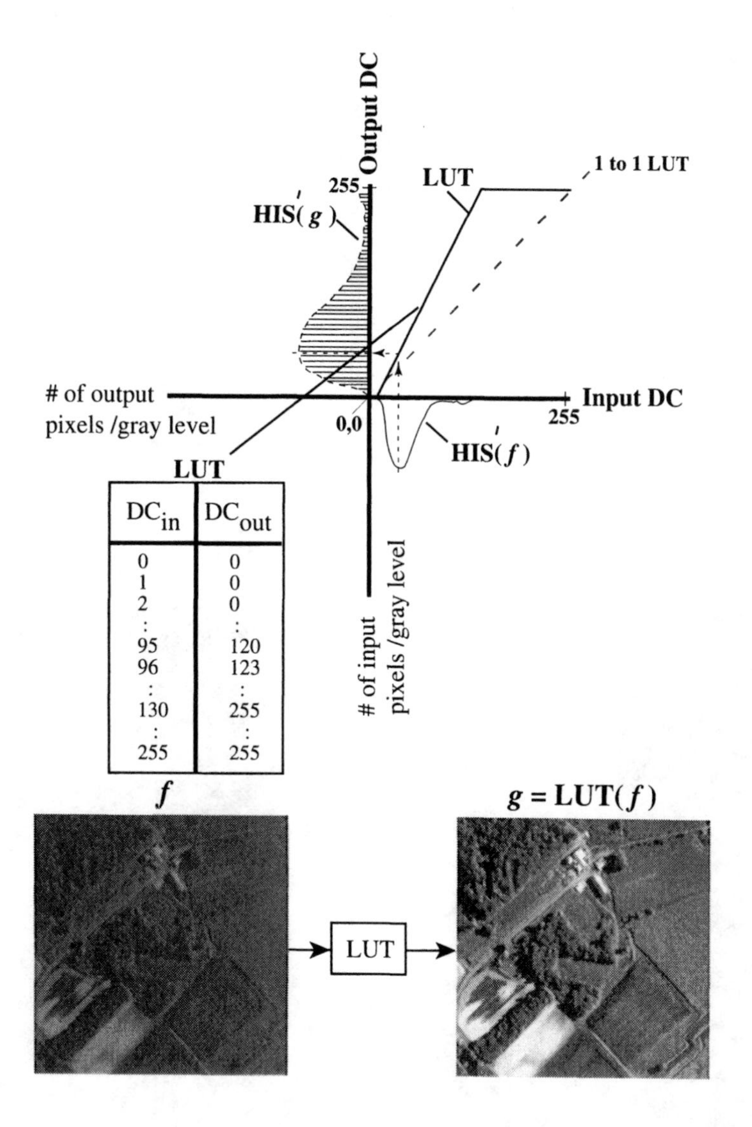

Figure 7.2 Lookup table concepts.

$$\mathrm{CDF}_n = \sum_{i=0}^{n} \mathrm{HIS}'(i) \tag{7.7}$$

For example, in an 8-bit image, the zero-th element would be the fraction of pixels in the image with DC = 0, the third element would be the fraction of the image with DC values between 0 and 3 inclusive, and the 255-th element would be 1 since it would be the fraction of the pixels with DC values between 0 and 255. Gonzalez and Wintz (1987) demonstrate that by multiplying the CDF vector by $\mathrm{DC}_{\mathrm{max}}$, converting the resultant values in the vector to integers, and using the integer vector as a LUT, an input histogram can be converted to an approximately equalized histogram (the approximation is due to the discrete nature of the integer data). This process, demonstrated in Figure 7.5, is commonly used as an initial pre-processor for image display. Although its effect is somewhat harsh for pictorial images, it is

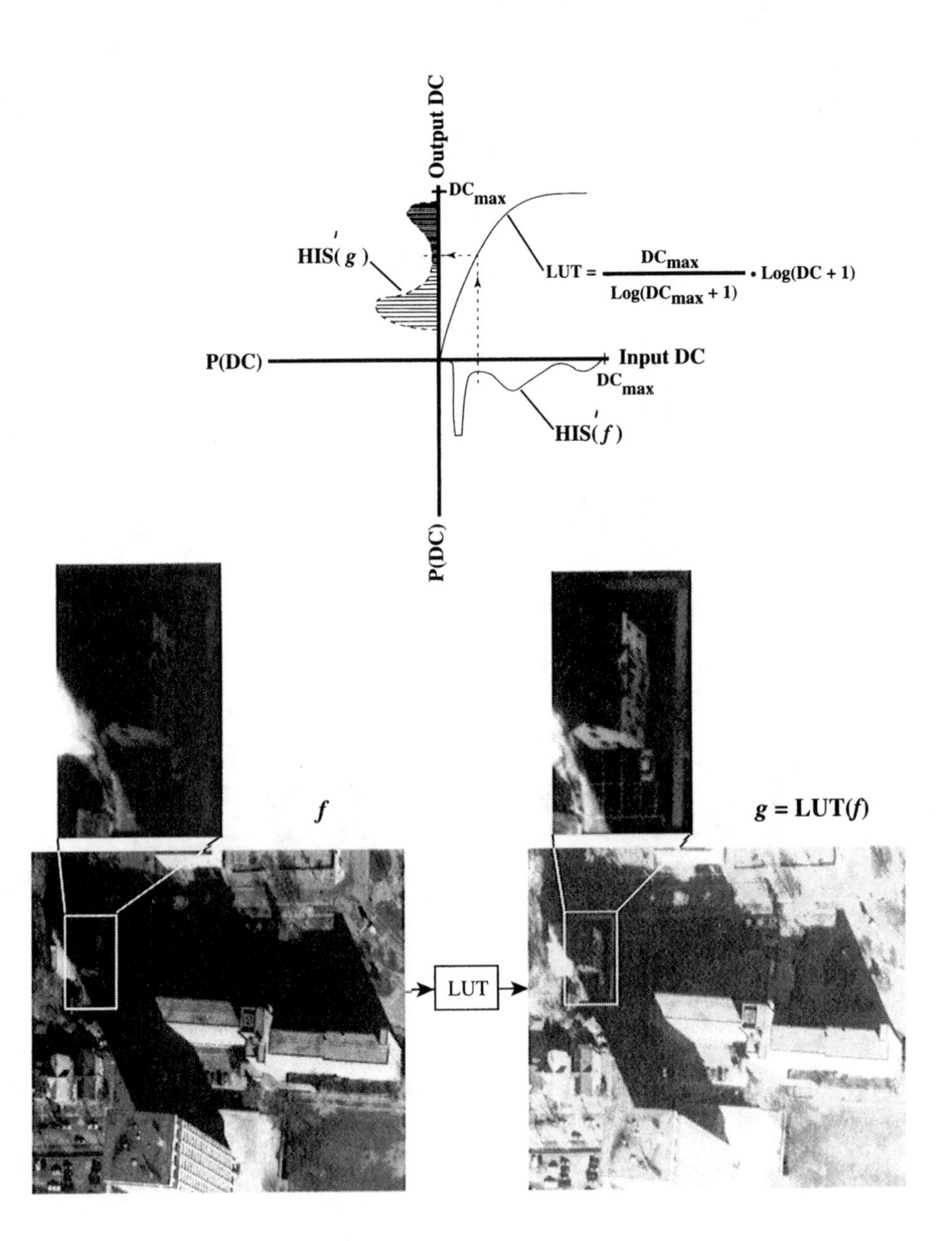

Figure 7.3 Logarithmic LUT used to enhance shadows in an air photo.

often acceptable for remotely sensed images.

It is clear from Figure 7.5 that the process can also be inverted to a good approximation by simply using the lookup table in reverse to convert the equalized image back to the original image. In this case, the inversion is only approximate when multiple binning takes place. This happens when two gray levels in the input image are assigned a single gray level in the output image. When reversed, the pixels that come from multiple bins are assigned to a single gray level or are randomly assigned to the two gray levels resulting in subtle changes

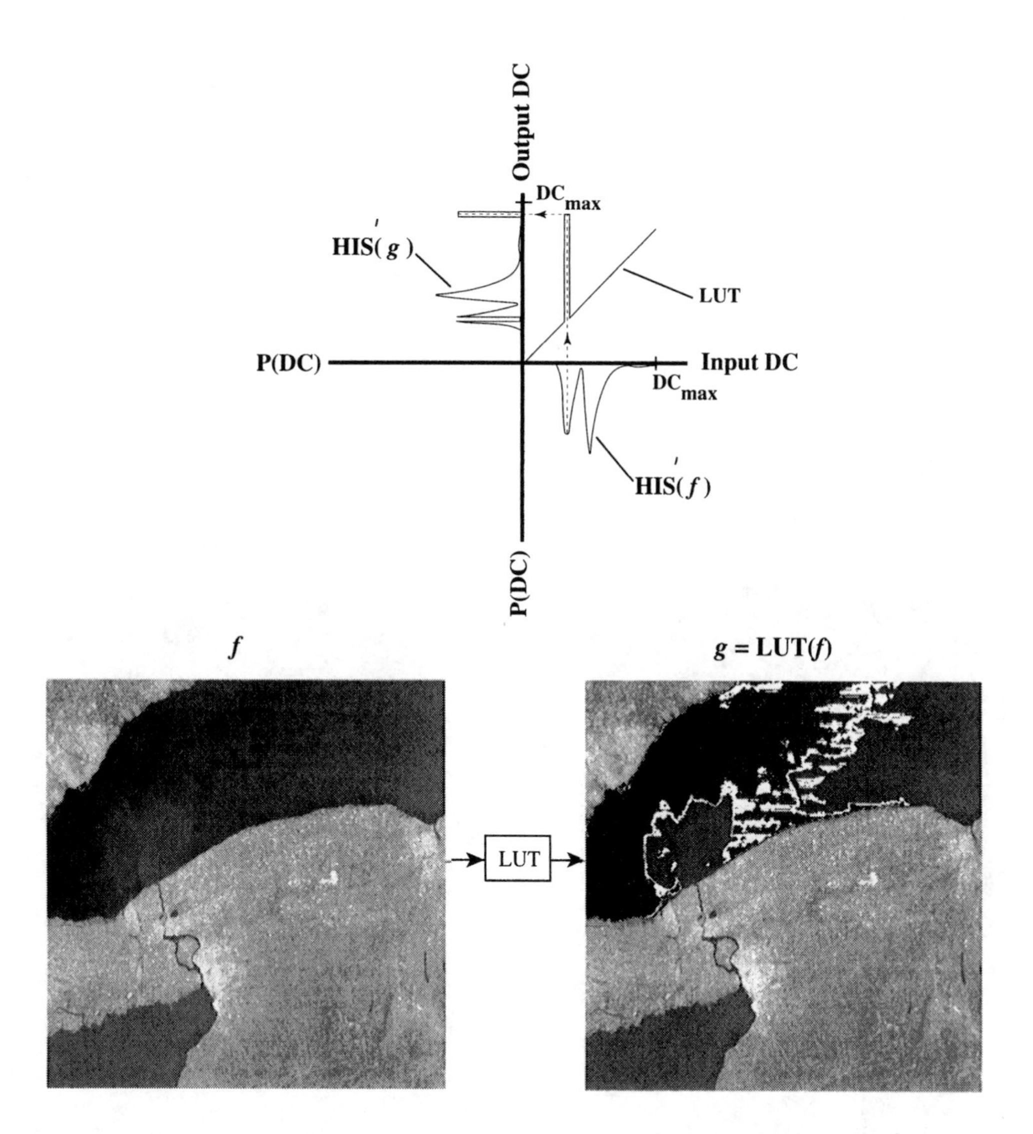

Figure 7.4 LUT used to isolate selected digital counts for use in generating isopleth maps. The input image is a Landsat Thematic mapper thermal band image of Lake Ontario.

in the image. We review this here not so much for the current discussion but to caution the reader that many gray-level manipulations are nonreversible and can result in loss of information due to multiple binning. Our interest in using the scaled CDF LUT in reverse is not to reverse the input image but to force one image to appear approximately the same as a second.

Consider two images of an identical scene taken with different sensor gain and bias or under different illumination levels. If the brightness distribution function (histogram) of the second image can be processed to match the distribution of the first, then the apparent differences between the images should be removed. Gonzales and Woods (1992) show that in the discrete case this probability matching or specification can be approximated using a process called *histogram specification*. This process is illustrated in Figure 7.6 where the histograms from two images of the same scene are computed along with their CDF's. We

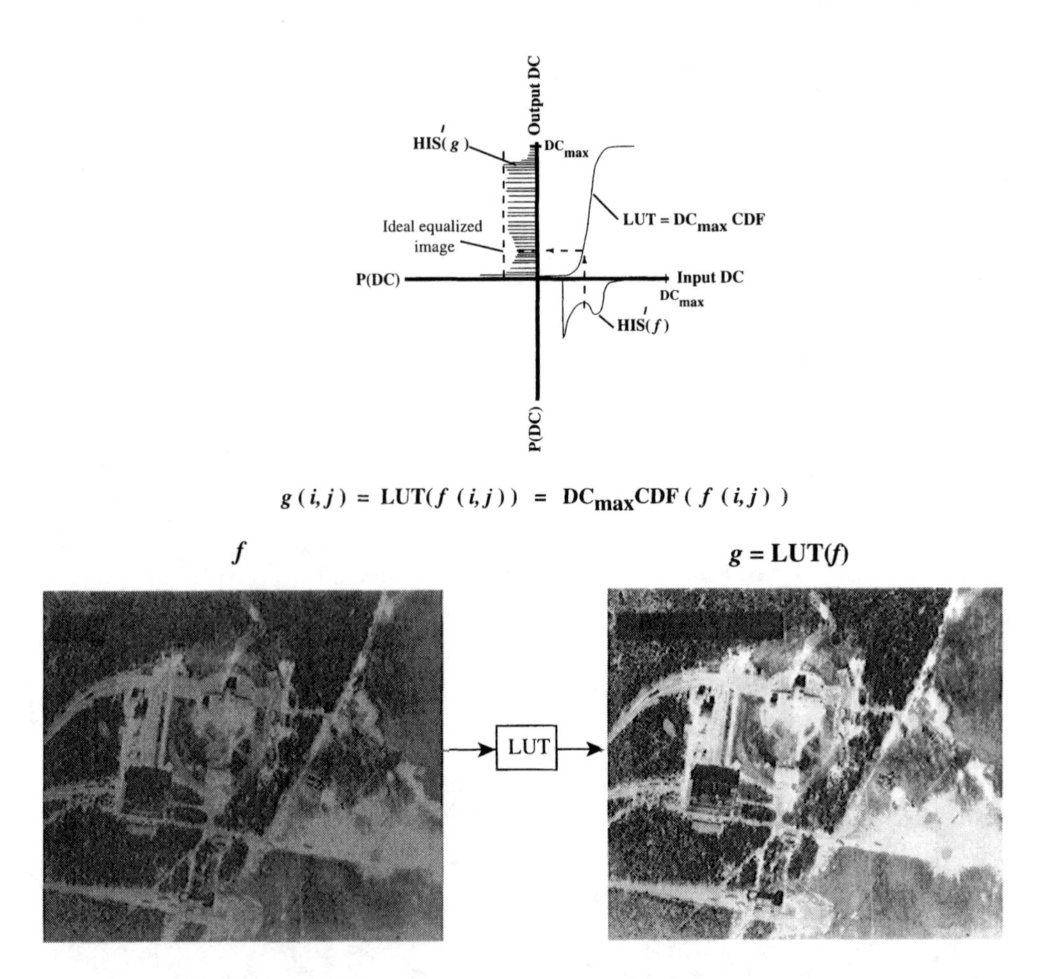

Figure 7.5 Histogram equalizing process. Image is a digitized U.K. reconnaissance photo of the V-2 rocket test facilty at Peenemunde acquired July 4, 1944. (Image courtesy of Autometric Inc.)

assume these images have nearly identical reflectance distributions and desire to force the lower contrast (day 2) image to have a histogram that matches the higher contrast (day 1) image in order to approximately remove illumination and sensor effects. This is accomplished by conceptually passing the low-contrast (day 2) input image through a LUT comprised of its scaled CDF, resulting in an equalized histogram as illustrated in Figure 7.5. This equalized image is then passed backwards through the scaled CDF of the high-contrast (day 1) image to force its histogram to match that of the high-contrast image. In practice, this is accomplished with a single LUT that combines the forward and reverse CDF LUT's. As seen in Figure 7.6, this process removes gross contrast differences between the images. However, in cases where the assumptions of identical reflectance distributions are seriously violated, false changes can be introduced. In this case, the water in the day-2 input image that was fairly clear has been forced to appear more turbid because the water on day 1 was turbid. Thus, as discussed in Section 6.5.2, care should be taken when using this approach as a quantitative normalization process.

We have implicitly assumed that the point operations presented here are applied uniformly over the entire image. They can also be applied in an adaptive fashion where the LUT is made a function of the local image region. For example, an algorithm can be designed that

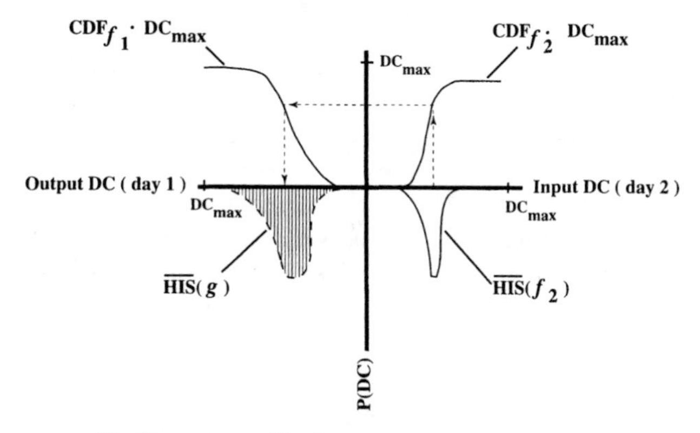

The histogram specification process can be expressed as

$$g(i,j) = DC_{\max} \cdot CDF_{f_1}^{-1}\left[DC_{\max} \cdot CDF_{f_2}\left(f_2(i,j)\right)\right]$$

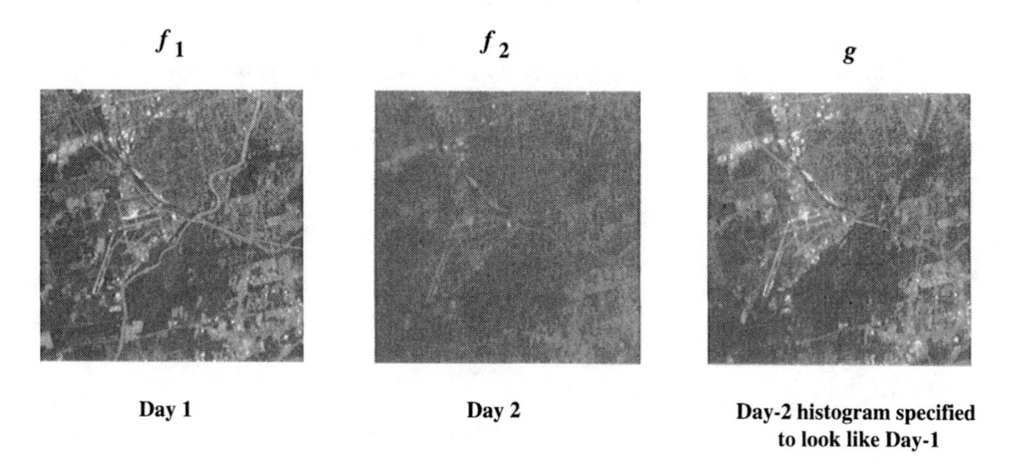

Figure 7.6 Histogram specification process.

increases contrast only in low-contrast regions of the scene. The local contrast is computed in the neighborhood of each pixel. If the contrast is less than some user-defined threshold, then the contrast is stretched about the mean of the local window. This results in an output image whose gray levels are unchanged, on average, but with enhanced contrast of localized areas of low contrast such as shadows and highlights. Such adaptive methods are not truly point operations, since the nature of the transform depends on the neighborhood around the point in question.

7.1.2 Neighborhood Operations-Kernel Algebra

The next class of DIP algorithms generates the gray value of the output pixel as an explicit function of some number of neighboring pixels. In most cases, neighborhood operations are designed to alter the sharpness (local edge contrast) of an image. Most of the common neighborhood operations of interest can be implemented using what is referred to as a *convolution kernel* (cf. Fig. 7.7). The kernel can be thought of as a small image window whose size defines the size of the kernel (most often square and with an odd number of pixels on each axis). Each pixel in the window is assigned a weight, and this array of values is defined to be the kernel. The kernel operation involves locating the center of the kernel over each input pixel location *(i,j)* and multiplying each value in the kernel times the DC value in the input image under the kernel. The sum of the products is defined to be the output pixel value *g(i,j)*. The kernel is said to be operating on the pixel under the center element of the kernel (hence the common use of odd-size kernels). The kernel is stepped sequentially over each pixel in the input image. Note that the edge pixels (i.e., where a portion of the kernel falls beyond the image) are generally either left undefined or one of a variety of methods can be used to fill the edges to maintain the overall image statistics. Schowengerdt (1983) shows that, ignoring edge effects, this process is mathematically the equivalent of the discrete correlation operation:

$$g(i,j) = \sum_{m=i-w/2}^{i+w/2} \sum_{n=j-w/2}^{j+w/2} f(m,n)h(m-i,n-j) = f(i,j) \star h(i,j) \tag{7.8}$$

where *m* and *n* are dummy variables of summation, *h* is the kernel, w is the size of the kernel, w/2 is a truncated integer value, and ★ denotes the correlation operation. The kernel is indexed over the range -w/2 to w/2, with the center location defined as *h*(0,0). If the values in the kernel are "flipped" about the *x* and *y* axis, then operating on the image with the flipped kernel is the mathematical equivalent of the discrete convolution operation from which it derives its name. The convolution operation can be expressed as:

$$g(i,j) = \sum_{m=i-w/2}^{i+w/2} \sum_{n=j-w/2}^{j+w/2} f(m,n)h(i-m,j-n) = f(i,j) * h(i,j) \tag{7.9}$$

where * denotes the convolution operation. For example, a 3 x 3 element correlation would be implemented mathematically as:

$$g(i,j) = \sum_{m=i-1}^{i+1} \sum_{n=j-1}^{j+1} f(m,n)k(m-i,n-j) \tag{7.10}$$

The convolution operation so fundamental to linear system theory can also be implemented using Eq. (7.10) by flipping the kernel values about *x* and *y*.

Let's examine the effect of various simple kernels to improve our intuitive understanding of kernel operations. Consider these simple 3 x 3 element kernels:

$$h_1(i,j) = \begin{bmatrix} 0 & 0 & 0 \\ 0 & 1 & 0 \\ 0 & 0 & 0 \end{bmatrix} \quad h_2(i,j) = \begin{bmatrix} 0 & 0 & 0 \\ 0 & k & 0 \\ 0 & 0 & 0 \end{bmatrix}$$

$$h_3(i,j) = \begin{bmatrix} 0 & 0 & 0 \\ 0 & 0 & 1 \\ 0 & 0 & 0 \end{bmatrix} \quad h_4(i,j) = \begin{bmatrix} 1 & 1 & 1 \\ 1 & 1 & 1 \\ 1 & 1 & 1 \end{bmatrix}$$

Kernel h_1 is the identity operation and simply replicates the input image. Kernel h_2 produces an image *k* times brighter than the input image (i.e., a contrast enhanced image for $k > 1$ or a contrast reduced image for $k < 1$). Kernel h_3 replaces each pixel with its neighbor to the right, resulting in an image that is shifted one pixel to the left. Note: If the convolution operation is performed Eq. (7.9) or the kernel flipped, the result would be a shift to the right. Kernel h_4 generates output pixels that are the sum of the input pixel and its eight nearest neighbors. The resultant image will be 9 times brighter than the input image and blurred

Image $f(i,j)$

7	9	45	21	39	45	29	10	20
18	20	41	37	11	19	23	13	18
12	29	9	16	16	10	18	24	16
3	7	5	36	25	21	23	17	
41	15	11	13	33	28	22		
10	33	18	22	29	43			
7	18	24	15	13				
12	31	27	17					

3 x 3 Kernel $h(i,j)$

1	1	1
1	2	2
1	2	1

3 x 3 Kernel flipped $h^F(i,j)$

1	2	1
2	2	1
1	1	1

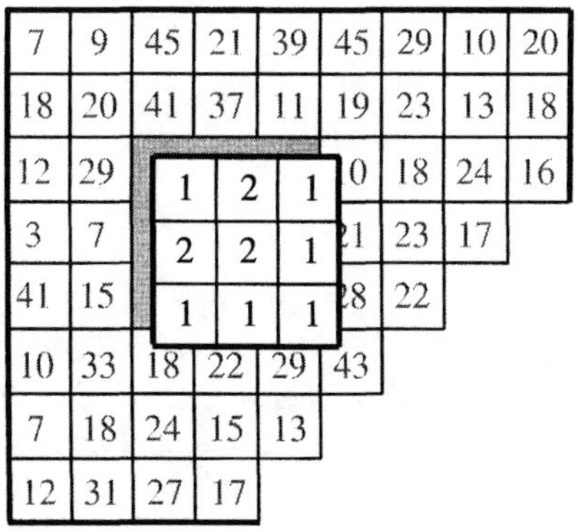

7	9	45	21	39	45	29	10	20
18	20	41	37	11	19	23	13	18
12	29	1	2	1	0	18	24	16
3	7	2	2	1	1	23	17	
41	15	1	1	1	8	22		
10	33	18	22	29	43			
7	18	24	15	13				
12	31	27	17					

$$g(4,4) = f(i,j) * h(i,j) = f(i,j) \star h^F(i,j)$$

$$g(4,4) = (1 \times 9) + (2 \times 16) + (1 \times 16) + (2 \times 5) + (2 \times 36) + (1 \times 25) + (1 \times 11) + (1 \times 13) + (1 \times 33) = 120$$

7	9	45	21	39	45	29	10	20
18	20	41	37	11	19	23	13	18
12	29	9	16	16	10	18	24	16
3	7	5	**120**	25	21	23	17	
41	15	11	13	33	28	22		
10	33	18	22	29	43			
7	18	24	15	13				
12	31	27	17					

Figure 7.7 Correlation (★) and convolution (*) operations with a kernel operator..

slightly. To avoid this increase in the magnitude of the output image, kernels are often scaled by their magnitude (the sum of the values in the kernel), so that on average the output image will span approximately the same range as the input image. The scaled version of h_4 is

$$h_5 = \frac{1}{9}\begin{bmatrix} 1 & 1 & 1 \\ 1 & 1 & 1 \\ 1 & 1 & 1 \end{bmatrix} = \begin{bmatrix} \frac{1}{9} & \frac{1}{9} & \frac{1}{9} \\ \frac{1}{9} & \frac{1}{9} & \frac{1}{9} \\ \frac{1}{9} & \frac{1}{9} & \frac{1}{9} \end{bmatrix} = \frac{1}{9} h_4(i,j) \tag{7.11}$$

A pixel in the output image will be the mean value of the corresponding input pixel and its eight nearest neighbors. The averaging operation "blurs" the image. In linear systems, it is referred to as a low-pass filter or a low-pass kernel, since it has little impact on low spatial frequencies (uniform areas) and filters out (blurs) high spatial frequencies (edges) (cf. Sec. 7.1.4 for additional treatment of filtering operations).

Kernel operations can also be used to locate or exaggerate edges by local differencing operations. Consider the following operators:

$$h_6 = \begin{bmatrix} 0 & 0 & 0 \\ 0 & -1 & 1 \\ 0 & 0 & 0 \end{bmatrix} \quad h_7 = \begin{bmatrix} 0 & 0 & 0 \\ 0 & -1 & 0 \\ 0 & 1 & 0 \end{bmatrix} \quad h_8 = \begin{bmatrix} 0 & +1 & 0 \\ +1 & -4 & +1 \\ 0 & +1 & 0 \end{bmatrix}$$

Kernel h_6 subtracts each input pixel from its neighbor on the right. The resultant image will have zero DC values if the two adjacent pixels have the same gray value, large positive values if the pixel to the right is significantly larger, and large negative values if the pixel to the right is significantly smaller than the input pixel. This operation is the equivalent of taking the discrete first derivative of the image with respect to x:

$$\frac{df}{dx} = \frac{\lim}{\Delta x \to 0} \frac{f(x+\Delta x, y) - f(x,y)}{\Delta x} = \frac{f(i+1,j) - f(i,j)}{1} = f(i,j) \bigstar h_6(i,j) \tag{7.12}$$

Since it is difficult to conceptualize and display negative gray levels, these images are usually biased by adding half the dynamic range of the input image [(DC_{max}+1)/2] to each pixel value such that a zero gray level becomes a medium gray. The output images detect vertical edges in the image, but reject all mean gray-level information. Kernel h_7 is the first derivative with respect to the y direction and will exaggerate horizontal edges. Kernel h_8 is the Laplacian operator and is the discrete equivalent of the sum of the second partial derivatives with respect to x and y, i.e.,

$$\begin{aligned} g = \nabla^2 f = \frac{\partial^2 f}{\partial x^2} + \frac{\partial^2 f}{\partial y^2} &= [f(i+1,j) - f(i,j)] - [f(i,j) - f(i-1,j)] \\ &+ [f(i,j+1) - f(i,j)] - [f(i,j) - f(i,j-1)] \\ &= f(i+1,j) + f(i-1,j) + f(i,j+1) + f(i,j-1) - 4f(i,j) = f(i,j) \bigstar h_8(i,j) \end{aligned} \tag{7.13}$$

This kernel can be thought of as subtracting the image from a blurred version of itself formed from averaging its four nearest neighbors. The resulting image highlights edges and isolated bright or dark pixels. Like the first derivative images, it is usually biased up for display because uniform regions would have zero DC values. The Laplacian is a form of high-pass filter because the low-frequency (mean DC level) information is lost (filtered out), and the edges or high frequencies are preserved and enhanced. Figure 7.8 provides an example of a variety of kernels and their impact on an image. The convolution operation is widely used in image processing because essentially any type of linear spatial filter can be implemented with a sufficiently large kernel. In addition, most optical processes (e.g., atmospheric and sensor degradation and blur due to detector sampling), as well as electronic processes (e.g., preamp frequency response effects), can be modeled using convolution (cf. Secs. 7.1.4 and 9.2 for more on filtering and modeling of images).

Several nonlinear image processing operations (e.g., median and mode) are also of interest to remote sensing. These operations also use a window centered about the pixel of interest to specify the neighborhood of the operation. The window steps from pixel to pixel in the same fashion as the correlation and convolution kernel operations. The pixels to be considered are indicated by 1's in the window. The output pixel value is the result of performing the operation on those pixel values that coincide with 1's in the window. An example of the mode operation is shown in Figure 7.9. This operation is commonly used to filter a class map image to remove random class values at class boundaries due to mixed pixel effects (cf. Sec. 7.2). The values under the 1's in the window are sorted to see which digital count value occurs most often (mode operation). When these digital counts represent class values, this operation will tend to replace isolated pixels or strings of pixels with the most common class in the neighborhood and to smooth the boundaries between classes. Note that this process can have detrimental effects as well. For example, if a critical class was resolved at about one or two GIFOV's, this operation may filter the targets out of the final class map.

The median filter is a nonlinear filter commonly used to remove "salt and pepper" noise in images. The median operation removes isolated pixels with very high and low values that can be introduced by certain types of electronic noise. The median filter is one type of rank order filter. A more general form for rank order filters is described by Hardle et al. (1993).

The majority of the image processing tools presented in the last two sections can be used to improve the appearance of an image, making it easier for an analyst to visually extract information. Many of them can also be used as preprocessing algorithms for machine analysis (e.g., noise reduction, edge delineation, etc.). Some of the neighborhood operations that are particularly useful as a preprocessor for machine-based statistical pattern recognition algorithms (treated in Sec. 7.2) are texture measures.

7.1.3 Structure or Texture Measures

Structure metrics are a class of neighborhood operations designed to characterize the variability (texture) in the neighborhood around a pixel. In general they utilize a window of some size that moves over the image in the same fashion as a kernel operation. The output pixels are formed from some measure of the texture in the window when it is centered over the pixel location. One of the simplest structure metrics is the range. For example, using a 3 x 3 window centered on $f(i,j)$, the output $g(i,j)$ would be the range (DC_{max}-DC_{min}) computed over the set of pixels comprised of $f(i,j)$ and its eight nearest neighbors. The output image would be bright in regions with structure and dark in regions with little structure. Another similar texture metric is the local standard deviation. Again, the output image pixel value $g(i,j)$ is simply the standard deviation of the pixels under a window centered on $f(i,j)$ in the input image. A large variety of texture metrics exists. Some are very direction-specific, some require fairly large windows and are aimed at characterizing low-frequency texture, others work best with small windows and are designed to differentiate high-frequency texture patterns. Some of these texture metrics are the co-occurrence metrics described by Haralick et al. (1973), run-length metrics discussed by Galloway (1975), and texture spectra described by Wang (1990). It is also common to use texture metrics isolated in the frequency domain

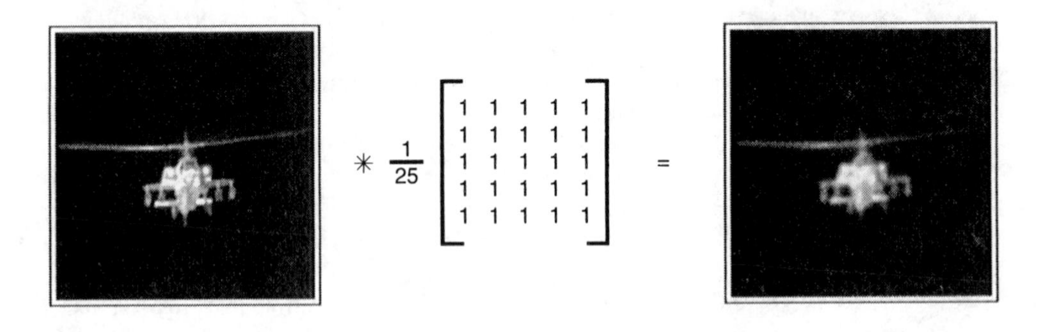

5 x 5 blur kernel (low-pass filter) used for noise reduction.

$$* \begin{bmatrix} 0 & 1 & 0 \\ 1 & -4 & 1 \\ 0 & 1 & 0 \end{bmatrix} +128 \quad =$$

Laplacian kernel (high-pass filter) to highlight edges.

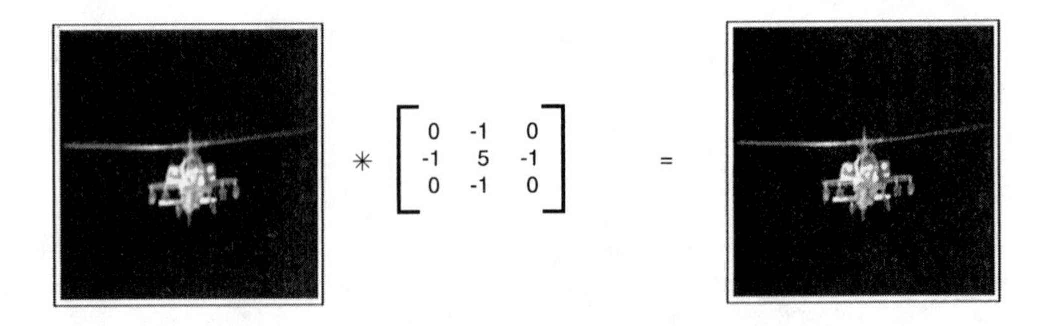

Image minus its Laplacian used to sharpen edges (high-frequency boost).

Figure 7.8 Examples of neighborhood operations employing kernel operator. Image is a MWIR image taken with a 2-D array sensors. (Image courtesy of Eastman Kodak.)

$$\star \begin{bmatrix} 0 & 0 & 0 \\ 0 & -1 & 1 \\ 0 & 0 & 0 \end{bmatrix} +128 \quad =$$

First-derivative image with respect to x

$$\star \begin{bmatrix} 0 & 0 & 0 \\ 0 & -1 & 0 \\ 0 & 1 & 0 \end{bmatrix} +128 \quad =$$

First-derivative image with respect to y

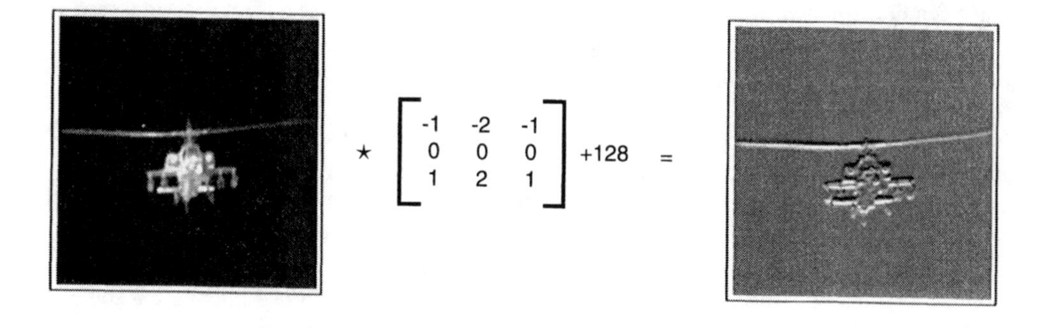

Sobel horizontal-edge detector

Figure 7.8 Examples of neighborhood operations employing kernel operator (continued)

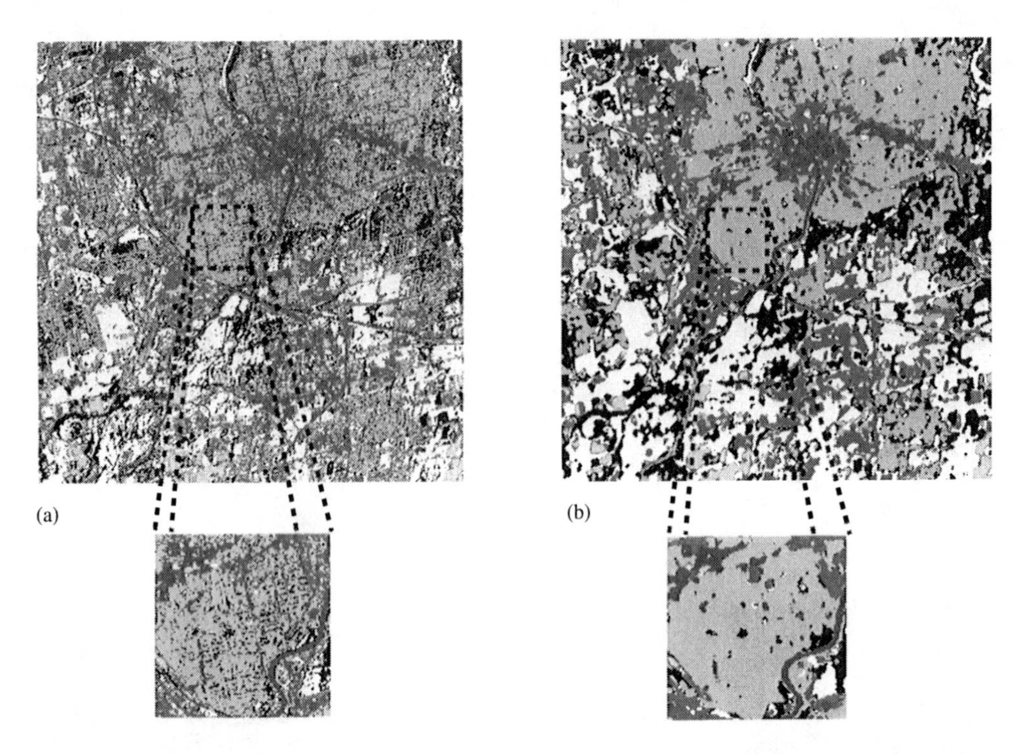

Figure 7.9 Mode filter operation and appliction to a class map image. (a) Original class map of a portion of a Landsat image of Rochester, New York. (b) Class map after application of a mode filter. See color plate 7.9.

as described by Stormberg and Farr (1986) (cf. Sec. 7.1.4 for more on frequency domain metrics). Figure 7.10 represents a sample set of images formed from texture metrics. The gray values in these images can be used in algorithms designed to classify material types on the basis of texture. For example, in a single-band image containing only trees and water, the mean gray level might overlap between trees and water. However, at certain resolutions, the local standard deviation in a 3 x 3 window of water pixels would be very small compared to a window containing forest pixels. In this case, a simple threshold of the standard deviation image can dramatically improve classification accuracy. In most cases, we will need to consider many texture metrics to aid in separating different material classes or to merge texture metrics with multispectral digital count values to provide greater differences between classes.

The statistical pattern recognition tools discussed in the next section (7.2) describe procedures for image classification using spectral and/or textural metrics. Because of the wide variety of possible textural metrics, it can be difficult to isolate those that are truly useful. Rosenblum (1990) describes a method for selecting texture features similar to the method described by Swain (1978) to select spectral bands for land cover classification. These band optimization methods (discussed in greater detail in Sec. 9.3) rely on maximizing the statistical separability between classes characterized using the Gaussian maximum likelihood theory developed in the next section (7.2).

7.1.4 Global Operations

Many convolution operations (such as low-pass filtering) require rather large kernels. At

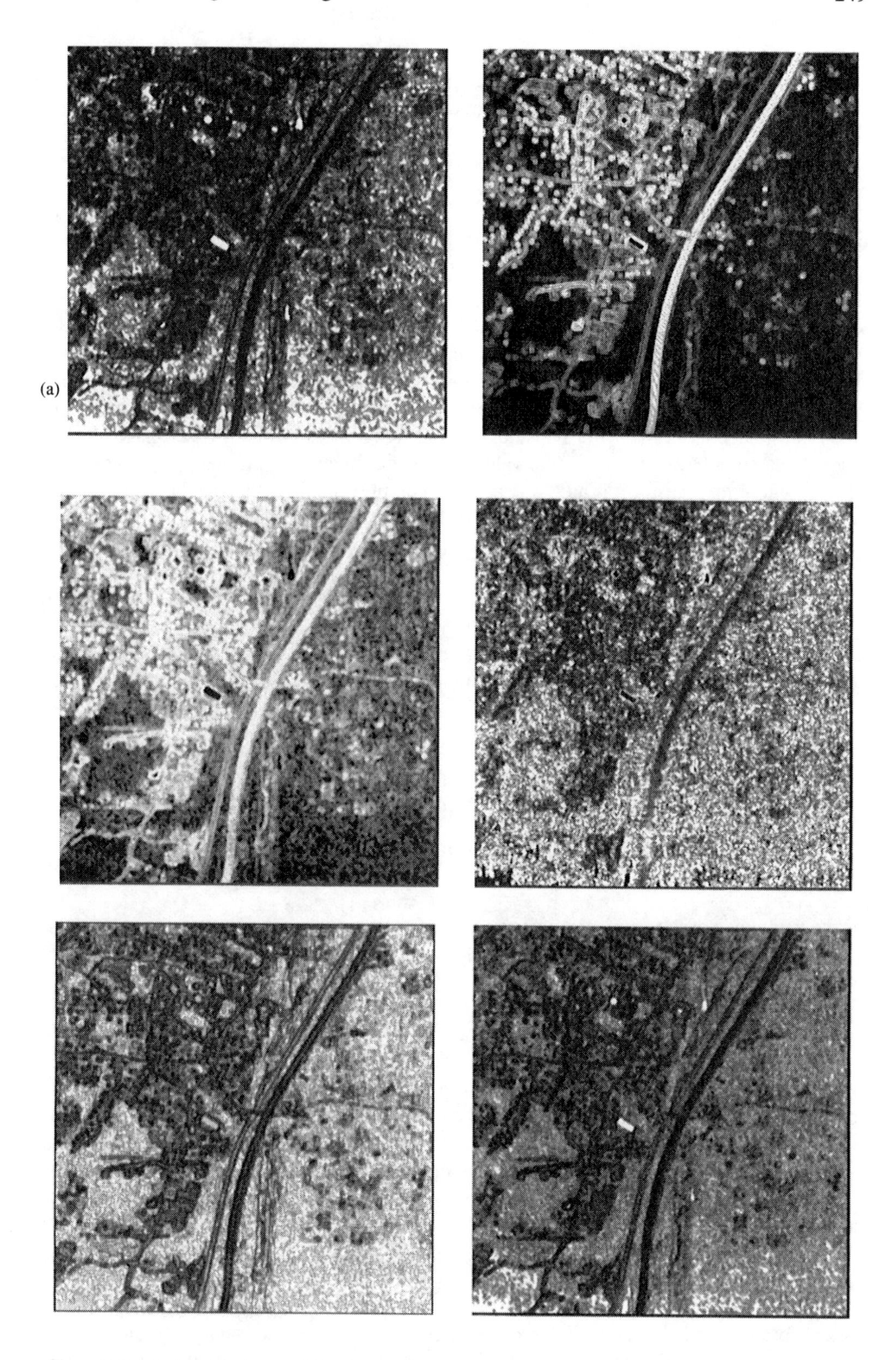

Figure 7.10 A digitized air photo (a) and several images representing a range of texture metrics or features. Note how different metrics accentuate different patterns or structures in the image.

some point, the number of computer operations becomes large compared to an equivalent approach of performing the filtering operation in the frequency domain (cf. Gonzales and Woods, 1992, for a more complete treatment of global operations). This involves transforming the entire image from the spatial domain into the frequency domain using the discrete Fourier transform, which can be defined for a square image as:

$$F(u,v) = \mathcal{F}(f(i,j)) = \frac{1}{N}\sum_{i=0}^{N-1}\sum_{j=0}^{N-1} f(i,j)e^{-2\pi i\frac{ui+vj}{N}} \tag{7.14}$$

where u,v are the spatial frequency indices in the transformed image oriented in the x and y direction, respectively, u/N, v/N are the spatial frequencies, $\mathcal{F}$ is the Fourier transform operation, N is the side dimension of the image in pixels, $i = (-1)^{1/2}$, and F is the complex frequency domain representation of the image. This transform represents all the information in the input image in terms of the sinusoidal spatial frequencies present in the image. It is a global operation because each output value $F(u,v)$ is a function of all the input values $f(i,j)$. The output of the transform is an N x N array of complex numbers that can be visualized by displaying the data as a pair of images, one being proportional to the real part of the Fourier transform and the other to the imaginary part, i.e.,

$$F(u,v) = \mathrm{Re}(F(u,v)) + i\mathrm{Im}(F(u,v)) \tag{7.15}$$

Alternatively, and more commonly, the data can be represented as the magnitude and phase derived from the Euler representation of complex numbers, i.e.,

$$F(u,v) = |F(u,v)|e^{i\phi(u,v)}$$

where

$$|F(u,v)| = \left[\mathrm{Re}(F(u,v))^2 + \mathrm{Im}(F(u,v))^2\right]^{\frac{1}{2}} \tag{7.16}$$

is the magnitude of the Fourier transform (also referred to as the *Fourier spectrum*), $|F(u,v)|^2$ is referred to as the *power spectrum*, and

$$\phi(u,v) = \tan^{-1}\left[\frac{\mathrm{Im}(F(u,v))}{\mathrm{Re}(F(u,v))}\right], \pi \geq \phi(u,v) \geq -\pi \tag{7.17}$$

is the phase. The power of the Fourier transform at any spatial frequency coordinate (u,v) is proportional to the amount of energy in the entire input image that was varying at (or contained) that frequency. From this perspective, the image is perceived to be made up of the sum of sinusoidal functions whose frequency magnitude and relative phase determines their impact on the final image. The spatial frequency interval in the output array can be expressed as:

$$\Delta u = \Delta v = \frac{1}{N\Delta x} = \frac{1}{N\Delta y}, \text{ if } \Delta x = \Delta y \tag{7.18}$$

where Δx and Δy are the sample intervals (i.e., distance between pixel centers assumed equal in this case) in x and y in the spatial domain representation of the image. We should recognize that the Fourier transform is a cyclic function with period N, so that the maximum frequency represented is the Nyquist frequency:

$$\frac{N}{2}\Delta u = \frac{N}{2}\Delta v = \frac{1}{2\Delta x} \tag{7.19}$$

For a 512 x 512 subimage of a Landsat TM scene with a 30-m GIFOV, the smallest non-zero spatial frequency sampled by the Fourier transform would be:

$$u = 1\Delta u = \frac{1}{512 \cdot 30 \text{ m}} = \frac{1 \text{ cycle}}{15{,}360 \text{ m}} \text{ or } 65 \cdot 10^{-6} \text{cycles / m} \tag{7.20}$$

i.e., approximately zero frequency. The highest frequency sampled is defined by the Nyquist limit of the sampling process and is

$$u = \frac{N}{2}\Delta u = \frac{512}{2} \cdot \Delta u = \frac{1 \text{ cycle}}{60 \text{ m}} \text{ or } 1.67 \cdot 10^{-2} \frac{\text{cycles}}{\text{m}} \tag{7.21}$$

It is important to recognize that the Fourier transform is a reversible process, so that we can convert the image information from the frequency domain to the spatial domain using an inverse Fourier transform of the form:

$$f(i,j) = \mathcal{F}^{-1}(F(u,v)) = \frac{1}{N}\sum_{u=0}^{N-1}\sum_{v=0}^{N-1} F(u,v)e^{+2\pi i\frac{(ui+vj)}{N}} \tag{7.22}$$

This means that if we operate on the image in the frequency domain, we can compute the inverse transform and take advantage of the impact the operation has in the spatial domain.

Regrettably, the mathematical underpinning of Fourier analysis and the intricacies of the links between optical processes and their mathematical representation in terms of frequency domain operations are beyond the scope of this book. The reader is urged to consider Bracewell (1978) or Papoulis (1962) for a detailed treatment of the Fourier transform, Gaskill (1978) or Goodman (1968) for treatment of the links between optical systems and frequency domain operations (linear systems), and Gonzalez and Woods (1992) for applications to digital image processing.

For the sake of brevity, our treatment of frequency domain operations will be largely pictorial in an attempt to provide the reader with an intuitive understanding. The terminology used, particularly for special functions, is adapted from Gaskill (1978). Many of the subtleties of the process (e.g., aliasing) must invariably be neglected by such a treatment, and the reader unfamiliar with linear systems is encouraged to delve deeper into the subject before relying heavily on frequency domain operations. Figure 7.11 shows an image and the various ways to represent the components of its Fourier transform. Note that the origin has been shifted from the top left corner to the center of the image as is customary in the treatment of optical systems. Also note that the periodic nature of the function is such that if the input image is real valued (which is always the case for digital images of optical intensity), then the magnitude of the Fourier transform will be symmetric about the origin in the zero-centered representation. In many remote sensing images, most of the visually interpretable data in the Fourier transform are contained in the magnitude. To save space, we will usually show only the power spectrum of the Fourier transform in the rest of this section. To help in visualizing the information in the Fourier transform, subsections of an image with particular characteristics can be treated as subimages and their transforms computed. This is shown for several windows in Figures 7.12 and 7.13. In Figure 7.13, we can see how the Fourier transforms of images of different land cover types appear significantly different. Stromberg and Farr (1986) suggested that these differences in the Fourier spectrum could be used to aid in

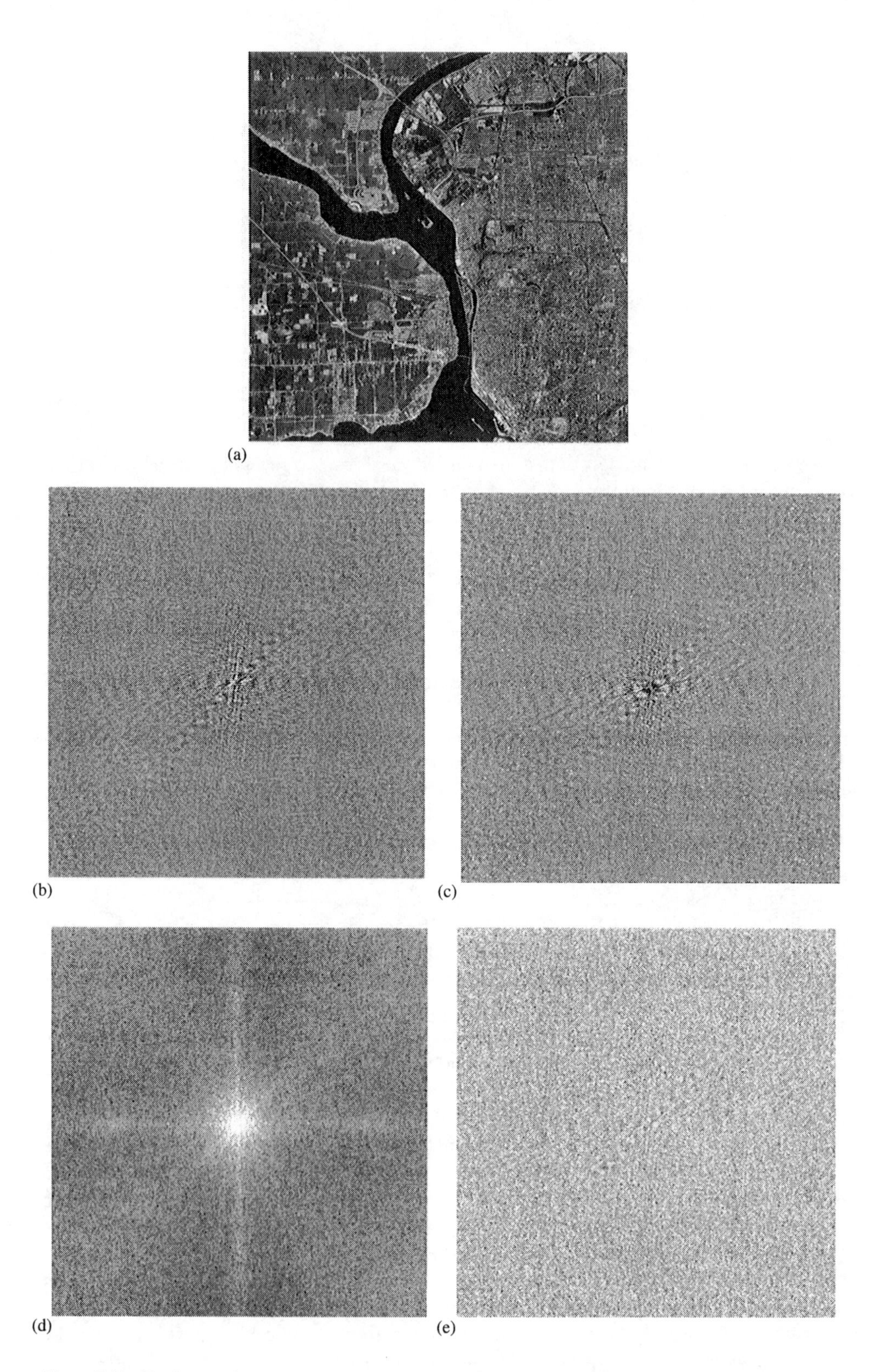

Figure 7.11 Fourier transform representation of an image. (a) Digitized air photo of Buffalo, New York and Niagara River (b) real; and (c) imaginary part of the Fourier transform; and (d) magnitude and (e) phase of the Fourier transform. Note that these images have been scaled for display purposes.

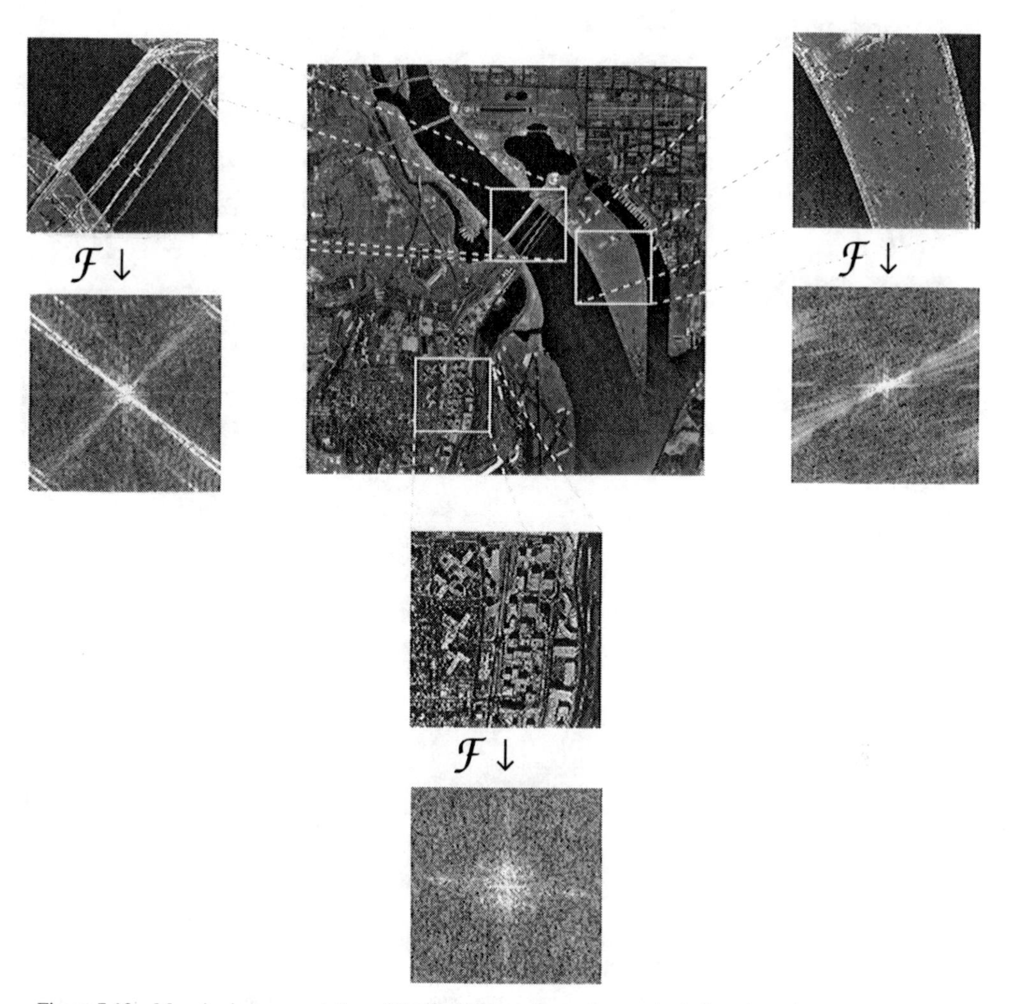

Figure 7.12 Magnitude representation of the Fourier transform of several windows in an image of Washington, D.C.

classifying features with different texture. Ehrhard et al. (1993) demonstrated that when the Fourier transforms were filtered (i.e., multiplied) by a set of bandpass filters, the inverse transforms could be used as texture metric images to successfully classify background land cover types using conventional statistical classifiers (cf. Sec. 7.2 for statistical classifiers).

Several important links exist between operations in the spatial domain and corresponding operations in the frequency domain. The one we are most concerned with is how frequency filters are related in the two domains. The convolution theorem from linear systems expresses this relation mathematically as:

$$f(i,j) * g(i,j) = \mathcal{F}^{-1}[F(u,v) \cdot G(u,v)] \tag{7.23}$$

$$f(i,j) \cdot g(i,j) = \mathcal{F}^{-1}[F(u,v) * G(u,v)] \tag{7.24}$$

where $f(i,j)$ and $g(i,j)$ are image arrays and $F(u,v)$ and $G(u,v)$ are their Fourier transforms. This theorem states that the convolution operation in the spatial domain is equivalent to mul-

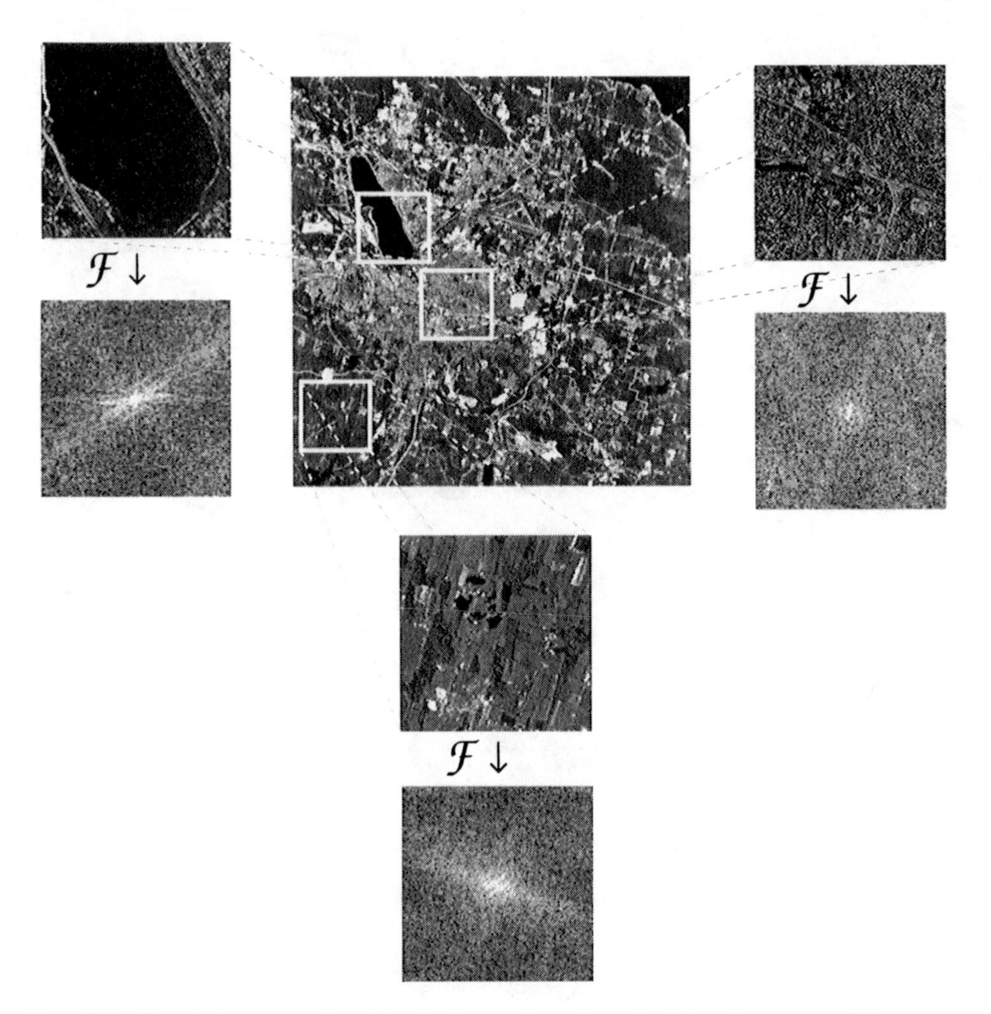

Figure 7.13 Magnitude representation of the Fourier transform of several windows in an image of Syracuse, New York.

tiplication in the frequency domain, and conversely that multiplication in the spatial domain is the equivalent of convolution in the frequency domain. Recall from Chapter 1 that we are assuming that the image chain can be reasonably approximated as a linear shift-invariant system.

Processing an image by changing the relative amount of energy in various spatial frequencies is most easily envisioned in the frequency domain. To facilitate this, we first introduce the concept of radial frequency to simplify our characterization of circularly symmetric filters. Let p be the spatial frequency associated with a radial distance from the origin, i.e.,

$$p = (u^2 + v^2)^{\frac{1}{2}} \tag{7.25}$$

For circularly symmetrical filters, we can then simply define the relative amount of filtration for each radial frequency and generate a two-dimensional filter by rotation about the origin, as shown in Figure 7.14. The value of the filter $H(u,v)$ at any point defines how much of the energy at that spatial frequency will be present in the output image. Where the output image is defined as

$$g(i,j) = \mathcal{F}^{-1}[F(u,v) \cdot H(u,v)] = \mathcal{F}^{-1}[G(u,v)] \tag{7.26}$$

For example, where the filter $H(u,v)$ is zero, so is the output amplitude $G(u,v)$. Values of $H(u,v)$ between 0 and 1 will reduce ("filter") the amount of energy, and values greater than 1 will increase the energy and exaggerate the frequencies affected. Figure 7.15 shows the impact of several frequency domain filters on remotely sensed images.

Clearly, filter design is conceptually simpler in the frequency domain. The number of elements in the kernel used to implement the filter in the spatial domain typically controls whether it is more appropriate to apply the filter by multiplication in the frequency domain (then back-transforming the filtered image) or by inverse-transforming the filter and convolving the image with the filter in the spatial domain. The ideal low-pass filter [Fig. 7.15(a)] eliminates all frequencies above the user defined cutoff frequency (p_0). The resultant image is blurred and has some ringing artifacts due to the abrupt cutoff of the filter. The Butterworth low pass filter (Fig. 7.15(b) has no visible ringing due to its gentler transition and has less blurring effects even with the same cutoff since some high-frequency data are allowed to pass. The steepness of the cut on (cutoff) is controlled by the order of the Butterworth filter [cf. Fig. 7.15(b) and (c)]. In many cases involving image enhancement, we are interested in a filter that amplifys ("boosts") the higher frequencies to compensate for the blurring in the imaging process that has reduced the high-frequency content in the image. Figure 7.15(d) shows the general shape of a boost "filter" and its sharpening impact on the image. Regrettably, most boost filters inevitably exaggerate high-frequency noise in the image, such that amplifying beyond a certain point degrades the appearance of the image.

Periodic noise in the image can often be filtered out so that it does not detract from the original image or from a boosted version in cases where the noise is subtle. An example of this process is shown in Figure 7.16. In this case, neither the original image [Fig. 7.16(a)] or a contrast enhanced version [Fig. 7.16(b)] shows any appreciable noise. However, when the original image is filtered with a boost filter to sharpen the edges and then contrast-enhanced [Fig. 7.16(c)], periodic noise patterns become visible. These subtle periodic noise patterns are common in most EO imagery as a result of slight detector-to-detector calibration differences, periodic noise sources in the processing electronics, or periodic noise in the readout and digitizing electronics. By transforming the original image into the frequency domain and then enhancing the contrast of the image of the magnitude of the Fourier transform, it is possible to identify the elevated energy content at certain frequencies that represent the unwanted noise [Fig. 7.16(d)]. By making a filter that rather quickly tapers to zero at these frequencies, it is possible to filter out the noise and leave the rest of the image largely unaffected [Fig. 7.16(e)]. When the boost filter is applied after the noise filter and the resulting image contrast stretched (after being transformed back to the spatial domain), the noise effects are removed [Fig. 7.16(f)].

These frequency domain operations (or their spatial domain equivalents, cf. Fig. 7.17) can be used to perform numerous image enhancement operations including blurring, sharpening, noise filtering, edge definition, patterned noise isolation, pattern isolation, and scene segmentation based on texture when coupled with other algorithms. In addition, one of their most useful functions is in modeling the performance characteristics of imaging systems relative to spatial image fidelity and in defining filters for reducing the apparent impact of spatial degradations along the imaging chain (cf. Sec. 9.1).

7.2 IMAGE CLASSIFICATION

One of the most common requirements in remote sensing is the need to segment or classify

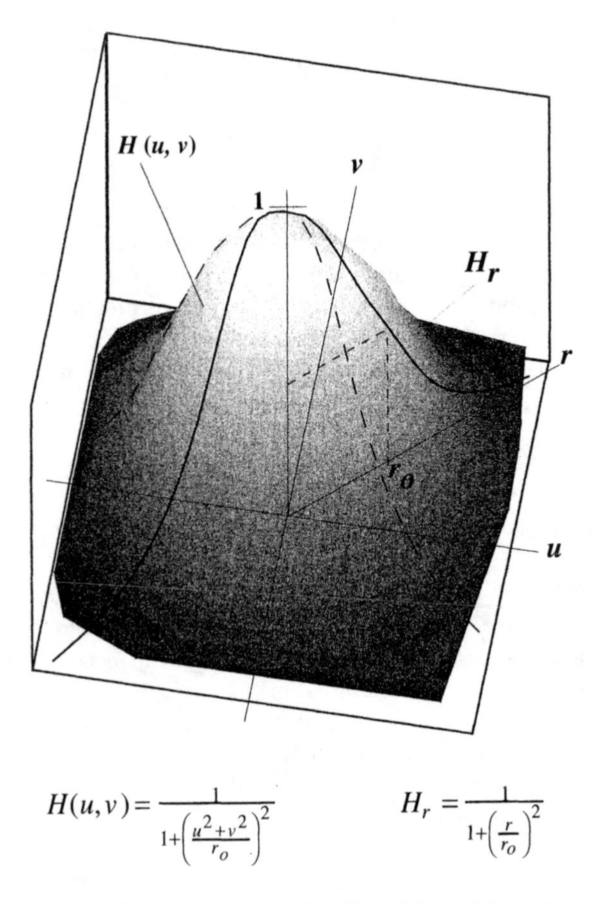

$$H(u,v) = \frac{1}{1+\left(\frac{u^2+v^2}{r_o}\right)^2} \qquad H_r = \frac{1}{1+\left(\frac{r}{r_o}\right)^2}$$

Figure 7.14 Use of radial frequency concept to simplify definition of circularly symmetric filters.

an image into land cover, material, or object classes. In some cases, this is an end in itself; in others, it is one step in a more involved process (cf. Sec. 7.5 on hierarchical classification). In most cases, the classification is performed with a man-in-the-loop on uncalibrated imagery, so we will concentrate this section on techniques used in these cases. Unsupervised classification methods are discussed briefly in Section 7.2.3, and the use of calibrated or normalized data are discussed in Section 7.2.2.5. A complete derivation of the multivariate statistics behind the Gaussian maximum likelihood classifier emphasized in Section 7.2.2 is beyond the scope of this treatment (cf. Morrison, 1967). To provide a stronger understanding of the multivariate process, we will provide a rigorous development of univariate classification methods in Section 7.2.1 and then extend that treatment, largely by analogy, to the multivariate (multispectral) case in Section 7.2.2.

7.2.1 Supervised Classification of a Monochrome Scene

In this section we will assume that a single spectral band digital image is to be segmented into classes by material or land cover type. We are going to use what is usually referred to as a *supervised*, or *man-in-the-loop*, process where the user identifies a sample of pixels of each type or class, and digital processing algorithms are used to assign all similar pixels to one of the classes. To simplify the process, assume that the image is comprised of l classes.

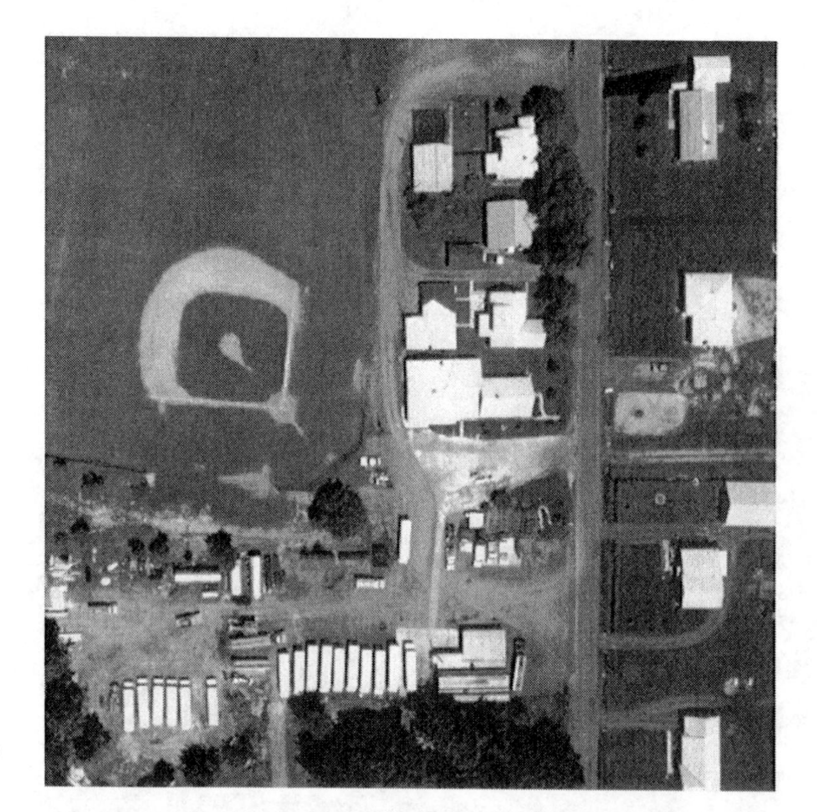

(a)

(b)

Figure 7.15 Application of several circularly symmetric filters to remotely sensed images. (a) Image, (b) is the magnitude of the Fourier transform.

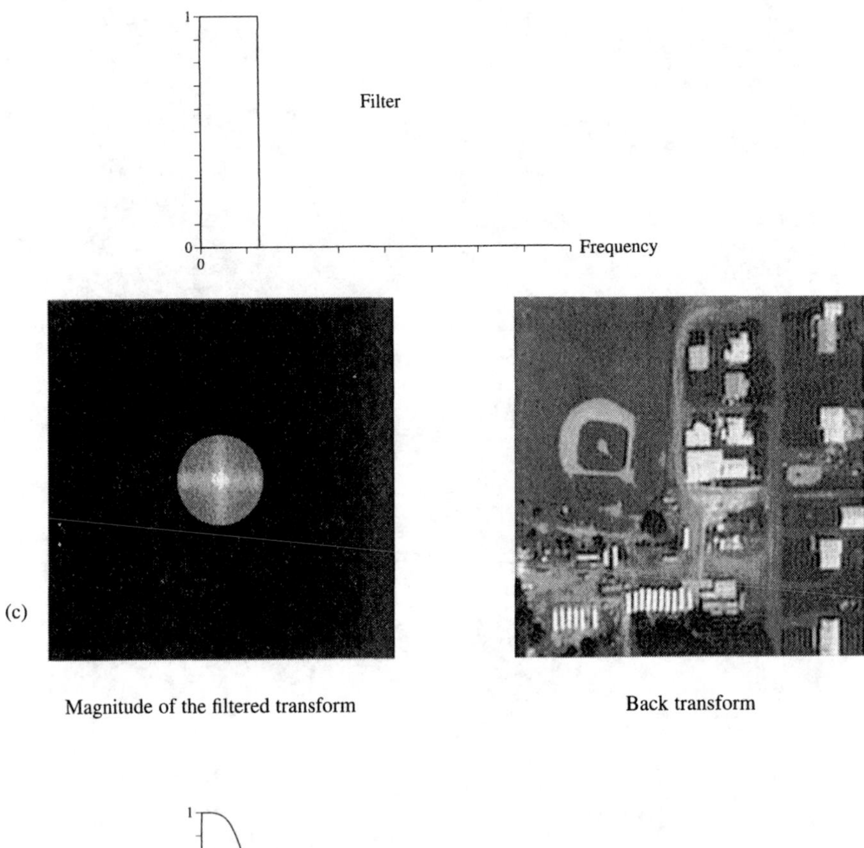

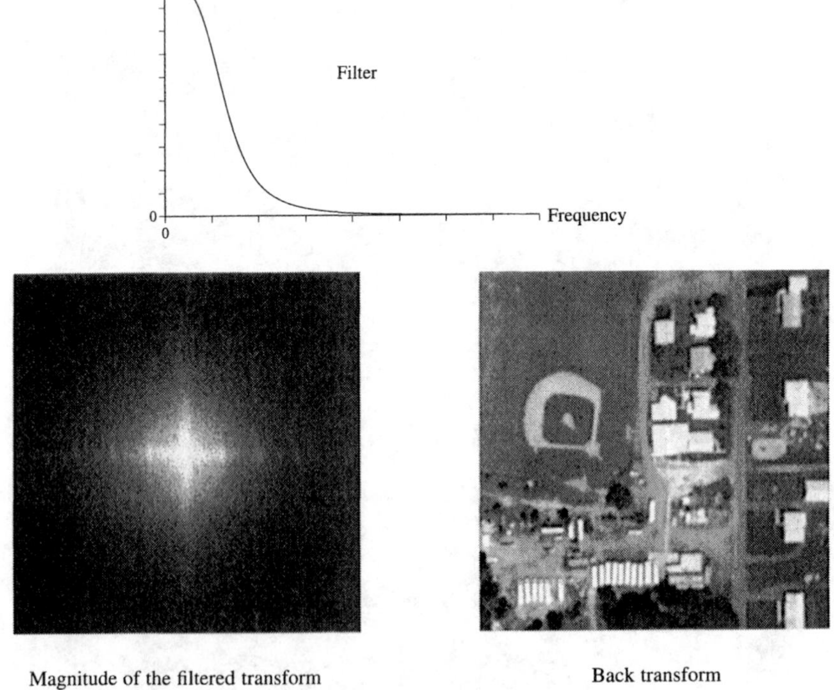

Figure 7.15 Application of several circularly symmetric filters to remotely sensed images (continued), (c) ideal low-pass filter (d) Butterworth low-pass filter.

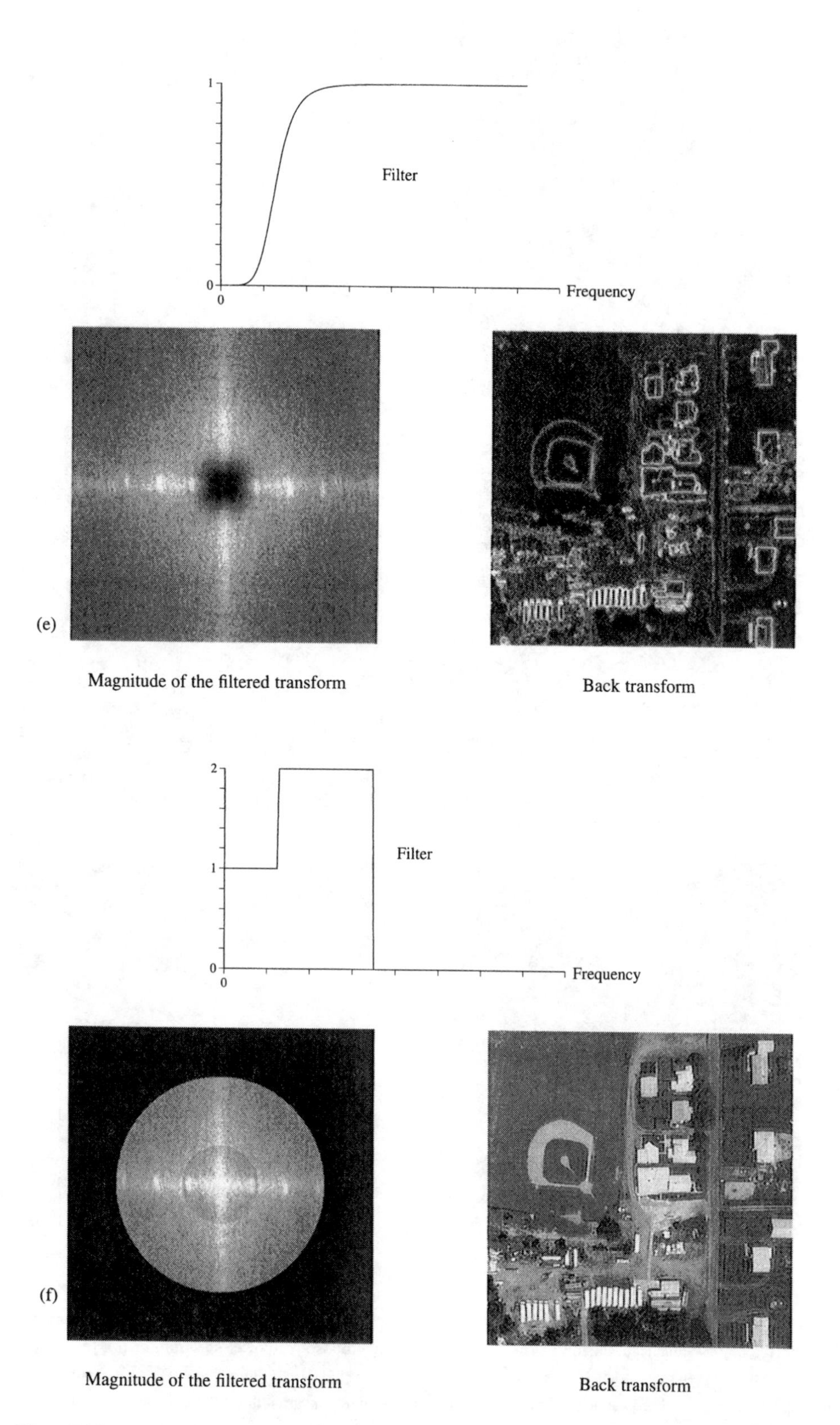

Figure 7.15 Application of several circularly symmetric filters to remotely sensed images (continued), (e) Butterworth high-pass filter and (f) high-frequency boost filter.

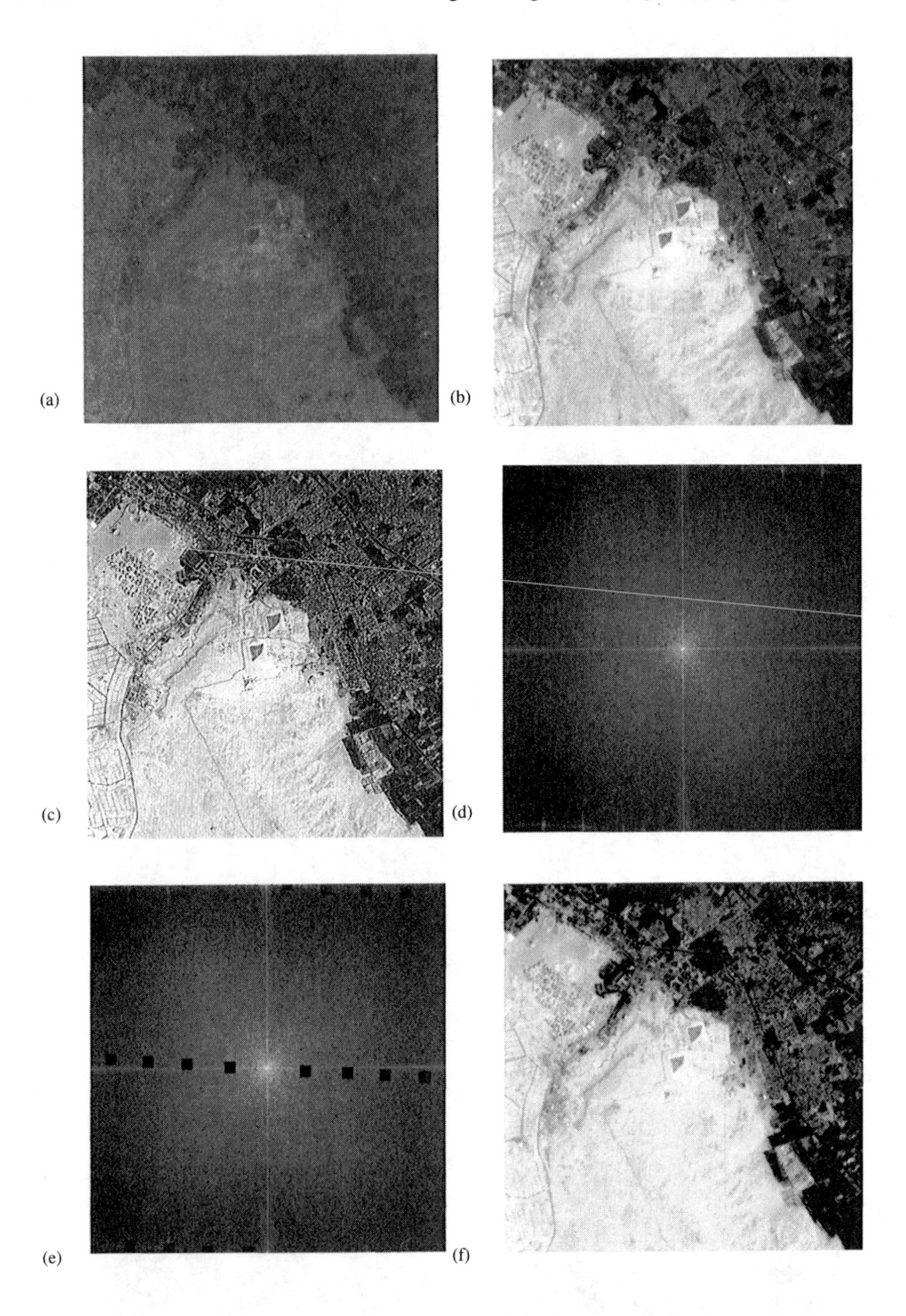

Figure 7.16 Filtering of noise in a SPOT image of the great pyramids: (a) original image, (b) contrast enhanced image, (c) contrast enhanced version of (a) after high-frequency boost, (d) power spectrum of (a), (e) power spectrum showing where frequencies associated with periodic noise have been filtered, and (f) contrast-enhanced and high-frequency boosted version of (a) after periodic noise filter was applied in the frequency domain.

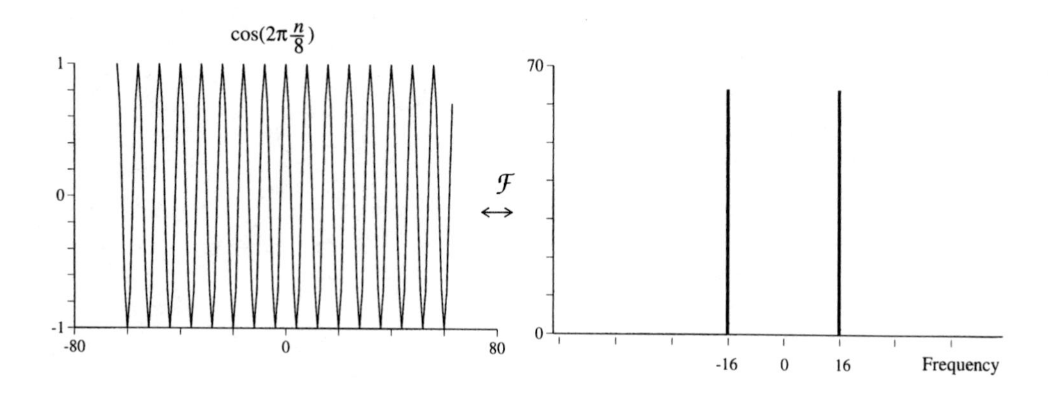

Cos function

Delta functions

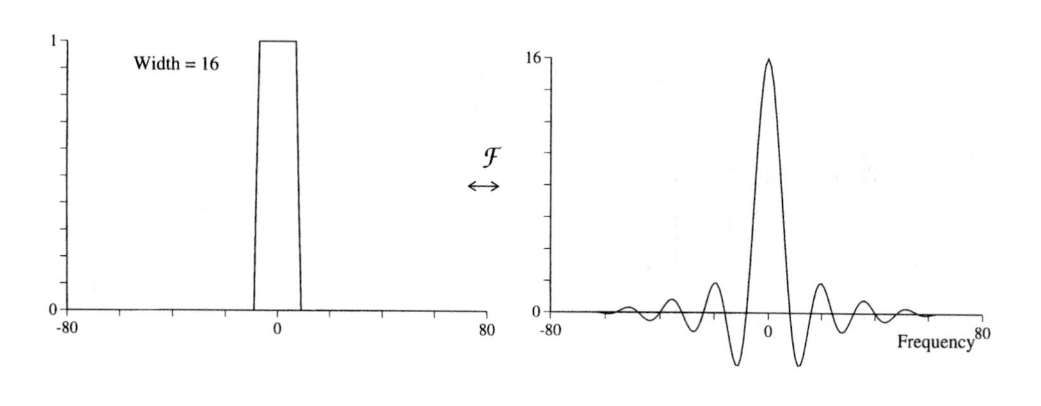

Rect function

Sinc function

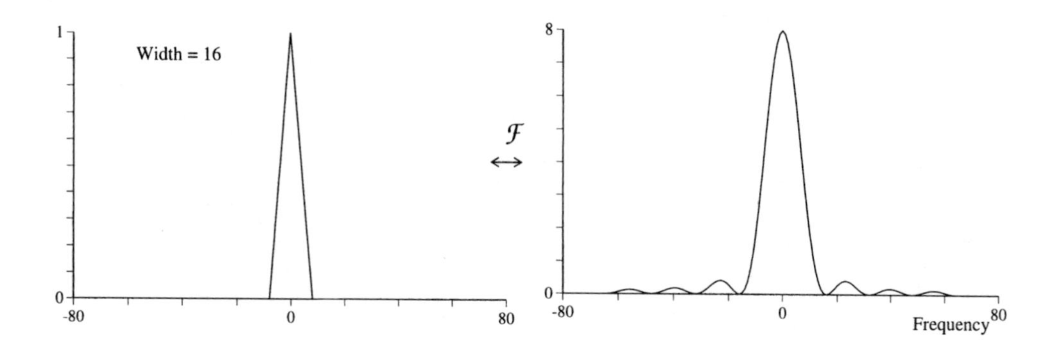

Tri function

Sinc^2 function

Figure 7.17 Fourier transform pairs.

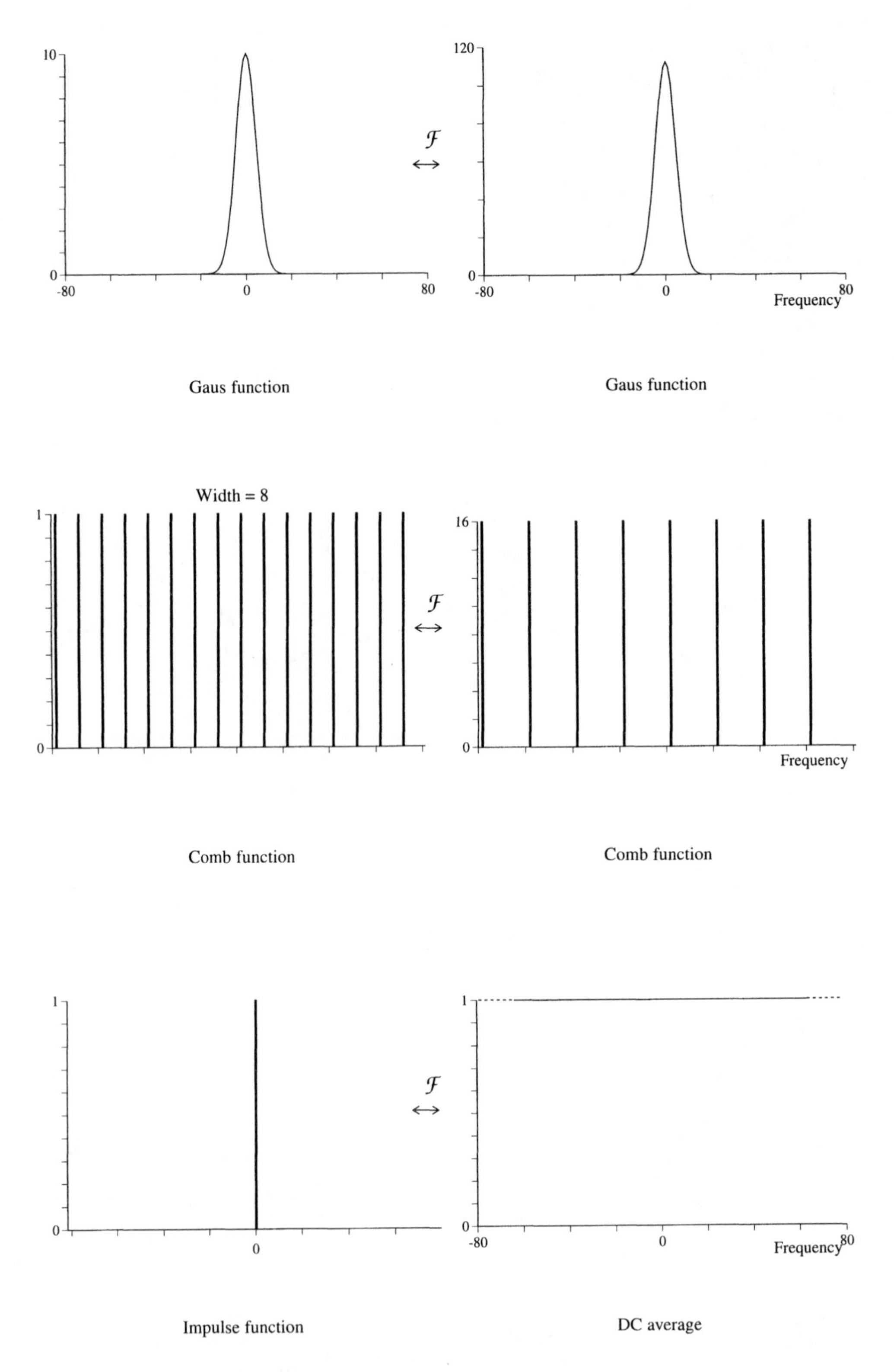

Figure 7.17 Fourier transform pairs (con't.).

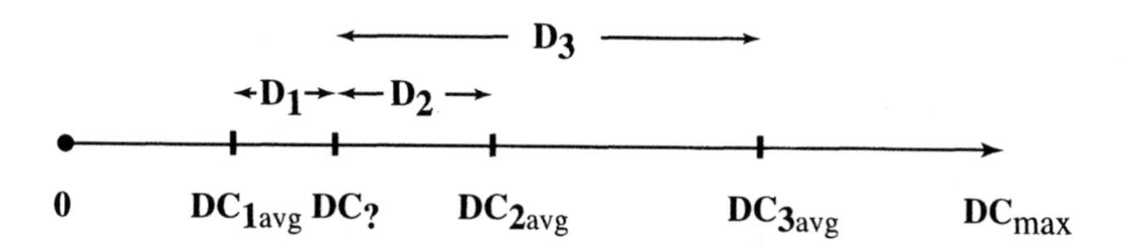

Figure 7.18 Use of the minimum distance to the mean classifier in a 3-class single-band image classification problem. Using the minimum distance to the mean classifier $DC_?$ is assigned to class 1 since D_1 is the smallest.

For clarity in the diagrams, we will set l equal to 3. We can think of this process as teaching or "training" the computer algorithms what each class "looks" like, and then using some attributes of the class to classify other pixels into the class they most "look" like.

7.2.1.1 Univariate Minimum Distance to the Mean

The simplest way to characterize what a group of pixels "looks" like in a monochrome scene is by their mean digital count. In Figure 7.18 we show the results of selecting several pixels from each class and computing the mean gray value of each class using

$$DC_{iavg} = \frac{\sum_{q=1}^{N} DC_{iq}}{N} \tag{7.27}$$

where DC_{iavg} is the mean of the ith class, DC_{iq} is the digital count for the qth sample in the ith class, and N is the total number of pixels in the sample set for that class. We have now "trained" the computer algorithm to what each class "looks" like as defined by the mean digital count of the samples. Any other pixel $DC_?$ must then be compared to the means to see which one it most "looks" like. In the simplest case, this can be defined by selecting the class whose mean digital count (DC_{iavg}) is the minimum distance (D) from the pixel to be classified $DC_?$ as defined by

$$D_i = \left[(DC_{iavg} - DC_?)^2\right]^{\frac{1}{2}} \tag{7.28}$$

where D_i is the distance of the pixel from the mean of the ith class. The pixel is assigned to the class (i) for which the D_i value is a minimum (hence the name of the method *minimum distance to the mean* or MDM). In the example shown in Figure 7.18, the unknown pixel would be assigned to class 1. In this way, every pixel in the scene can be classified. It is also possible to use threshold values so that pixels too far from any class mean are assigned to a class of unidentified pixels.

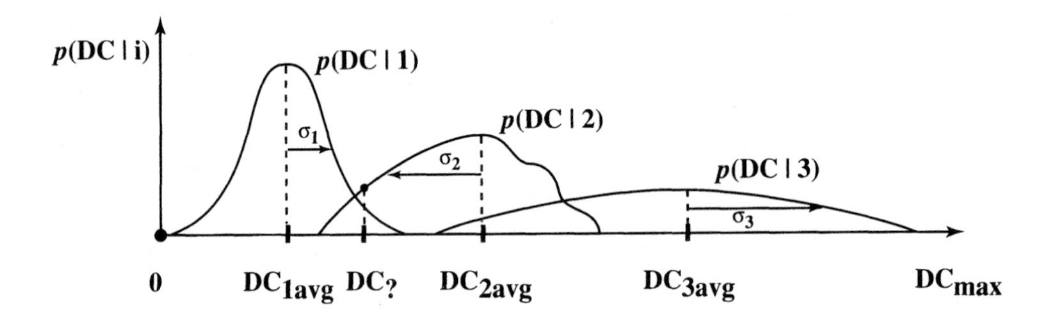

Figure 7.19 Use of simple statistical distances to develop classifiers that are sensitive to variance in the data.

7.2.1.2 Simple Statistical Measures of Distance

A limitation of the minimum-distance-to-the-mean method is illustrated in Figure 7.19. In this case we have shown not only the mean of the sampled values but also a normalized histogram $p[(DC|i)]$ of the sampled values for each class and a measure of the variation in the sampled values in the form of the standard deviation from the class mean (σ_i). In this case, we see that pixels belonging to class 1 show very little variation from the mean compared to classes 2 or 3. A pixel having the same digital count $DC_?$ as used in the minimum distance to the mean example would intuitively be classified not into the closest class (class 1 in this case) but into the class that has the higher *conditional probability*. The normalized histogram is referred to as the conditional probability $[p(DC|i)]$ because it describes the probability of a digital count (DC) occurring subject to the condition that we are sampling from the *i*th class. Large sample sizes are required to characterize the shape of the normalized histogram. As a result, the shape is often assumed to be approximately Gaussian. Then the standard deviation, together with the mean, define the shape of the normalized histogram. Thus assigning the pixel to the class with the highest conditional probability $[p(DC|i)]$ can be approximated by assigning the sampled pixel to the class whose mean is the fewest number of standard deviations away from the gray value of the unknown pixel. This simple statistical distance can be expressed as

$$D_{\sigma i} = \frac{\left[\left(DC_{iavg} - DC_?\right)^2\right]^{\frac{1}{2}}}{\sigma_i} \tag{7.29}$$

where σ_i is the simple standard deviation of the samples belonging to class i. This classifier agrees with our intuitive reaction to data with different variances, stating that we assign the pixel to the class from which it is separated by the fewest number of standard deviations (class 2 in the example in Fig. 7.19). Thus by accounting for variance within the class, we've introduced a statistical measure of distance that should improve our classification accuracy relative to the simple Euclidean distance metric used in the MDM method.

As shown in Figure 7.19, the use of the conditional probabilities (or an approximation based on the number of standard deviations), to make classification decisions makes intuitive and logical sense when the number of pixels in each class is approximately the same. However, if we anticipate that the number of pixels per class will differ significantly across classes, we might want to adjust our classification algorithm. For example, assume that we

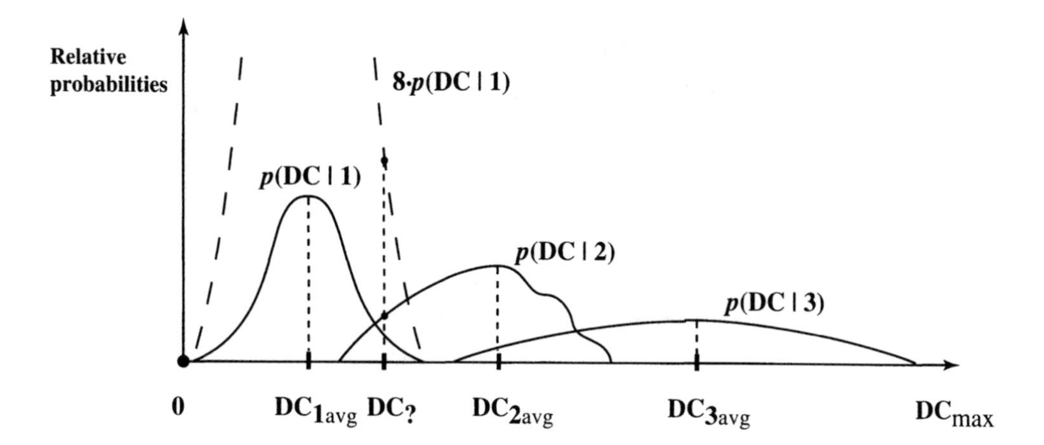

Figure 7.20 Effect of a priori probabilities on class decisions. Use of the a priori probabilities will change our decision about what class pixels with a digital count of $DC_?$ belong to. In this case, many more pixels with digital count $DC_?$ are in class 1 than class 2, so we will be correct more often if we assign pixels with this digital count to class 1.

know our example scene contains roughly equal numbers of forest and urban pixels but 8 times as many water pixels. This foreknowledge of the probability of a randomly sampled pixel being in a given class is called the *a priori probability* [$p(i)$]. In our case, the a priori probabilities are 0.8, 0.1, and 0.1 for the water, forest, and urban classes, respectively. This means that if we want to plot the relative heights of the unnormalized histograms rather than the normalized histograms in Figure 7.19, we would multiply each curve by the a priori probability. In our case, since we are only interested in relative heights of the curves, we could achieve the same end by multiplying the conditional probability for water (class 1) by 8, leaving the others unchanged (cf. Fig. 7.20). This accounts for the fact that there are eight times more water pixels than forest or urban. From Figure 7.20, we see that there will be both class 1 and class 2 pixels with digital count $DC_?$. However, when we factor in the relative number of class 1 pixels, it is clear that a pixel with a digital count value of $DC_?$ is much more likely to be in class 1 than in class 2 or class 3.

7.2.1.3 Univariate Gaussian Maximum Likelihood Classifier

The use of a priori probabilities can be formalized using Bayesian probability theory to define the a posteriori probability that a pixel with digital count DC belongs in class i:

$$p(i|\mathrm{DC}) = \frac{p(\mathrm{DC}|i)\,p(i)}{p(\mathrm{DC})} \tag{7.30}$$

where $p(\mathrm{DC})$ is the probability of the digital count occurring anywhere in the image (i.e., it is the normalized histogram of the entire image). The classification process simply involves computing the a posteriori probability of the sampled digital count value $DC_?$ being in each class and assigning pixels with that digital count to the class that yields the highest a poste-

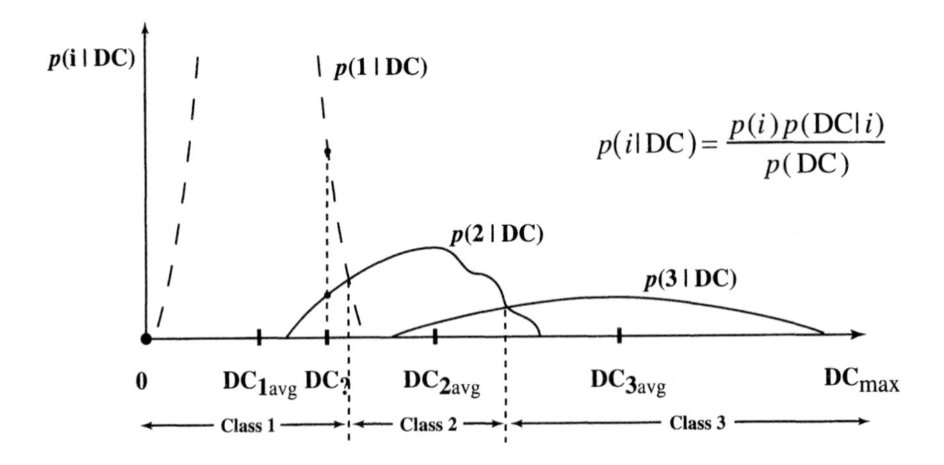

Figure 7.21 Application of probability theory to univariate classification. Using maximum likelihood theory we would assign a pixel to whatever class has the highest a posteriori probability; $p(i|\mathrm{DC})$. In our case, a pixel with a digital count of $\mathrm{DC}_?$ would be assigned to class 1.

riori probability (cf. Fig. 7.21). In practice, $p(\mathrm{DC})$ is the same for all classes. As a result, $p(\mathrm{DC})$ scales all the a posteriori probabilities, but it will not change their rank ordering, and so for convenience it is not included in the classifier. The a priori probabilities can be obtained from general knowledge of scene content or from a previous class map of the region [in some cases, a simple minimum-distance-to-the-mean classifier is run first to estimate the $p(i)$ values]. If no information is known about a priori probabilities $p(i)$, they can simply be assumed to be equal. The conditional probabilities can be estimated from the normalized histograms generated from the training data. This presents a problem in terms of the amount of data to be stored in the algorithm and the amount of training data required to generate a good estimate of the true class histogram. In a univariate case (such as we are considering here), the histogram is not too large (e.g., for an 8-bit image, it would only be 256 values for each class). However, in the multispectral cases we will consider shortly, these histograms are multidimensional and require massive amounts of data to accurately estimate their shape and to store the resulting estimate.

In cases where the shapes of the class histograms can be assumed to be approximately Gaussian, this process can be greatly simplified. The shape of the histogram, and therefore the conditional probability, can be estimated from the mean and the standard deviation of the training data according to:

$$p(\mathrm{DC}|\mathrm{i}) = \frac{1}{\sqrt{2\pi\sigma_i^2}} \mathrm{e}^{-\frac{(\mathrm{DC}-\mathrm{DC}_{\mathrm{iavg}})^2}{2\sigma_i^2}} \tag{7.31}$$

We can then define a class discrimination metric (D'_i) by combining Eqs. (7.31) and (7.30) and ignoring the $p(\mathrm{DC})$ term. This yields:

$$D'_i = p(\mathrm{DC}|i)p(i) = \frac{p(i)}{\sqrt{2\pi\sigma_i^2}} e^{-\frac{(\mathrm{DC}-\mathrm{DC}_{iavg})^2}{2\sigma_i^2}} \tag{7.32}$$

For each pixel, the D'_i values for each class are computed, and the pixel is assigned to the class with the highest D'_i value. According to Eq. (7.30) and our analysis, this process will optimize the probability of correctly classifying the pixel under the assumption of Gaussianly distributed data (hence the name for this classifier, *Gaussian maximum likelihood* or GML). Note that for simplicity, we can redefine the discriminant metric by computing the logarithm of D_i:

$$D''_i = \ln D'_i = \ln[p(i)] - \frac{1}{2}\ln(2\pi) - \ln(\sigma_i) - \frac{(\mathrm{DC}-\mathrm{DC}_{\mathrm{iavg}})^2}{2\sigma_i^2} \tag{7.33}$$

or

$$D_i = \ln D'_i + \frac{1}{2}\ln 2\pi = \ln[p(i)] - \ln(\sigma_i) - \frac{(\mathrm{DC}-\mathrm{DC}_{\mathrm{iavg}})^2}{2\sigma_i^2} \tag{7.34}$$

Taking the natural logarithm and adding a constant will change neither the rank ordering of the discriminant metrics nor any resultant decisions about class assignment. Inspection of Eq. (7.34) indicates that only the last term is a function of the digital count and that this term is proportional to the magnitude of the squared value of the number of standard deviations from the pixel gray value to the mean. Thus for a given class, reducing the number of standard deviations increases the discriminant value in agreement with our intuitively derived simple statistical-distance metric (Eq. (7.29). Furthermore, if the a priori probabilities are the same and the standard deviations are the same, Eq. (7.34) reduces to:

$$D_i = A_0 - \frac{(\mathrm{DC}-\mathrm{DC}_{\mathrm{iavg}})^2}{2\sigma^2} \tag{7.35}$$

where k is the number of classes, σ is the common standard deviation for all classes, and $A_0 = \ln(1/k) - \ln\sigma$. In this case, D_i is maximized by minimizing the Euclidean distance (i.e., the problem reduces to an MDM classification).

In summary, if the data can be assumed to be Gaussian, the GML classifier described by Eq. (7.34) will minimize the number of misclassified pixels. In cases where the data are not Gaussian, we would have to use a discriminant function based on maximizing the a posteriori probabilities [cf. Eq. (7.30)] to minimize errors. While this is possible in univariate cases, it becomes operationally difficult to characterize effectively the conditional probabilities as the number of variables (spectral bands) increases. As a result, it is often useful to select classes whose brightness distributions are approximately Gaussian (e.g., a bimodal class could be split into two unimodal approximately Gaussian classes).

To this point, we have assumed that our goal has been to minimize overall classification error. This may not always be the case. For example, consider the error regions indicated in Figure 7.22. In this example, class 1 is a target class that is for some reason more important than the other classes (e.g., a crop of high market value). From the perspective of class 1, the error regions can be separated into errors of commission (false alarms) and errors of omission (failure to identify a true target-false negative). From the perspective of class 2 (i.e., if class 2 is designated as the priority or target class), these error regions would be reversed. In a case where it is very critical to find all occurrences of a target (e.g., the target is toxic leachate from a waste dump), we might want to shift the decision point in Figure 7.22 to the

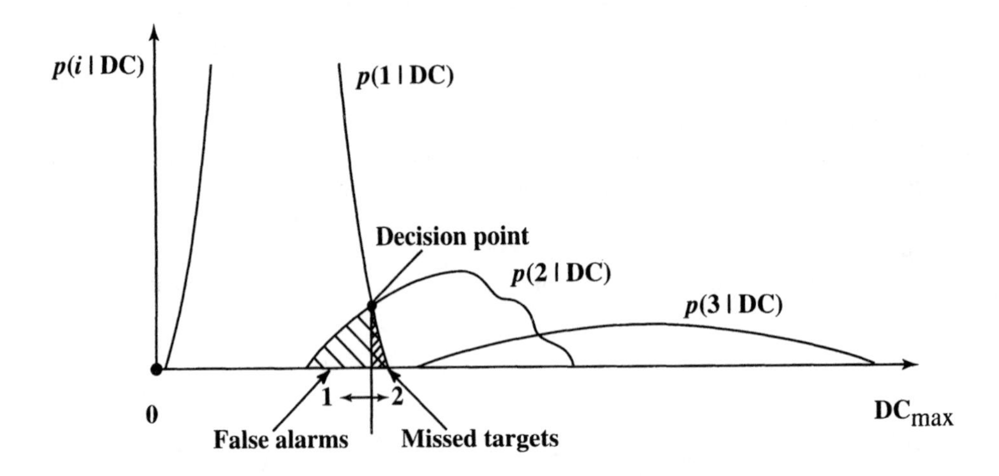

Figure 7.22 Adjustment of decision bounds to control false alarms on misses.

right. This would increase the number of false alarms but decrease the number of misses. The overall effect would be to increase the number of misclassified pixels. Moving the decision point to the left would have the opposite effect of decreasing false alarms and increasing the number of targets missed. The overall increase in errors might be acceptable if there were a high cost associated with false alarms.

In scenarios such as these, we realize that we want to minimize not the number of misclassified pixels but the cost or loss associated with the overall classification. This can be done by assigning loss functions or weights to each type of classification (cf. Duda and Hart, 1973). For example, let $l(i|j)$ be the loss or cost of assigning a pixel to the ith class when it actually belongs to the jth class. Then the risk or loss associated with assigning a pixel to class i will be related to its digital count by the conditional risk given by

$$R(i|\mathrm{DC}) = \sum_{j=1}^{k} l(i|j)\, p(j|\mathrm{DC}) \tag{7.36}$$

where k is the number of classes in the image. We minimize risk by assigning the pixel to the class that has the minimum conditional risk. In most cases, the most difficult part of this process is defining the loss matrix. In some cases, this is done based on an economic assessment, and dollar values can be used to define the cost of each type of misclassification. In other cases, costs are estimated in terms of relative loss or "pain" levels associated with different types of misclassification. Figure 7.23 shows a loss matrix for our simple three-class problem where we have defined the water (class 1) to have a high cost associated with false alarms (false positives) in case 1 and a high cost of missing a target (false negatives) in case 2. If we then choose to use one of these loss matrices to minimize our cost using Eq. (7.36), we will make more overall errors in classification but minimize our cost as defined by the matrix. The final classification can be very sensitive to the relative magnitudes of the loss matrix, and the user should carefully evaluate how the loss values are defined. In many cases, there is too little data to develop good loss functions, and the misclassification error is simply minimized by maximizing the a posteriori probability.

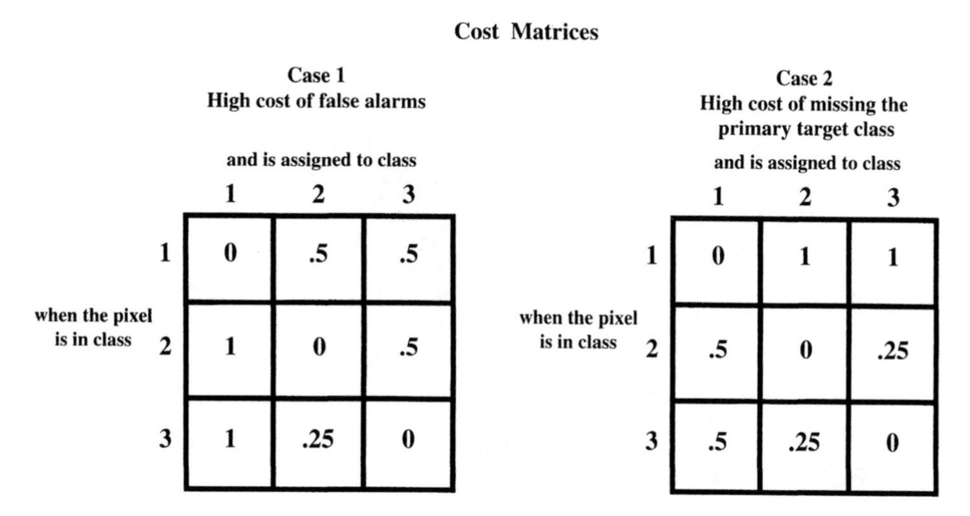

Figure 7.23 Example cost matrices for a three class scenario: (a) Case 1 is for a high cost of false alarms, with class 1 as the target class. (b) Case 2 is for a high cost of missing the target class.

7.2.2 Supervised Multispectral Image Classification

The single-band classification case described in the previous section is seldom used in practice because only limited class separability is achieved when only one spectral band is available. However, it is fairly straightforward to extend the principles to many spectral channels. If there are reasonably unique spectral signatures, then the classifiers can provide separation of classes. To keep the illustrations clear, we will generally illustrate only the two-spectral-band case, although the mathematical solutions will be in terms of l spectral bands. Consider a set of training data illustrated in Figure 7.24, where the three classes are plotted with different symbols for two spectral bands. It is clear that if these values were projected onto either axis, discrimination using a single band would result in considerable overlap. Yet, in the two-band case, there appears to be fairly clear separability between the classes. The objective of this section is to define a set of classification algorithms that can take advantage of this apparent separability using multiple spectral bands.

7.2.2.1 Multivariate Minimum Distance to the Mean Classifier

To begin, we define for each pixel a column vector (**X**) made up of its digital count values (DC) in each of l spectral bands, i.e.,

$$\mathbf{X} = \begin{bmatrix} DC_1 \\ DC_2 \\ \vdots \\ DC_\ell \end{bmatrix} \tag{7.37}$$

where **bold** letters will be used to indicate vectors and matrices. Similarly, the center of a cluster of multispectral sample points can be defined by the multivariate mean (**M**) for the class using a column vector comprised of the mean digital count values in each band for samples from class (i) according to:

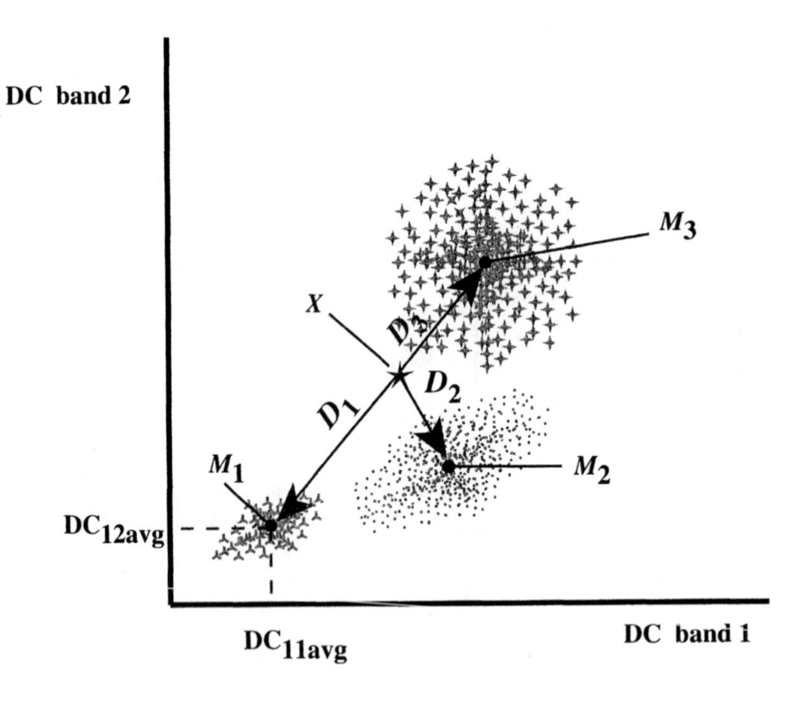

Figure 7.24 Multispectral classification using a minimum distance to the mean classifier.

$$\mathbf{M}_i = \begin{bmatrix} DC_{iavg1} \\ DC_{iavg2} \\ \vdots \\ DC_{iavg\ell} \end{bmatrix} \tag{7.38}$$

From Figure 7.24, we see that the simple Euclidean distance between **X** and $\mathbf{M}_i$ in a one-dimensional space is:

$$D_i = \left[(\mathbf{X} - \mathbf{M}_i)^2\right]^{\frac{1}{2}} \tag{7.39}$$

We could then implement a multivariate minimum-distance-to-the-mean (MDM) classifier by computing D_1 or, more practically, D_i^2 values for each pixel relative to the mean vector ($\mathbf{M}_i$) for each class and assign the pixel to that class with the minimum-squared distance (D_i^2). This classifier, like its univariate equivalent, is very easy to implement and can be coded to run rapidly on today's computers. However, it ignores the variability in the data as illustrated in Figure 7.25. The pixel **X** in Figure 7.24 would be classified into class 2 by an MDM classifier, yet our intuition would assign it to class 3 based on the high variability in class 3 and the relatively low variability in band 2 for class 2. Recall that in the univariate case, we used a simple statistical distance employing the number of standard deviations to account for variance.

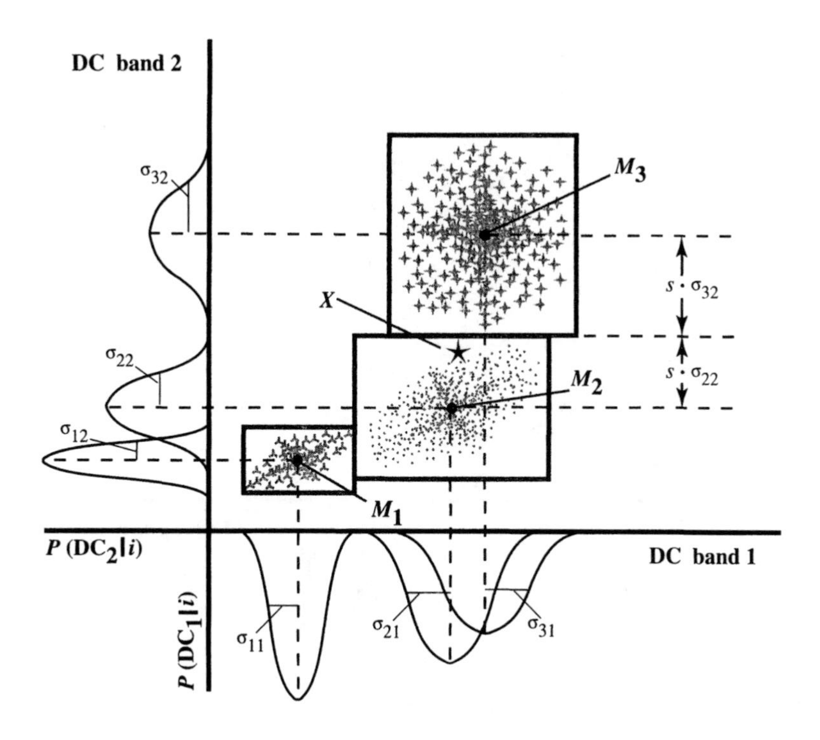

Figure 7.25 Parallelepiped classifier using equal numbers of standard deviations for the boundaries. To locate the boundary in band 2 between class 2 and class 3, we equalize the number of standard deviations (s•σ) from the mean of each class to the boundary.

7.2.2.2 Parallelepiped Classifiers

In the multivariate case, the variance in the data can be included in a classifier by locating boundaries in the classification space with parallelepipeds, as shown in Figure 7.25. Any vector that falls within the one-dimensional parallelepiped associated with class i is assigned to that class. In two dimensions, these parallelepiped boundaries could be defined interactively on a computer screen. However, as the number of dimensions (spectral bands) increases, this approach becomes more cumbersome, as well as yielding unsystematic results. In general, some simple formula is used to define the boundary locations. One method that draws on the simple statistical distance is to place the parallelepiped boundaries at equal numbers of standard deviations from the means, as illustrated in Figure 7.25. The basic parallelepiped method does not account well for the spectral shape of the sample distributions, as illustrated in Figure 7.26. Here it is clear that the pixel **X** would be classified into class 3 by the parallelepiped method, but our intuition would indicate that the pixel is more likely a member of class 2. Our intuition is responding to the elongated shape of the class 2 data, which is described by the correlation between bands 1 and 2 for class 2. We would like to identify a classifier that accounts for the shape of the spectral distribution.

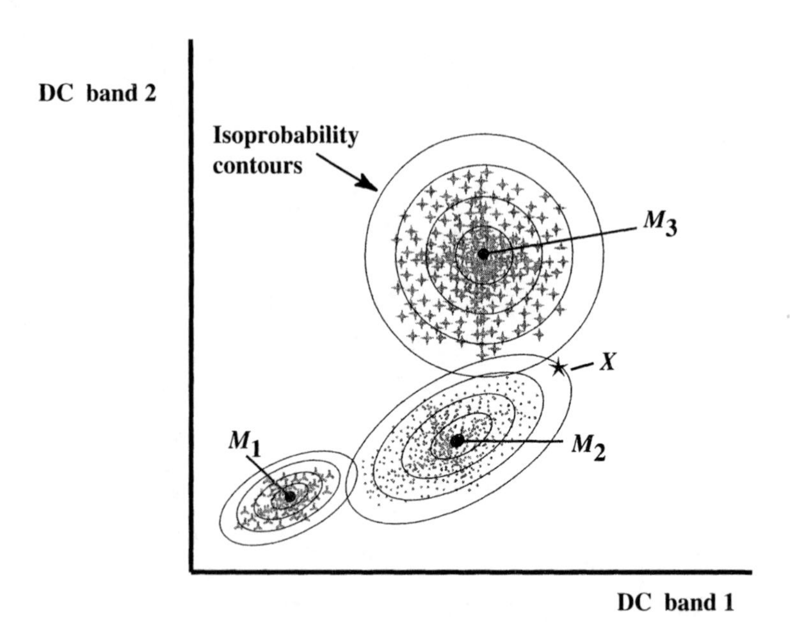

Figure 7.26 Classification using a multispectral Gaussian maximum likelihood classifier.

7.2.2.3 Multispectral Gaussian Maximum Likelihood

One way to include the shape of the class sample distribution is to use the probability theory introduced in Section 7.2.1. The a posteriori probability of a pixel with a spectral vector **X** belonging to class *i* is:

$$p(i|\mathbf{X}) = \frac{p(i)p(\mathbf{X}|i)}{p(\mathbf{X})} \tag{7.40}$$

where the definitions of the probability terms are the same as introduced previously, only now they are with respect to the spectral vector **X** instead of the scalar digital count value (DC). The conditional probability $p(\mathbf{X}|i)$ can be estimated from the normalized multidimensional histogram for each class as illustrated for a two-band case in Figure 7.27. The simple probability of **X** occurring in the image, $p(\mathbf{X})$, (while not needed in the classifier as we saw earlier) is the value of the normalized multidimensional histogram of the entire image. The pixel is assigned to whatever class has the highest a posteriori probability for the spectral vector **X** associated with the pixel. While very attractive from a theoretical standpoint (i.e., this classifier will minimize classification errors by fully incorporating the spectral shape of the sampled pixels in each class), this approach is cumbersome to implement. It requires a very large set of training pixels to characterize adequately a multidimensional probability surface, and it is difficult to store a description of such a surface unless it is very wellbehaved.

In cases where the surface is wellbehaved, it is possible to approximate a solution to Eq. (7.40) by using a multivariate Gaussian maximum likelihood classifier. This classifier is based on the assumption that the conditional probability $p(\mathbf{X}|i)$ for each class can be approximated by a multivariate normal distribution. This is the multivariate equivalent of the assumption of Gaussian behavior for the conditional probability $[(p(\mathrm{DC}|i)]$ that was made in developing the univariate GML classifier. If the conditional probabilities are normal (i.e., if they can be represented by a multivariate Gaussian distribution), then they can be expressed

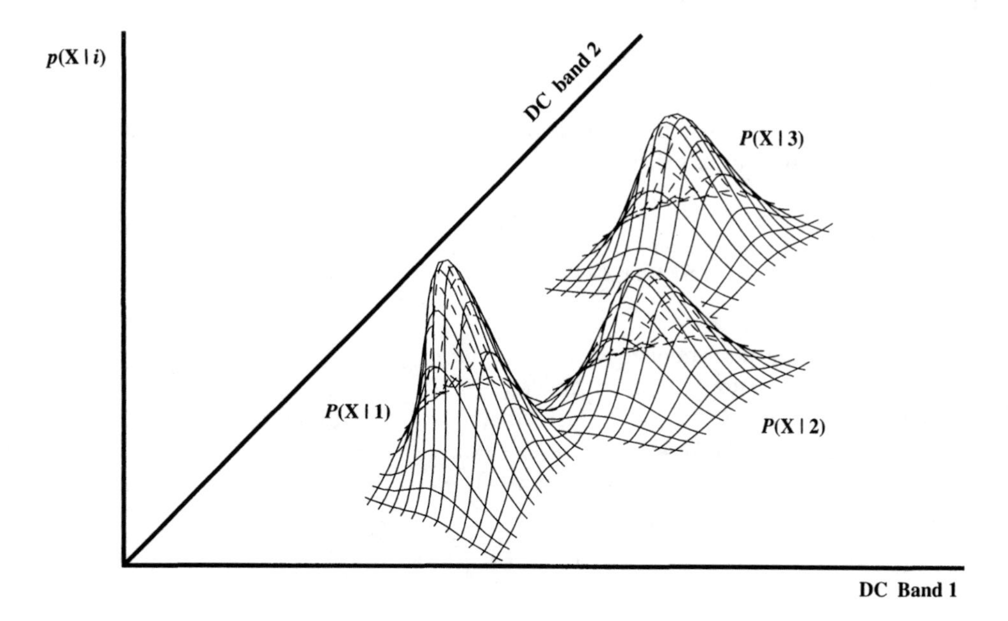

Figure 7.27 Conditional probability surface (multidimensional normalized histogram) for a two-band, three-class case.

as:

$$p(\mathbf{X}|i) = \frac{1}{(2\pi)^{\ell/2}|\mathbf{S}_i|^{1/2}} e^{\left[-\frac{1}{2}(\mathbf{X}-\mathbf{M}_i)'\mathbf{S}_i^{-1}(\mathbf{X}-\mathbf{M}_i)\right]} \tag{7.41}$$

where ℓ is the number of spectral bands, $\mathbf{S}_i$ is the covariance matrix for class i, $|\mathbf{S}_i|$ is the determinant of $\mathbf{S}_i$, $\mathbf{S}_i^{-1}$ is the inverse of $\mathbf{S}_i$, and $(\mathbf{X}\text{-}\mathbf{M}_i)'$ is the transpose of $(\mathbf{X}\text{-}\mathbf{M}_i)$. The covariance matrix is a matrix of dimension ℓ x ℓ comprised of the band-to-band spectral covariance values for the class i.e.,

$$\mathbf{S}_i = \begin{bmatrix} \sigma_{i11} & \sigma_{i12} & \cdots & \sigma_{i1\ell} \\ \sigma_{i21} & \ddots & & \\ \vdots & & & \\ \sigma_{i\ell 1} & & & \sigma_{i\ell\ell} \end{bmatrix} \tag{7.42}$$

where the covariance between spectral bands m and n for the ith class is defined as:

$$\sigma_{imn} = \sum_{q=1}^{N} \frac{(\mathrm{DC}_{im}(q) - \mathrm{DC}_{im\mathrm{avg}})(\mathrm{DC}_{in}(q) - \mathrm{DC}_{in\mathrm{avg}})}{N-1} \tag{7.43}$$

where N is the number of pixels in the sample set.

The location of the multivariate normal distribution for a class is fully characterized by the mean vector ($\mathbf{M}_i$), and the shape of the distribution by the covariance matrix ($\mathbf{S}_i$). Duda and Hart (1973) point out that the iso-probability contours are hyperellipsoids centered on

$\mathbf{M}_i$ and having a constant Mahalanobis distance from $\mathbf{M}$. The square of the Mahalanobis distance of $\mathbf{X}$ from $\mathbf{M}_i$ is defined as:

$$d_i = (\mathbf{X} - \mathbf{M}_i)' \mathbf{S}_i^{-1} (\mathbf{X} - \mathbf{M}_i) \tag{7.44}$$

Combining Eqs. (7.40) and (7.41) yields the a posteriori probability for $\mathbf{X}$ belonging to class i if the data are normally distributed:

$$p(i|\mathbf{X}) = \frac{p(i)}{p(\mathbf{X})(2\pi)^{\ell/2}|\mathbf{S}_i|^{1/2}} e^{\left[-\frac{1}{2}(\mathbf{X}-\mathbf{M}_i)'\mathbf{S}_i^{-1}(\mathbf{X}-\mathbf{M}_i)\right]} \tag{7.45}$$

We can define a discriminant function for the multivariate GML classifier using the same simplifying steps used in the univariate case. The decision regarding which class has the maximum value of $p(i|\mathbf{X})$ is independent of $p(\mathbf{X})$, so this term is dropped from Eq. (7.45). Taking the natural log will further simplify the expression and not change the rank order of the discriminant function, which is then defined as:

$$D_i = \ln[p(i)] - \frac{\ell}{2}\ln(2\pi) - \frac{1}{2}\ln|\mathbf{S}_i|$$
$$-\frac{1}{2}(\mathbf{X} - \mathbf{M}_i)'\mathbf{S}_i^{-1}(\mathbf{X} - \mathbf{M}_i) \tag{7.46}$$

Training data are used to generate estimates of the mean vector ($\mathbf{M}_i$) and covariance matrix ($\mathbf{S}_i$) for each class. The a priori probabilities are estimated from previous data or from a simple preliminary classification. From this information, the discriminant value for any unknown pixel can be computed from its spectral vector ($\mathbf{X}$) using Eq. (7.46). The pixel is then assigned to whichever class yields the maximum D_i value. The process is repeated for each pixel until the entire scene is classified. This will yield a classified scene that has the maximum number of correctly classified pixels based on the sample data available and the assumption of Gaussian distributions for the class populations. When something other than a minimum error solution is required, the risk functions discussed in Section 7.2.1.3 can be applied in exactly the same way as in the univariate case.

Before leaving this subject, we should point out some caveats involved in using GML classifiers (many of these caveats apply to any classifier):

1. The user needs to make sure that the training data set is robust enough to characterize fully the class [cf. Fig. 7.28(a)]. For example, a forest class should include samples from different slopes, aspects, stand types, and densities. In addition, the user must ensure that there are sufficient data to estimate adequately the values of $\mathbf{M}$ and $\mathbf{S}$. Swain (1978) points out that $\ell + 1$ samples are a theoretical minimum (just enough to get you into trouble), but that $10\,\ell$ is closer to a practical minimum, and $100\,\ell$ is a desirable objective.

2. The user must ascertain that the data are approximately Gaussian. In particular, any multimodal classes should generally be split into separate classes for classification and merged into a common class afterwards [cf. Fig. 7.28(b)]. For example, forest classes on east-and-west facing slopes may be significantly different, and forcing a unimodal Gaussian assumption can cause confusion with another class.

3. Many GML software packages allow the user to assume that the class covariance matrices are similar and to use a pooled covariance matrix for all classes. This

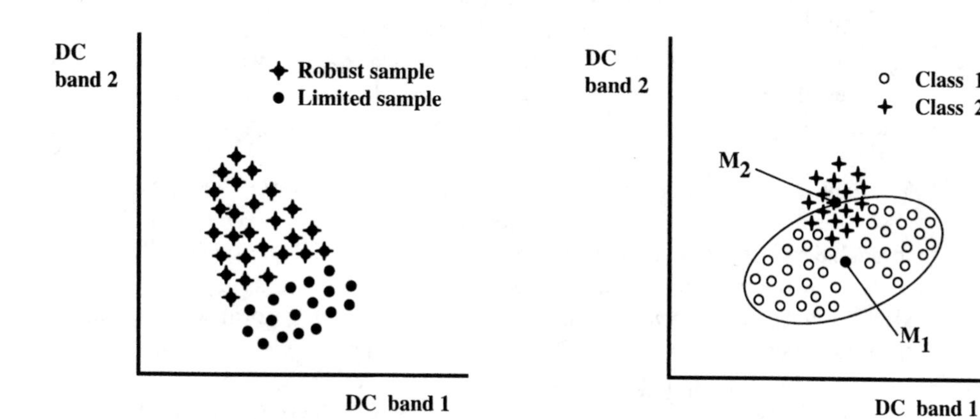

(a) Effect of limited sample set in training process; note that both the mean and covariance can be seriously in error

(b) Effect of failing to separate a bimodal distribution in Class 1 into two classes

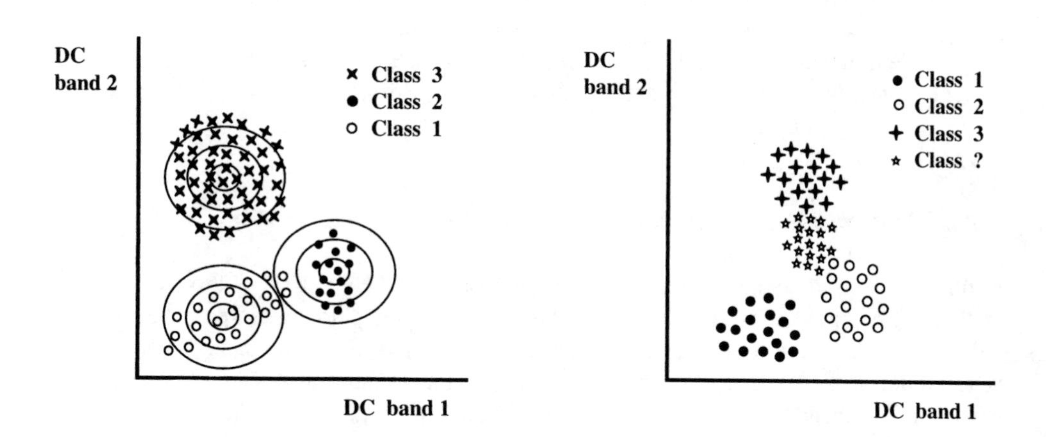

(c) Effect of using a pooled covariance matrix; note that the isoprobability curves are all the same except for location

(d) Effect of a class not included in the training process

Figure 7.28 Considerations in using a GML classifier.

assumes that the distributions are all the same shape and simply displaced by the mean vectors [cf. Fig. 7.28(c)]. This can lead to serious misclassification error if the assumption is false, which it generally is.

4. The user must ensure that all classes are included in the training process, or

untrained classes will be grossly misclassified [cf. Fig. 7.28(d)]. To some extent this effect can be mitigated by using classification thresholds that will assign a pixel to an unknown class if the discriminant function fails to exceed a user defined threshold. This is often useful in isolating small numbers of unusual materials and isolating pixels at boundaries. These "mixed pixels" have spectral vectors comprised of mixtures of two material classes and can be seriously misclassified by a GML classifier.

5. The quality of the classifier should be evaluated prior to its use. There are several ways to evaluate the quality of a classifier. The first is to simply run the training data through the classifier and produce a confusion matrix as shown in Figure 7.29. This shows how the training data would be classified by the classifier they trained. It is referred to as a *dependent data set* and can yield inflated performance estimates if the training data are not robust. It is often supplemented by also running the classifier on data from training sites that were not included in the data used to train the classifier. This independent data set is a good check of the robustness of the classifier. If both data sets yield comparable results, this is a good estimate of how well the classifier will perform on the uniform target areas that are typically used for training (cf. Fig. 7.30). In many cases a better estimate of performance is obtained by having the computer randomly select and highlight pixels on an interactive display. The user identifies the class the pixel belongs to. The pixel is also classified by the classifier and the result entered into a confusion matrix. This process is continued until a good statistical estimate is obtained (typically 50 pixels per class yields a good estimate of performance). In many cases, this test is run after the initial classification so that equal numbers of pixels from each class can be randomly presented to the user. Figure 7.29 contains typical confusion matrices for dependent, independent, and randomly selected data sets. Note that the random classifier generally shows significantly poorer performance because it will include samples from transition regions and mixed pixels.

7.2.2.4 Multivariate Classification using Texture Metrics

Multivariate classifiers are generally used on multispectral data with from 2 to approximately 10 or 12 spectral bands. The classifiers, however, are "blind" to the source of the data and will work on any data set that looks like spatially registered multiband data. Thus, one can use a monochrome image and several texture metric images (such as described in Sec. 7.1.3) to form a multiband image that can then be classified (cf. Fig. 7.31). Alternatively, texture metric images can be combined with multispectral data to improve classification accuracy. Care must be taken when using GML classifiers to ensure that the Gaussian approximations are valid.

7.2.2.5 Limitations of Conventional Multispectral Classification

A major limitation of multispectral classifiers, as they are normally used, is that the classifier is scene-specific and cannot be used on any other scene. This is because the classifier is based on raw digital count values and has the sensor and atmospheric calibration effects incorporated into the classifier. In some sense, this is a beneficial attribute since we don't need calibrated data to build a classifier, but it limits the applicability of the classifiers. Thus, it is often not cost effective to build extremely elaborate classifiers only to discard them after a single use. Two alternatives exist to this disposable classifier approach. The first is to cal-

% classified as class

Actual class	1	2	3	4	5
1	99.3	0.0	0.7	0.0	0.0
2	0.0	86.0	12.8	1.2	0.0
3	0.0	0.2	99.8	0.0	0.0
4	2.5	10.4	7.0	80.0	0.1
5	0.0	0.0	0.0	0.0	100.0

(a) Dependent data set

% classified as class

Actual class	1	2	3	4	5
1	97.3	1.5	1.1	0.0	0.0
2	0.2	74.0	12.2	1.5	12.1
3	0.0	4.1	95.9	0.0	0.0
4	2.4	1.6	6.4	89.6	0.0
5	0.0	0.0	0.0	0.0	100.0

(b) Independent data set

% classified as class

Actual class	1	2	3	4	5
1	93.5	0.0	2.2	0.0	4.3
2	0.0	29.7	65.7	0.8	3.8
3	12.3	0.0	87.4	0.3	0.0
4	0.0	0.0	1.4	98.6	0.0
5	0.0	0.0	0.0	0.0	100.0

(c) Randomly selected data set

Figure 7.29 Typical confusion matrices for: (a) dependent data, (b) an independent data set, and (c) a randomly selected data set.

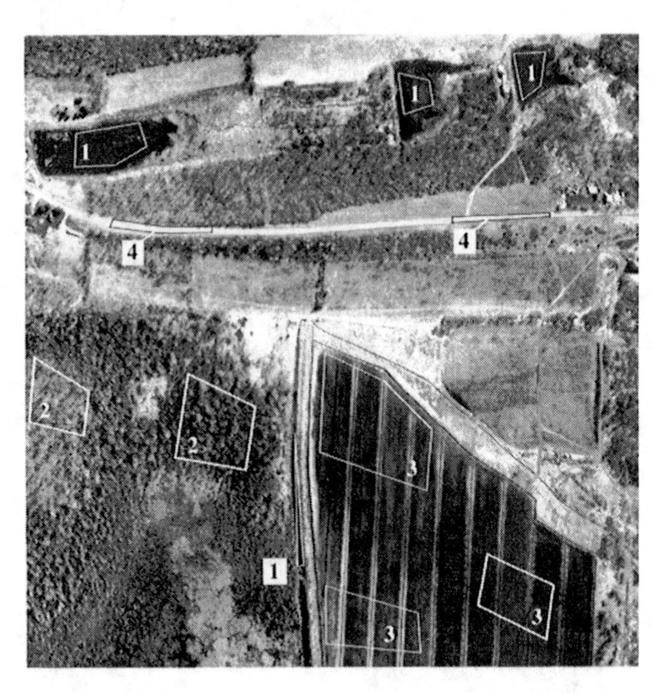

Class 1 : Open water
Class 2 : Forest
Class 3 : Agricultural muckland
Class 4 : Roads
•
•
•

Figure 7.30 Training data sets for input to a classifier.

ibrate scenes into reflectance units and to build reflectance-based classifiers. These classifiers could be reused and made more and more sophisticated because they are based on stable physical units. The limitation of this approach is that most of the operational atmospheric calibration techniques have residual errors large enough to introduce serious classification errors. The second approach is to use multidate radiometric normalization techniques. In this approach, a standard supervised classifier is constructed using imagery acquired on one day. Images of the same scene on other days are radiometrically transformed using one of the methods discussed in Section 6.5.2 such that the day-2 DC values vary with reflectance in the same way as the day-1 values. The day-1 classifier can then be directly applied to the transformed day-2 image. Using the method of Schott et al. (1988), it was demonstrated that under certain conditions update scenes normalized with the pseudo-invariant feature technique could be classified with a day-1 classifier nearly as well as with a classifier built specifically for the day-2 images. However, extension of classifiers to multiple days using normalization is usually limited to short time spans between acquisitions (e.g., several weeks) or to images within a few weeks of anniversary dates (e.g., one year ± several weeks) to avoid significant seasonal changes in the reflectance spectra.

7.2.3 Unsupervised Multivariate Classifier

In some cases, it is useful to have the computer sort out which pixels have similar characteristics (e.g., spectra) rather than to try to force the pixels into a class based on our culturally driven sense of their similarities. This is done using an unsupervised classifier. Duda and Hart (1973) describe several approaches to unsupervised classification. We will only present the simplest and most commonly used method here. It is often referred to as the *k*-means approach and requires only one piece of user-supplied data. The user must specify the num-

Figure 7.31 Example of the image shown in Figure 7.10 classified using texture metrics images. See color plate 7.31.

ber of classes (k) in the image. The k-means algorithm then attempts to locate the mean vector ($\mathbf{M}_i$) for each of the k classes. Normally the image data are subsampled by selecting every qth pixel in the image to reduce the data volume. Next, k initial estimates of the location of the mean vectors in the ℓ-dimensional spectral space are made at random. If we designate these initial estimates as $\mathbf{M}''_i$ (cf. Fig. 7.32), then we can tentatively assign each pixel to a class based on how close (minimum distance to the mean) it is to the mean vectors. The mean of all the pixels tentatively assigned to the ith class becomes our new estimate of the class mean ($\mathbf{M}'_i$). The sample pixels are tentatively reassigned using the new class means, and the procedure repeats in this fashion until the class means no longer change (i.e., the change is less than some threshold). At this point, the tentative means are assumed to be good estimates of the class mean vectors ($\mathbf{M}_i$). All pixels in the image can then be assigned to a class using the MDM classifier. Alternatively the pixels assigned in the last iteration to the ith class can be used to generate estimates of the other statistics for the class (e.g., the covariance matrix), and a parallelepiped or GML classifier can be used.

The resultant classes will be indicative of the natural spectral clusters in the data. They may or may not correspond to land cover or material classes as we normally think of them. For example, in high-resolution airphoto data, a shadow class may develop or a class that is a combination of shadow and water. In some cases, classes are formed with no obvious common characteristic. Because of this limitation, unsupervised classification is often used as a preprocessor for other algorithms. For example, when attempting to locate training regions for supervised classification, it is often useful to know where natural spectral groupings occur. If an unsupervised classifier clearly delineates forests on east- and-west facing slopes as two classes, then it is a good idea to make them two classes in the supervised classifier.

In practice, it is often useful to run an unsupervised k-means classifier with the initial estimate of k slightly greater than expected to "see" what types of spectral groupings natu-

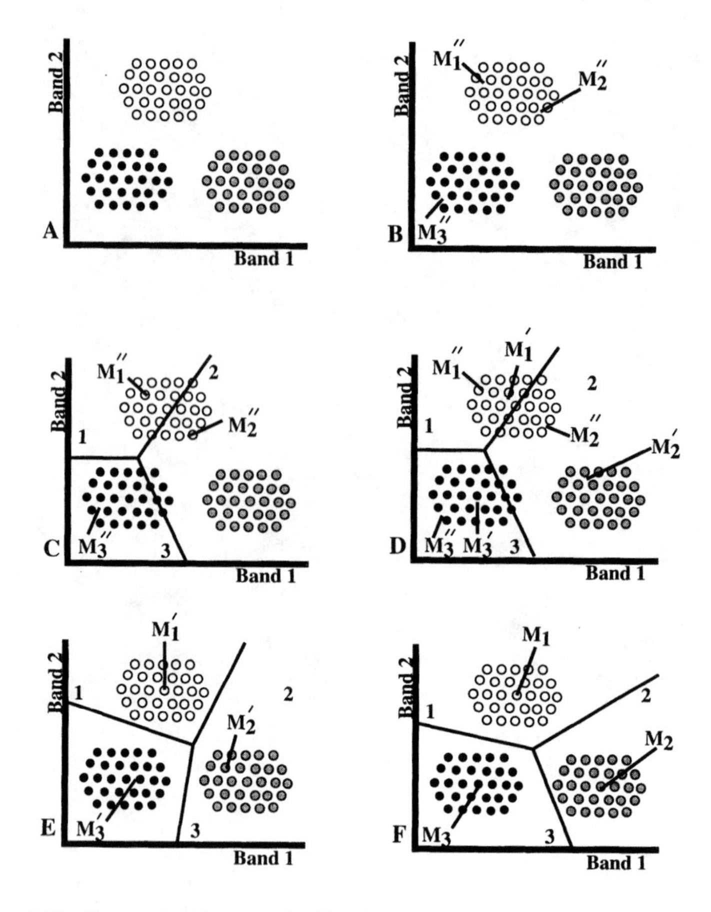

Figure 7.32 Unsupervised *k*-means classifier illustrating a case with two bands and three classes.

rally occur in the image. These "extra" classes often identify spectral features that would confuse a supervised classifier (e.g., clear and turbid water as two separate classes) or point out classes that will be very difficult to separate spectrally.

As with the supervised classifiers, it is possible to mix texture and spectral data in an unsupervised classifier. For example, if an initial run with only spectral data showed that two classes were going to be difficult to segment, it would be useful to consider whether texture metrics would provide increased separability. A second unsupervised classifier could be run with texture metrics included to try to differentiate the classes of interest.

7.3 IMAGE TRANSFORMS

In most cases, we tend to assume that more spectral bands will yield a better classifier when processing multispectral data. In cases where we have only a few bands, this is often a good assumption. However, more are not always better for two reasons: processing time and inter-band correlation. In the first case, if we add too many bands, the processing time to run multispectral image analysis algorithms can become staggering. The second reason is that both spatial (texture) and spectral data tend to be highly correlated. Adding another band of data that is highly correlated with previous bands may add little new data and only increase the noise and the processing time, possibly even reducing the effectiveness of the algorithm.

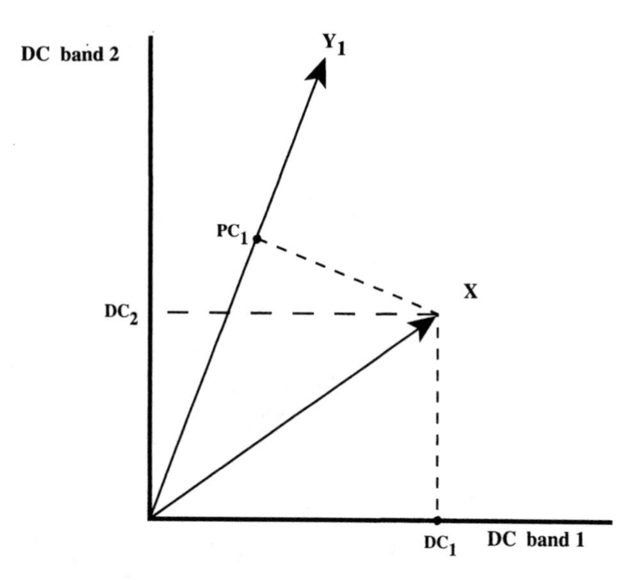

Figure 7.33 Projection of the vector X onto the Y_1 axis to yield the magnitude of the projected value (PC_1.).

Image transforms can be used to overcome some of these limitations. These transforms are designed to redefine the multivariate feature vectors into a feature space where the features are more decorrelated. Therefore, fewer features can carry the information needed for processing algorithms such as multispectral classifiers. These transforms are also often used to attempt to reduce some of the variations that result from processing uncalibrated data. In many cases, the transforms tend to normalize out some of the atmospheric and sensor calibration effects.

The most common transform is the principal component transform that is specifically designed to decorrelate the data and maximize the information content in a reduced number of features. Each feature value in the transformed data set is a linear combination of the features in the input data set. If **X** is the vector comprised of ℓ digital count values corresponding to the ℓ features (most often spectral bands) in the input image, then an output feature (PC) can be defined as:

$$\mathrm{PC}_1 = \mathbf{Y}_1'\mathbf{X} = y_{11}\mathrm{DC}_1 + y_{12}\mathrm{DC}_2 \cdots y_{1\ell}\mathrm{DC}_\ell \tag{7.47}$$

where PC_1 is the "brightness" value of the first principal component feature, and $\mathbf{Y}_1$ is the first principal component vector comprised of ℓ weights ($y_{11}, y_{12} \dots y_{1\ell}$,), i.e.,

$$\mathbf{Y}_1 = \begin{bmatrix} y_{11} \\ y_{12} \\ \vdots \\ y_{1\ell} \end{bmatrix} \tag{7.48}$$

Recall from vector algebra that Eq. (7.47) can be thought of as projecting the vector **X** onto the vector $\mathbf{Y}_1$. The vector $\mathbf{Y}_1$ can be thought of as a new axis, and PC_1 is the magnitude of the vector **X** along the axis defined by the vector $\mathbf{Y}_1$ (cf. Fig. 7.33).

There are ℓ principal component features, so we can define a principal component feature vector **PC** for each pixel by transforming the input feature vector through the transform matrix **P** according to:

$$\mathbf{PC} = \mathbf{P'X} \tag{7.49}$$

where

$$\mathbf{PC} = \begin{bmatrix} PC_1 \\ PC_2 \\ \vdots \\ PC_\ell \end{bmatrix} \tag{7.50}$$

and **P** is comprised of ℓ principal component column vectors (**Y**), i.e.,

$$\mathbf{P} = \begin{bmatrix} \underset{\mathbf{Y}_1}{\begin{pmatrix} y_{11} \\ y_{12} \\ \vdots \\ y_{1\ell} \end{pmatrix}} \underset{\mathbf{Y}_2}{\begin{pmatrix} y_{21} \\ y_{22} \\ \vdots \\ y_{2\ell} \end{pmatrix}} \cdots \underset{\mathbf{Y}_\ell}{\begin{pmatrix} y_{\ell 1} \\ y_{\ell 2} \\ \vdots \\ y_{\ell\ell} \end{pmatrix}} \end{bmatrix} \tag{7.51}$$

The discussion to this point is valid for any transformation matrix **P**. What makes the principal component transform unique is that the new axes ($\mathbf{Y}_i$) are orthogonal and are selected such that the first principal component axis is the axis that will exhibit the most variance when the input data vectors are projected onto it. The second principal component is the axis orthogonal to $\mathbf{Y}_1$ that exhibits the next largest variance, etc. Formally, P is defined such that it diagonalizes the covariance matrix, i.e.,

$$\mathbf{P'SP} = \begin{bmatrix} \lambda_1 & 0 & 0 & \cdots & 0 \\ 0 & \lambda_2 & & & \\ 0 & 0 & & & \\ 0 & & & & \lambda_\ell \end{bmatrix} \tag{7.52}$$

where λ_i is the variance of the ith principal components feature values (PC_i), and the λ values are arranged such that $\lambda_1 > \lambda_2 \ldots > \lambda_\ell$. The columns of P are the ordered eigenvectors ($\mathbf{Y}_i$) corresponding to the ordered eigenvalues (λ_i) of the image covariance matrix. We can generate a transformed image with ℓ bands where each band is comprised of PC_i values. The first band, or feature, in the multiband principal component image will account for most of the variance in the data, with decreasing amounts in the remaining bands. The latter bands in a many-band image tend to contain mostly random noise. Figure 7.34 shows a principal component transformed six-band Landsat TM image. All the data have been contrast stretched to enhance the visual representation of the data. The actual percentage variance in each principal component band is shown in the figure. This is typical of Landsat TM scenes with most of the image data contained in the first three or four principal components.

Multispectral classification algorithms can be run using only the first few principal component images, greatly reducing run times. In some cases, simpler classifiers can be used because the transform process, by selecting axes with high variance, will tend to increase the separability of classes. This can be seen for a simple case in Figure 7.35. Principal component transformed images can also be used for data compression in storing and transmitting multispectral data, since the latter PC bands can often be completely neglected and intermediate bands require fewer bits per pixel because of their low variance.

Care must be taken when using the principal component transform. Principal compo-

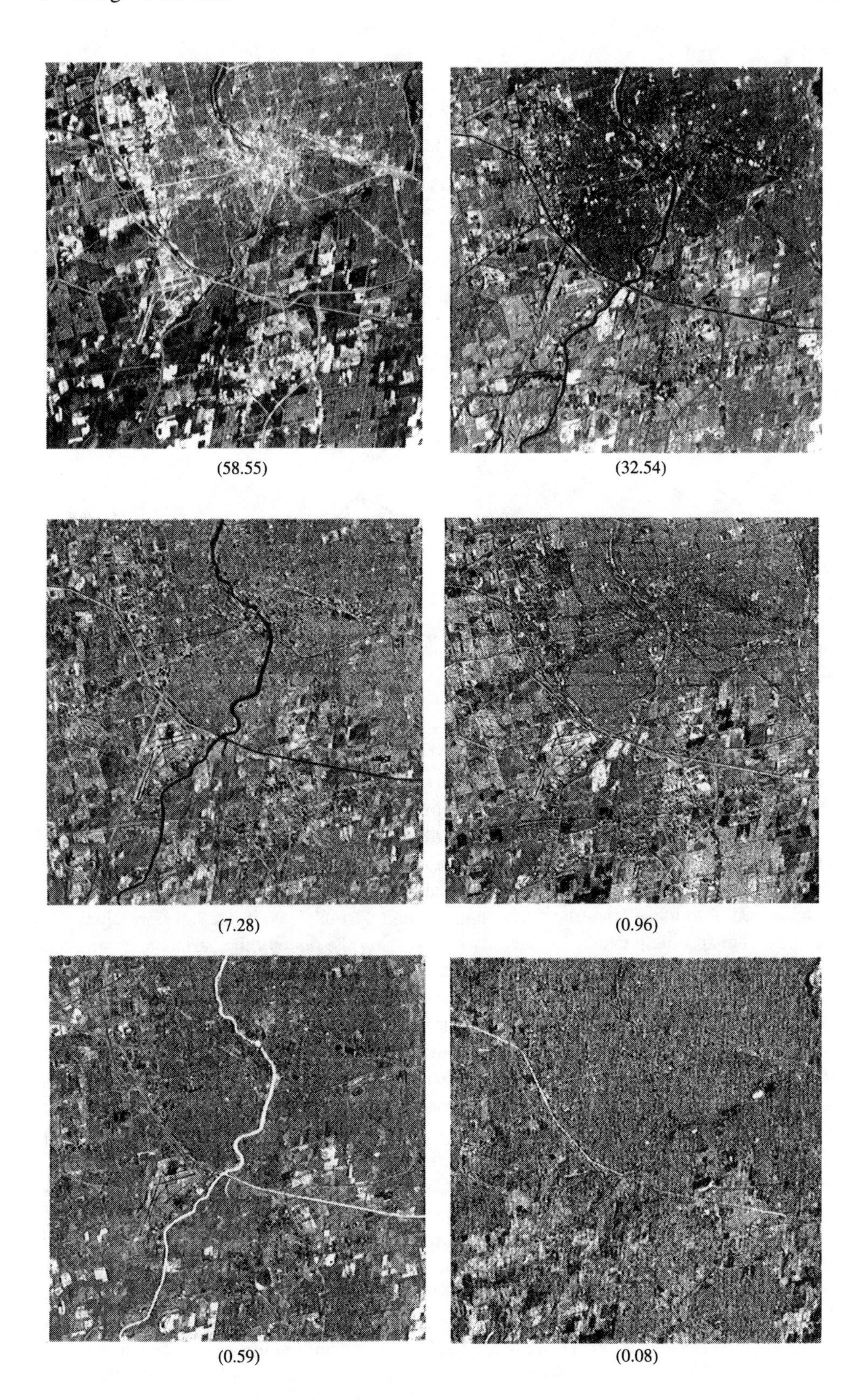

Figure 7.34 Principal component images from a six band TM scene of Rochester, New York. Percentage of the variance contained in each PC band is shown in parentheses.

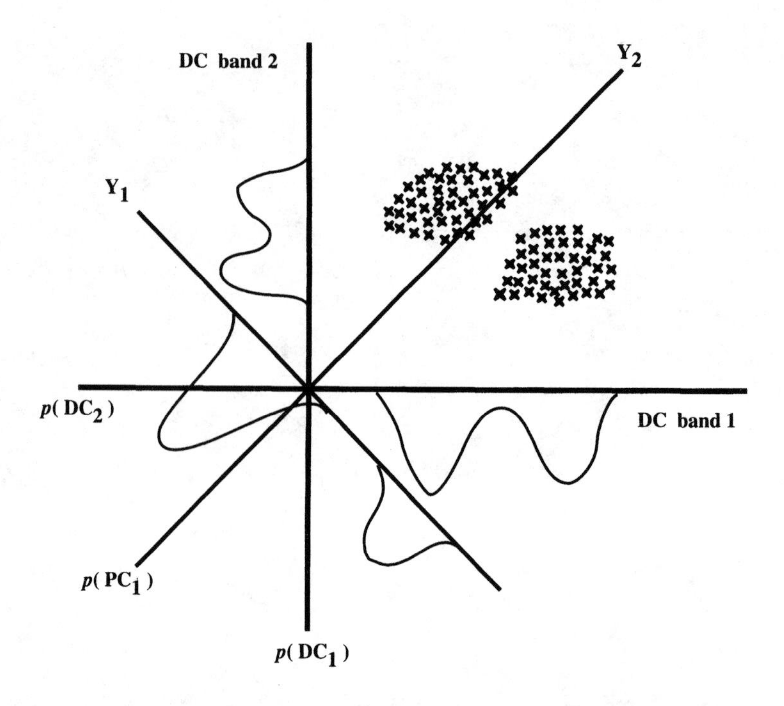

Figure 7.35 Principal component axis showing how increased separability can be realized by maximizing the variance on the projected axis.

nent analysis essentially equates variance with information. Thus, high-variance bands will tend to dominate the first few principal components, with low-variance bands appearing in the later PC bands regardless of information content. For example, in Landsat TM imagery, the thermal band contains unique information not depicted in the other bands. However, because the thermal band usually has low variance, the data from the thermal channel tends to be mixed with noise from the other channels and partially lost in the later principal components. This problem can be avoided by normalizing the data by the variance in each band before performing the principal component analysis.

A disadvantage of the principal component data is that the **PC** vectors are scene-dependent and often have no obvious correlation with a physically interpretable concept. Thus, image interpretation of principal component images can be difficult. Furthermore, a PC_2 image of one scene may represent different information than a PC_2 image from another scene. On the other hand, principal component images of multidate scenes of the same area tend to be correlated with each other and to some extent can be processed in the same way (i.e., some degree of normalization is included in the transform process). As with most multidate approaches, the interpretation of the PC transforms will change if there is a significant change in the information content in the scene (e.g., crop development).

A limitation of the principal component transform can arise if there are a small number of pixels associated with a critical class. Since the transform is based on whole-image statistics, such a class may not be well separated in the transform space. This problem can be addressed using the cononical transform, which is designed to identify a new coordinate space that maximizes the separation between class means and minimizes the dispersion within the classes (cf. Schowengerdt, 1983). Clearly, this transform requires information about the mean and covariance matrix for each class and is, therefore, only applicable when training data are already available.

An alternative transform approach was developed by Kauth and Thomas (1976). They defined a transform space that was designed to improve the analysis of agricultural scenes using Landsat MSS data. The *tasseled-cap* transform (named after the shape of the population distribution) is designed to project the data along a set of axis where the first three axes correspond roughly with the brightness of soils (brightness axis), the vegetation biomass (greenness axis), and the senescence of vegetation (yellowness axis). The tasseled-cap transform is sensor-specific (assuming fixed sensor gain) and is designed to transform the input data to a feature space where the features are more directly correlated with an application parameter and more intercomparable over time. While not an optimized transform, it has the advantage from the user's standpoint of using a constant precomputed transform matrix. It must, however, be recognized that the transform was designed for particular types of scenes and is sensor-specific.

7.4 IMAGING SPECTROMETER DATA ANALYSIS

Many of the multispectral data analysis approaches treated to this point become extremely unwieldy when tens or hundreds of spectral bands are available. As a result, new image processing approaches are evolving to take advantage of imaging spectrometer data (cf. Lee and Landgrebe, 1993). One set of approaches to analysis of imaging spectrometer data is to run preprocessing software to reduce the number of bands to a point where conventional multispectral algorithms can be used. In many cases, the correlation between bands (particularly adjacent bands) is so high that the higher dimensions of the data do not carry significantly more data. The principal component type of transform is one possible approach to reducing the dimensionality of the data. However, unless many bands are maintained, the subtleties in the data can be lost by a principal component transform. Another approach is to select which subset of the many bands available is most relevant to the particular application of interest. This can be done with real data or simulated data using reflectance spectra and atmospheric propagation models. Some form of metric is identified against which an optimization can be performed. For example, in Section 9.3.1 methods are described that maximize the statistical distance between the means of all classes. Combinations of ℓ spectral bands are considered until an optimum set is identified. This type of approach requires the user to have identified the application and the algorithm ahead of time so that the data can be optimized against the particular algorithm. It also requires real or simulated data characteristic of the actual data on which to run the band selection optimization routine. It will also often require that training subsets or some other segmentation of the data be available to the optimization routine and can involve long run times to select the optimum set. On the positive side, in most cases the band selection routines can be run in advance so that only those spectral bands particular to an application need to be recorded or transmitted by the sensor, and in general band selection need only be performed for each scenario or application.

A second set of algorithms for processing multispectral data draws heavily on the detailed spectral content of the data and uses all, or nearly all, of the spectral bands. A good example of this is the spectral mixing model approach using end members described by Smith et al. (1990a, b), and Roberts et al. (1993). This approach has not been used so much to perform initial classification but rather to refine classifications and to look for contaminants or trace contributions to a spectrum. In its simplest form, it assumes that the observed spectral vector, comprised of digital count values, is a linear combination of the digital count that would be observed from pure pixels of the materials in a pixel. Put more simply, it assumes that most pixels are mixtures of two or more distinct material types called *end members*. The end-member spectral vectors can be combined in a linear fashion to yield the observed spectral vector. The weights used in the linear combination are the fractions of the end members present in the pixel. This can be expressed mathematically as

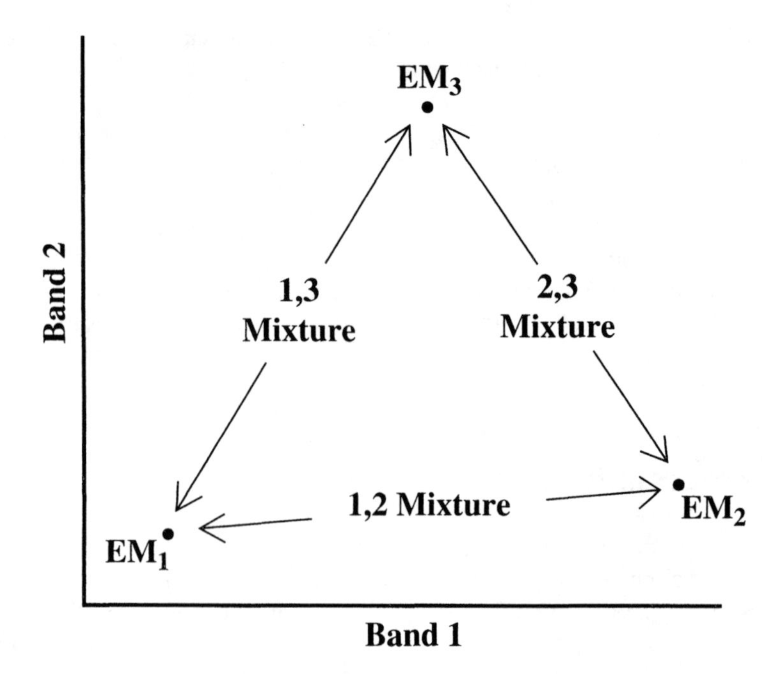

Figure 7.36 Spectral mixing models: Mixture of any two end members (EM) must fall on the line between them. Mixtures of three must fall in the space bounded by the lines.

$$\mathbf{DC} = \sum_{j=1}^{n} f_j \mathbf{DC}_j + \mathbf{e} \tag{7.53}$$

where **DC** is the vector comprised of observed digital count values in all spectral bands of interest, $\mathbf{DC}_j$ is the vector of spectral digital count values contributed by the jth end member, f_j is the fraction of the jth end member in the pixel, n is the number of end members assumed present in the pixel, and **e** is a spectral error vector representing the difference between the observed vector and the modeled vector. The model can be constrained such that the fractional values (f_i) must be positive and their sum must equal 1. Equation (7.53) can be solved for each pixel by minimizing the residual error vector using least-squares methods. Generally, only a few end members are assumed present in any pixel, and the possible end members are restricted by a presegmentation of the scene (e.g., a pasture area might have only grass and soil end members). The model assumes that the end members represent extremes in spectral space and that the mixed pixel will be located between these spectral extremes. In the two-element case, this would be along the line connecting the end members as shown in Figure 7.36. Because of the large number of spectral bands available with multispectral and hyperspectral systems, this approach can be quite effective if the end members can be adequately identified. Because most instruments with many spectral bands have limited spatial resolution, it is often difficult to identify "pure" end-member pixels. This is particularly important with materials like soil or understory vegetation in a forested scene. We know they are contributing to the mixing process, but it can be very difficult to obtain an in-scene estimate (i.e., a pure pixel).

To overcome this limitation, the spectral mixing model approach commonly uses laboratory or field reflectance spectra for the end members. If we assume that the digital count in each band is a simple linear function of reflectance, as discussed in Chapter 4, then the reflectance in each band can be expressed as

$$r_i = g_i \mathrm{DC}_i + o_i \tag{7.54}$$

where r_i is the reflectance in the ith band, g_i includes band-dependent sensor gain and atmospheric source and attenuation effects, o_i includes sensor bias and atmospheric path radiance effects, and DC_i is the digital count in the ith band. On rearranging, we see that we can express the digital count in each band as

$$\mathrm{DC}_i = (r_i - o_i)/g_i \tag{7.55}$$

and we recognize that each DC_i value is an element of the spectral end-member vector $\mathbf{DC}_j$. To take advantage of laboratory reflectance spectra, the calibration values (g_i and o_i) must be computed for each band. This is most commonly done using a simple two-point empirical line calibration for each band as described in Section 6.2.1. The reflectance values for at least two objects (one dark and one bright) in band i are regressed against observed DC_i values for those objects to yield g_i and o_i values in each band [cf. Eq. (7.54)]. These scene-derived calibration values can then be used for any reflectance spectra to generate effective "pure pixel" spectral end-member vectors ($\mathbf{DC}_j$) for use in Eq. (7.53). These values will be valid as long as the calibration values (g_i and oi) are approximately constant over the scene. When this is not the case, additional calibration or correction procedures must be employed. Smith et al. (1990 a, b) describe the use of an additional end-member spectra referred to as *shade*, which, when included as an end member in the mixing models, can be used to account for illumination and shadow variations on a pixel-by-pixel basis.

The output of the spectral mixture analysis includes image maps showing the fraction of the various end members that are determined to be present in each pixel. Roberts et al. (1993) point out how the residual vector (**e**) can be used to help in identifying the presence in a pixel of contributions from spectra not included as end members. A plot of the residual vector will exhibit peaks when the end-member spectra have absorption signatures not observed in the actual spectra. On the other hand, the residual vector will be negative when the observed spectra have absorption characteristics not modeled by the end members. This approach can be very useful when applied to imaging spectrometer data where detailed spectra are available.

7.5 HIERARCHICAL IMAGE PROCESSING

Many image processing algorithms cannot be run effectively in a single direct application to an image. In most cases, the more sophisticated algorithms are designed to be applicable only to a subset of the image data. For example, the NDVI algorithm is only meaningful in vegetated regions. An algorithm designed to measure the amount of damage to a particular crop based on correlation with field reflectance data will yield false signatures on other crops, as well as on any other targets in the scene. To take advantage of these target-specific algorithms, a hierarchical approach to image processing is often required. For example, in the crop damage case we might run the following sequence of algorithms. First the scene would be calibrated to reflectance using ground truth panels. Second, a GML classifier could be trained and run to isolate agricultural fields from other land cover types. Third, a land cover classifier, including texture metrics, could be trained and run on just the agricultural fields to isolate the particular crop of interest. Next, a post-classification algorithm (mode filter) could be used to clean up the classifier. Finally, the crop damage algorithm can be run on the reflectance values for just those pixels that correspond to the crop for which the algorithm was designed. This hierarchical approach allows us to build more and more specialized algorithms based on a priori knowledge about specific pixels, (i.e., only operate on agricultural fields or only a particular crop). In this way the algorithms can become very specialized without becoming extremely complex.

7.6 REFERENCES

Bracewell, R.O. (1978). *The Fourier Transform and Its Applications.*, 2d ed., McGraw-Hill, NY.

Duda, R.O. & Hart, P.E. (1973). *Pattern Classification and Scene Analysis.* Wiley, NY.

Ehrhard, D.G., Easton, R.L. Jr., Schott, J.R., & Comeau, M.J. (1993). "Frequency-domain texture features for classifying SAR images." Proceedings of SPIE's OE/Aerospace and Remote Sensing, 1993, Vol. 1960, No. 03, pp. 21–32, Orlando, FL.

Fu, K.S. (1974). *Syntactic Methods in Pattern Recognition.* Academic, NY.

Galloway, M.M. (1975). Texture classification using gray level run lengths. *Comp. Graphics and Image Proc.*, Vol. 4, pp. 172–79.

Gaskill, J.D. (1978). *Linear Systems, Fourier Transforms and Optics.* Wiley, NY.

Gonzales, R.C., & Wintz, P. (1987). *Digital Image Processing.* 2d ed., Addison-Wesley, Reading, MA.

Gonzalez, R.C., & Woods, R.E. (1992). *Digital Image Processing.* Addison-Wesley, Reading, MA.

Goodman, J.W. (1968). *Introduction to Fourier Optics.* McGraw-Hill, NY.

Haralick, R.M., Shanmugam, K., & Dinstein, I. (1973). Textural features for image classification. *IEEE Transactions on Systems, Man, and Cybernetics*, Vol. SMC-3, No. 6.

Hardie, R.C., & Boncelet, C.G. (1993). LUM filters: a class of rank-order-based filters for smoothing and sharpening. *IEEE Transactions on Signal Processing*, Vol. 41, No. 3, pp. 1061–76.

Kauth, R.J. & Thomas, G.S. (1976). "The tasseled cap - a graphic description of the spectral-temporal development of agricultural crops as seen by Landsat." Proceedings of the Symposium on Machine Processing of Remotely Sensed Data, Purdue University, W. Lafayette, IN.

Lee, C., & Landgrebe, D. A., Analyzing High Dimentional Multispectral Data, *IEEE Transactions on Geo Science and Remote Sensing*, Vol. 31, No. 4, pp. 792 800.

Morrison, D.F. (1967). *Multivariate Statistical Methods.* McGraw-Hill, NY.

Papoulis, A. (1962). *The Fourier Integral and its Applications.* McGraw-Hill, NY.

Pratt, W.K. (1991). *Digital Image Processing.* 2d ed., Wiley, NY.

Richards, J.A. (1986). *Remote Sensing Digital Image Analysis, An Introduction.* Springer-Verlag, Berlin.

Roberts, D.A., Smith, M.O., & Adams, J.B. (1993). "Green vegetation, non-photosynthetic vegetation and soils in AVIRIS data." *Remote Sensing of Environment*, Vol. 44, pp. 255-69.

Rosenblum, W. (1990). "Optimal Selection of Textural and Spectral Features for Scene Segmentation." Masters thesis, Center for Imaging Science, Rochester Institute of Technology, Rochester, NY.

Rosenfeld, A. & Kak, A.C. (1982). *Digital Picture Processing.* Volume 2d, ed., Academic, Orlando, FL.

Schott, J.R., Salvaggio, C. & Volchok, W.J. (1988). "Radiometric scene Normalization using pseudoinvariant features." *Remote Sensing of Environment*, Vol. 26, No. 1, pp. 1-16.

Schowengerdt, R.A. (1983). *Techniques for Image Processing and Classification in Remote Sensing.* Academic, Orlando, FL.

Smith, M.O., Ustin, S.L., Adams, J.B., & Gillespie, A.R. (1990a). "Vegetation in Deserts: I. a regional measure of abundance for multispectral images." In *Remote Sensing of Environment*, Vol. 31, pp. 1-26.

Smith, M.O., Ustin, S.L., Adams, J.B., & Gillespie, A.R. (1990b). "Vegetation in deserts: II. environmental influences on regional abundance." In *Remote Sensing of Environment*, Vol. 31, pp. 27-52.

Stromberg, W.D. & Farr, T.G. (1986). "A fourier-based textural feature extraction procedure." *IEEE Transactions on Geoscience and Remote Sensing*, Vol. GE-24, No. 5, pp. 722-31.

Swain, P.H. (1978). "Fundamentals of pattern recognition in remote sensing," In P.H. Swain and S.M. Davis, eds. *Remote Sensing: The Quantitative Approach*, McGraw-Hill, New York, NY.

Tou, J.T. & Gonzalez, R.C. (1974). *Pattern Recognition Principles.* Addison-Wesley, Reading, MA.

Wang, L., & He, D. (1990). "A new statistical approach for texture analysis." *Photogrammetric Engineering and Remote Sensing*, Vol. 56, No. 1.

CHAPTER

8

Information Dissemination

One of the most important and often overlooked links in the image chain is the distribution of the information to the user. In many cases, the information is provided in a form where only an imaging expert can use it (i.e., the user sees meaningless blobs) or without sufficient support data to justify meaningful decisions (e.g., a class map with no information on the class accuracies). Entire fields of science are evolving around the topics of information display and scientific visualization. Remote sensing scientists need to take advantage of advances in these fields to assist in the distribution of remotely sensed information. We will touch here on only a few points related to basic image display issues and methods for merging various forms of spatial information both for greater information extraction and improved image display.

8.1 IMAGE DISPLAY

In many cases, our output product will be the raw image, or more likely a processed version of the raw image, with the final interpretation done by the user. We must assume that in general the user is an applications specialist, not a remote sensing or imaging expert. If the data are improperly displayed, the user may well assume the information is not available from the image data rather than question the display or processing procedures. It is incumbent on the remote sensing expert to close the final links in the image chain by making sure that the information in the image is properly presented to the user.

In general, there are two media for image display. *Soft copy* refers to images displayed on video or computer monitors. *Hard copy* refers to images printed on photographic transparency or print media (e.g., photographic media). We need to recognize the capabilities and limitations of the display devices, as well as the capabilities and limitations of the human visual system, which is the next link in the image chain. Before we discuss the display media, we will review a few critical features of the visual response system (for a more thor-

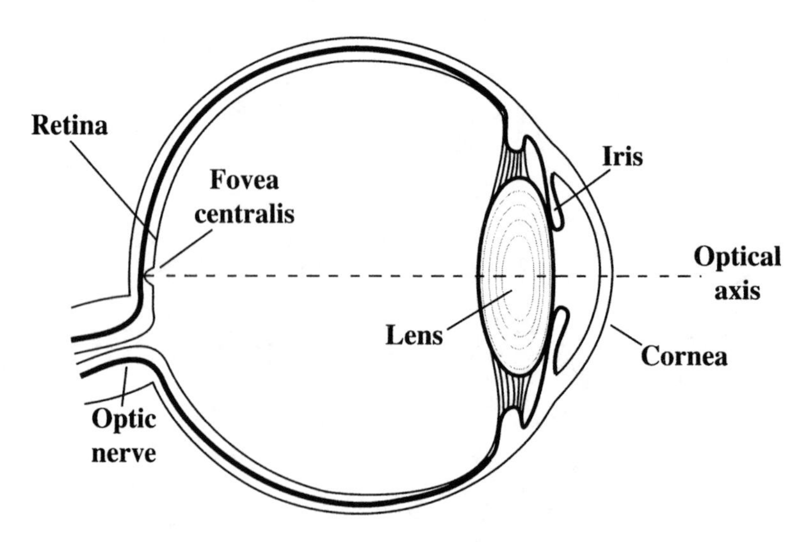

Figure 8.1 Elements of the human eye.

ough treatment of visual systems in this context, refer to Levine, 1985).

First (referring to Fig. 8.1), recall that the eye is made up of a simple lens and two arrays of detector elements on the focal plane (retina). Concentrated in a region (fovea) near the visual axis is an array of detectors (cone cells) with each element sensitive to either red, green, or blue flux in a manner conceptually similar to the color CCD cameras illustrated in Figure 5.31(a). The visual response system interpolates the spectral samples into what appears to be a continuous color image. The remainder of the retina is covered with monochrome receptors (rods) that provide low light response (scotopic response) and peripheral vision. In general, we are concerned with focused observations with adequate lighting (photopic response) where the cells in the foveal region are the dominant receptors.

Visual response is expressed in terms of perceived brightness. The visual response function is extremely adaptive, allowing us to see over many orders of magnitude of illumination levels. However, when adapted to a complex reference level (such as when viewing a scene), visual response is approximately logarithmic and covers a range of about 2.2 log units (cf. Gonzales and Wintz, 1987). This has three important implications. First, the adaptive nature of visual response means that the eye-brain system is sensitive to differences in flux but not readily able to discern (measure) absolute levels. It is important, therefore, that we make sure that changes in digital count are properly translated into discernible changes in brightness since these are so important to visual analysis. Second, the logarithmic nature of the response means that it takes a larger change in flux in bright regions to generate the same perceived change in brightness that a smaller change would generate in a dark region. Finally, the dynamic range limit at any adaptation level means that the visual system can only perceive a limited radiometric range at a reference adaptation level. If a scene is displayed over a larger brightness range (3.3 log units), the eye would typically adapt to a midlevel, and variations in the brightest and darkest regions would be lost to the visual system.

Clearly, the size and spatial sampling of the visual receptors introduce visual resolution limits typically referred to as *acuity*. An average observer's angular resolution is limited to about 1.7 line pairs per milliradian. This angular resolution can easily be scaled to display resolution when the distance to the display is known i.e.,

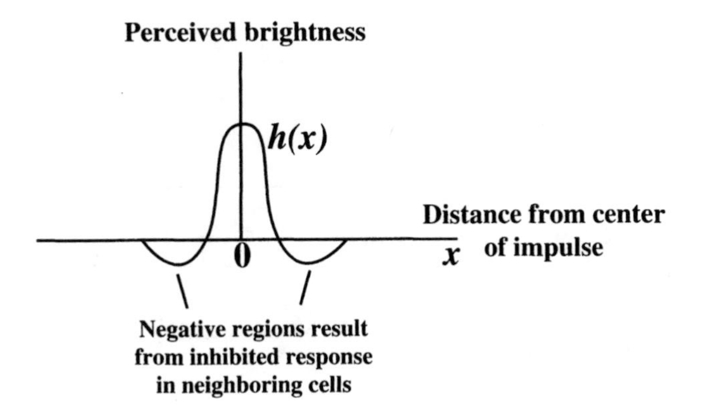

Figure 8.2 Simplified point spread function associated with lateral inhibition.

$$\frac{\text{\# line pairs}}{\text{mm}} = \frac{\text{\# line pairs}}{\text{milliradian}} \cdot \frac{1}{H} \tag{8.1}$$

where H is the distance from the observer to the display. For example, at 0.5 meters we would resolve approximately 3.4 line pairs per millimeter, which is better than a standard computer monitor (i.e., we would be limited by the monitor not the observer). On the other hand, if we were observing hard copy with a resolution of 10 line pairs/mm at this same distance, we would be limited by the visual system, and some of the fine detail would not be observed by the user.

This brings us to another limitation of the visual system, namely, closeness of focus. The near point (closest point where the eye can comfortably focus) varies with individuals and often with age, but generally is about 0.2 meters or more. Thus one can only slightly improve the spatial detail resolved by the eye by moving the image closer. After that, some form of magnification is required.

Unlike photographic and most electro-optical systems, the response of the visual receptors are a function of mean brightness levels, previous illumination conditions, and illumination on adjacent cells. The nerve cells that process the signals from the receptor cells (rods and cones) interact with surrounding cells in a number of ways. The lateral connection of nerve cells in the eye, coupled with additional processing in the visual cortex result in a variety of unexpected perceptual effects (cf. Cornsweet, 1970). The subtle details of visual response effects are too numerous for this treatment, so we will restrict ourselves to the most dominant effect. One of the strongest interactions is called *lateral inhibition*. Each cell reduces the sensitivity of adjacent cells when it is excited (i.e., when the receptor is illuminated). The spatial response to a point source in a cell structure exhibiting lateral inhibition is shown in cross section in Figure 8.2. This is called the *point spread function* of a sensor system (cf. Sec. 9.2). The perceived brightness can be approximated by the convolution of the point spread function with the illumination field onto the retina. Figure 8.3 shows the effect of convolving the point spread function of Figure 8.2 with a step change in brightness. The net effect of lateral inhibition is to exaggerate edges (the Mach band effect). This exaggeration of edges in the lowest level of visual processing is part of the overall cognitive process that keys heavily on edge elements. Thus, in displaying image data for visual analysis, we want to make sure we preserve, or possibly even exaggerate, edges to make them fully accessible to the image analyst's cognitive processes.

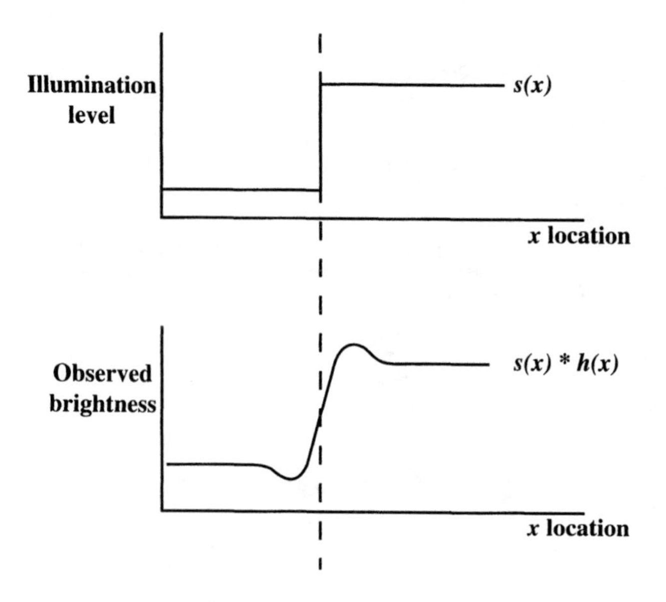

Figure 8.3 Convolution of the point spread function of Figure 8.2 with a step in illumination.

To this point we have ignored color vision which is also a science in its own right. Our main concern will be in recognizing that the visual system responds to a range of spectral stimuli that can be described in a variety of color spaces (cf. Robertson, 1992). The CIE has defined one of the most commonly used color spaces based on the use of tristimulus values derived from color-matching function designed to approximate the effective response of the human visual system (cf. Wyszecki and Stiles, 1982). These color-matching functions (plotted in Fig. 8.4) are not the actual response functions of the red, green, and blue photo receptors but a set of weighting functions (based on the CIE 1931 Standard Observer) incorporating the average observer's perceived response to spectral radiance. By cascading the color-matching functions together with the incident spectral radiance, three tristimulus values are derived according to:

$$X = k\int L_\lambda x'(\lambda)d\lambda$$
$$Y = k\int L_\lambda y'(\lambda)d\lambda$$
$$Z = k\int L_\lambda z'(\lambda)d\lambda \qquad (8.2)$$

where X, Y, and Z are the tristimulus values, x', y'; and z' are the color-matching functions shown in Figure 8.4; k is a constant used for units conversion; and the integral is over the visual response range. The Y tristimulus value is proportional to the overall luminance. For convenience of plotting color in a two-dimensional space, the x and y chromaticity coordinates are defined as

$$x = \frac{X}{(X+Y+Z)}$$

and

$$y = \frac{Y}{(X+Y+Z)} \qquad (8.3)$$

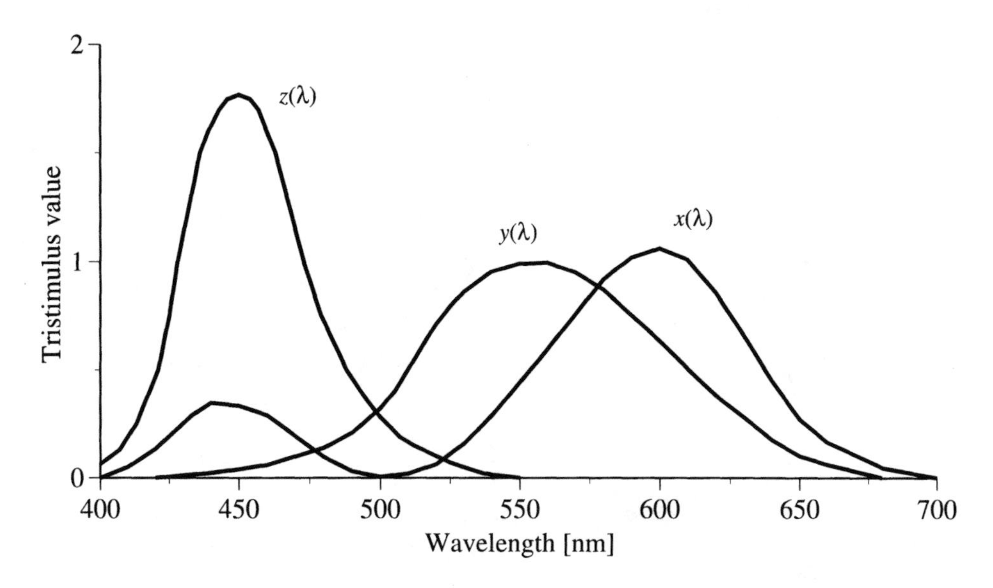

Figure 8.4 Color-matching functions.

with any perceived color uniquely specified by (Y, x, y), where Y represents the luminance, and x and y provide what we think of as the color characteristics (hue and saturation). The color space is then characterized by plotting x versus y as shown in Figure 8.5. This is known as the CIE 1931 chromaticity diagram. The horseshoe-shaped region is constructed by computing the chromaticity coordinates for monochromatic flux at the indicated wavelengths and connecting the points to form the horseshoe shape referred to as the spectrum locus. The *spectrum locus* delimits the range of visually perceived color. This region, inside the spectrum locus, is referred to as the visual *color gamut.* Any given display system will be capable of displaying some subset of this gamut, as shown in Figure 8.6. Thus, the color gamut of the display system may limit the amount of information we can provide to the visual system and influence how we choose to display it. It is also important to recognize that all points in a color gamut are not perceived as unique colors. Surrounding any point in the chromaticity diagram, there is a region where changes in the chromaticity values are not noticeable as color changes. For display purposes we want to make sure that if we want to show steps in color space, the step sizes must be such that they exceed these just noticeable difference (JND) steps that were characterized in chromaticity units by MacAdam (1942).

When displaying images for hard or soft copy analysis, we need to factor in the visual response function. To accomplish this, the relationship between digital count and radiance from soft-copy displays needs to be calibrated. In the case of hard-copy displays, the relationship is usually between digital count and reflection or transmission density. For visual assessment, we would generally like to have this relationship approximately log linear with brightness (or linear with density since this is a log function). We also need to ensure that significant changes in digital count (i.e., changes greater than the overall system noise) translate into changes in brightness that are greater than the just noticeable differences (JND's) of the visual system. In cases where the display system is not readily adjusted, this is often best accomplished by adjusting the digital count driving the display using a lookup table approach (cf. Sec. 7.1.1). An example of this is shown in Figure 8.7 where a density vs. DC calibration curve is shown in the first quadrant and a lookup table used to linearize the relationship between image digital count and density is developed in the remaining quadrants. Similar

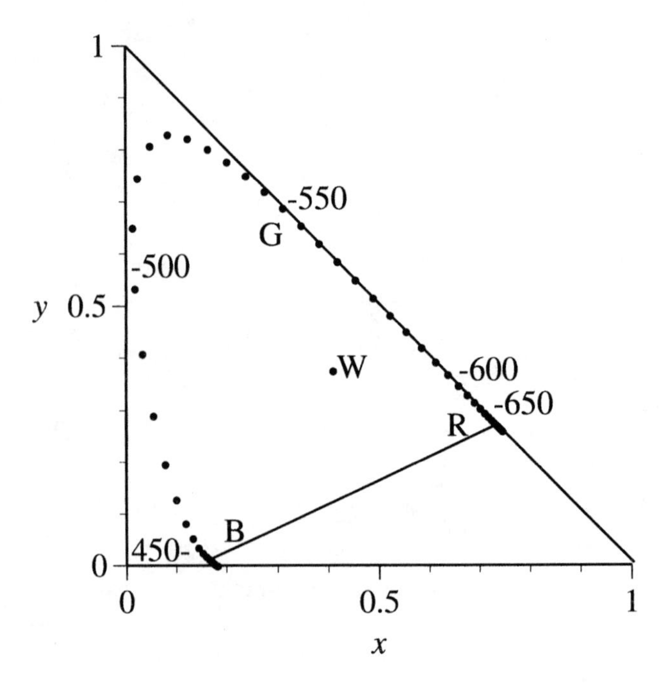

Figure 8.5 CIE 1931 chromaticity diagram; R, G, B, and W are the red, green, blue, and white regions, respectively.

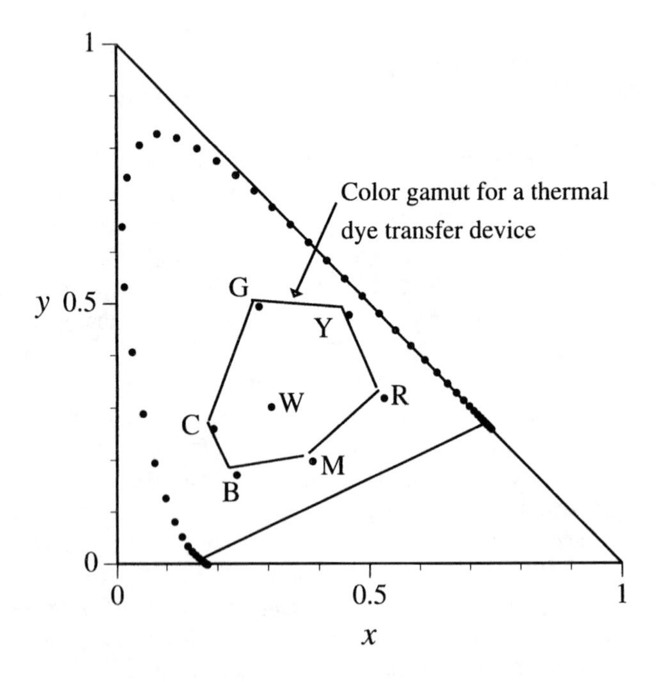

Figure 8.6 Chromaticity diagram showing the color gamut of an output device. Only colors inside the color gamut boundaries can be produced by this device. (Data courtesy of RIT's Munsell color lab.)

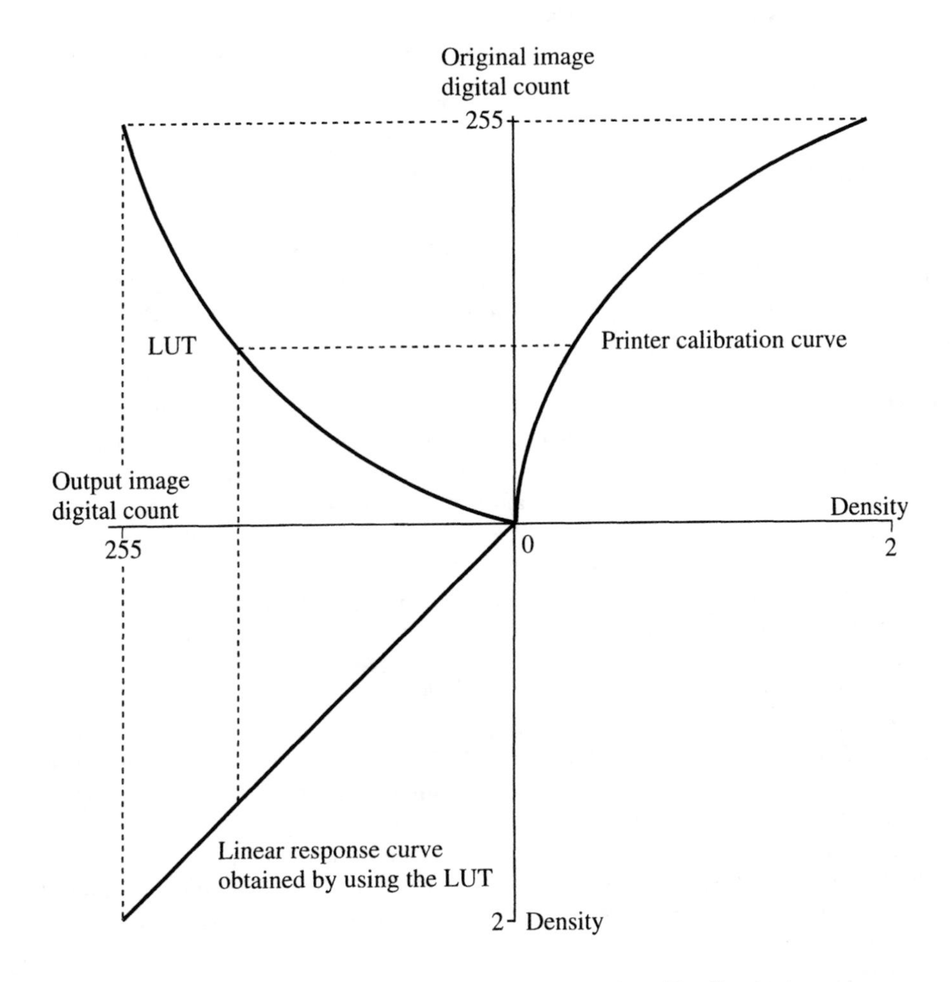

Figure 8.7 Calibration and correction of output device using a lookup table to linearize the output.

display corrections need to be developed for color displays to ensure that the available color gamut is effectively used and to make sure that we don't try to display outside the color gamut. The background lighting conditions at the analyst's station can be very critical here, as they can have significant impact on the appearance and interpretation of color, as well as affecting the ability to detect subtle gray-level variations. Where possible these illumination factors should be included in the calibration of the output device (cf. Hunt, 1987).

Resolution effects will be discussed in more detail in the next chapter. So for the moment, we will restrict our concern to ensuring that the observer is able to "see" all the resolution available in the data if required. For soft-copy displays, this means that the user should be able to zoom in to a point where the visual resolution is better than the resolution of the image displayed. Similarly, for hard-copy display, the observer should be able to magnify the display to a point where visual resolution is no longer a limiting factor. The reader should note, however, that most analysis will not be done at these extreme resolutions, and the synoptic perspective afforded by lower resolutions and large fields of view is also highly desirable. Furthermore, providing the user with a capability to observe the image at magnifications much beyond where the visual resolution matches the display resolution can introduce unwanted and distracting artifacts (e.g., grain noise in the film can begin to be interpreted as image information). The ideal system would allow the user the full synoptic perspective at the maximum resolution. A practical compromise is to provide full synoptic coverage at reduced resolution and an ability to zoom in on localized regions to a point where the visual resolution is the limiting factor.

8.2 THEMATIC AND DERIVED INFORMATION

In many cases, the user needs information derived from the image rather than, or in addition to, the image itself. This is commonly thought of as thematic information where the image data has been processed to represent some form of spatial theme(s), such as land cover, vegetation stress, or water quality. Here again, it is critical that we ensure that any processed information is available to the analyst. For example, we will often color code themes (e.g., a land cover map), and we need to ensure that the colors selected are within the color gamut of the output device and clearly differentiable (i.e., several JND's apart in color space). Furthermore, for both esthetics and ease of interpretation, it is often valuable to select color or gray shades in thematic maps that are intuitively meaningful to the user (e.g., blue water).

Another important and often overlooked aspect of conveying thematic information to the user is to recognize that decisions must be based on this information. It is often important to provide not only the results (e.g., the thematic map) but also information on the quality of the results so the user can decide how heavily to weigh them into any decision. This can be done, for example, by providing a confusion matrix of classification accuracies on independent data (cf. Sec. 7.2). In many cases where statistics are used in the processing algorithm (e.g., a GML classifier), it is even possible to provide confidence data on a pixel-by-pixel basis. For example, in a GML classifier the a posteriori probabilities can be calculated and scaled so that they can be presented in image form with brightness varying with probability.

Recognize that the user needs this same type of information even when the output is in the form of numerical or tabulated data. For example, suppose an algorithm identifies 20% of a study region as forest and 17% of the forest area as dead or severely stressed due to an insect infestation. If possible, we should also provide the user with information on the confidence of these estimates, e.g., the study region is $20.0 \pm 0.5\%$ forested and of that forested region $17 \pm 8\%$ is dead or severely stressed. As a result, it is important for the remote sensing scientist to always understand the algorithms being used and to perform sufficient analysis to determine what confidence should be placed on any results.

8.3 OTHER SOURCES OF INFORMATION

It is all too easy for experts to become myopic and neglect sources of information other than those with which they are most familiar. The remote sensing scientist is most often involved in extracting information from images. This is one very critical element of spatial information analysis. However, there are many other possible sources of information that may be helpful in addressing an information analysis problem. The whole field of geographic information systems (GIS) has evolved in recent years to address the problems of combining and analyzing spatial information. In this section, we will discuss a few of the fundamentals of GIS and point out other sources of information and analysis tools a remote sensor should look to for assistance. A more thorough treatment of GIS concepts can be found in Goodchild and Kemp (1990).

8.3.1 GIS Concepts

Geographic information systems (GIS) loosely refer to both databases and data processing tools used to analyze spatial data. Normally the data are tied to some geographic reference system, so that each piece of information refers to a point or a region on the earth. In a GIS, many types of information exist that describe spatially distributed variables. For any location covered by the GIS, all the attributes (the value of the spatial variables) can be accessed, or conversely any attribute can be mapped or displayed as an image. Most importantly, algorithms can be implemented that allow processing of any or all of the attributes in order to search for certain combinations or relationships.

8.3.1.1 Spatial Data

Before we worry too much about GIS processing logic, we should briefly consider some of the forms of spatial data that may be of use in remote sensing and image analysis. To begin, each image is itself a set of spatial data that could be attributed on a pixel-by-pixel basis to the location on the ground where the pixel would be projected. Similarly, any image-derived data could be attributed to little parcels of real estate (ground sample spots) in the same fashion (e.g., land cover type). Another way to think about this process is to imagine the boundaries of thematic regions projected onto a map and the enclosed region assigned a theme (e.g., land cover type). This introduces the two primary GIS data formats: raster (pixelized data) and vector (line segments and enclosed polygons), as illustrated in Figure 8.8. Different images or data derived from different images (e.g., thermal, multispectral, radar) can all be combined by having the data spatially registered to each other or to a common geographic coordinate system. Other types of spatial data can also be used to characterize the same piece of real estate represented by the image data. Any type of map data, for example, can be merged into a common GIS based on geolocation. This could include soils maps, road maps, and topographic data in the form of elevations, or derived data in the form of slope and aspect maps. It is also possible to include geo-referenced point or line information such as the location of a pollution source or a telephone line with an information file describing the attributes of the point or line. For most of us used to working with images, the database is most easily thought of in raster form with all of the features (or layers) in the database conceptualized as spectral bands or feature bands in a many layered image (cf. Fig. 8.9). It is important to recognize that the resolution of the layers need not all be the same. The lower-resolution data can be resampled to a higher pixel density when necessary for comparison with the higher-resolution data.

8.3.1.2 Registration and Resampling

The formating and processing of GIS data rely very heavily on our ability to transfer spatial data into a common coordinate system (registration) and to resample the data so that we can easily access and process information from the same spatial location simultaneously. To begin to understand image registration, we will first take the case of two images of the same region. For convenience, we assume that the images are of relatively flat terrain so that image distortion due to terrain elevation is negligible. The images may, however, have different resolution, rotation, view angle, etc., so that pixel locations in one have no clear relationship to pixel locations in another. We desire a way to relate the geometric coordinate system in one image to that of the other image and eventually to transform one of the images so it has a common coordinate system with the other.

In the case of images, it is possible to use arbitrary coordinate systems, with pixel row

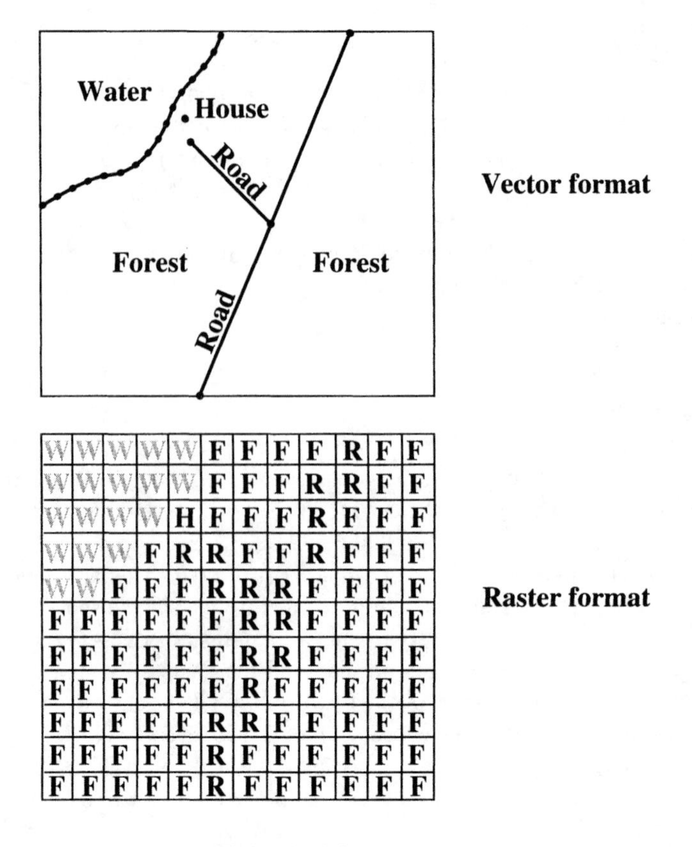

Figure 8.8 GIS data formats.

and column values representing the location of pixel centers in a pair of Euclidean coordinate systems, as indicated in Figure 8.10. We use x', y' to designate points in the primary or target coordinate system of the reference image and x,y to designate points in the sample image coordinate system. We would like to take the sample image and warp it so that it registers with the reference image. To accomplish this, we need to know the relationship between the two coordinate systems. We assume that the systems can be approximately related by a least-squares fit to a polynominal of the form

$$x = a_0 + a_1 x' + a_2 y' + a_3 x' y' + a_4 x'^2 + a_5 y'^2 \cdots + \varepsilon_x$$

$$y = b_0 + b_1 y' + b_2 x' + b_3 y' x' + b_4 y'^2 + b_5 x'^2 \cdots \varepsilon_y \tag{8.4}$$

where ε_x and ε_y are the residual errors after the transform. For simple distortions between the two images, a good transform can be achieved with a fairly low-order transform. For example, only the zero-th–order terms a_0 and b_0 are needed for a simple shift of the origin, and the first two terms (a_1, a_0, b_1, and b_0) are needed for a combined scale adjustment and shifting of the origin. Higher-order terms are required to account for rotation, skew, and keystoning effects due to acquisition and perspective differences (cf. Schowengerdt, 1983). A coordinate transform of the form of Eq. (8.4) can only be assumed valid for images or portions of images that have good internal rectilinear geometry. For example, scanner images with residual roll distortion should be corrected for roll before use of Eq. (8.4). In scenes with considerable topographic variation, the simpler forms of Eq. (8.4) are only approxi-

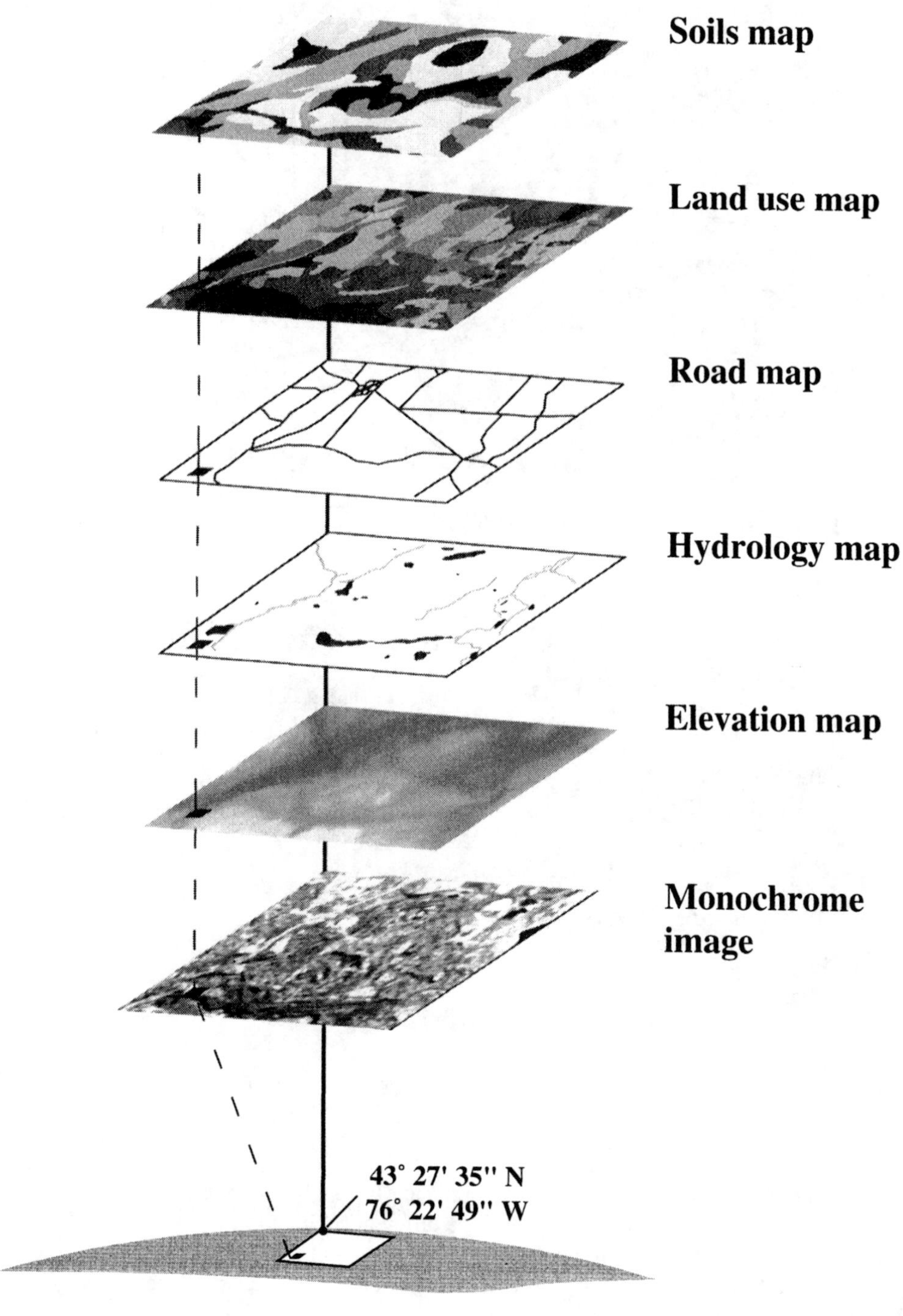

Figure 8.9 Layers in a GIS database. (Data courtesy of the Mapping Science Laboratory, SUNY-ESF, Syracuse.)

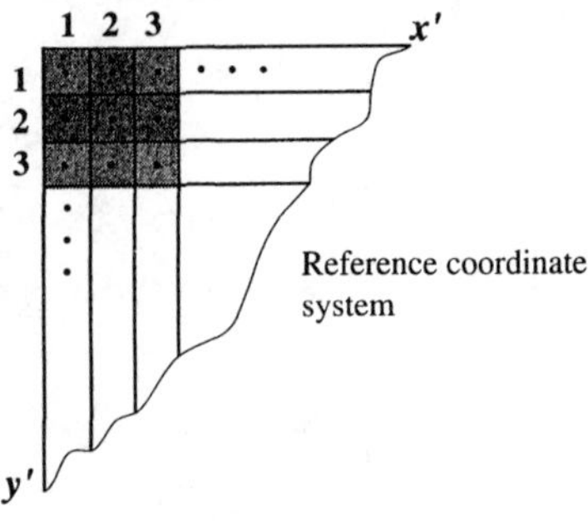

Control point coordinates

	x', y'			x, y	
A':	52	382	A:	9	91
B':	358	254	B:	332	157
C':	574	540	C:	344	518

(a) Target on reference image and coordinate system

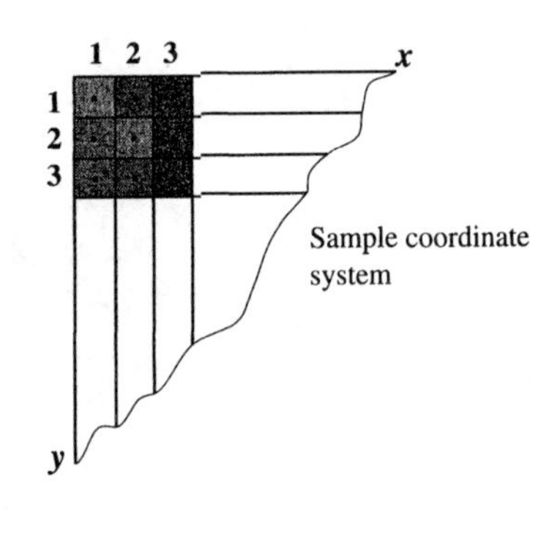

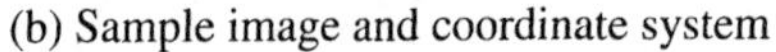
(b) Sample image and coordinate system

Figure 8.10 Coordinate systems for geometric registration.

mately valid for surfaces that approximate a plane. In rough terrain, piecewise approximations must be applied across the image using a different solution to Eq. (8.4) for each region (cf. Fig. 8.11). To avoid error in boundary regions, Eq. (8.4) must be truncated to four terms if quadrilaterals are used, or three terms if triangles are used. This produces a less precise but unique solution at all boundaries between solution regions.

In order to apply Eq. (8.4), common objects must be uniquely located in each image. The x,y and x',y' values of these control points generate the input data to a least-squares regression that is used to solve for the a_i and b_i coefficients. As with any regression solution, the input data should cover the entire solution space, and in general, the solution should not be extended beyond the sample space. This is particularly true when higher-order terms are used in the solution for a whole-image transform. Severe distortion can occur beyond the

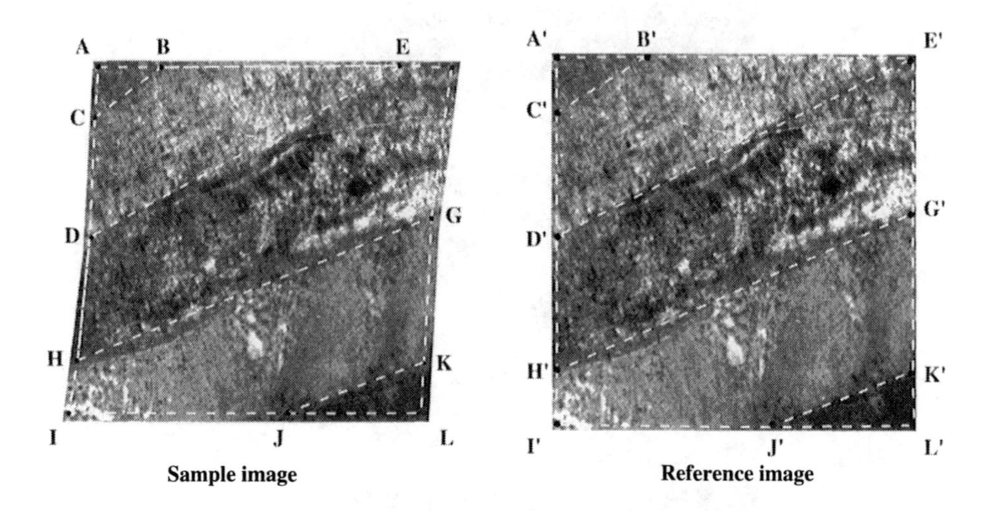

Figure 8.11 Use of quadrilateral regions in a piecewise geometric rectification for regions with high-terrain relief.

sample space (for this reason lower-order solutions should be used whenever possible, and the user should make sure that sufficient control points are selected to ensure a robust solution).

The control points used for input to a coordinate transform can be manually located in each image. However, this can be a tedious, time-consuming task. A number of methods have been developed to automate the selection of control point pairs. One of the most straightforward involves the use of correlation; cf. Eq. (7.8). A small window in one image is selected as a correlation kernel, and its normalized correlation with the other image is computed as:

$$g(i,j) = \frac{1}{K_1 K_2} f(i,j) \star h(i,j) \tag{8.5}$$

where K_1 is the sum of the values in the correlation kernel, K_2 is the sum of the values in the image under the correlation kernel, $f(i,j)$ is the image digital count and $h\ (i,j)$ is the correlation kernel. Where the kernel passes over the corresponding region in the search image, a maximum in the normalized correlation should occur. The original coordinates of the center of the correlation kernel and the location of the maximum in the normalized correlation image become a pair of control points.

Regrettably, the correlation value is sensitive to scale and rotation, so the images must be roughly scaled and rotated in order to use this approach. Most satellite images have had first-order geometric corrections applied so that this is often not a concern. To reduce the likelihood of finding a false maximum and to speed up the process, the search region for the correlation is often restricted to a region delimited by the expected error in nominal registration between the two images. For example, if the two-sigma (two standard deviations) pointing error in a satellite translates to 30 pixels in the image, we would expect a pair of control points on two images from the satellite to be located within ±60 pixels of each other (i.e., the images should be registered to first order to within 60 pixels). Thus, the search window need only be 60 x 60 pixels rather than the entire scene. Furthermore, the correlation kernel (control point) can be selected to have high contrast and sharp edges to improve the sharpness of the correlation peak and, therefore, the precision of the conjugate control point location. An

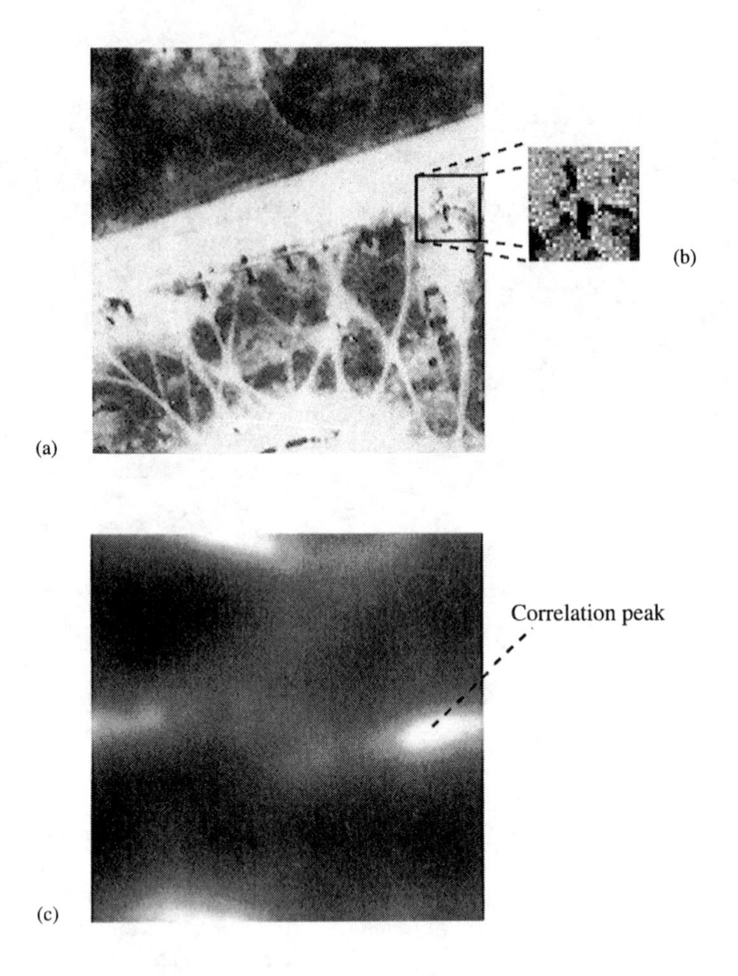

Figure 8.12 (a) Search window, (b) image kernel, and (c) correlation surface. The peek in the correlation shows where the image kernel matches the brightness structure in the search window.

example of a control point kernel, search window, and correlation surface is shown in Figure 8.12. Many control points could be automatically located in this fashion and used as input to a regression solution to an equation of the form of Eq. (8.4).

Once the geometric relation between the two images is known, a method for mapping the image values from one coordinate system to the other must be defined. For each x',y' pixel center in the reference coordinate system, we must determine what value to assign from the input sample image. Using Eq. (8.4) we can locate the x,y location in the input image coordinate system corresponding to x',y'. Generally this x,y value will not be an integer value falling exactly on a pixel center. A simple method of selecting the appropriate digital count value is to select the value of the pixel center closest to the x,y location. This method of nearest neighbor resampling can be easily implemented and is computationally very quick. The nearest-neighbor resampling method can introduce staircase artifacts at edges, as shown in Figure 8.13. This effect can be reduced using a simple bilinear resampling approach, as illustrated in Figure 8.13. Bilinear resampling reduces the edge artifacts but tends to blur the image slightly. Also, since it depends on four neighboring pixels, it runs slightly longer than the nearest-neighbor method.

The resampling process can be thought of as convolving the image with a kernel that is

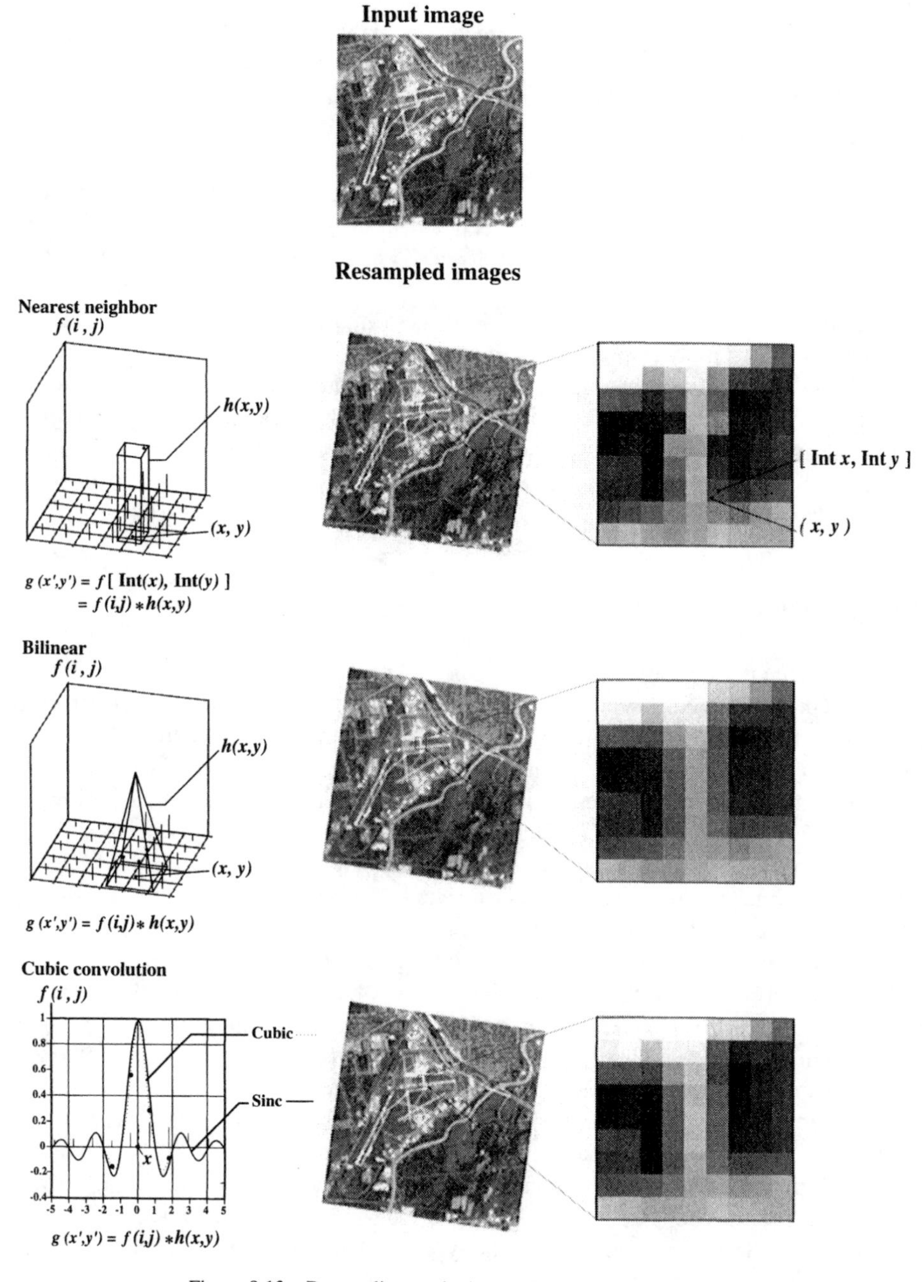

Figure 8.13 Resampling methods used in image registration.

centered at each consecutive x,y location in the input image corresponding to the x',y' values in the output image. The nearest-neighbor kernel would have a base of 1 pixel by 1 pixel and unit height. Thus it will always just return the value of the 1 pixel located under the kernel. The bilinear interpolator is a four-sided pyramid kernel of base 2 pixels x 2 pixels and central height of unity. Linear systems theory indicates that the ideal sampling kernel would leave all frequencies unaffected (i.e., a rect function in frequency space), which means the convolution function should be a sync function centered at the (x,y) sampling location [cf. Eq

(9.9) and (9.10)]. Use of the sync function is impractical as it would require sampling the entire image, as shown in Figure 8.13. Instead, it is approximated with a cubic polynominal fit that uses the nearest 4 pixels in each direction as shown in one dimension in Figure 8.13. This cubic convolution resampling approach is a good approximation to the ideal sinc function producing less blur than the bilinear interpolator, but at the cost of increased run time (cf. Park and Schowengerdt, 1983).

It is important to recognize that any resampling other than nearest neighbor introduces some blurring of pixel radiance values and some mixing of spectral signatures in a multispectral image. For this reason, where it is practical, it is often desirable to perform radiometric calculations before geometric transforms. For example, a multispectral land cover classifier could be run on the raw image, and then the class map image could be transformed.

To this point we have concentrated on image-to-image registration. However, exactly the same principles apply to warping an image to a map. The control points in this case are map features that can be clearly identified on the image, and the reference coordinate system is the map coordinate system. To perform the resampling, a grid size is defined in the map coordinate system and the image location computed for each grid center, just as in image-to-image resampling. Any of the resampling methods can be used. Using this method any image or digital map data can be transformed to a common coordinate system.

8.3.1.3 GIS Processing Logic

Once the spatial data are assembled and warped to a common geometric coordinate system, a host of GIS processing tools can be applied to the data. If the data are in raster form, all the image processing tools discussed in Chapter 7 can be applied, since the data can now be fashioned to look like multispectral image data. In addition, a number of logical operations can be performed to facilitate combining and extracting information from the various layers in a GIS. Two of the most useful of these operations that were not introduced in Chapter 7 are region growing and Boolean logic. Region growing is the simple process of locating all the grid cells (pixels) within some distance (number of pixels) of a feature. For example, from a rasterized hydrographic map (or a land cover map derived from multispectral imagery) we could find all pixels that were coded as water. We could then add a new layer to the GIS comprised of all pixels within 300 meters of a water course simply by having the computer locate all pixel centers within x pixels (x = 300 meters/grid cell size) of the water pixels. Similarly, we could make a map of distance ranges (e.g., 0 to 100 m, 100 to 200 m, etc.) from any feature of interest located in any of the GIS layers. The GIS layers can then be combined using combinations of simple algebra and Boolean logic (e.g., and, or, not) to extract whatever data are required.

A full treatment of GIS processing logic is beyond the scope of this book, so we will restrict ourselves to one simple example to provide the reader with some insights into the possibilities of GIS processing. For simplicity, we will assume that all the data are preregistered and in raster format and that all image processing will use raster processing methods, as these will be similar to the image processing tools introduced in Chapter 7. Table 8.1 lists the layers that are available in the example GIS database and how they are coded. For this example, our interest is in locating all possible sites within the study area that meet a set of site criteria listed in Table 8.2. Figure 8.14 shows the steps that would be used to generate some intermediate layers in the GIS and the logic that could be used to combine the layers to isolate suitable pixels. Finally, a screening algorithm is run to determine if the number of contiguous pixels is sufficiently large to meet the area requirements for the site.

One of the greatest advantages of a GIS system is that once the database has been populated, the data can be combined in many different ways in support of diverse interests. For example, commercial, transportation, environmental, agriculture, public health, and public

Table 8.1 Layers in an Example GIS Database

Source	Feature(s)	Code
Landsat TM Classification map	Landuse Water features	Agriculture = 1 Streams = 2 Extractive industry = 3 Forest = 4 Wetlands = 5 :
USGS quadrangle Digitized manually	Roads	Roads = 1 Other = 0
Soil conservation service Digital soils maps	Soil type	Type A = 1 Type B = 2 :
USGS digitial terrain Elevation data	Elevation	1 m = 1 2 m = 2 3 m = 3 :

utilities engineers can all access the same database in different ways to address applications-specific questions. The information derived from a GIS may include images, maps, data tables, location indices, or simply decisions (e.g., no suitable site). In all cases the display philosophy of making sure the data are in a format meaningful to the user and easily understood, and whose limitations are clearly delineated, should be a major design consideration.

8.3.2 Data Bases and Models

In addition to spatial databases, analysis of remotely sensed data increasingly relies on other sources of data and modeling tools to increase our ability to extract information. Some of the data are simply tabulated databases. For example, there are a variety of databases containing laboratory or field reflectance spectra for various material types. Some of these databases also include thermal and dielectric properties that can be useful in thermal infrared and radar image analysis. In many cases, the databases require analytical or stochastic models to make the support data useful to the image analyst. For example, the thermal properties of materials would typically only be of use if a thermal model were available to translate materials information along with meteorological variables into temperature estimates (cf. Sect. 10.2.2). Radiation propagation models such as LOWTRAN and MODTRAN discussed in Chapter 6 are another example where atmospheric databases and models are combined to provide the image analyst with tools to extract better information about a target's temperature or reflectivity. There are also a whole host of application-specific models that relate

Table 8.2 Sample Criteria Required for a Site to Be Located Using the Example GIS Database of Table 8.1

Feature	Criteria	
Water	Site must be : > 500 feet from any water feature	
Roads	Best suitability :	< 2000 feet from a road
	Moderate suitability:	> 2000 feet from a road
Landuse	Sites must be located on either:	
		Agriculture
		Forest
		Extractive industry
Soils		Soil associations
	Best suitability :	1, 3, 6, 11, 12
	Moderate suit. :	7, 23, 35, 45
	Not suitable :	all others
Slope	Best suitability :	0% - 4%
	Moderate suit. :	4% - 12%
	Not suitable :	> 12%

reflectance or temperature data to phenomena of interest. For example, Suits (1972) describes a model relating canopy reflectance to the structure, layering, and individual leaf reflectance in a vegetation canopy, and Gordon et al. (1983) describe a model relating water reflectance to the concentration of various parameters in the water.

While we will not delve into these specific models here, they reflect a growing trend toward image analysis that focuses on information extraction by coupling conventional image processing with physical models of the scene and the imaging process. This coupling more closely relates the final image parameters to the phenomena that caused them. This provides the user more ability to understand not only what was imaged but also in many cases, why it appears as it does (i.e., is the change in appearance of the vegetation due to a species change, a change in canopy density, a change in slope and azimuth, or a change in vegetation vigor?). This model-based image analysis essentially continues the image chain all the way to the application-specific phenomenology of interest. We will return to this topic again in Chapter 10 where we discuss a particular set of models aimed at synthetic image generation.

8.4 IMAGE FUSION

Loosely interpreted data fusion is the process of merging data from different sources to improve our knowledge of a scene. Thus most GIS- and model-based analysis could be construed to be data fusion. In this section, we want to concentrate on image fusion where two types of image data are somehow combined to form a new type of image data that ideally contains more interpretable information than could be readily accessed from the two image sets separately. Often fusion involves merging data from different sensors with different resolutions to attempt to improve the interpretability of the lower-resolution system.

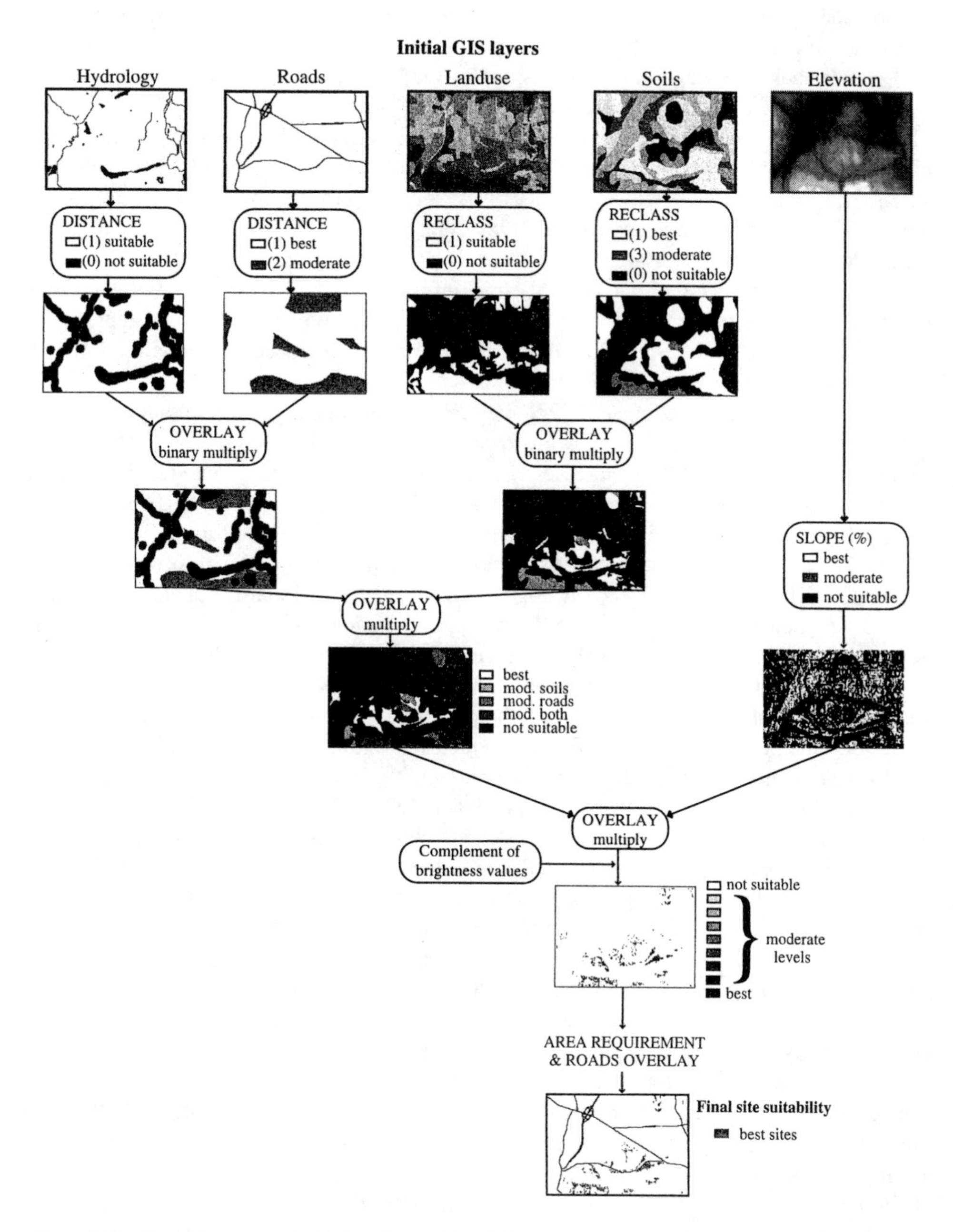

Figure 8.14 Logical steps used in site location problem defined in Table 8.2 for GIS data set listed in Table 8.1.

Haydn (1982) used this approach to improve the interpretability of LWIR images from the Heat Capacity Mapping Mission (HCMM) sensor. The LWIR images had a 500-m GIFOV. The HCMM image was geometrically registered and resampled to the GIFOV of a Landsat MSS image of the same region (80-m GIFOV). The data were "fused" by having one band of the MSS data represent intensity and the thermal band from HCMM represent hue in an intensity hue and saturation (IHS) color space. This color space was chosen because the visual-cognitive system tends to treat intensity (brightness), hue (color), and saturation (purity of color or how pastel the shade is) as roughly orthogonal perceptual axis.

While it may be very difficult to decide the relative change in redness, greenness, or blueness between two colors, it is much easier to say what the relative intensity, hue, and saturation are. Haydn took advantage of this by displaying the high spatial resolution data from MSS as intensity, since the visual system's spatial cues are most sensitive to intensity variations. The thermal data were displayed as hue with saturation left at a fixed level. This is accomplished by a simple transformation of the data from IHS space to the red, green, and blue (RGB) space used in most display devices (cf. Conrac Corporation, 1985). The resulting image "appears" to be a high-resolution thermal image, since the visual system's spatial cues are triggered by intensity variations, and thermal variations are interpreted by the variation in hue (color). There is actually no more thermal information, but by fusing it in this fashion, it is more readily interpreted. Figure 8.15 shows an example of this approach applied to a Landsat TM image where the 30-m GIFOV data from the red spectral band is used to control intensity, and the 120 m LWIR data controls the hue.

Chavez et al. (1991) compared three methods for fusing a high-resolution panchromatic (pan) band image (e.g., SPOT), with lower-resolution multispectral band images (e.g., SPOT Multispectral or TM). In all cases, the multispectral data are registered and resampled to the pixel size of the high-resolution pan band. The first method uses the IHS transform described above by taking three TM bands as RGB inputs and transforming to IHS color space. The resulting intensity image is found to be roughly proportional to the pan image brightness. The pan image is then histogram equalized to the intensity image and used to replace the intensity in the IHS image. The "high resolution" IHS image can then be back-transformed to RGB space for display. The resulting image appears to be a high-resolution three-band multispectral image. This method works reasonably well if the three input bands are highly correlated with each other and the pan band. When this is not the case, the substitution of the pan band for intensity can produce substantial shifts in the radiometry of the hybrid scene. This method is also limited in that only three bands can be processed at a time. A second method using the principal components transform has many similarities to the IHS transform method, but all the bands can be operated on simultaneously. In this case, the multispectral data are transformed into principal component (PC) images (cf. Sec. 7.3). The first PC image is often roughly proportional to the pan brightness. Again, a histogram equalization and substitution approach are used. The PC images with the high-resolution histogram equalized pan band substituted for PC band 1 are then back-transformed to produce the hybrid high-resolution multispectral bands. This method yields results similar to the IHS transform. The images appear to have a higher resolution and are reasonably correct in terms of brightness when the first PC image and the pan band are highly correlated. However, when this is not the case, the radiometry of the hybrid image can be significantly different than in the original image. The third method was suggested by Schowengerdt (1980) and uses quite a different approach. In this case, the high-resolution image is high-pass filtered, leaving an image with only the high-frequency edges. This image is summed with each band of the low-resolution MS image to yield a hybrid high-resolution image. This method is attractive in that the resultant image appears to have all the resolution of the pan image, and since low-frequency regions are left unchanged, the mean level radiometry is preserved. The limitation of this approach comes into play if the edge information in the pan band is not highly correlated with the edge information in one or more of the multispectral bands. In this case, while the image may "look" nice, false edges may be introduced or true edges neglected.

Most of the methods presented thus far were primarily focused on enhancing the appearance of the hybrid high-resolution image to facilitate visual image exploitation. Another class of algorithms exist which are primarily focused on fusion techniques designed to preserve the radiometric integrity of the high-resolution hybrid image for quantitative or machine exploitation. The visual improvement associated with these methods is comparable to the methods described above, but comes as a by-product of the algorithm's attempt to gen-

Red channel of TM (intensity)

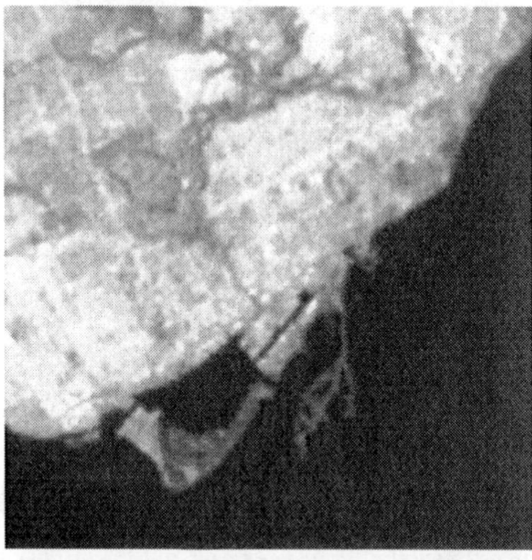

LWIR channel (hue)

Color-coded fused image (IHS to RGB)

Figure 8.15 Image fusion of the red channel of TM with the LWIR channel using the IHS transform. See color plate 8.15

erate a radiometrically correct estimate of what a high-resolution scene would look like. All of these methods must, of course, use some method of estimating how the high-resolution radiance is distributed in the hybrid scene. This cannot be known exactly, since even with the high-resolution pan image, there is too little information for a perfect reconstruction. As a result, this is an active area for ongoing research since many estimation methods exist and their utility may be both scene- and application-dependent.

One of the most straightforward methods of fusing multispectral data with higher-resolution panchromatic data relies on the assumption that there is some degree of correlation between the multispectral band and the pan band brightness values. The method operates on super-pixels that have GIFOV's the size of the original multispectral (MS) pixels. Each super-pixel is comprised of subpixels that are resampled to the same GIFOV as the pan band (cf. Fig. 8.16). The hybrid high-resolution pixel brightness can then be expressed as

$$\mathrm{DC}'_i = \frac{\mathrm{DC}_i}{DC_{\mathrm{pavg}}}\mathrm{DC}_p \tag{8.6}$$

where DC'_i is the digital count of the hybrid output subpixel value in the ith spectral band, DC_i is the digital count of the super pixel in the ith spectral band, DC_{pavg} is the mean digital count in the pan band for the subpixels contained in the low-resolution super-pixel, and DC_p is the digital count in the pan band for the high-resolution pixel of interest. The algorithm implicitly assumes that $\mathrm{DC}_i/\mathrm{DC}_{pavg}$ represents a fixed proportional relationship that is approximately constant over all subpixels in the super-pixel.

This method is very attractive in that on average over a hybrid super-pixel, it must provide exactly the same radiance as in the original super-pixel, i.e.,

$$\mathrm{DC}'_{i\mathrm{avg}} = \frac{\sum_{j=1}^{N}\mathrm{DC}'_{ij}}{N} = \frac{\sum_{j=1}^{N}\left(\frac{\mathrm{DC}_i}{\mathrm{DC}_{\mathrm{pavg}}}\mathrm{DC}_{pj}\right)}{N} = \frac{\mathrm{DC}_i}{\mathrm{DC}_{\mathrm{pavg}}}\frac{\sum_{j=1}^{N}\mathrm{DC}_{pj}}{N} = \mathrm{DC}_i \tag{8.7}$$

where DC'_{iavg} is the mean digital count over the hybrid super-pixel in band i, and the sum is over the N subpixels in the super-pixel. This means that on average at the resolution of the original MS imagery, the radiometry is preserved exactly. Pradines (1986), Price (1987), and Munechika et al. (1993) all used methods similar to this to enhance TM or SPOT MS data using the 10-m SPOT panchromatic band. This approach yields both radiometrically and visually improved images. It is, however, subject to error when the assumptions are violated. This occurs when the MS band and the pan band are only weakly correlated or negatively correlated. It also occurs even with well-correlated bands if the correlation ratio ($\mathrm{DC}_i/\mathrm{DC}_{pavg}$) is not a constant over the super-pixel, as might be the case for a mixed pixel (i.e., a super-pixel comprised of subpixels representing two or more material classes). Munechika et al. (1993) describe improvements to the baseline methods that incorporate adaptations for mixed pixels, and Price (1987) suggests corrections for correlation effects. Braun (1992) combined some of these approaches by developing algorithms that adapt to both poor correlation and mixed pixel effects. Both Munechika et al. (1993) and Braun (1992) used MS land cover classification accuracy to evaluate the quality of the hybrid images and showed that the hybrid images provided at least as good, and often significantly better, scene classification than the original scenes. The "apparent" visual improvements in the MS imagery can be seen in Figure 8.16.

The subject of data fusion will continue to grow in importance as we generate higher spectral resolution sensors of modest resolution and attempt to apply the data to spatial phenomena at or below the GIFOV of the spectrometer images. Since the limited radiometric flux density will make high-resolution spectrometers very difficult and expensive to build,

TM false color IR

Fused false color IR

Spot panchromatic

Figure 8.16 Fusion of multispectral TM data with geometrically registered and resampled SPOT data. See color plate 8.16

fusion will be an interim solution. Regrettably, most fusion techniques make the image "appear" better, since the eye keys on the increased spatial structure whether or not it is radiometrically correct. Braun (1992) points out the need for quantitative, machine-based, image fidelity metrics to aid in evaluating the relative merits of image restoration algorithms (fusion can be considered a subset of the overall image restoration problem) and points out that most of the conventional metrics (e.g., classification accuracy and RMS error) are inadequate.

8.5 KNOW YOUR CUSTOMER

This chapter has emphasized ways to combine data and prepare it for dissemination to a later stage of analysis or to final users. These last stages of the image chain are often the most neglected. The remote sensing community must be careful to recognize that if the information passed to the next stage of analysis is not in a form and in terms (e.g., units) that are readily understood, then the information effectively does not exist. Furthermore, for users to obtain full value from the information, they must be able to assess its integrity. This means that the data must be analyzed in such a fashion that the confidence or error limits on the final data must be known and the limitations (e.g., robustness) of each approach clearly delineated. In order to prepare these final output products effectively, the remote sensing specialist must become familiar with the end user's requirements and final applications.

8.6 REFERENCES

Braun, G.J. (1992). "Quantitative evaluation of six multi-spectral, multi-resolution image merger routines." M.S. thesis, Center for Imaging Science, Rochester Institute of Technology, NY.

Chavez, P.S., Sides, S.C., & Anderson, J.A. (1991). Comparison of three different methods to merge multiresolution and multispectral data: Landsat TM and SPOT panchromatic. *Photogrammetric Engineering and Remote Sensing*, Vol. 57, No. 3, pp. 295-303.

Conrac Corporation (1985). *Raster Graphics Handbook*, 2d ed., Van Nostrand Reinhold, NY.

Cornsweet, T.N. (1970). *Visual Perception*. Academic, NY.

Goodchild, M.F., & Kemp, K.K. eds. (1990). "NCGIA core curriculum." National Center for Geographic Information and Analysis, University of California.

Gordon, H.R., Brown, D.K., Brown, J.W., Evans, O.B., Broenkow, R.H.B., & Broenkow, W.W. (1983). Phytoplankton pigment concentraitons in the Middle Atlantic Bight: comparison of ship determinations and CZCS estimates. *Appl. Opt.*, Vol. 22, pp. 20-36.

Haydn, R. (1982). "Multidisciplinary investigations on HCMM data over Middle Europe and Morocco…Southern Germany and Marrakech, Morocco." NTIS No. 82N24588.

Hunt, R.W.G. (1987). *The Reproduction of Colour in Photography, Printing & Television*, 4th ed., Fountain Press, Tolworth, England.

Levine, M.S. (1985). *Vision in Man and Machine*. McGraw-Hill, NY.

MacAdam, D.L. (1942). Visual sensitivities to color differences in daylight. *JOSA*, Vol. 32, No. 5, pp. 247-274.

Munechika, C., Warnick, J.S., Salvaggio, C., & Schott, J.R. (1993). Resolution enhancement of multispectral image data to improve classification accuracy, *Photogrammetric Engineering & Remote Sensing*, Vol. 59, No. 1, pp. 67-72.

Park, S., & Schowengerdt, R.A. (1983). Image reconstruction by parametric cubic convolution. *Computer Vision, Graphics and Image Processing*, Vol. 20, No. 3.

Pradines, D. (1986). "Improved SPOT image size and multispectral resolution." Proceeding of the SPIE, Vol. 660, pp. 98-102.

Price, J.C. (1987). Combining panchromatic and multispectral imagery from dual resolution satellite instruments." *Remote Sensing of Environment*, Vol. 21, pp. 119-128.

Robertson, A.R. (1992). Color perception. *Physics Today*, pp. 24-29.

Schott, J.R. (1993). Contribution to *The Focal Encylopedia of Photography*, 3d ed, Edited by L. Stroebel and R. Zakia, Focal Press, Boston, MA.

Schowengerdt, R.A. (1983). *Techniques for Image Processing and Classification in Remote Sensing*. Academic, Orlando, FL.

Schowengerdt, R.A. (1980). Reconstruction of multispatial, multispectral image data using spatial frequency content. *Photogrammetric Engineering and Remote Sensing*, Vol. 46, No. 10, pp. 1325-34.

Society for Information Display, Palisades Institute for Research Services, Inc., 201 Varick Street, Suite 1006, NY.

Suits, G.H. (1972). "The calculation of the directional reflectance of a vegetative canopy." *Remote Sensing of Environment*, Vol. 2, No. 117.

Wyszecki, G., & Stiles, W.S. (1982). *Color Science*. Wiley, NY.

CHAPTER

9

Weak Links in the Chain

In much of our discussion to this point, we have simplified many of the equations and concepts in order to introduce and deal with the fundamental or governing equations. In doing so, we recognize that there are times when the simplifications will be less valid than desired, and that in general noise and/or error sources should become part of the equation. In analyzing an individual image or image product, we have to realize that it is the net result of the entire image chain: acquisition conditions, sensor system, processing procedures, and display or analysis. The end product will be limited by errors or degradations introduced all along the image chain and, like any chain, will be no stronger than its weakest link. This factor motivates us to understand the subtleties of the image chain, because it will provide insight into sources of limitations in the images, how much confidence to place where, and what processing techniques are viable, and may improve the potential for information extraction from the image. The same image chain analysis can show the system analyst where the weak links exist so that improvements are focused where they can do the most good. Making a strong link even stronger has little impact on the quality of a chain, whereas even the slightest improvement in the weakest link impacts the entire chain. In this section, we will focus on spatial and radiometric fidelity issues that have been simplified to this point for clarity of presentation, but which are often sources of weakness in the image chain.

We will continue to assume that most of the image chain can be approximately modeled as a linear shift invariant (LSI) system as described in Chapter 1. This is generally a good assumption for EO sensors and can typically be used, though more cautiously, for the image output links of the chain. However, as discussed in Section 9.1.1, the LSI approximations are less applicable if there are visual links in the image chain and may have no meaning for non-imaging links (e.g., landcover tables, decision trees, etc.). The reader should recognize that the LSI approximations are widely used and quite robust, but that their validity should be evaluated for any system under study.

9.1 RESOLUTION EFFECTS (SPATIAL FIDELITY)

In this section, we want to look in more detail at the factors that impact the spatial resolution of the final image and to introduce a formalism for characterizing spatial image fidelity and cascading it through the system.

9.1.1 Spatial Image Fidelity Metrics

In earlier chapters, we introduced some simple methods for characterizing the spatial resolution of imaging systems (e.g., IFOV, GIFOV, and just-resolvable line pairs in a tribar target). The detector spot-size metrics like IFOV and GIFOV provide intuitive estimates of resolution but imply a detector-limited system. The discernible contrast of a tribar measure is more satisfying in that it incorporates end-to-end system effects but only provides performance information about the highest resolution, and then in subjective form. To overcome this limitation, the concept of the contrast or square-wave transfer function (CTF) is used as a measure of image fidelity. The CTF is an objective measure and characterizes image quality over all the spatial frequencies in the image. It is based on the concept of image contrast (C) or square-wave modulation defined as (cf. Fig. 9.1):

$$C = \frac{A - B}{A + B} \tag{9.1}$$

where A and B are the observed brightness for white and dark bars in a square-wave target. For example, given a printed image of a tribar target on the ground acquired with an imaging system, we would like the image to reproduce exactly the modulation on the ground. Figure 9.1 shows an example of the radiance measured from cross sections of a ground target and the corresponding radiance measured from an image (in arbitrary units). The contrast transfer (CT) is defined as the ratio of the contrast of the input and output images. Since the absolute magnitude of the image contrast will be a result of the tone reproduction process, the contrast transfer is often normalized to unity at zero spatial frequency, either by extrapolation or by using the lowest spatial frequency as a reference. As a result, we will often be interested in relative contrast transfer (i.e., relative to zero spatial frequency) under the implicit assumption that most imaging systems reproduce low frequencies very well (which they do), and we are interested in how well higher frequencies are reproduced relative to the low-frequency "ideal." This concept of contrast reproduction as a function of the spatial frequency of line pairs in a tribar target is captured in the CTF as illustrated in Figure 9.1. The normalized contrast is measured for several linepair patterns in an image of a tribar target (note that these could be measurements from a print, transparency, digital image, or soft-copy display). The normalized contrast is then plotted as a function of the pattern frequency, and a curve fit through the data provides the CTF. The value of using the CTF rather than the frequency of the just-resolvable line pairs as a resolution metric is shown in Figure 9.2. If we assume that a relative contrast of 0.1 is where an observer could just distinguish the lines, then systems A and B illustrated in Figure 9.2 would have the same performance against the simple cutoff metric. However, using the CTF metric, it's clear that system A performs much better in the midrange frequencies and very similarly at higher frequencies. If all other considerations are equal, system A would be a much better system.

While having obvious intuitive appeal, the CTF metric is not commonly used because of the greater mathematical flexibility available through the use of the sinusoidal modulation transfer function (MTF). The MTF is defined in essentially the same fashion as the CTF, except that sinusoidally varying brightness functions are used as input (sine waves in cross section). The MTF tells how well a sinusoidally varying brightness of a given frequency will be reproduced by the imaging system. Again, the modulation (peak minus trough divided by

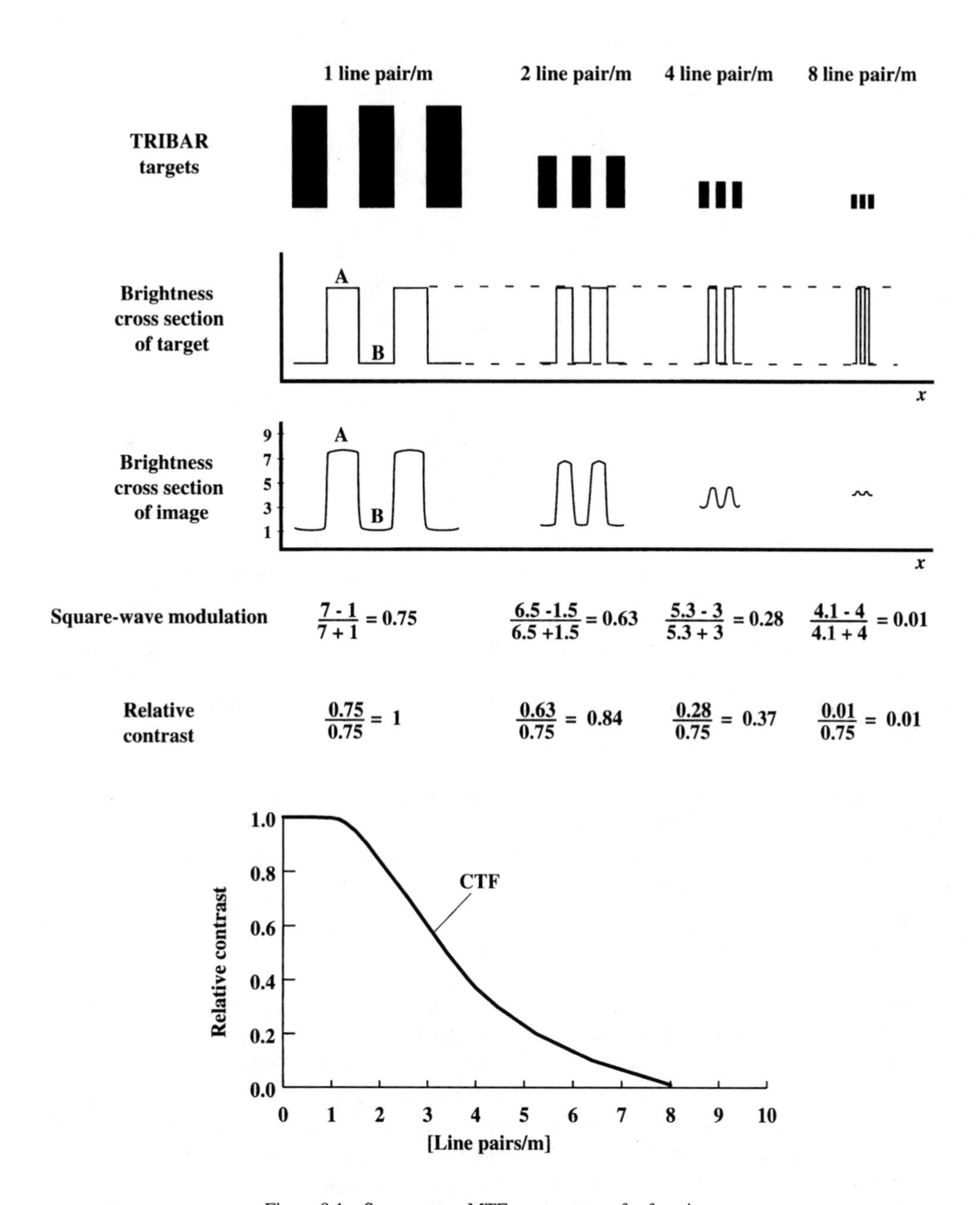

Figure 9.1 Square-wave MTF, contrast transfer function.

peak plus trough) is normalized usually by the modulation at zero frequency. In cases where low-frequency response differs significantly from unity, the modulation recorded by the system can be carefully measured and compared with the modulation in radiance of the scene to compute the actual low-frequency MTF. The MTF carries essentially the same information as the CTF in terms of its intuitive appeal, while having a number of properties that make it easy to manipulate and solve for using linear systems theory. For example, the two-dimensional inverse Fourier transform of the MTF is the system impulse response or point spread function (PSF). Recall that the PSF is the response of the system to a point source of radi-

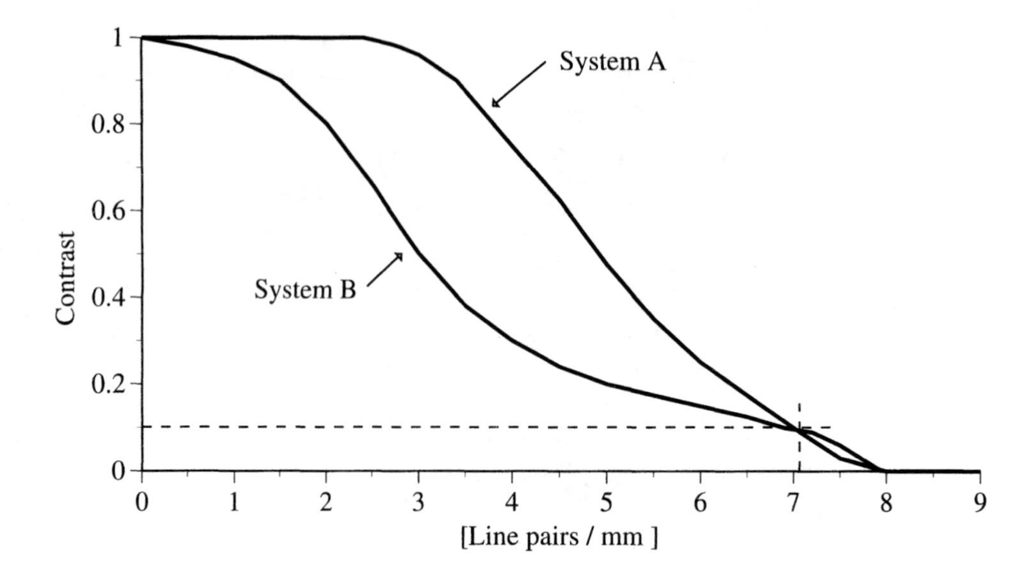

Figure 9.2 Comparison of the CTF's for two hypothetical systems.

ance (mathematically the response to a delta function). Conceptually, this is the shape of the blurred image of a point source (i.e., the blur spot). The PSF is also often used as a measure of resolution by measuring its full width at half the maximum (FWHM), as illustrated in Figure 9.3. When the systems PSF is projected onto the ground, its FWHM is often referred to as the *ground spot* or *ground sample size*.

For simplicity in this treatment, we will often assume that the system can be treated as separable in x and y. Gaskill (1978) shows that in this restricted case the one-dimensional impulse response along the x axis corresponds to the one-dimensional inverse Fourier transform of the MTF along the u axis, i.e.,

$$h(x,0) = \mathcal{F}^{-1}(H(u,0)) \tag{9.2}$$

where $h(x,0)$ is the profile of the PSF along the y axis, $H(u,0)$ is the profile of the MTF along the u axis (i.e., the profiles along corresponding axes form one-dimensional transform pairs), and the variables and nomenclature are those introduced in Chapter 7. In most cases, the separability assumption is a good approximation for many EO systems. However, a full two-dimensional treatment should be undertaken when detailed quantitative results are required. Gaskill (1978) points out that if the PSF is circularly symmetric, the Hankel transform can be used to compute the PSF from a radial slice through the system MTF.

Coltman (1954) showed that the MTF and CTF can be calculated from each other according to:

$$\begin{aligned} \mathrm{CTF}(u) &= \frac{4}{\pi}\left[\mathrm{MTF}(u) - \frac{\mathrm{MTF}(3u)}{3} + \frac{\mathrm{MTF}(5u)}{5} + \frac{\mathrm{MTF}(7u)}{7} + \cdots\right] \\ \mathrm{MTF}(u) &= \frac{\pi}{4}\left[\mathrm{CTF}(u) + \frac{\mathrm{CTF}(3u)}{3} - \frac{\mathrm{CTF}(5u)}{5} + \frac{\mathrm{CTF}(7u)}{7} + \cdots\right] \end{aligned} \tag{9.3}$$

where u is spatial frequency [cycles/mm]. This yields a CTF approximately 10% higher than the MTF in the midfrequencies, with comparable values at the highest and lowest frequencies.

Rather than solving for the MTF by studying many sine wave inputs, it is often more

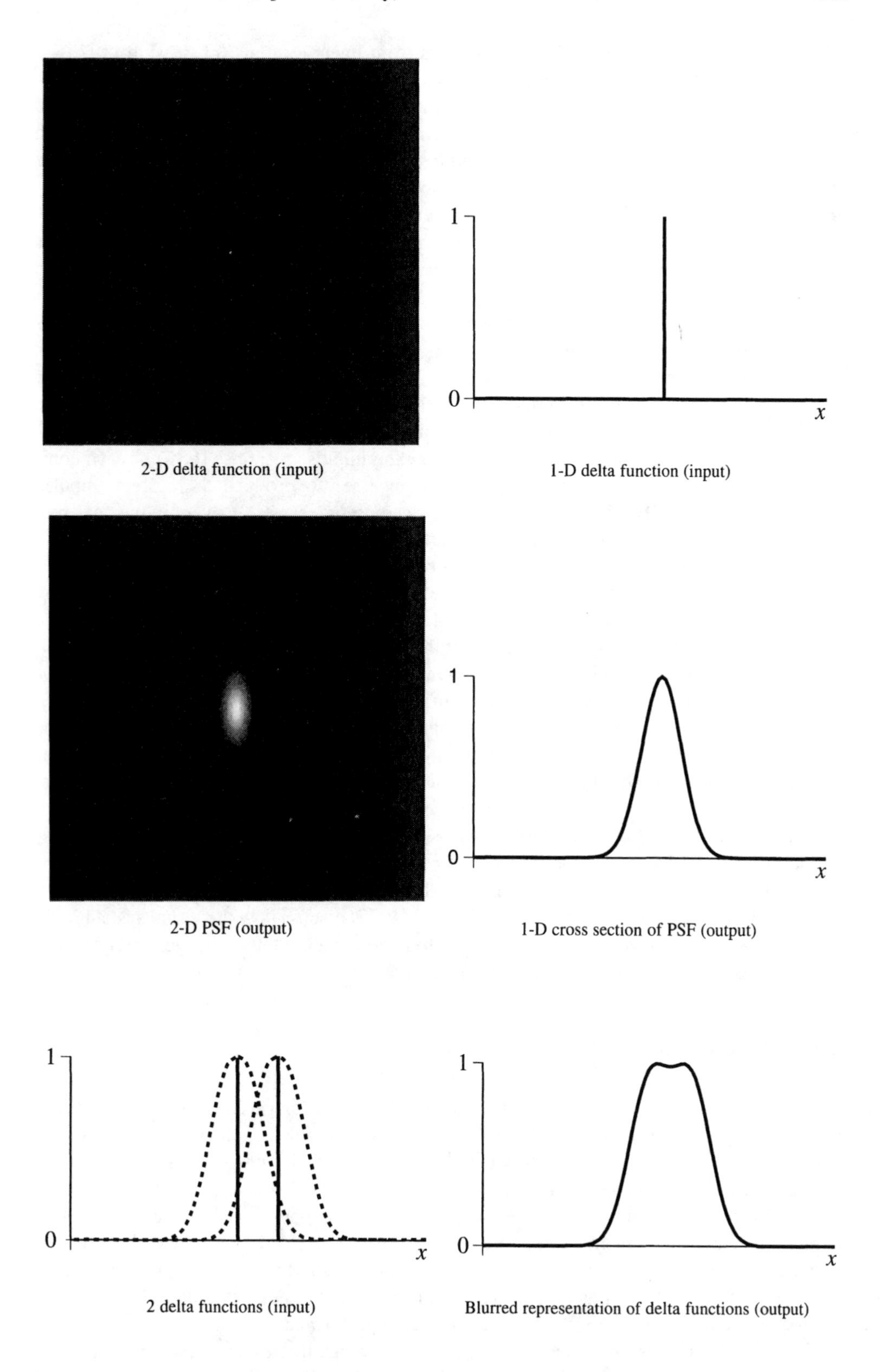

Figure 9.3 Point spread function (PSF) as a measure of resolution.

convenient to take advantage of some concepts from linear systems (cf. Gaskill (1978) or Goodman (1968) for a more thorough treatment of these concepts). Recall that a step function or edge can be thought of as being constructed by summing many sine waves of varying frequency and amplitude. A perfect edge contains all frequencies. If we introduce a knife edge (i.e., a step function) as the input to an optical system, then the image along the perpendicular to the edge is the edge spread function. The derivative of the edge spread function with respect to position is the line spread function (LSF) in that direction. Furthermore, one line through the 2-D MTF is the Fourier transform of the LSF, as illustrated in Figure 9.4. [Note that for separable systems LSF (x) = PSF (x)] The interrelationship of the image quality metrics plus the ease of combining the effects of components of the imaging system on the overall system MTF makes this the most common metric for describing the spatial resolution of imaging systems. The cascading principles of linear systems tell us that the MTF's of the components multiply to yield the system MTF (strictly speaking, this is only valid for incoherent systems where phase effects can be ignored). Recall from Section 7.1.4 that the convolution theorem states that multiplication in the frequency domain is equivalent to convolution in the spatial domain. Thus, if we know the impulse response (PSF) of each component of a system, they could be convolved together to produce the system impulse response function (PSF). We will utilize this cascading process extensively in the next section to study the impact of the image chain on the overall system MTF.

It is important to recognize that in many cases a human image analyst is the final consumer of the image. From one perspective, we are preparing the image for consumption by the analyst. However, I think it is often useful to consider the analyst as another link in the image chain. From either perspective, it is of the utmost importance to recognize the spatial frequency response of the visual system. Figure 9.5 shows an estimate of the contrast sensitivity of the visual system with the maximum response occurring characteristically upward from the lowest frequencies. The contrast sensitivity can be thought of as being analogous to the MTF. The visual spatial frequency response can be considered to be a weighting factor that determines the importance of the spatial frequency to visual analysis. For example, in the simplest case, it is clear that the midrange frequencies are the most important, and that the loss of the lowest frequencies will have less impact than degradation of the midrange frequencies. This is due to the adaptive nature of the visual system, which causes it to work with relative rather than absolute gray levels. This reduces the importance of absolute gray levels for visual perception. Granger and Cupery (1972) discuss how the visual system MTF and the image system MTF can be combined to generate a combined image quality metric for visual assessment called the *subjective quality factor* (SQF).

9.1.2 System MTF

In this section, we want to assess how various links in the image chain impact the spatial image fidelity of the final image. In particular, we want to evaluate how the MTF concept can be used to characterize and assess the spatial characteristics of the image chain. From this standpoint, each link of the system is treated as a potential source of image degradation due to blurring. We generally assume that the world is composed of objects of interest that exhibit all spatial frequencies, and an ideal system would perfectly reproduce all those frequencies in the final image (i.e., have an MTF of unity at all observable frequencies). Each link can then be thought of as a filter that attenuates some frequencies of interest as characterized by the MTF of the link. The overall system performance is then the product of the individual MTF's. From the image chain standpoint, we are interested in the performance of the entire chain so that we understand the characteristics of the final image. We also are interested in determining the weak links so that we know where to concentrate our efforts at improving the system.

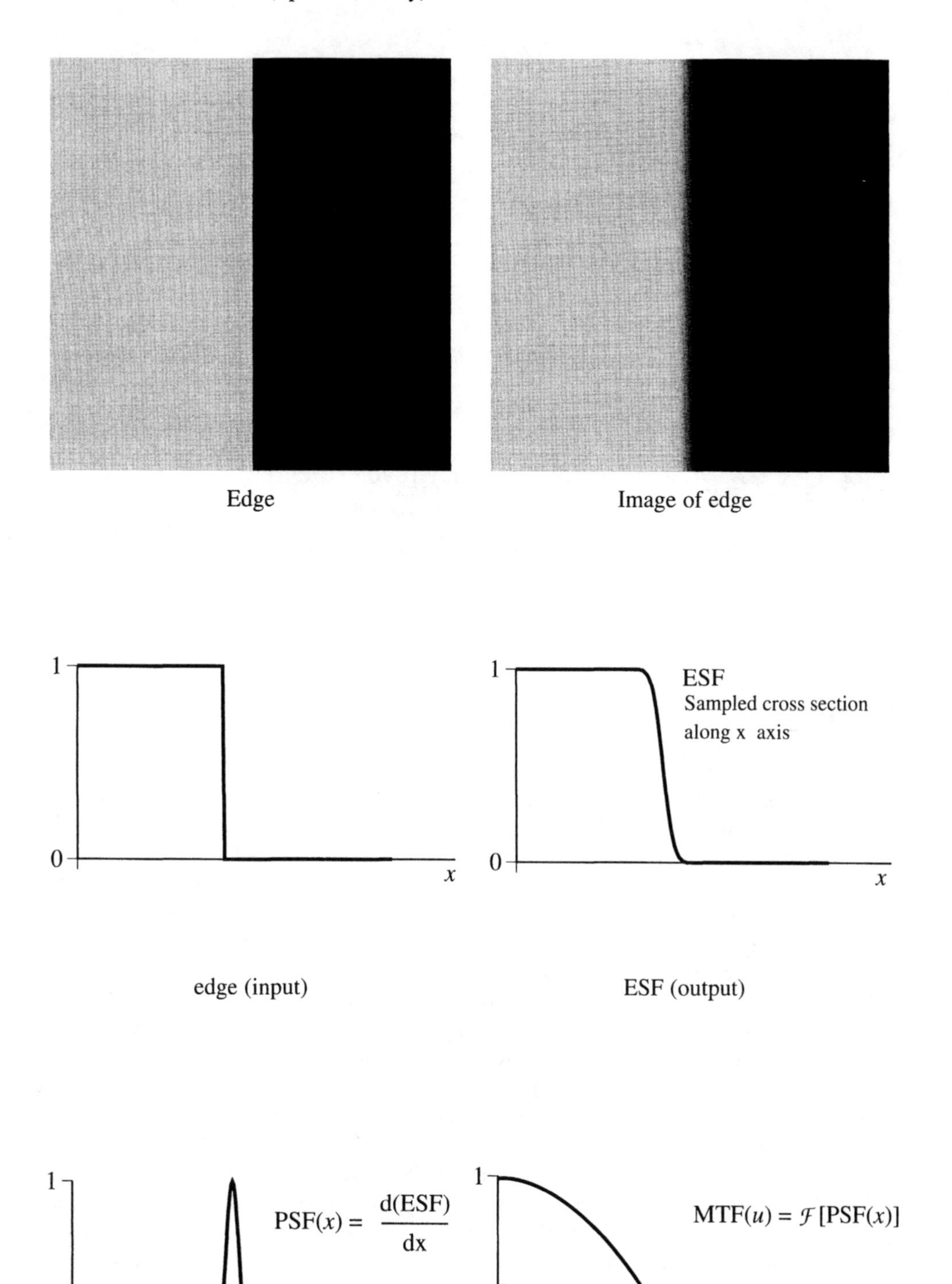

Figure 9.4 Relationship between the point spread function (PSF), edge spread function (ESF), and the modulation transfer function (MTF) of a system.

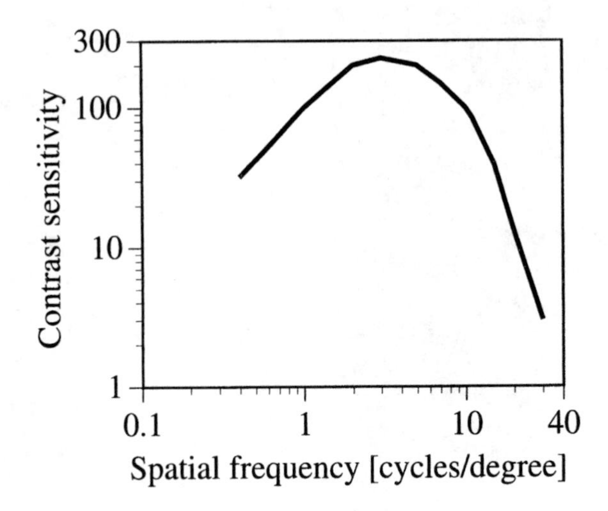

Figure 9.5 Contrast sensitivity of the human visual system. (Adapted from Hall and Hall, 1977.)

9.1.2.1 Example System

The next several sections (9.1.2.2 to 9.1.2.10) will examine the impact of various links in the image chain on the MTF of the entire chain. In each subsection, some example calculations will be included for an example system. For this treatment, we want to take a hypothetical system of modest complexity and examine how the components might impact the final system performance. The components of an actual system would vary in complexity, but the approach would remain fundamentally the same. The image chain of interest in this case is illustrated in Figure 9.6. We are analyzing a single-channel airborne pushbroom imager that utilizes an onboard digital recording system. The system specifications of interest are listed in Table 9.1. The image is geometrically rectified on the ground and printed as a transparent photographic image using a thermal dye diffusion film writer. From the point of view of the image analyst, we might tend to emphasize the image fidelity on the final transparency (i.e., at the end of the chain). However, the entire chain must be understood if we begin to look at tradeoffs to improve the image fidelity. For example, can we use an alternative film writer to improve the image fidelity or by using the same device write out a smaller portion of the image and achieve better fidelity over a smaller area? If we ship the imager overseas to be used with a different image processing system, what kind of performance can be expected? In this case, at a minimum we need to know the performance of the airborne imager as a unit, the imager with the digital ground processing taken together, and the film writer together with the prior links (i.e., the entire chain).

9.1.2.2 Atmospheric Effects on Spatial Image Fidelity

The first link that must be considered is distortion to the wavefront reaching the optical system due to atmospheric turbulence and scattering of flux from adjacent pixels into the pixel of interest (i.e., effects from type I photons in Fig. 3.1). The distortion due to atmospheric turbulence results from density fluctuations in the lower atmosphere causing variations in the

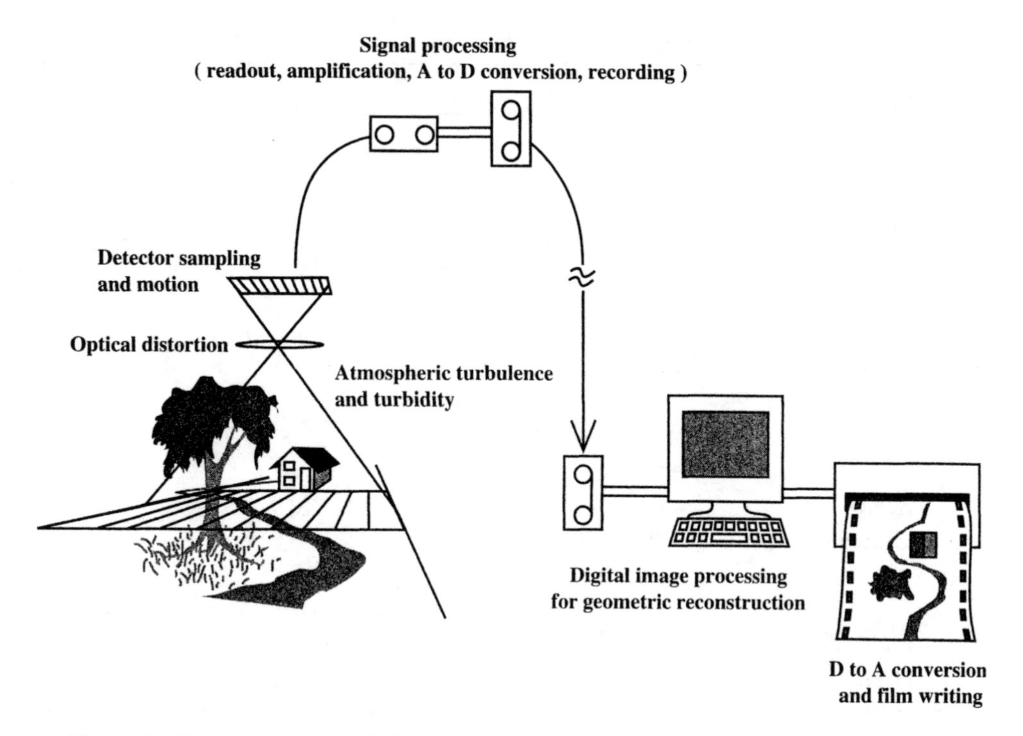

Figure 9.6 Components of a hypothetical imaging system for use in a spatial image quality assessment.

index of refraction of the air that in turn induce random deviations in the wavefront. For our purposes, we can normally think of the aggregate of these statistical fluctuations as a blur. Rees (1990) suggests that for vertical viewing, the angular dispersion will be of the order of $3 \cdot 10^{-6}$ radians. Or, assuming the effective height of the atmosphere (scale height) for turbulence to be 6666 m, the blur spot due solely to turbulence will be the product of the angular dispersion and the scale height. Thus, even when viewing down from space, the PSF due to atmospheric turbulence will be quite small (of the order of 0.02 m in diameter). This should not imply that turbulence can always be neglected. It can become very important for long paths in the lower atmosphere such as encountered by a forward-looking infrared (FLIR) system ((cf. Beland, 1993) for a more thorough treatment).

The reduction in contrast due to the scattering of flux from adjacent pixels into the line of sight is a function of the turbidity of the atmosphere. Kaufman (1982) shows that this adjacency effect can be used to generate an MTF of the atmosphere using atmospheric radiative transport models. His results are in agreement with Pearce (1977), who used Monte Carlo simulations of the paths of the individual photons through radiative transport models and found that for low-to-moderate turbidity levels, the MTF is decreased but remains fairly constant over a wide range of spatial frequencies representing target scales from a few to a few hundred meters (cf. Fig. 9.7). Thus, over a wide range of conditions, the dominant effect would appear as an overall reduction in scene contrast and would often be accounted for to the first order by mean-level corrections for atmospheric transmission and path radiance. However, some residual effects persist and will be more pronounced under hazy conditions and near-large high-contrast boundaries, i.e., large land-water interfaces. This effect does not fall off rapidly at higher frequencies, and often goes unnoticed since it does not appreciably

Table 9.1 Specifications of Example System

Description	Parameter	Value
F number	$F\#$	5.6
Focal length	f	0.56 m
Lens diameter (pupil)	d	0.1 m
Bandpass	$\Delta\lambda$	0.4 - 0.7 μm
Nominal wavelength	λ_{test}	0.55 μm
Side dimension of detector		
Element	ℓ	15 μm
Nominal flying height	H	2000 m
Number of pixels in array	N	2048
Scale	s	$2.8 \cdot 10^{-4}$
Ground instantaneous field of view	GIFOV	0.05 m
Exposure time	t	$1 \cdot 10^{-3}$ sec
Field of view	FOV	±1.6°
Ground speed	W	90 mph

General: The system is an airborne push-broom scanner employing a silicon-based linear array detector and conventional refractive optics.

impact the appearance of the image. However, it can impact precise radiometric calculations and should not be neglected under the conditions indicated above (i.e., large high-contrast boundaries).

The units of measure of spatial resolution or spatial frequency response will often be different for the different components of a system. To cascade the MTF's together, we need to convert all information to common units such as cycles per unit distance on the ground, which are used here. Pearce's data are already in this form. To convert the MTF due to atmospheric turbulance into similar units, we assume that the point spread function due to averaging of the randomly refracted beam can be represented as a Gaussian blur spot with a full width at half maximum of 0.02 m. The MTF is estimated by computing the Fourier transform of the PSF. Fig. 9.8, shows how this would roll off the atmospheric MTF but only at very high frequencies. The overall MTF of the atmosphere would have the general shape indicated by the atmospheric MTF curves shown in Figure 9.8 and is obtained by multiplying the MTF due to turbulence by the MTF due to turbidity. This is the general shape of this function, but it will depend on the viewing conditions and the composition and temperature of the atmosphere.

9.1.2.3 Optical Effects on Spatial Image Fidelity

Assessment of Figure 9.8 tells us that, from a spatial image fidelity standpoint, the image should reach an airborne or space-based system largely undistorted by the atmosphere except

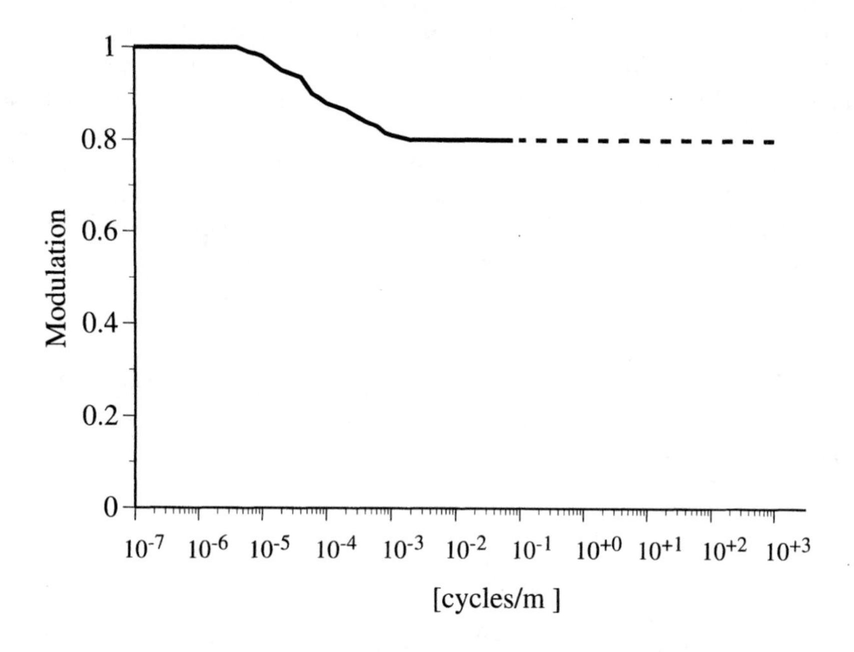

Figure 9.7 Modulation transfer function of the atmosphere due to turbidity effects (after Pearce, 1977).

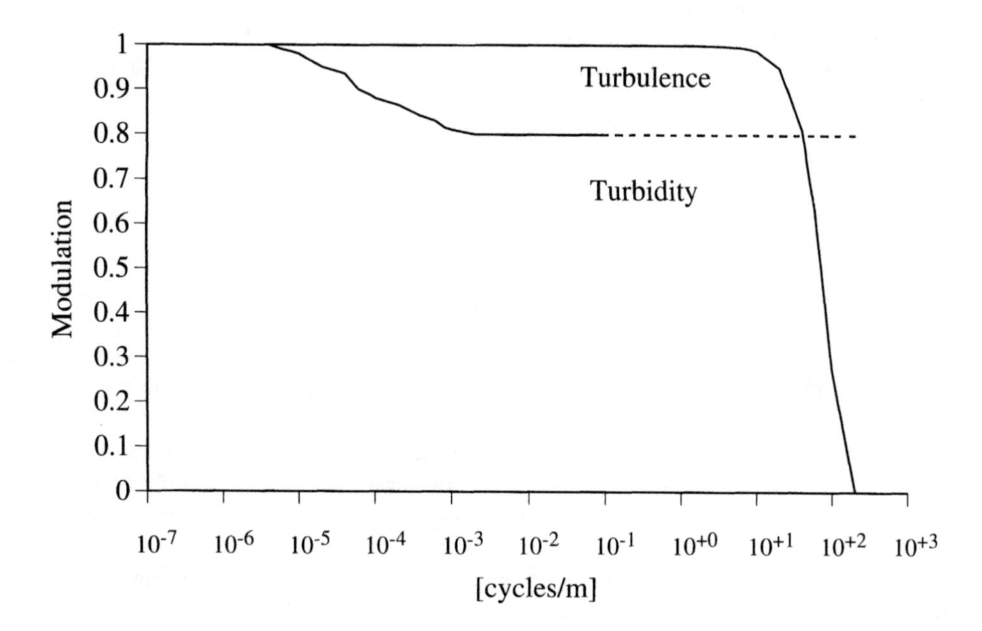

Figure 9.8 MTF of the atmosphere due to the combined effects of turbidity and turbulence.

at extremely high frequencies (spot sizes of the order of 2 cm). The next link to be considered is the optical system, which for convenience we will treat as a unit including all windows, mirrors, and lens elements. Most on-axis remote sensing optical systems can achieve near diffraction-limited performance, making it relatively easy to calculate their expected performance to first order. For wider fields of view, aberrations (coma in particular) can become important, and a more detailed assessment of the lens design is required to determine expected performance. For a diffraction-limited incoherent system, the MTF is proportional to the correlation of the pupil function of the optical system with itself (i.e., the autocorrelation of the pupil function). For a simple optical system with an unobstructed circular pupil, the autocorrelation of the pupil function looks approximately like a circularly symmetrical trianglular function. The high-frequency cutoff for incoherent irradiance is given by

$$u_{max} = \frac{d}{\lambda f} = \frac{1}{\lambda F\#} \tag{9.4}$$

where u_{max} is the maximum spatial frequency [cycles/mm], λ is the wavelength, d is the aperture diameter, f is the focal length, and $F\#$ is the f number of the optical system. The PSF of the optics, obtained by taking the Fourier transform, is the well-known Airy disc pattern (cf. Fig. 9.9). The diameter of the Airy disc (to the first zero) is given by

$$D = 2.44\frac{f\lambda}{d} = 2.44\lambda F\# \tag{9.5}$$

where D is the diameter of the Airy disc. The central peak in the Airy disc (area enclosed by the first zero) contains 84% of the energy from a point source. The radius of the Airy disc is commonly used as a measure of the resolving power of optical elements according to the Rayleigh criterion. This is the minimum distance on the focal plane that must separate the images of two point sources (e.g., stars) so that they appear as visually distinct (i.e., the peak of one Airy disc will fall on the first zero of the second; cf. Hecht, 1987). The MTF of the optical system is typically computed and expressed in terms of cycles per millimeter at the focal plane. This can be converted to cycles per unit distance on the ground simply by multiplying the spatial frequency by the scale factor, i.e.,

$$(u') = (u \cdot s) = \left(u \cdot \frac{f}{H}\right) \tag{9.6}$$

where u' is the spatial frequency expressed in cycles per unit length on the ground, u is spatial frequency expressed in cycles per unit length (typically mm) on the focal plane, s is the scale, f is the focal length, and H is the range of the sensor to the target (i.e., height above ground level for a vertical viewing system). The MTF due to diffraction limited optics for a clear aperture F5.6 system flown at 2000 m is shown in Figure 9.10 ($d = 0.1$ m, $\lambda = 0.55$ μm).

As discussed in Chapter 5, many remote sensing systems use Cassegrain style optics that have a central obscuration. The pupil function for a Cassegrain system has the general appearance shown in Figure 9.9. The MTF, obtained from the autocorrelation of the pupil function, and the PSF (Fourier transform of the MTF) for a typical Cassegrain system are also shown in Figure 9.9. Note that for comparison, the diameter of the primary and the effective focal length in the Cassegrain are the same as in the clear aperture example. The MTF of the Cassegrain has a characteristic hipped shape showing more filtering of the low- to midrange frequencies and slightly better performance than the unobscured system at higher frequencies (note, however, that the cutoff frequency would be the same for both optics). As a result, the central peak in the PSF of the Cassegrain is slightly narrower, with more of the power shifted to the first lobe.

For the hypothetical clear aperture system, we see from Figure 9.10 that the optics have approximately the same spatial image fidelity as the atmosphere for this very high-resolution system (projection of the Airy disc onto the ground yields a spot size of 0.03 m). For any system of more modest resolution, the high-frequency atmospheric effect can usually be considered negligible.

9.1.2.4 Effect of Sampling by the Detector on Spatial Image Fidelity

In most system designs, the physical size of the detector is the limiting spatial factor in the optical radiation collection, because its dimensions are often larger than the blur spot (PSF) of the optics. To characterize the impact of the detector on spatial image quality, we need to briefly review sampling theory since the EO detector samples the scene. The sampling theorem states that the highest spatial frequency that can be reproduced by an ideal sampling system [i.e., where the sampling is performed by a series of unit area delta functions with spacing Δx referred to as $1/\Delta x$ comb $(x/\Delta x)$] is

$$u_{max} = \frac{1}{2\Delta x} \tag{9.7}$$

where u_{max} is the Nyquist frequency and Δx is the distance between sample centers. The MTF in an idealized system would be unity for frequencies less than the Nyquist frequency and zero for frequencies higher than Nyquist. Note that this assumes that the image does not contain frequencies higher than the Nyquist frequency. Higher frequencies must be filtered prior to sampling to avoid aliasing. Figure 9.11 shows a simplified one-dimensional representation of the sampling process. For linear array EO systems, Δx is the distance between detector centers. For scanning systems, Δx is the effective distance the detector has moved between samples (i.e., the distance the image has moved across the focal plane between sample acquisitions). In a push-broom system, for example, if we define the along-track direction to be y, then the sample interval Δy will be defined by the period of the sample acquisition. In the across-track direction x, the sampling interval Δx will be defined by the distance between detector centers in the array. In practice, the detectors do not sample at a point, but occupy some finite dimension. In this case, the sampling process can be represented as a convolution of the detector impulse response with the image which is then sampled by the comb function, i.e.,

$$g(x) = [f(x) * h(x)] \cdot \frac{1}{\Delta x} \mathrm{comb}\left(\frac{x}{\Delta x}\right) \tag{9.8}$$

where $g(x)$ is the sampled image, $f(x)$ is the input image brightness, $h(x)$ is the impulse response of the detector (which is normalized to unit magnitude), $1/\Delta x$ comb $(x/\Delta x)$ is the sampling function, which is unity for integer values of $x/\Delta x$ and 0 elsewhere. A rectangular detector can be treated as a rectangular averaging process in each dimension having a one-dimensional impulse response defined by

$$h(x) = \frac{1}{|\ell|} \mathrm{rect}\left(\frac{x - x_0}{\ell}\right) \tag{9.9}$$

where rect$((x\text{-}x_0)/\ell)$ is a function equal to unity over the range $x_0\text{-}\ell/2$ to $x_0+\ell/2$, equal to 1/2 at $x_0 \pm \ell/2$, and zero elsewhere. The MTF due to the detector blurring is then given by the Fourier transform of the impulse response (PSF) of the detector; in one dimension

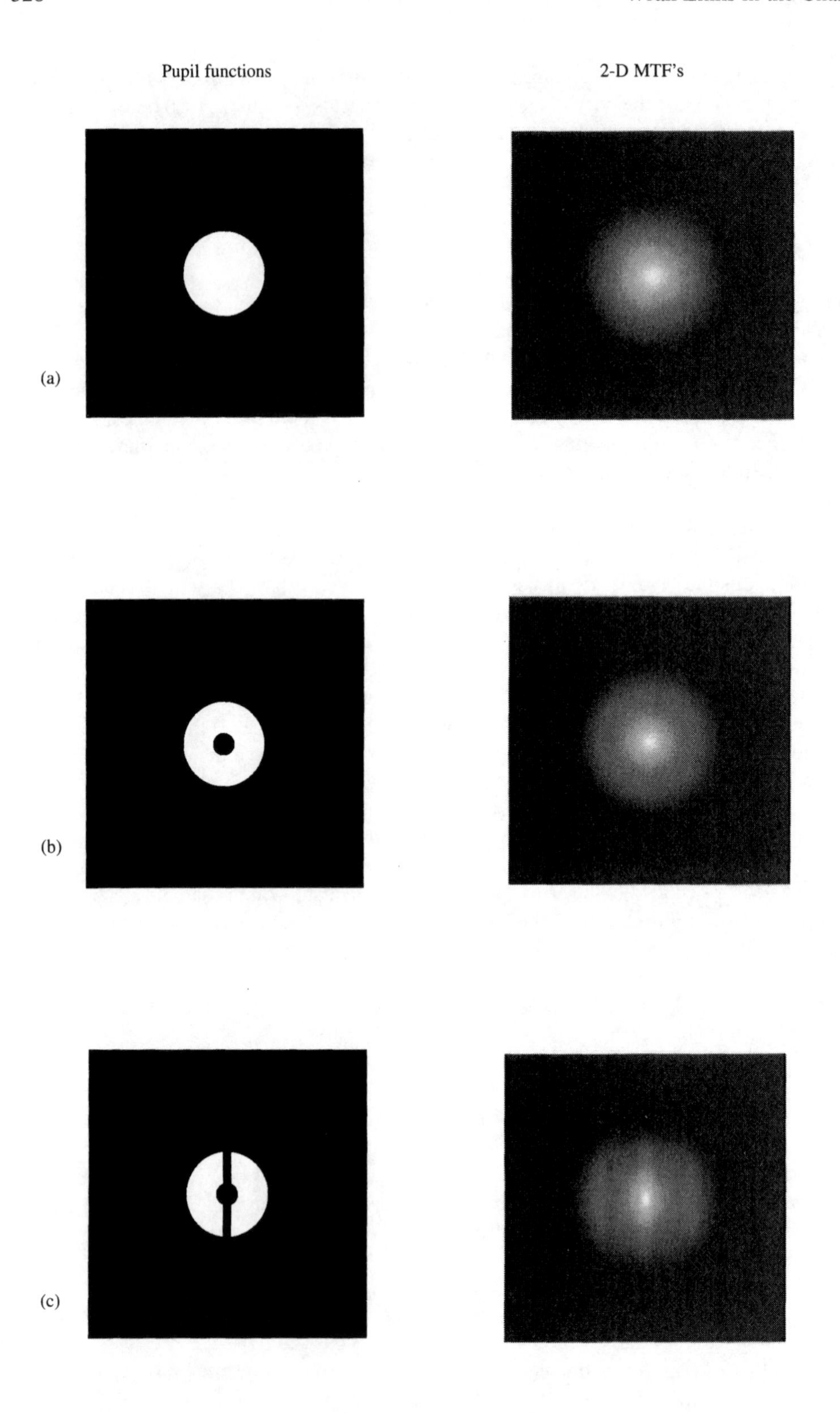

Figure 9.9 Pupil functions, 2-D MTF's, 1-D MTF's, and PSF's for some typical optical systems.

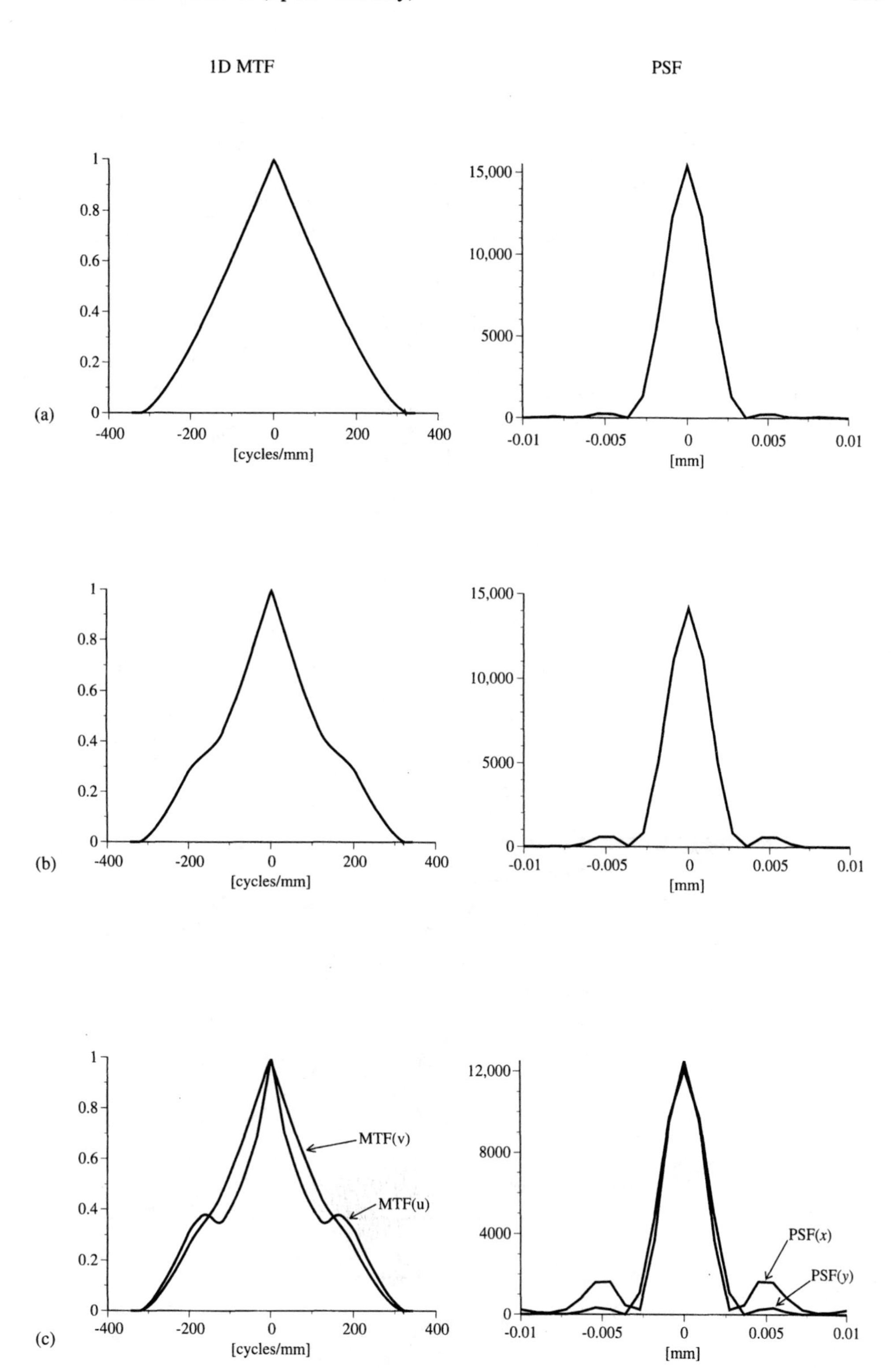

Figure 9.9 (continued)

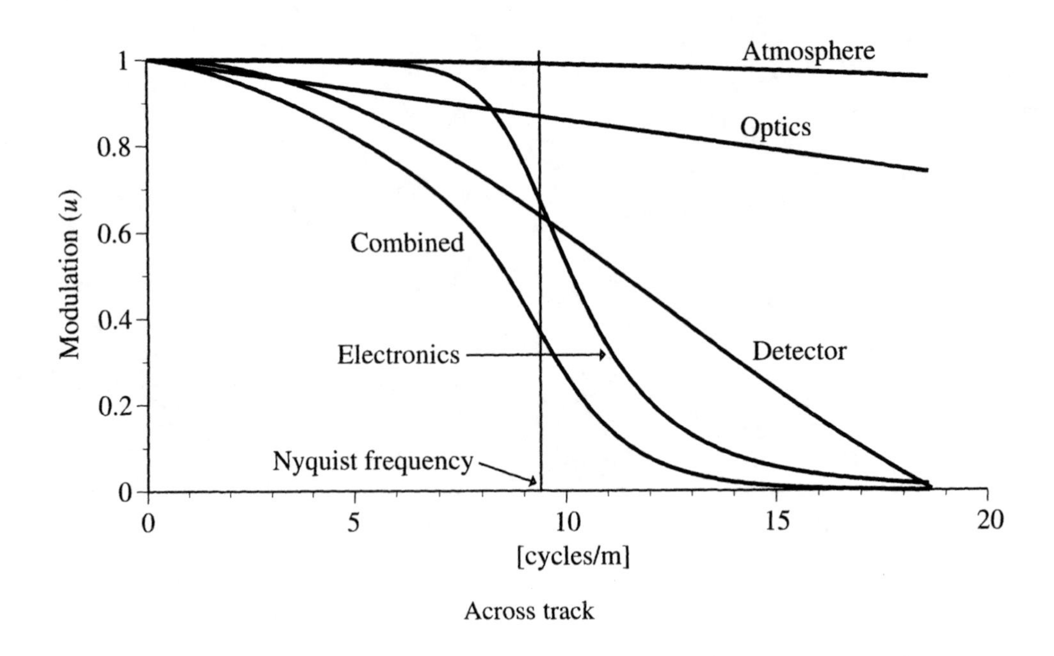

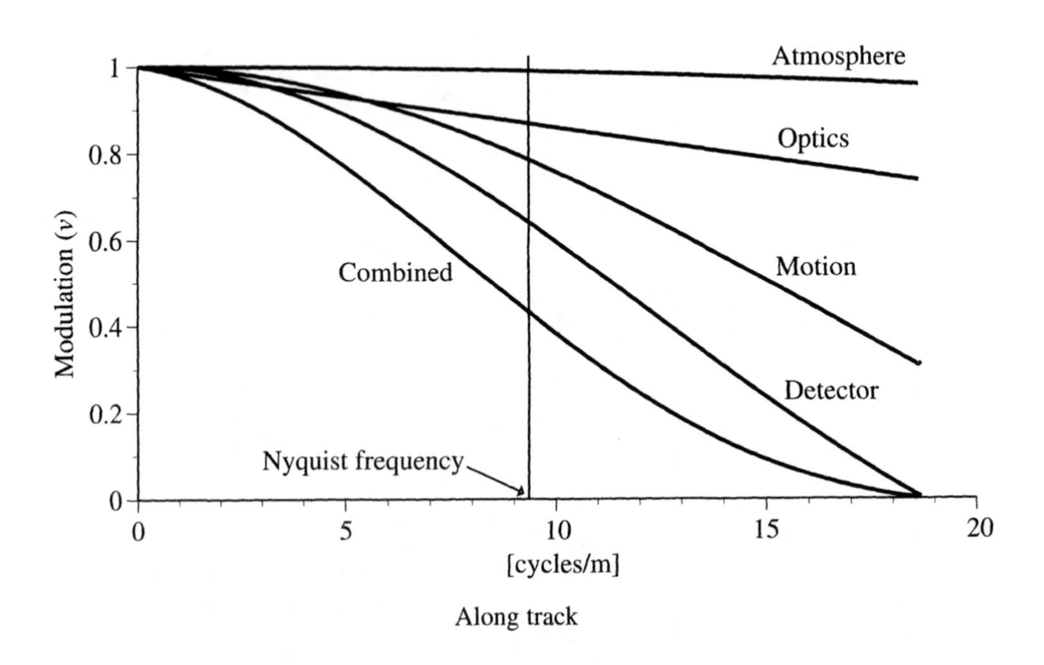

Figure 9.10 Cascaded MTF's of the atmosphere, optics, detector, detector motion, and electronics for a hypothetical system.

$$\mathrm{MTF}(u) = \mathcal{F}\left[\frac{1}{|\ell|}\mathrm{rect}\left(\frac{x}{\ell}\right)\right] = \frac{\sin(\pi \ell u)}{\pi \ell u} \equiv \mathrm{sinc}\ \ell u \tag{9.10}$$

In many cases, ℓ is approximately equal to the detector spacing (Δx) in the across-track and the along-track directions. This results from using adjacent square detectors. The shape of the MTF for this case is shown in Figure 9.11. In some cases, particularly in the across-track direction for push-broom imagers or in either dimension for staring arrays, the effective detector size does not completely fill the space between detector centers. In this case, ℓ is less than Δx (more closely approximating the ideal delta function), and the MTF is improved as shown in Figure 9.11 for the case where $\ell = 0.5\Delta x$. The downside of this is that fewer photons are detected by low-fill factor detectors (i.e., detectors whose sensitive area is less than their physical size), thus reducing the signal to noise.

It is important to recognize that to this point we have been able to assume that the PSF's of the links and, therefore, the MTF's were symmetric in x and y. In many cases, due to detector shape or differential sampling schemes in the x and y directions, the PSF's and MTF's are asymmetric. While ideally we would always treat the full two-dimensional functions in all our analysis, the mathematics and graphic representation of the results can become cumbersome. For this reason it is common practice to just present the x and y or u and v components of the PSF's and MTF's for simplicity. The user must evaluate the characteristics of the particular system under study to determine when more detailed assessment is warranted (e.g., when the system response is not separable in x and y).

For our example case, we are going to assume that we have fully filled square detectors on 15-μm centers and that they are sampled in the along-track direction at time intervals that correspond to 15 μm of image motion, so that we should have symmetry in x and y. Using a linear array of 15-μm square detectors in our hypothetical push-broom system, we would expect to be resolution-limited by the detector. The detector dimension (ℓ = 15 μm) is approximately twice the Airy disc diameter ($D = 2.44\ \lambda F\# = 2.44 \cdot 0.55\ \mu\mathrm{m} \cdot 5.6 = 7.5\ \mu\mathrm{m}$). The impact on the MTF scaled to ground units is shown in Figure 9.10. We see that for this system the detector limits the high-frequency response. However, the MTF of the optics and detector are both significant contributors to the reduced fidelity at higher frequency.

9.1.2.5 Motion Effects on Spatial Image Fidelity

Another factor that must be considered in the detector sampling process is the time interval over which the sensor is integrating the flux. Because the image is moving across the detector during this process, there is additional blurring in the motion direction (i.e., along-track for a push broom or across track for a scanner). This process can be represented by an impulse response function in the form of a rect. The rect would have a base as long as the distance the image moves during the integration time. For example, in the system we are discussing, if the aircraft ground speed (W) is 90 mph (40 m/sec) and the integration time (t) is $1.0 \cdot 10^{-3}$ sec, then the PSF in the y direction due to motion is

$$\mathrm{PSF}(Y) = \frac{1}{|Wt|}\mathrm{rect}\left(\frac{Y}{Wt}\right) \tag{9.11}$$

and the PSF is expressed in ground coordinates. The MTF in terms of spatial frequencies at the image plane (v) is

$$\mathrm{MTF}(v) = \mathcal{F}\left[\frac{1}{|sWt|}\mathrm{rect}\left(\frac{y}{sWt}\right)\right] = \mathrm{sinc}(vsWt) \tag{9.12}$$

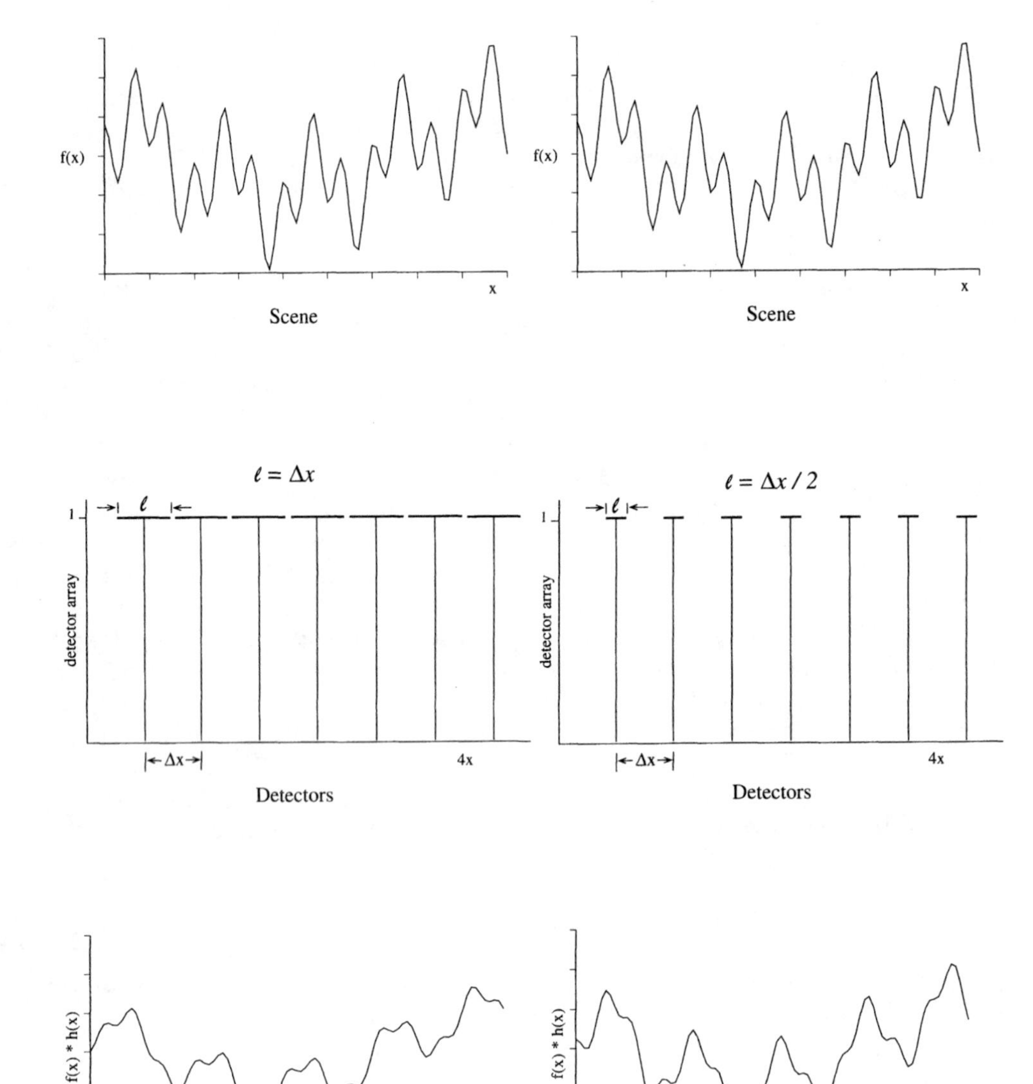

Figure 9.11 Detector sampling concepts and MTF's.

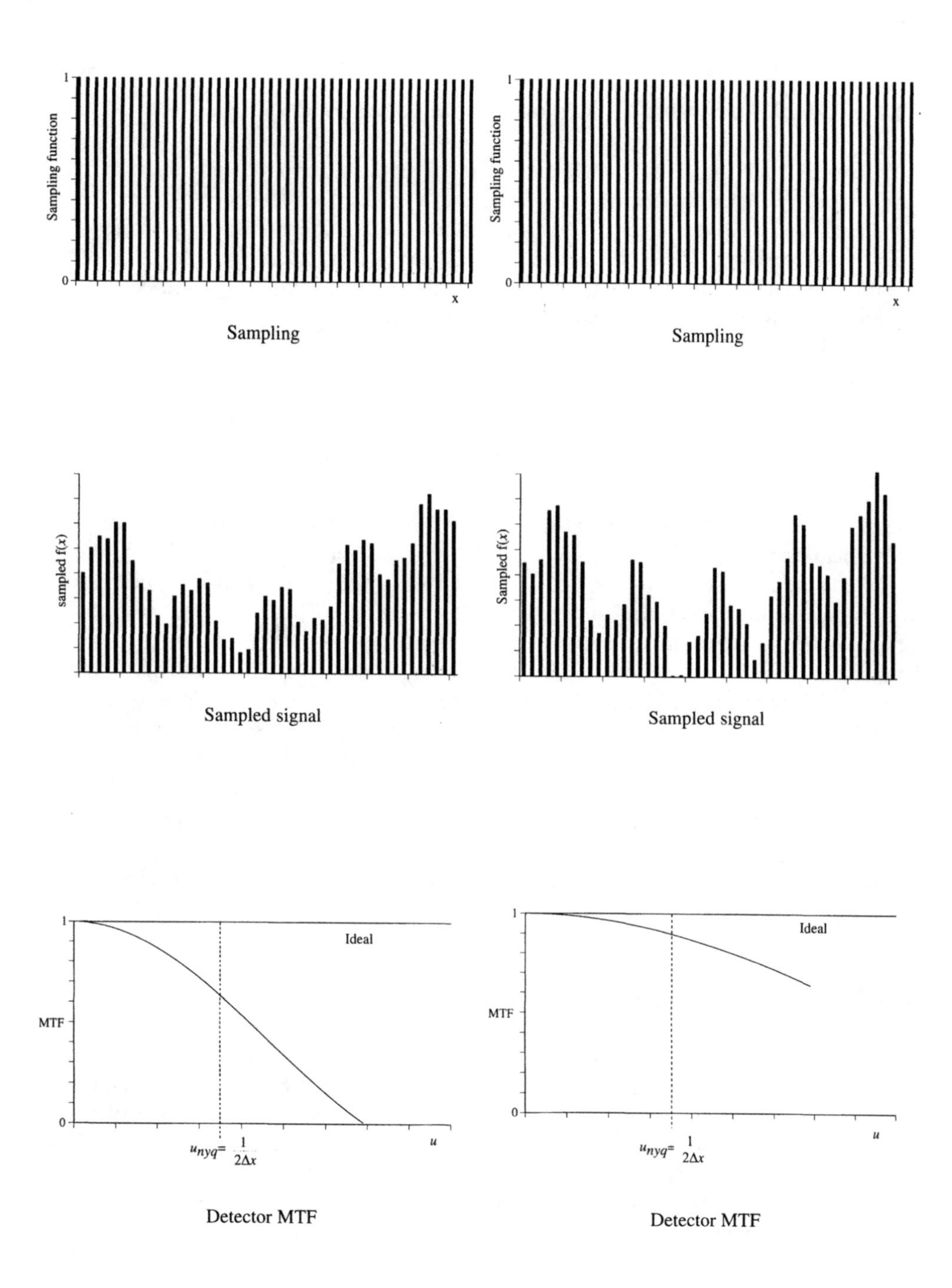

Figure 9.11 (continued)

where s is the scale factor, $sWt = 11$ μm, and $Wt = 0.04$ m. This has been scaled to ground coordinates for plotting in Figure 9.10. Note that in this case the blur due to motion is quite substantial: nearly as large as the blur due to the size of the detector.

9.1.2.6 Signal Processing Effects on Spatial Image Fidelity

In the across-track direction, a different set of phenomena introduce incremental signal degradation. These include blurring of the signal due to less-than-perfect charge transfer out of the linear array and less-than-perfect amplification and digitizing of the frequencies in the electronic processing. These effects can be combined and represented by the MTF of the electronic readout and processing (MTF_e). Since this is an electronic signal, it is normally expressed initially in terms of temporal frequencies. The MTF_e can be converted to spatial frequencies on the focal plane based on the detector readout rate and the image scale if required, i.e.,

$$u = u_e\left[\frac{\text{cycles}}{\text{sec}}\right]\cdot t_c\left[\frac{\text{sec}}{\text{pixel}}\right]\cdot\frac{1}{\Delta x}\left[\frac{\text{pixels}}{\text{mm}}\right] = u_e\,{}^{t_c}\!/_{\Delta x}\left[\frac{\text{cycles}}{\text{mm}}\right] \tag{9.13}$$

where u_e [cycles/sec] is the temporal frequency of the electronics associated with detector readout and electronic processing including the A-to-D converter, t_c [sec/pixel] is the clocking time of the readout electronics, and Δx [mm/pixel] is the spacing between pixel centers on the array. In the hypothetical system we are considering, each line of data is transferred simultaneously to a data storage buffer employing First In, First Out (FIFO) technology and then read out at a slower rate to minimize degradation of the MTF. The electronic amplifier is designed to cut off at the Nyquist limit to reduce aliasing. The final MTF due to the readout and processing electronics is shown in Figure 9.12 in terms of the spatial frequency at the focal plane, and in Figure 9.10 in terms of spatial frequency on the ground. Note that the spatial frequency effects due to the electronics adversely impact the MTF, but in this case they are not as significant as the motion effects in the y dimension. This is shown in Figure 9.12 where the two-dimensional point-spread function due to the size of the detector, motion in y, and electronic readout, electronic processing, and sampling in x are shown. This asymmetric PSF is typical of most electro-optical systems.

9.1.2.7 Sensor System MTF

The final MTF of the imaging system, including the atmosphere, is shown in Figure 9.10. This is obtained by cascading (multiplying) the other curves in Figure 9.10 to obtain the effective sensor MTF with respect to $Y(\text{MTF}(v'))$ and with respect to $X(\text{MTF}(u'))$. An analysis of these curves shows that the spatial frequency response is dominated by the size of the detector, but in this case the motion, electronics, and optical effects are all significant. This is an example of a system where the GIFOV is not a particularly good indicator of resolution. If we wanted a simpler way of characterizing the resolution than the PSF or MTF, we could use the full width at half maximum (FWHM) of the PSF scaled to the ground (the GSS) as shown in Figure 9.13. Alternatively, Slater (1980) suggests use of the effective instantaneous field of view (EIFOV), which is derived from the frequency where the MTF is 50% of maximum $u_{0.5}$:

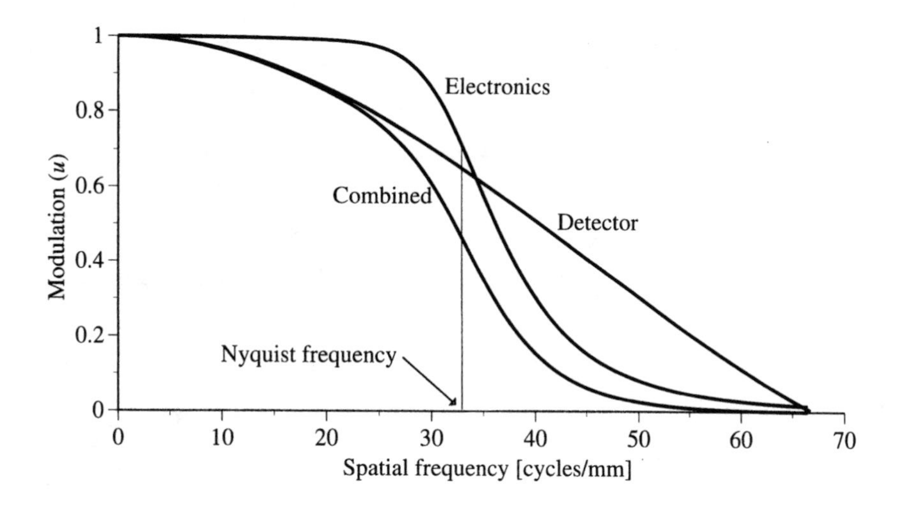

Across track

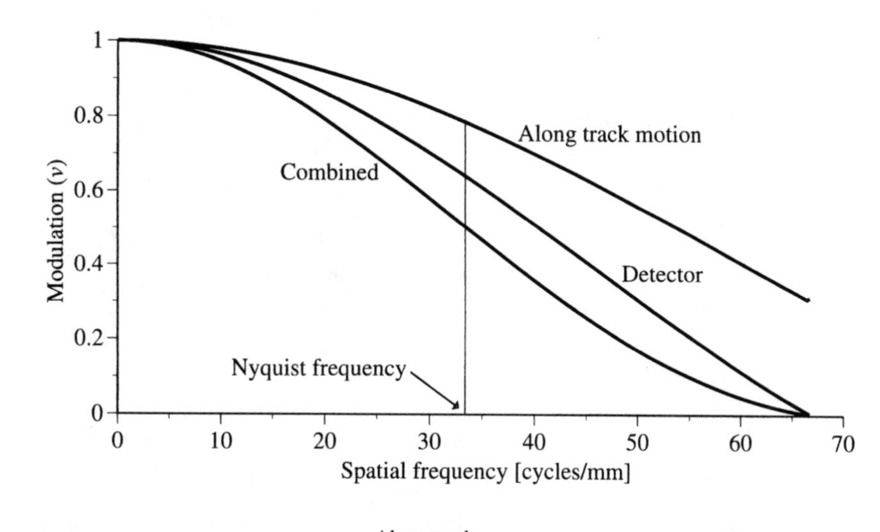

Along track

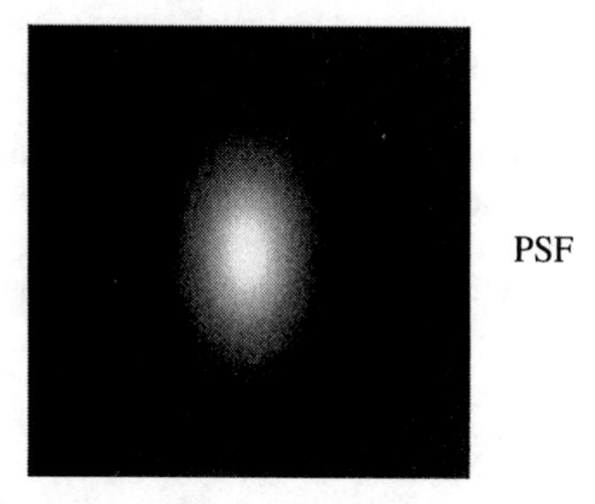

PSF

Figure 9.12 Effects of motion and detector readout on image MTF and 2D PSF.

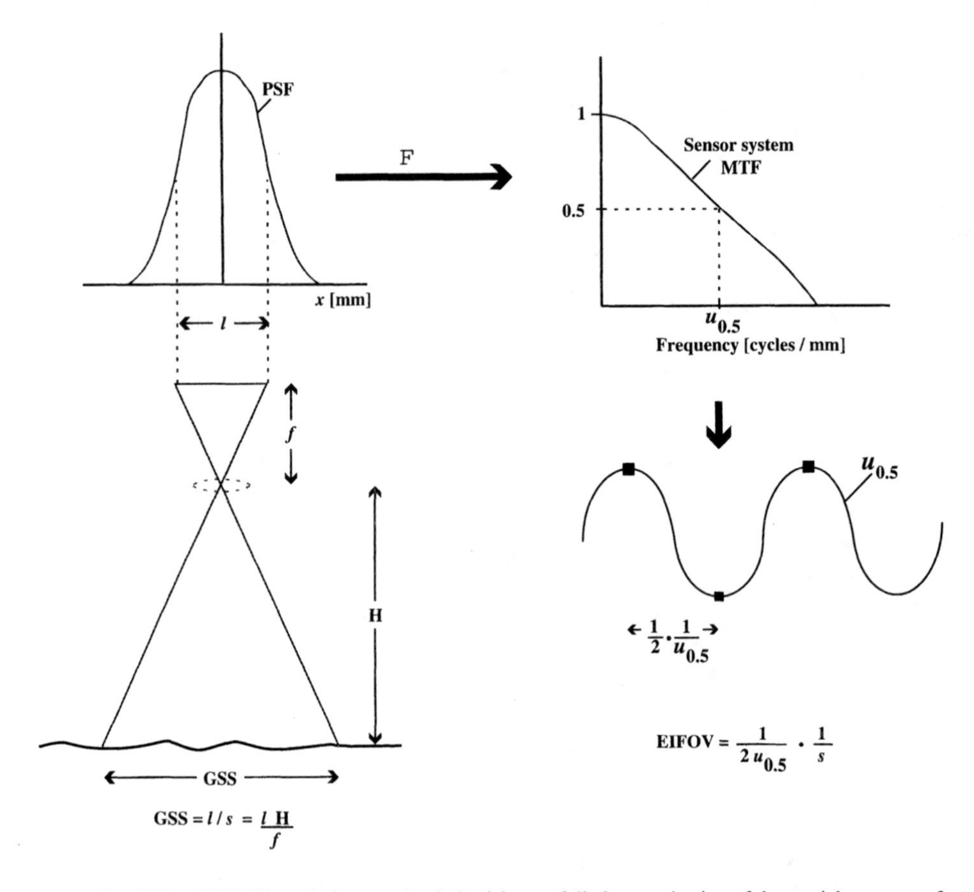

Figure 9.13 GSS and EIFOV resolution metrics derived from a full characterization of the spatial response of an imager.

$$\mathrm{EIFOV} = \frac{1}{2u_{0.5}s} \tag{9.14}$$

where s is the scale factor. Note that both the GSS and the EIFOV can and should be defined in both the along-track and across-track directions because of the asymmetry of the EO imaging process.

These full-system resolution metrics are much better estimates of system performance than simple measures such as GIFOV. However, for many systems the detector remains by far the dominant factor in determining resolution, such that the GIFOV is a good estimate of performance, and for this reason remains widely used because of its simplicity. It should be recognized that while all the resolution metrics introduced are attempting to characterize common performance characteristics and will result in similar values, they should only be directly compared to performance metrics derived in a similar fashion (i.e., EIFOV to EIFOV, not GIFOV to GSS).

9.1.2.8 Effect of Image Processing on Spatial Image Fidelity

For assessment and comparison of sensor performance, analysis of the spatial frequency performance of most systems stops here. However, from an end user's standpoint, the image the sensor "sees" typically is not the final product, and other factors must also be considered. In a well-designed image chain, one would expect that little or no incremental degradation would take place. However, this is not always the case, and only a full image chain assessment will adequately characterize the final image fidelity.

The next major step in the chain is digital image processing. All possible processing operations to which an image may be subjected are far beyond the scope of this effort, so we will treat them in the aggregate for simplicity and mention only two of the most common processes explicitly. Many images, particularly satellite images, are transmitted to ground receiving stations. This process can degrade the spatial frequency content of the image. If the data are still in an analog form, this degradation will be a function of the bandwidth and performance characteristics of the transmission channel. However, most data are in digital form such that degradation due to transmission is much less likely. With today's error correction schemes, digital data can be transmitted with essentially no degradation to the signal.

The most common problem associated with image transmission is that the growing volume of data due to increased spatial resolution, coverage, and number of spectral bands may exceed the bandwidth of the downlink systems. To compensate for this, various forms of data compression are employed. When lossless methods do not provide enough compression, lossy approaches are often employed. The most primitive form of lossy compression is to transmit only some of the data. For example, in a multispectral system, if we know the data are to be used for water quality studies, and three near infrared bands are collected by the sensor, we might choose to only downlink one of the NIR channels. This saves channel capacity so that all the visible channels could be transmitted in a lossless fashion. Since nearly all the information content for water quality is in the visible region and the single IR channel would provide input for atmospheric correction, this might be considered an acceptable "lossy" approach. This approach may be among the most acceptable methods of lossy data compression as we evolve to sensor systems with tens or hundreds of channels. On the other hand, in many cases we may not know the eventual use of the data, and so we want to preserve as much of the full data set as possible. In this case, more conventional forms of lossy data compression are employed that attempt to minimize the distortion in the imagery. Rabbani and Jones (1991) describe a number of these approaches. In general, these lossy compression methods tend to degrade high frequencies in an image and may introduce artifacts. In general, the dominant effect of compression schemes can often be characterized to first order as a filter impacting the spatial frequencies in the image (i.e., the effect of the compression can be estimated by an MTF). There are cases where the impact of the compression on the image is much more difficult to characterize, and changes can be introduced of which the user is unaware and may inadvertently interpret as real phenomena. For this reason, the user is cautioned to evaluate fully the characteristics of a compression algorithm before analyzing compressed data. For our purposes in this discussion, we will limit our considerations to the impact of the compression as a filter on the spatial frequencies by assuming that the compressor has a characterizable MTF.

The other image processing step applied to essentially all images is some form of image reconstruction. This process may occur at a central ground station or distribution center or it may be performed by the individual analyst. In general, it is the step where the geometric sampling of the image is taken into account. It may involve corrections for roll, pitch, and yaw in an airborne system, earth rotation effects in a satellite system, or registration to a ground coordinate system. For some systems (e.g., radar) this can be a very involved process; for others it may involve only simple line shifting and/or image rotation. In all cases, there is some impact on the image fidelity which can be characterized by the MTF of the reconstruction (or resampling) process. In addition, a number of other image processing steps may be performed prior to display or printing of the image (e.g., radiometric corrections, requantization of gray levels, noise suppressions algorithms, etc.). For convenience, we will treat all of these effects simultaneously by referring to the cascaded product of their MTF as the *processing MTF* and recognize that it may be asymmetric with respect to the along- and across-track directions. We should also point out that in many cases it is difficult to develop an analytical form for the MTF of these processes, and so empirical methods are employed.

For the hypothetical example system we are considering, there is no data compression, and the primary image reconstruction step would be shifting of lines to correct for aircraft roll. The imaging time is assumed to be short compared to the pitch-and-yaw periods of the aircraft, so that only roll distortion is significant. Furthermore, the frequency of the aircraft roll is low enough that only line-to-line motion (not pixel-to-pixel) need be considered. The line shifting and resampling will result in a slight degradation that can be minimized by the use of cubic convolution resampling, as described in Section 8.3.1.2.

9.1.2.9 Effect of Image Output Devices on MTF

The fidelity of the image presented to the analyst may be further impacted by the display process itself. If the image is interpreted from a monitor, the MTF of the monitor must be evaluated under the viewing conditions to be used for analysis. If the image is to be interpreted from hard copy, then the MTF of the film writing system must be evaluated. Let's take this second case for our example. If the image were to be viewed unaided, then we would need to factor the angular resolution of the visual system into our final calculations. In this case, we will assume that the analyst will be using optical magnification to assist in viewing the images and that this will introduce no significant degradation to the image fidelity. Thus the MTF of the film writer will be the only factor under consideration.

First consider the case where 2048 pixels of each row of the digital scene are written with a continuous tone film writer onto an 8- x 10-inch transparency. We will assume that 7- x 9-inches are actually available for the image. The 2048 pixels in each line would normally be written across the width of the image using the 7 inches addressable, and 2633 (2048•9/7) lines could be written using the 9 inches addressable along the length of the film. However, the electronics that drive the film writer can introduce spatial frequency degradation. More importantly, the individual spots (pixels) will not be perfectly reproduced due to spreading of the illuminating source in the film writer and diffusion in the film. Overall, the film writer might have an MTF such as illustrated in Figure 9.14. When scaled to the ground, this yields the MTF shown in Figure 9.15. This would indicate that for this overall system the film writer was the limiting factor. On the other hand, if we were willing to look at only half of the image width at a time written over the whole output image area of the film writer, then the MTF of the film writer would not change, but the image written to it would not contain the high frequencies that are so heavily filtered by the film writer. This comes about because when we scale the film writer MTF to the ground, the spatial frequencies are shifted upward by a factor of two when the image scale is doubled. This significantly reduces the adverse effects of the film writer on image fidelity, as shown in Figure 9.15. Obviously other tradeoffs are involved in determining whether writing half a line of the image at a time is an acceptable procedure.

9.1.2.10 MTF of the Entire Image Chain

Figures 9.10, 9.12, and 9.15 show how the various links in the imaging chain can be ana-

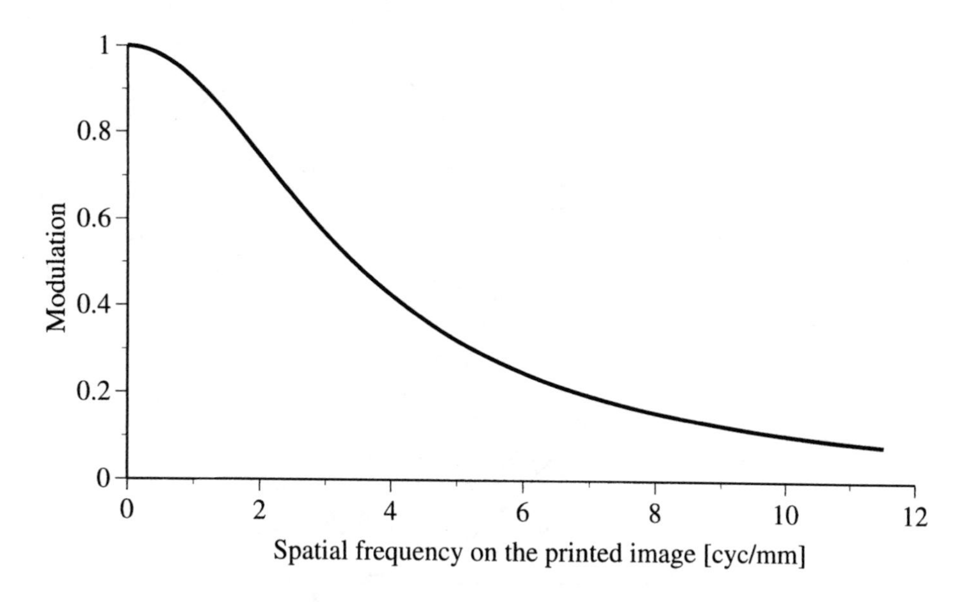

Figure 9.14 MTF of a hypothetical film writer.

lyzed, not only to see the net effect on spatial frequencies but also to identify the weak links in the image chain. This list will indicate where efforts to improve system performance would be most productive. In the case treated here, the film writer is by far the weakest link, and improvements elsewhere in the chain would have had almost no impact on final image fidelity. By simply changing the way the film was written, a significant improvement in spatial image fidelity could be achieved. However, the productivity on the unit is reduced by four, and materials costs quadrupled. This solution would be only an interim step until a higher-performance film writer could be brought on line. This section has discussed how image chain analysis of the spatial frequency response of a system can be used in image fidelity assessment and system performance analysis. It is also a powerful tool in end-to-end system design, not only in evaluation of expected system performance but also to avoid over-engineering components that are not limiting factors in system performance.

9.1.3 Measurement of, and Correction for, MTF Effects

In the previous section, we described how the MTF can be used to characterize the spatial image fidelity of components and entire image chains. In this section we will briefly discuss how the MTF of a system can be measured. It is always best to assess the overall spatial fidelity of a system after final assembly or at least as far down the assembly stage as is physically possible to ensure that the integrated system matches design expectations.

A variety of approaches may be used to perform this type of evaluation. For our purposes we will consider only two simple methods to illustrate the type of analysis that can be performed. The first method uses step functions (edges) where the brightness drops at a straight edge from a uniform high brightness level to a uniform low brightness. The second method will involve the use of a line source.

To begin, let's consider the case of evaluating the MTF of a staring sensor system using a two-dimensional array. We assume that like most remote sensing systems, the sensor has a fixed focus at infinity, so that we would like to evaluate the MTF for the image of an object

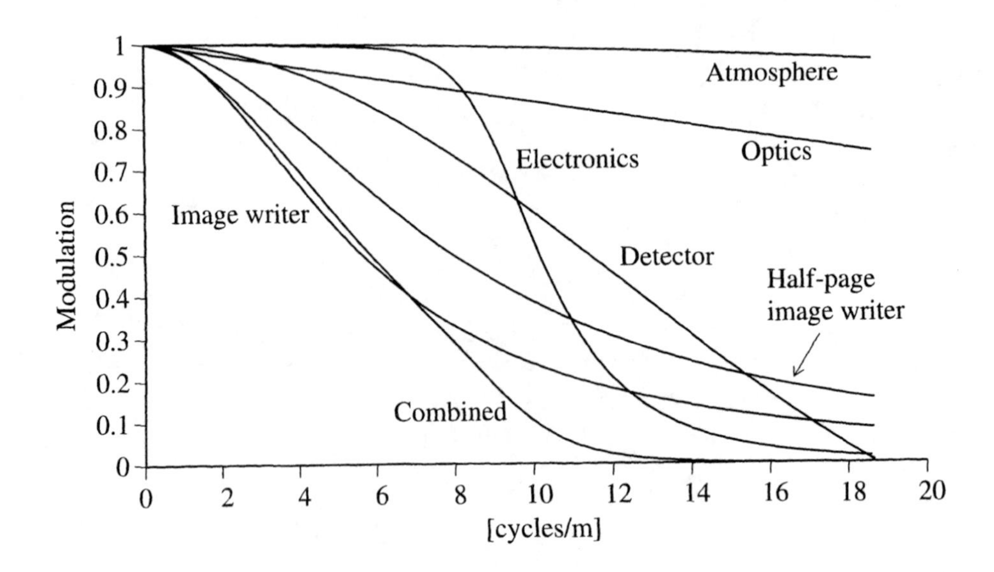

Across track

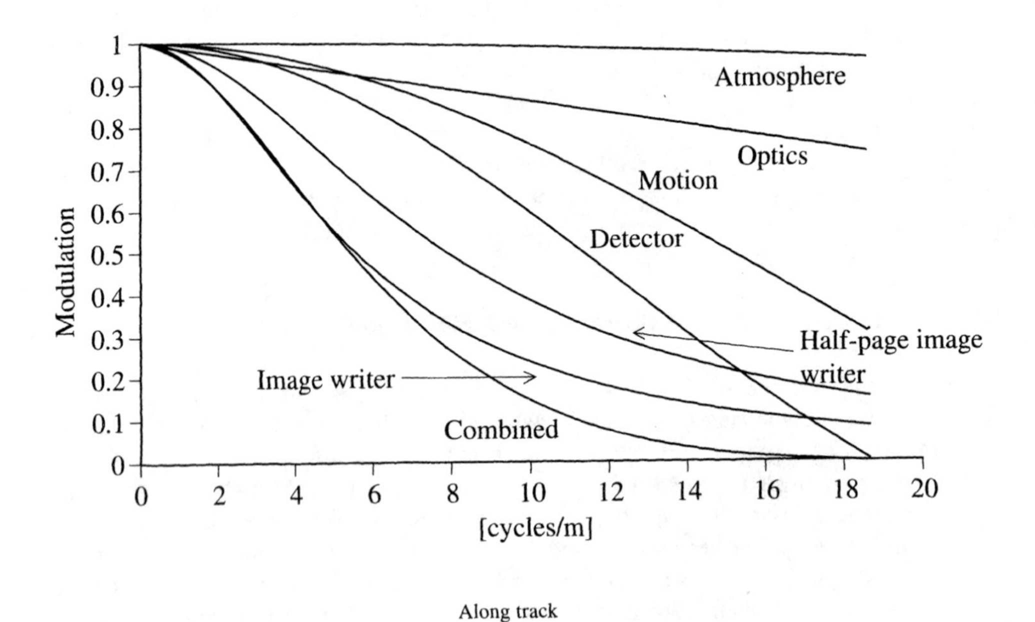

Along track

Figure 9.15 MTF of the overall image system.

that appears to be at infinity (i.e., the rays incident on the first components of the sensors are approximately parallel). This is achieved in the laboratory with a collimator as shown in Figure 9.16. An object at the focal plane of the collimator will appear to be located at infinity to a sensor looking into the collimator. If we introduce a knife edge as the object, then an image of the knife edge should be a step function. In fact, it is a blurred step and can be analyzed as described in Section 9.1.1 to yield the MTF of the entire system along one direction (cf. Fig. 9.4). In this case, the measured MTF would be the product of the sensor MTF and the test system (i.e., the collimator). The MTF of the collimator can be treated as known from previous tests (e.g., of distant point sources), so the sensor MTF is then

$$MTF_s = \frac{MTF_m}{MTF_c} \tag{9.15}$$

where MTF_s is the sensor MTF along the axis perpendicular to edge, MTF_m is the measured MTF, and MTF_c is the MTF of the collimator. This process must be repeated several times at different edge locations to avoid any artifacts due to the location of the projection of the edge onto the individual detectors in the array (phase effects) and the results averaged. This is most often done with a drive stage that steps the edge across the focal plane in step sizes a small fraction of the detector size. This method is sensitive to noise effects because of the derivative operation employed in generating the LSF. Wherever possible, averaging is employed to minimize the adverse impact of noise [e.g., long, narrow sampling slits oriented perpendicular to the edge are scanned over the edge to generate an estimate of the edge spread function (ESF)]. This entire process must, of course, be repeated for the perpendicular direction so that any differences between the along-track and across-track direction are determined. To avoid singularities, the MTF of the collimator should be significantly larger than zero for all spatial frequencies relevant to the sensor. The entrance aperture of the collimator must be at least as large as the sensor entrance aperture (larger for scanning systems). Ideally, an end-to-end sensor test such as this should include all sensor components, including signal conditioning, A-to-D conversion, and recording where applicable. Clearly, the signal should also be sampled along the processing chain so that the component MTF's can also be determined.

Warnick (1990) used a more elegant procedure suggested by Foos and Fintel (1990) to determine the MTF of a digital imaging sensor. This method used a test setup similar to that in Figure 9.16. However, the edge was introduced at an angle to the sensor array, as shown in Figure 9.17. The pixel locations in the vicinity of the edge in the digital image were projected onto a perpendicular to the edge and their brightnesses plotted as a function of the distance from the edge as determined by the projected value. The resultant sampled edge spread function was analyzed using the method of Tatian (1965) to generate the MTF of the measurement set up along the direction perpendicular to the edge. This method has the distinct advantage of reducing the number of experimental measurements required, since phase effects are included in the single measurement. This approach also has the advantage of generating an ESF with reduced noise because many points are included in the estimate, and aliasing can be reduced by sampling at numerous points along the ESF. The final sensor MTF calculation must still use Eq. (9.15) to remove any effects due to the collimator.

This same general approach can be used to compute the MTF of push-broom and scanning sensors. The MTF in the along-track direction can be difficult to measure unless a precise translation stage is used to scan the edge across the sensor. For many sensors, it may be possible to perform much of this end-to-end evaluation using a test range. For example, in calibrating the RIT airborne line scanner, we often locate the scanner on a test pad on the roof of one building such that the scan mirror sweeps perpendicular to the roof of a second building. With the scanner located on a rotating stage, panoramic images can be acquired, as shown in Figure 9.18. Building edges can then be used as step functions to generate the sen-

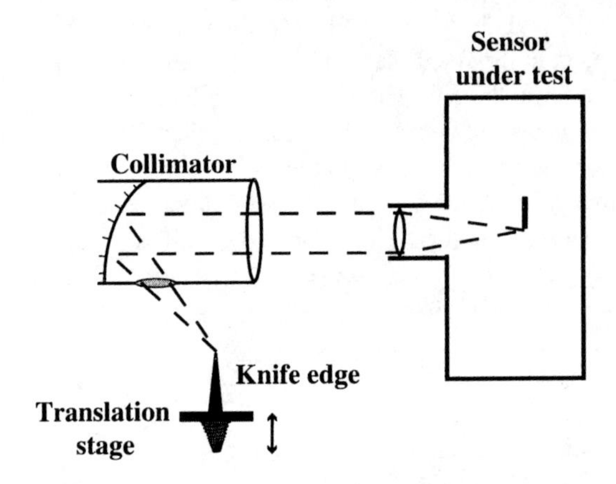

Figure 9.16 Test setup for measuring the MTF of an imaging sensor.

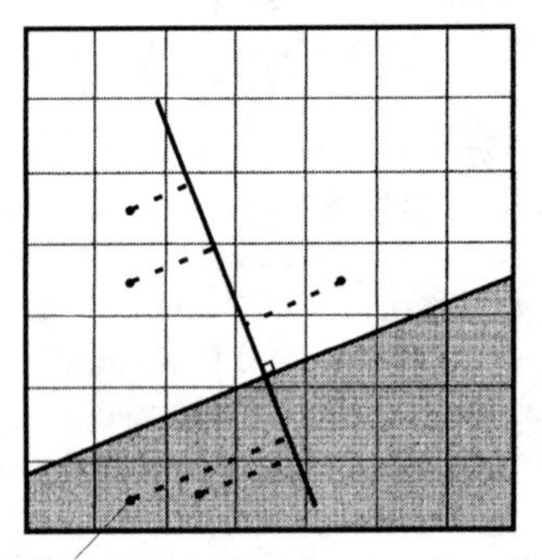

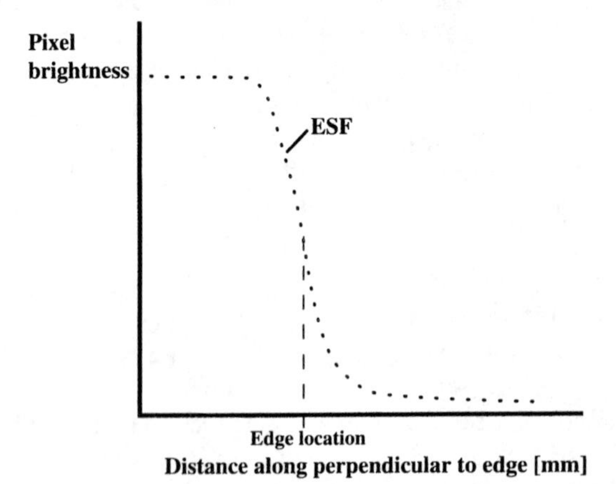

Figure 9.17 MTF measurements using an edge rotated slightly from the detector axes.

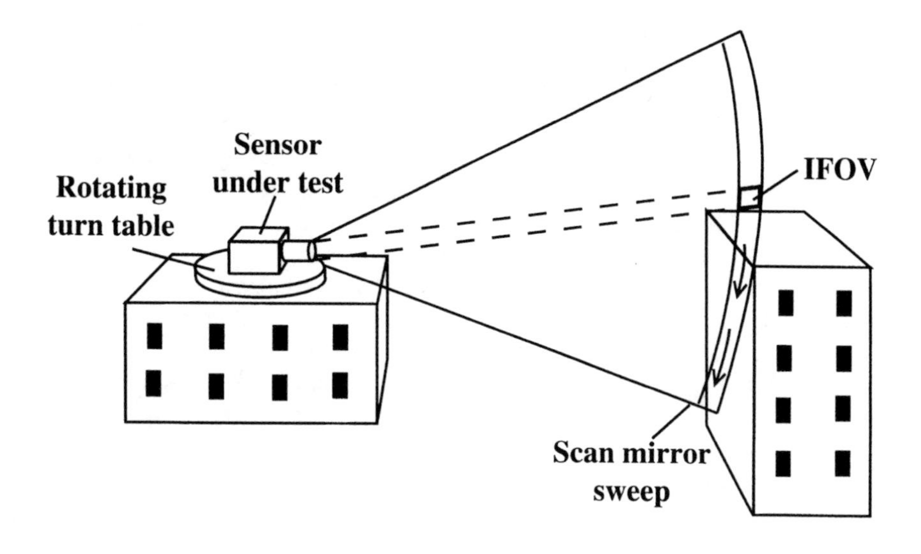

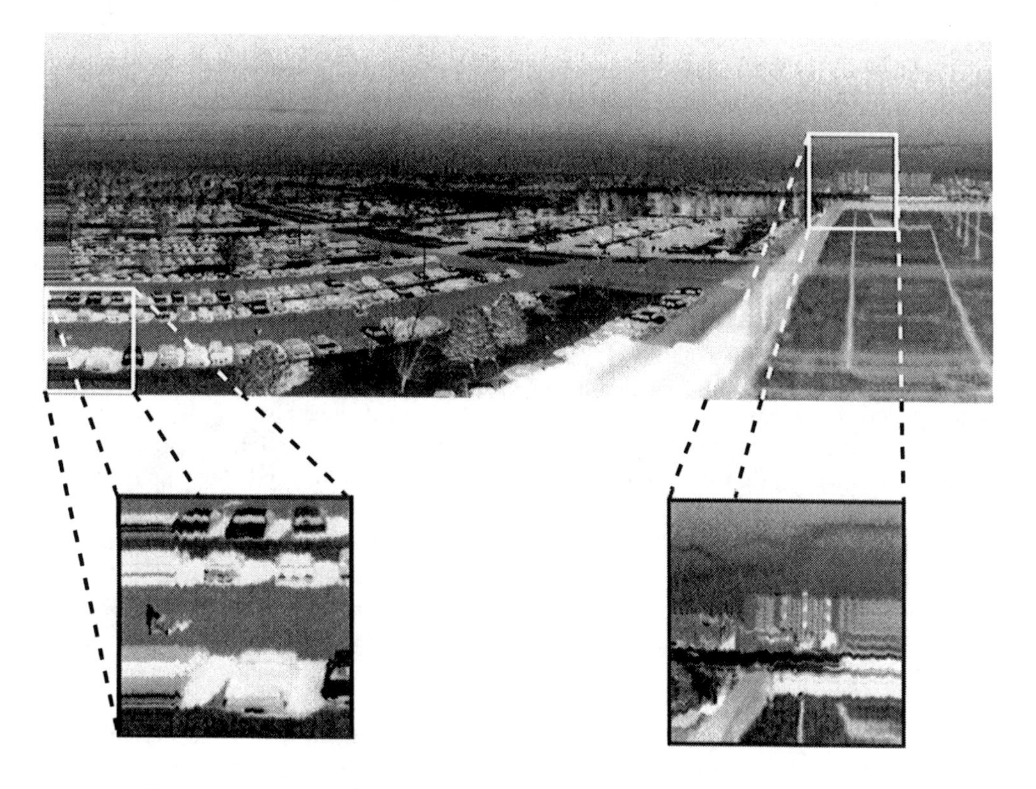

Figure 9.18 Use of a test range for assessment of sensor MTF. (LWIR image courtesy of RIT's DIRS lab.) Note that image at far left is before turntable rotation, and horizontal image is due to the motion of the pedestrian.

sor MTF. It is also possible to install test targets (e.g., tribars) on the roof or face of the adjacent structure for use in resolution studies. For this approach to be useful, the range distance must be comparable to the operational imaging distance so that changes in the PSF due to differences in focal depth are negligible. Also the edges of the buildings must have uniform backgrounds and be good approximations to straight edges at the relevant resolutions. For example, the RIT line scanners tested in this manner were 1- or 2-milliradian systems usually flown at 0.4 km. The buildings used as targets were approximately 0.44 km away, which would result in GIFOV's of 0.44 or 0.88 m. The variation in the edge of the buildings was of the order of 0.015 m and should introduce no error in the computations.

The other source of error in this measurement is turbulence due to the long horizontal path. Because the IFOV of this device is large compared to turbulence effects under low turbulence conditions, this effect is often negligible. For operational testing, measurements were used only when a visual assessment of turbulence was negligible for angular resolutions significantly higher than those of the sensor under test. Watkins (1991) describes a range target and procedures used to characterize sensor performance and performance degradations under conditions when turbulence is important. The use of a test range as an alternative or supplement to laboratory bench tests is particularly attractive for sensors with large entrance apertures or large fields of view where the collimators can become very expensive and the physical size of the test arrangement becomes quite cumbersome.

The one feature of the sensor MTF that is usually not tested by either the bench or range tests is blur due to the forward motion of the platform. This is normally treated as simple linear motion and cascaded with the along-track MTF to generate the final MTF of the sensor package.

For some systems, it can be difficult to generate a step function that is both very straight and has uniform high and low brightness. This can be a particular problem in the thermal infrared where the contrast is usually generated by thermal differences or by emissivity differences with a hot or cold surround. To avoid the effort of trying to ensure uniformity, an alternative target is often employed. This is simply a very thin resistive wire located at the focal plane of the collimator. The wire is heated by running a current through it. If a perfect image of the wire would be small compared to the PSF of the system, then the actual image can be treated as a line spread function LSF (an image of a mathematical line source having no width and infinite extent). The LSF image can be analyzed as illustrated in Figure 9.4 to yield the MTF of the sensor.

We have emphasized approaches for characterizing the spatial frequency response of the entire sensor. In general, each component is analyzed in the same fashion. Known signals (sine wave, step, etc.) or image representations of a well-defined function are input to a link in the chain (e.g., film writer, image reconstruction algorithm), and the output signals (images) are analyzed by treating the output signal (image) as a convolution of the input signal with the impulse response (PSF) of the link under test. The impulse response of the link is then determined by deconvolution, which is most easily done in the frequency domain using Eq. (9.15) or by using simple input functions, so that the approach illustrated in Figure 9.4 can be used. The overall system performance can be obtained by cascading the component MTF's, or the original input images can be passed through the entire image chain and the final images deconvolved to yield the image chain MTF.

For most airborne or space-based systems, even range tests must be treated merely as a baseline assessment, with final performance determined under flight conditions. In many cases, it is very difficult and expensive to perform detailed in-flight assessment of spatial image fidelity, so the normal procedure is to generate detailed performance data before flying the instrument. In-flight programs are then used to verify that no changes have occurred due to flight conditions or launch stresses. For airborne systems with reasonably high resolution, the flight tests can be simple replications of the laboratory or range tests employing similar targets on the ground at scales appropriate to the imaging conditions. For lower-res-

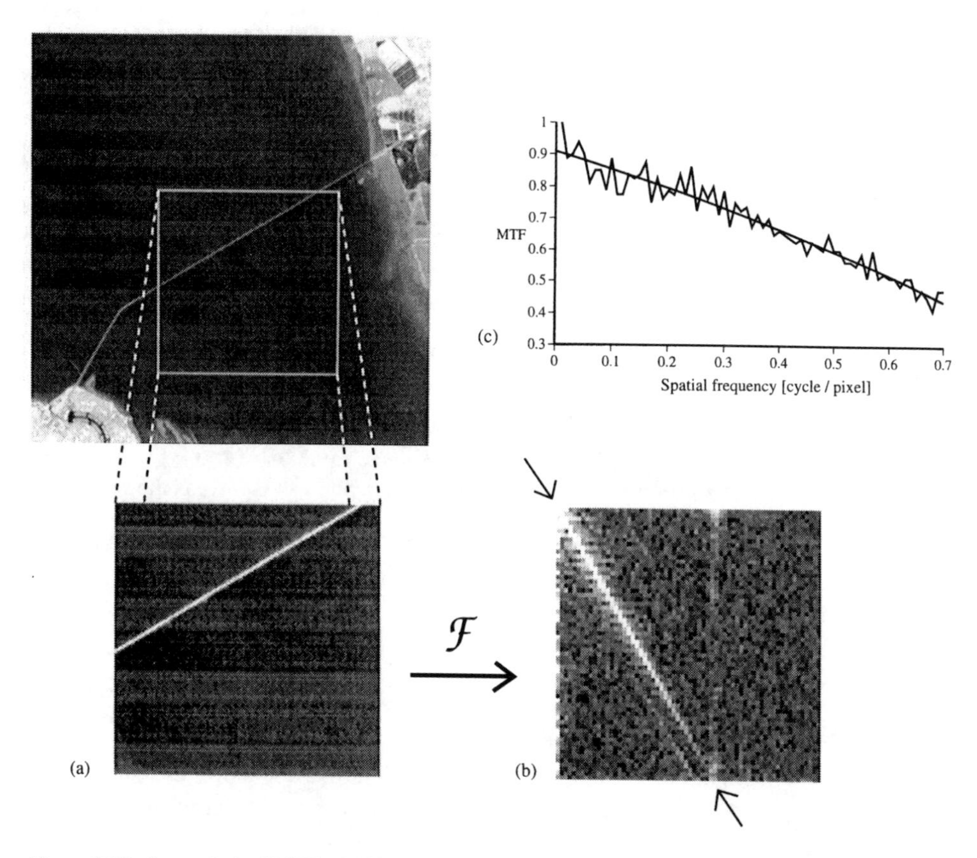

Figure 9.19 Image derived MTF. a) 128 x 128 pixel sub image of a TM image of the San Mateo Bridge, b) 1 quadrant of the Fourier transform of a) showing the modulus and the frequency dimension sampled, c) resulting sensor system MTF showing noise levels and smooth polynomial approximation (after Schowengerdt et al. 1985)

olution systems or in the thermal infrared where target costs can become substantial (although the line target is still often affordable), the use of ground targets can become quite cumbersome. Under these conditions, it is often possible to use serendipitous acquisition of targets that approximate the type of well-behaved patterns (lines, steps, points, line pairs) that would be used as test targets. Schowengerdt et al. (1985) describe two variations on this approach for calibration of the postlaunch spatial frequency response of the Landsat TM. In one case, a long, narrow bridge over a uniform water background is used as the input target (cf. Fig. 9.19). The brightness distribution of the bridge (the input function) can be simply defined along an axis perpendicular to the bridge. The deconvolution process can be accomplished in the frequency domain by division, i.e.,

$$\mathrm{MTF_s}(\nu) = \frac{\mathrm{MTF_m}(\nu)}{\mathcal{F}[o(\ell)]} \tag{9.16}$$

where $\mathrm{MTF_s}(\nu)$ is the MTF of the sensor with respect to spatial frequency (ν) along the axis perpendicular to the bridge, $\mathrm{MTF_m}(\nu)$ is the modulus of the Fourier transform of a subset of the TM image that only contains the bridge and the water background (the modulus is sampled along the axis corresponding to spatial frequencies v), and $o(\ell)$ is the function that describes the brightness variation of the bridge as a function of distance ℓ along the axis perpendicular to the long axis of the bridge. This approach suffers from two limitations. The first is that the target is a rather unusual feature not readily available in most images, and the second is that the method does not allow for easy separation of the along- and across-track MTF since the result will normally be a composite value. To overcome these limitations, Schowengerdt et al. (1985) demonstrated an approach using high-resolution aircraft data to assess the MTF of lower-resolution satellite data. In effect, one assumes that the high-resolution image contains all relevant spatial frequencies and can be treated as the object. The resultant image, after registration and gray-level matching, can be expressed in the frequency domain as

$$\mathrm{MTF_m}(u,v) = O(u,v) \cdot R(u,v) \cdot \mathrm{MTF_s}(u,v) \tag{9.17}$$

where $\mathrm{MTF_m}(u,v)$ is the modulus of the Fourier transform of the satellite image, $O(u,v)$ is the modulus of the Fourier transform of the registered aircraft image, $R(u,v)$ is the MTF associated with the resampling process required for registration, and $\mathrm{MTF_s}(u,v)$ is the sensor system MTF. The resultant sensor $\mathrm{MTF_s}$ can then be found from

$$\mathrm{MTF_s}(u,v) = \frac{\mathrm{MTF_m}(u,v)}{O(u,v) \cdot R(u,v)} \tag{9.18}$$

This process is usually quite noisy, and several noise reduction steps are necessary (as discussed in the reference) to yield final MTF values. To be effective, the aircraft imagery must be taken in essentially the same spectral band, at the same time, and from the same perspective as the satellite to ensure the validity of the assumption that it contains the same spatial frequencies as the scene. As a result, while this method yields a good assessment of the post-launch performance, it can be expensive and cumbersome to acquire the necessary data. Wherever possible, the use of naturally occurring targets is still a very attractive alternative, particularly for operational verification of the ongoing performance of a sensor.

Once the sensor or system MTF is known, it is often desirable to attempt to compensate for the degradation in high-frequency response that is common in most systems. Conceptually this is quite simple. An inverse filter is applied to the observed image by dividing its Fourier transform by the transfer function of the sensor, resulting in a boost of the high frequencies. The inverse transformed image should, therefore, appear sharper and more closely resemble an undegraded image. In operation, noise in the process often results in the presence of frequencies with zero amplitude in the computed transfer function, thus causing the process to break down. Even where methods are used to avoid zero values in the transfer function, division by small amplitude values tends to exaggerate periodic noise in the imagery, thus producing visual distractions in the corrected images. More advanced methods can be employed that include models of the image noise in the design of the inverse filter (cf. the treatment of image restoration by Gonzalez and Wintz, 1987 or Pratt, 1991). With the speed of today's image processors, ad hoc solutions are often developed simply by interactively applying high-frequency boost filters until a visually pleasing result is obtained. Figure 9.20 shows an image processed using a boost filter to enhance the high spatial frequencies. Because these filters are designed to enhance high frequencies, there is often no requirement for preserving radiometry or imagewide statistics. Consequently, the filters are usually implemented in the spatial domain using relatively small convolution kernels. In any situation where image restoration or boost filters are used, caution should be taken in any ensuing radiometric analysis of the image data to ascertain the impact of the filters on image radiometry. In many cases, it is advisable to restrict radiometric assessment to analysis of

(a)

(b)

Figure 9.20 Effect of a high-frequency boost filter: (a) original image and (b) image after high-frequency boost.

relatively large objects before any high-frequency restoration is attempted. By avoiding measurements near edge boundaries, the adverse effects of blurring on the radiometric values can largely be avoided. The boosted images can then be used for visual assessment or algorithms that use texture or spatial analysis.

9.2 RADIOMETRIC EFFECTS

In this section, we want to look at how radiometric image fidelity can be characterized through the image chain. Like spatial image fidelity, radiometric image fidelity can also be characterized in many ways and at many points along the image chain. A major distinction involves questions of absolute versus relative radiometry. In absolute calibration, we are concerned with how closely the measured value can be matched to a set of external reproducible standards. In relative calibration, the primary interest is internal consistency (i.e., does a change of 10 units always represent the same effect even if we don't know how many absolute units those 10 units represent?). When speaking of sensors, this is often thought of in terms of radiance. A sensor calibrated in absolute units typically would have internal sources of known absolute radiance, so each reading could be converted to an absolute value. A sensor with relative calibration might consistently return the same value when exposed to a specific radiance level, but the absolute value of that radiance could not be inferred. Alternatively, a system with a lesser degree of relative calibration might be able to determine only that a change of so many radiance units occurred between point A and point B, but not what the radiance level was. In general, as discussed at length in Chapter 6, the calibration of the sensor is only one step in the process, since our interest is typically rooted in how well we can measure parameters like ground temperature or reflectance. The image chain thus includes all radiation propagation effects, sensor effects, and analytical procedures discussed in Chapters 4, 5, and 6. In analyzing the mean-level radiometric performance along the image chain, we would have to assess all the contributions to the governing equation and their impact on the final measurements using error propagation modeling, as discussed in Section

4.6.2. We will not reiterate these issues here except to recall that the discussions in those chapters implicitly assumed a unit MTF response for all measurements. Clearly, as we saw in the last section, readings on small objects or near edges will not accurately represent the radiance that should be associated with those targets (small here is usually measured in terms of several EIFOV's).

Our discussion in previous chapters often assumed that the measured values were simply a function of radiance as characterized by the governing equations of Chapters 4, 5, and 6. We recognize, however, that nature is not this simple, and there are variations about the mean-level values described by the governing equations. We collectively refer to these deviations as *noise*, and in this section we will look briefly at the propagation of noise through the image chain.

9.2.1 Noise

Noise is perhaps best characterized as our uncertainty as to whether an individual measurement represents the value of a parameter. As discussed in Section 5.3, it is generally characterized in terms of the deviation from a mean value. Perhaps more importantly, in terms of image chain analysis, uncorrelated noise sources tend to add in quadrature such that

$$n_T = (n_1^2 + n_2^2)^{1/2} \tag{9.19}$$

where n_T is the total noise and n_1 and n_2 are the noise levels from two uncorrelated noise sources. The nature of quadrature addition is such that the largest noise source(s) drastically dominates the total noise. Thus, when trying to reduce the noise level of the entire imaging system, it is important to have identified the magnitude of the noise sources in each link so that mitigating measures are applied where they will have some impact.

When considering noise sources, we recognize that the various sources of noise that influence the signal level from the detector can be treated collectively as one source of noise (n_d). The components of detector noise were introduced previously in Section 5.3. It is useful to recall that we can think of this noise in the measured units (i.e., voltage or current) in terms of radiometric input units (i.e., noise equivalent power) or in terms of target parameters (i.e., noise equivalent temperature difference or noise equivalent reflectance difference). The preamplifier and conditioning electronics can also be sources of noise (n_a). For analog systems, the record and playback systems will be additional noise sources (n_r). In digitizing the signal, the continuous input signal is quantized into bins. which results in additional uncertainty in the signal level that is characterized as quantization noise (n_q). Oppenheim and Schafer (1975) point out that, when the quantization value is at the center of a bin of width b, and the input values are uniformly distributed, the standard deviation in quantization noise n_q can be expressed as

$$n_q = b / \sqrt{12} \tag{9.20}$$

Depending on the image chain, additional noise sources will exist and must be included in the computation of the overall system noise (n), i.e.,

$$n = (n_d^2 + n_a^2 + n_r^2 \cdots + n_q^2)^{1/2} \tag{9.21}$$

In general, the noise values will be expressed in terms of electronic signal levels (amps or volts). The concept of noise-equivalent input units can still be used to convert overall system noise into intuitive values (i.e., change in apparent temperature or reflectance) using the principles described in Section 5.3.

These standard noise concepts are generally thought of as variations about a mean level. However, in terms of the impact of noise on sensor performance, more complex issues come into play having to do with the interaction of noise levels, the spatial structure of the noise, and the spatial response (MTF) of the system. This introduces the problem that all noise is not equal, as illustrated in Figure 9.21 where the same image is shown with two sources of noise having the same RMS noise levels. In image (b), the noise is random, and while the image is degraded, it is not nearly as distracting as in (c) where the same RMS noise is shown as a patterned or correlated noise. Thus we see that the distribution of the spatial frequency of the noise can be a factor in the impact of the noise. This is a visual analysis problem that also impacts machine-aided exploitation where algorithms will be more sensitive to noise at certain frequencies.

The relative impact of noise levels and spatial frequency distribution on image fidelity is very often a function of the procedures to be used in the image analysis. A detailed assessment of these complex interactions is beyond the scope of our interest here. However, we do want to point out that the image fidelity issues are often addressed by relying on an empirical metric. For example, in the case of thermal infrared systems, a common metric is the minimum resolvable temperature difference (MRTD; cf. Holst, 1993). This is the temperature difference that must exist between a set of four bar targets (assumed blackbodies) for the bar pattern to be just visually discernible. The MRTD is specified under a set of test and observation conditions (e.g., target scale, viewer conditions, etc.). The MRTD of a system as a function of spatial or angular frequency will be a function of noise levels, noise structure, and the MTF of the entire system. Thus, as a figure of merit, it considers all of the important features. It has the disadvantage of not being fully modeled, so that the MRTD under other conditions must be obtained by experiment, which can be a tedious process.

Other performance-based metrics have been proposed for other sensors and/or applications. In general, they tend to be application specific and can be modeled as a complex function of the noise level, noise power spectrum, and system MTF. We will restrict ourselves here to stating the need to characterize these functions as indicators of image fidelity, recognizing that in general a complete set of data is needed to characterize image fidelity and that only in restricted cases can image fidelity be simplified to a single metric.

9.2.2 Noise Artifacts

The type of periodic noise shown in Figure 9.21 is usually caused by electronic or mechanical variations (or their harmonics) in sensor systems that induce noise in the detector or preamplifier. Other sources of periodic, correlated, or patterned noise are caused by detector-to-detector calibration differences. In linear arrays and whisk-broom-style detectors, the digital count out of each detector is a function of the incident radiance. The radiance-to-digital-count conversion is normally computed separately for each detector. Any slight error in calibration can result in the output digital counts for that detector being slightly high or low (or having slightly more or less gain). This will result in a periodic striping in the image, which will be most apparent in uniform areas or when the contrast has been highly stretched. This effect is very pronounced when a detector fails. This striping will be in the across-track direction for scanners and in the along-track direction for push-broom systems.

Another type of periodic noise occurs associated with groups of lines. This often results from slight gain or bias differences in groups of detectors. For example, in a bow-tie scanner, the detectors may be calibrated at the start of each line, and then a slight bias drift (droop) can occur across the line. For example, in the Landsat TM, all 16 detectors in each reflective band are calibrated against on-board references at the start of each line. Because the mirror scans in both directions, any droop effects occur on opposite sides of the image in groups of 16 lines. This results in a banding structure in the across-track direction, which can become

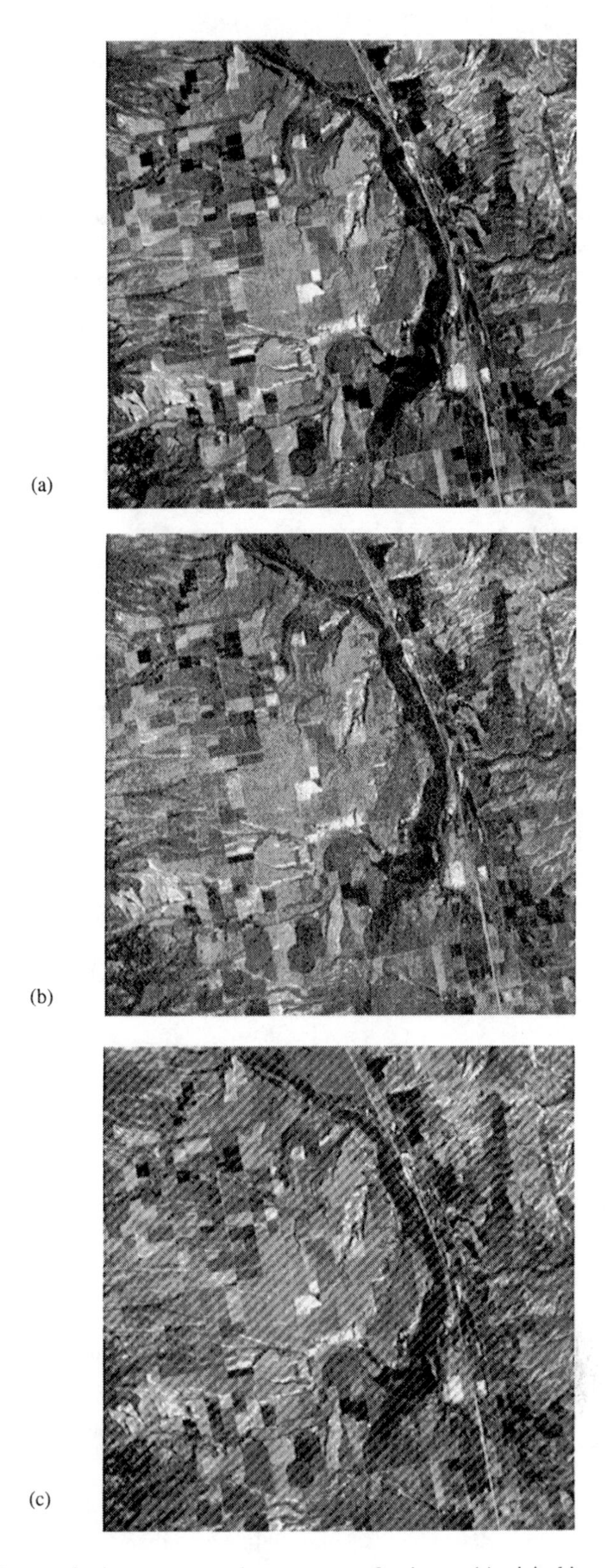

Figure 9.21 Impact of noise structure on the appearance of an image: (a) original image, (b) image with random noise added, (c) image with periodic noise added. The RMS noise levels of the noise added in (b) and (c) are the same.

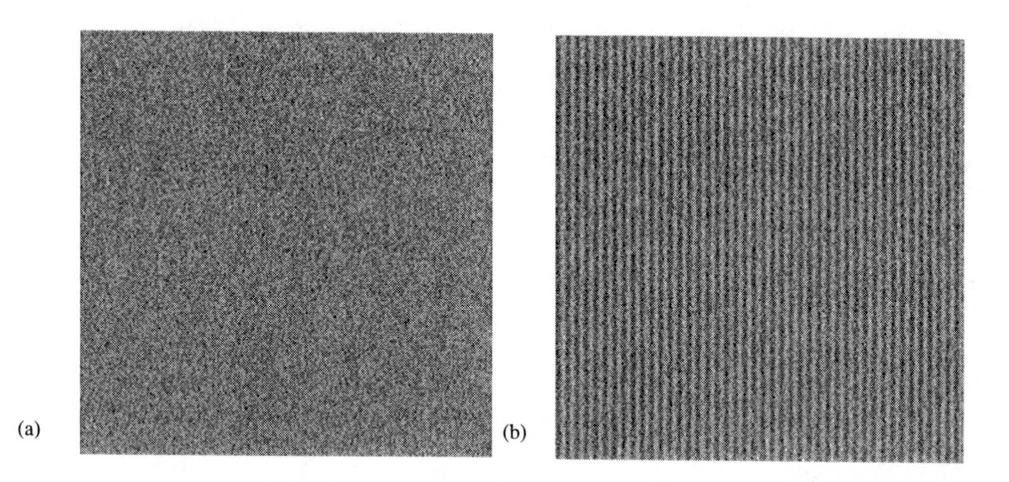

Figure 9.22 Examples of noise structure: (a) random noise, (b) structured noise.

apparent in uniform regions when the contrast is stretched. Similarly, in a linear array made up from many elements (cf. Fig. 5.29), the readout electronics may be slightly different for each element, causing small gain or bias differences between the groups of detectors in the elements. This will result in an along-track banding pattern (cf. Fig. 9.22).

A number of other artifacts often exist in raw sensor imagery. Most of these are very sensor-specific and require the development of sensor-specific correction algorithms to reduce or eliminate the effects. For example, in linear array systems it is common to see streaking or smearing after a very bright target. This occurs in the readout direction because the charge is not completely transferred out of the cell and some is left to be added in with the succeeding charge packet. This results in a bright streak behind very bright objects. Some of these artifacts can be very difficult, or even impossible, to characterize, to the point where the impact can be successfully modeled for removal from the imagery. In the next section, we will briefly treat some of the more straightforward methods for reducing noise artifacts.

9.2.3 Approaches for Correction of Noise and Periodic Structure in Images

With very large detector arrays, individual detector failures become highly probable. When failures occur, several correction options exist. The most straightforward is to simply replace the affected line with an adjacent line from a working detector. This results in some spatial artifacts, but does not change the radiometric value of any of the pixels. An alternative method is to use one of the interpolators discussed in Section 8.3.1.2. These interpolators will reduce the spatial artifacts, but may introduce radiometric artifacts in the interpolated line.

One method to reduce the effects of detector-to-detector variation in a whisk-broom or bow-tie system is to use the imagewide histograms for each detector (e.g., for the Landsat TM this would be the histogram made up of every sixteenth line of raw data). Over many, many lines, the mean and standard deviation of these detector specific histograms should be the same. Any slight difference can be attributed to gain or bias calibration differences between the detectors. These calibration differences can be corrected by forcing the histograms to be the same, using the linear histogram matching algorithms described in Section 6.4 for the PIF scene normalization process. The values for each detector are adjusted for gain and bias differences from some norm (either the detector with maximum contrast or the imagewide histogram). A similar approach can often be used for certain types of banding. Histograms of pixels from a common subelement in a linear array are matched to another element's histogram using linear histogram specification (i.e., a gain and bias correction). Because subelements can be very long, the scene structure can sometimes induce real variations in the histograms. To avoid this affect, only homogenous regions of the scene are used in generating each element's statistics.

To this point, we have concentrated on approaches for dealing with image artifacts. There are also methods available for reducing the effects of more random noise, but they often have some negative attributes. The baseline methods rely on the fact that the random noise in a signal will be reduced by averaging multiple samples. The noise reduction being proportional to $(1/N)^{1/2}$, where N is the number of samples averaged. Thus, one way to reduce noise in an image is to simply replace every pixel with the average in an N pixel neighborhood. The resulting image should have the noise reduced by $(1/N)^{1/2}$. This process is easily implemented using a convolution kernel, as discussed in Section 7.1.2. However, the process acts like a low-pass filter and will blur the output image. As discussed in Section 7.1.2, if the noise is more of a salt-and-pepper type (random spikes and dropouts), a nonlinear filter (e.g., median) may be more effective, although this too will introduce some artifacts. Rather than average spatially and suffer the resultant blur, it is sometimes possible to average multiple images of the same scene. For example, if the sensor is a rapid framing system that produces many images over a short period, multiple frames can be averaged together. The random noise will again be reduced by $(1/N)^{1/2}$. This approach can be very effective if there is little relative motion between the scene and the sensor over a period of many frames. In this case, the successive images will be in registration, and no spatial blurring will occur. Figure 9.23 illustrates some of the tradeoffs associated with spatial and temporal averaging for noise reduction.

It is also possible to take advantage of time averaging in the design of systems that are not of the rapid framing style. This alternative method uses multiple detectors in the along-scan direction of a scanning system or in the along-track direction of a push-broom system (cf. Sec. 5.4). As the image moves across the detectors, it is sampled by the first detector, and then the exact same image point on the ground is sampled a moment later as it is swept over the next sensor. By using several detectors in the image motion direction, the same point on the ground is imaged several times. These sampled values can then be averaged to reduce noise. This process is referred to as *time delay and integration* (TDI) because each sampled signal is held a short time (while the image advances over the detectors) and is then summed with the signal from each succeeding detector. The final integrated value should have its noise level reduced by $(1/N)^{1/2}$; where N is the number of detectors in the TDI process. This method has the advantage that with carefully controlled timing circuits, there should be little or no spatial blurring. Since the time between successive detector acquisitions is very small, even moving objects will normally have very limited blur. The negative side to this approach is the additional cost and complexity in the sensor design. The (TDI) process is normally implemented on board the sensor (often right on the focal plane) to avoid recording multiple images. It is possible, where recording bandwidth is available, to record images from each successive detector and shift and average the images during ground processing to achieve the TDI effect.

In many cases, external sources of electronic or mechanical noise that are induced into the detector or preamp will result in periodic noise patterns. When these patterns are identified in preflight conditions, they can often be eliminated by improved shielding and isolation of the detector, preamplifier, or noise source. Regrettably, these noise sources often develop in flight and must be removed in postprocessing. If the noise is periodic, it will cause localized increases in the power spectrum of the image. The particular frequencies can be identified by noise spikes in the power spectrum and filtered in the frequency domain using a very

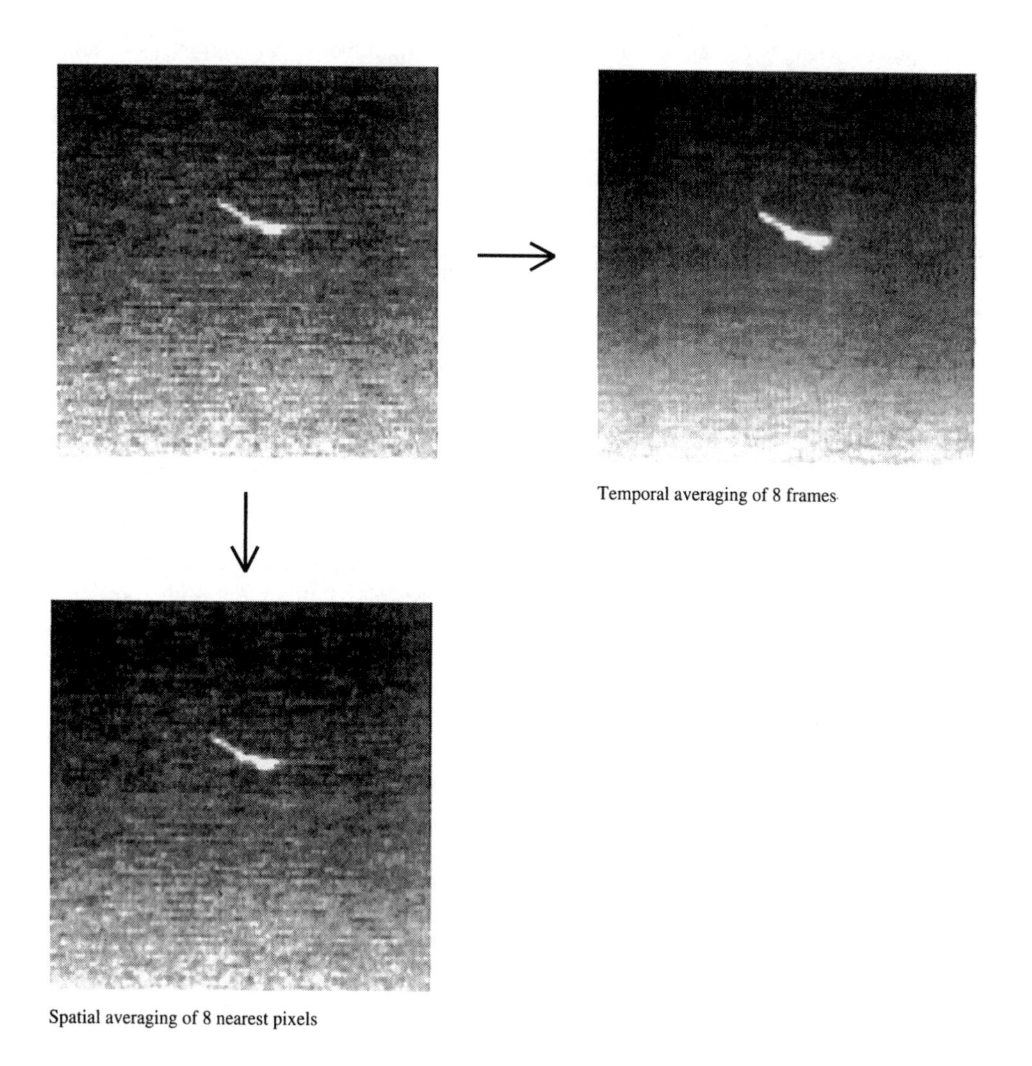

Figure 9.23 Effects of spatial and temporal averaging.

localized band rejection filter as discussed in Section 7.1.4 and illustrated in Figure 7.16. This will result in loss of information at the filtered frequencies, so the filters should be kept small and used with caution.

9.3 SPECTRAL AND POLARIZATION EFFECTS

A potential problem source in any imaging chain, but one that is taking on increasing importance as we move to multispectral analysis tools, is a shift in the system spectral response after ground spectral calibration. This can be a simple shift in band centers such as might result in a slight mechanical deviation in a grating in a spectrometer, or it can be due to a change in the bandpass of filters such as that reported for narrow band interference filters by Flittner and Slater (1991). It is believed that the interference filters used for band selection undergo mechanical changes in the space environment that resulted in changes in the spacing of the layers in the filter. The net result of these changes is a shift in the band edges. If these spectral changes go unnoticed, they can introduce radiometric calibration errors, atmospheric correction errors, and misinterpretation of spectral signatures. The relative importance of these changes will depend a great deal on the calibration techniques used and on where the spectral shift occurs relative to spectral structure in the target or the atmosphere (i.e., a very small change at the edge of an atmospheric window could have a major impact, whereas a large change in the middle of a window or in a spectral region where the target and backgrounds were slowly varying would have limited impact). It can be very difficult to detect and characterize these changes in satellite systems. In most cases, a careful analysis of well-known targets with spectral structure near the band edges must be performed to identify and characterize the problem in instruments with bandpass filters. In the case of spectrometers, errors are often easier to detect if the edges of atmospheric absorption structure are sampled. Many of these absorption features are well characterized (cf. Sec. 3.5). They can be used to identify errors in spectral calibration, and the spectrometer recalibrated in an iterative fashion by fitting the observed spectra to the absorption line structure (cf. Green et al., 1994). Since many spectrometers are designed to study slight variations in line spectra, or don't sample across a major absorption feature, more involved on-board or in-flight verification methods may be required. As early as 1973, earth-based laser beams were imaged by space platforms (cf. Piech and Schott, 1975), and this practice may be required in the postlaunch characterization of the spectral calibration of satellite systems if on-board spectral sources are not available.

9.3.1 Feature/Spectral Band Selection

For systems with more than a few spectral bands, problems arise associated with the high volumes of data. In many cases, this becomes a data storage or transmission problem (i.e., we are producing data faster than we can move it down the image chain). In other instances, it is an analysis problem produced by having so much data to process that the algorithms run too long. In the case where the data are to be used in multivariate classifiers, it is often not necessary, or even desirable, to process all of the data. This is particularly the case when feature bands (spectral bands) are highly correlated. In this instance, adding more bands to a classifier can actually reduce classification accuracy. Several methods have been suggested to attempt to isolate optimal or near-optimal subsets of features for use with GML classifiers. Swain (1978) describes three measures referred to as *divergence*, *Jeffries-Matusita distance*, and *transformed divergence*. These are all measures of class separability, which, when averaged over all classes, yield a measure of the discrimination quality of a group of spectral features. By comparing all possible combinations of subsets of the spectral features (i.e., which 4 out of 12 available bands), the one that produces the highest quality metric can be selected. Only the reduced subset of bands are then used in the overall image classification. Swain (1978) indicates that if the classes can be assumed to be normally distributed, then the divergence between class i and class j can be expressed as

$$\mathrm{Div}_{ij} = \frac{1}{2} tr\left[(\mathbf{S}_i - \mathbf{S}_j)(\mathbf{S}_i^{-1} - \mathbf{S}_j^{-1})\right] + \frac{1}{2} tr\left[(\mathbf{S}_i^{-1} + \mathbf{S}_j^{-1})(\mathbf{M}_i - \mathbf{M}_j)(\mathbf{M}_i - \mathbf{M}_j)'\right] \tag{9.22}$$

where Div_{ij} is the divergence between class i and class j and tr [A] indicates the trace of the matrix A; the rest of the terms were defined in Section 7.2. The first term in the equation for divergence characterizes the difference between the covariance matrices (i.e., the difference in the shape of the distributions), and the second term is a measure of the normalized statistical distance between the means. Overall, the divergence is a measure of the separability between class i and j. The average divergence is a weighted summation over all pairwise combinations of classes, i.e.,

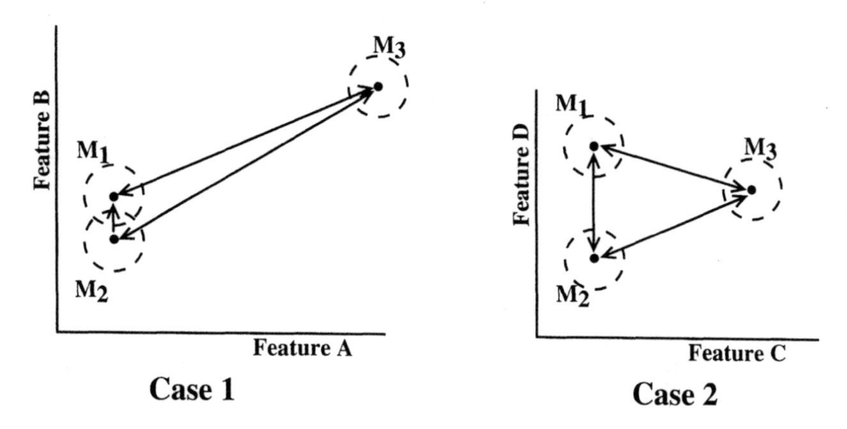

Figure 9.24 Limitation of simple divergence. A feature set that has overlap in clusters but has one cluster center very far from all the others can appear better than a set with approximately the same separation in all centers.

$$\overline{Div} = \sum_{i=1}^{k}\sum_{j=1}^{k} \rho(i)\rho(j)\text{Div}_{ij} \tag{9.23}$$

where the weights are simply the a priori probabilities. The average divergence of each possible subset of ℓ spectral bands can then be computed, and the subset that yields the largest average divergence will tend to yield the best separability. A problem arises, however, when one combination of bands generates classes with very large divergence values for some classes and small values for others, and a second generates modest divergence values for all classes, as illustrated in Figure 9.24. Clearly, the second case represents a better overall pairwise selection of features. This indicates that increasing the pairwise divergence has a diminishing return. Swain (1978) indicates that this limitation is overcome by the Jeffries-Matusita distance, but at considerable computational cost. A more commonly used heuristic approach is the transformed divergence expressed as

$$\text{Div}_{ij}^{T} = 2\left[1 - e\left(-\text{Div}_{ij} / 8\right)\right] \tag{9.24}$$

This has the characteristic of exponential saturation of the divergence measure and scales the transformed divergence over the range 0 to 2 (Mausel et al., 1990), in evaluating separability measures used a scaling factor of 2000 rather than 2 and this factor is widely used). The average transformed divergence can then be computed by substituting the transformed divergence in Eq. (9.23).

Rosenblum (1990) describes a similar metric that was used to separate texture-based image features (cf. Sec. 7.1.3), as well as select spectral bands based on reflectance spectra in sensor design studies. This metric, referred to as *thresholded separability*, uses a ratio of the statistical distance to a user-defined maximum distance value. The ratio is then thresholded, so that any value greater than 1 is set equal to 1. The separability measure of class j from class i takes on the intuitively appealing form of

$$\text{Sep}_{ij} = \frac{|\mathbf{S}_i| + (M_j - M_i)'\mathbf{S}_i^{-1}(M_j - M_i)}{\text{Sep}_{\max}} \tag{9.25}$$

where the numerator looks very much like the GML discriminant function between class means for the case of equal a priori probabilities (cf. Sec. 7.2) and the denominator is chosen to ensure very good, but not extreme, separability between class means. The threshold operation ensures that the problem of excessive separability in one class does not overwhelm the final metric (cf. Fig. 9.24). The overall separability can then be defined as

$$\overline{\text{Sep}} = \sum_{i=1}^{k}\sum_{j=1}^{k} \text{Sep}_{ij} \tag{9.26}$$

and the subset of features that maximizes the overall separability would be chosen as near optimal. Rosenblum found the separability measure performed comparably to the transformed divergence with improved speed.

All of the feature selection methods presented in this section are scenario-dependent. This means that the features (e.g., spectral bands) chosen are nearly optimal for classification of the target classes used in the optimization study. In order to use this approach in deciding which bands to downlink or record, robust studies of optimum band combinations would need to be performed for various scenarios (i.e., various target classes) and imaging conditions.

9.3.2 Polarization Issues

To this point, we have ignored the effects of polarization on the imaging process. Implicitly, we have assumed that either the flux incident on the detector was randomly polarized or the sensor was insensitive to polarization effects (i.e., flux at any polarization would yield the same signal level). These assumptions are often invalid, and in some cases may lead to substantial errors. For example, grating spectrometers tend to be very polarization-sensitive, and most EO imaging systems tend to have some polarization-dependent sensitivity, unless great effort has gone into minimizing this sensitivity. As a result, if the radiance reaching the sensor is polarized, it will cause different signal levels depending on the orientation of the polarization. Simply put, a polarization-sensitive sensor calibrated with randomly polarized flux will be out of calibration for anything but randomly polarized flux. One way to deal with this problem is to design polarization-insensitive sensors and to characterize the polarization sensitivity so that any error due to residual sensitivity can be tracked. An alternative approach is to attempt to use the polarization state as a signature. In cases where the energy matter interactions at the earth's surface induce a change in the polarization state of the incident flux, we can use this change (which will be a function of the makeup and condition of the surface) as a signature. In order to take advantage of this type of feature, our sensor must be able to discriminate variations in the polarization state of the incident flux. One simple way to do this is to orient cross-polarized filters in front of matched sensors. Ideally, orientation of the filters is varied to maximize the difference between the two images formed. Figure 9.25 shows an example of a pair of images obtained with a polarization-sensitive system. In this case, the system consisted of two conventional film cameras with polarization filters on each. The filters are oriented at right angles to each other and the cameras synchronized for image acquisition (cf. Duggin et al., 1989). Differences between the image pairs are then a function of the polarization effects of the scene elements (note that image-to-image normalization is usually performed first to account for cross-calibration effects and polarization effects in the atmospheric parameters).

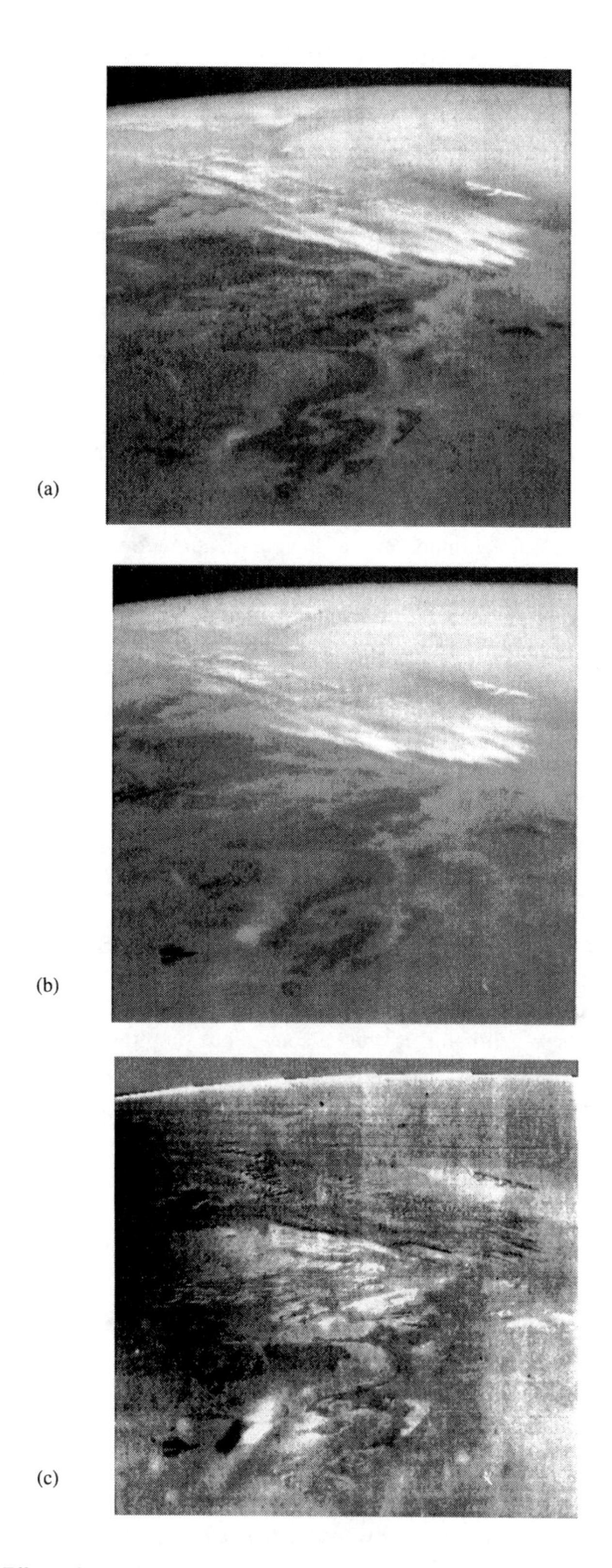

Figure 9.25 Effects of scene-induced polarization differences: (a) and (b) are space shuttle images cross-polarized relative to each other, and (c) is a polorization difference image. (Images courtesy of M. Duggin SUNY CESF, Syracuse, New York.)

The degree of terrain-induced polarization difference will be a function of wavelength, surface roughness (relative to the wavelength) and sun-object-sensor angle. Rough surfaces will tend to randomly scatter the flux inducing random polarization, while smoother surfaces will tend to more selectively scatter relative to polarization. This difference in the amount of energy reflected into perpendicular or parallel components will tend to be rather small when the incident flux is near normal and become more exaggerated at grazing angles (cf. Hecht, 1987). The images shown in Figure 9.25 were taken at a low view angle to enhance polarization effects. Note that in the difference image, smooth textured surfaces will tend to have higher differences and, therefore, appear brighter in the image.

The extent to which polarization is an issue is very much a function of viewing geometry, target structure, and wavelength (cf. Whitehead, 1992). Rough targets viewed from nadir will tend to have little polarization, so that the impact of the polarization sensitivity of the sensor on calibration is reduced. On the other hand, for quantitative measurements of smooth surfaces from an oblique perspective, great care must be taken to ensure that the sensor is not polarization-sensitive or that the polarization effects are fully known (which is extremely difficult).

Travis (1992) points out the importance of the polarization effects of the atmosphere on polarization imaging. Rayleigh scattering introduces significant polarization that is strongly a function of the scattering phase angle. He suggests that polarization effects in the visible region are dominated by atmospheric effects, with terrestrial polarization becoming more important as we move into the near IR where the mean albedo (particularly over land) is higher and Rayleigh scatter appreciably reduced. The presence of aerosols tends to reduce the polarization induced by Rayleigh scatter. Travis (1992) suggests that multispectral polarimetric imagers may be able to use this reduction in polarization to characterize the aerosol content. These data would be of considerable value to both climatologists interested in studying the atmosphere and remote sensing scientists interested in reducing atmospheric effects.

9.4 SPATIAL, SPECTRAL, AND RADIOMETRIC TRADEOFFS

For the most part we have tended to treat the radiometric and spatial resolution image chains as separate entities. While this is often a convenient simplification, it is also useful to recognize that these image chains are fully interwoven. This is most easily seen if we think about observing a scene where the radiance is varying at some spatial frequency ν. Using the G# concept developed in Eq. (5.7), the variation in flux onto the detector in the absence of MTF effects should be

$$\Phi(\nu) = \int \frac{L_\lambda(\nu) A_0 d\lambda}{\text{G\#}} \tag{9.27}$$

where A_0 is the detector area [cm^2]. In fact, Eq. (9.27) is only valid if the spatial MTF is unity, which for most systems would only be true at very low frequencies. The actual expected variation in flux $\Phi'(\nu)$ should more rigorously be expressed as

$$\Phi(\nu) = \Phi'(\nu) = \text{MTF}(\nu) \int \frac{L_\lambda(\nu)}{\text{G\#}} A_0 d\lambda \tag{9.28}$$

where for convenience we have assumed that the system MTF is constant over the spectral bandpass of the detector, and the (′) superscript is dropped for clarity. The variation in the signal (S) observed when viewing a scene whose radiance is varying at a spatial frequency ν could then be expressed as

$$S(v) = \text{MTF}(v)\int \frac{R(\lambda)L_\lambda(v)}{\text{G\#}} A_0 d\lambda \tag{9.29}$$

where $R(\lambda)$ is the sensor's spectral responsivity.

However, Eq. (9.29) is only valid in the absence of noise ($N(v)$). The actual expected signal S' would be

$$S(v) = S'(v) = \text{MTF}(v)\int \frac{R(\lambda)L_\lambda(v)}{\text{G\#}} A_0 d\lambda + N(v) \tag{9.30}$$

where the ($'$) superscript is again dropped for clarity. The signal-to-noise can then be estimated as

$$\frac{S(v)}{N(v)} = \frac{\text{MTF}(v)\int \frac{R(\lambda)L_\lambda(v)A_0}{G\#}}{N(v)} d\lambda \tag{9.31}$$

For convenience, we have expressed the parameters in Eqs. (9.27) to (9.31) as a function of a single spatial frequency dimension (v). Clearly, for imaging systems these variables are often more effectively addressed in terms of the spatial frequencies (u,v) with respect to the x and y axis [i.e., $S(v)$ becomes $S(u,v)$ etc.].

In the formalism we introduced in Chapter 7, Eq. (9.30) would be expressed as

$$G(u,v) = H(u,v) \cdot F(u,v) + N(u,v) \tag{9.32}$$

where $G(u,v)$ is the frequency domain representation of the observed image (i.e., the 2-D Fourier transform of the image), $H(u,v)$ is the 2-D MTF of the entire image chain leading up to the observed image, $F(u,v)$ is the frequency domain representation of the image in the absence of any MTF effects, and $N(u,v)$ is frequency domain representation of the noise. We choose to use the representation in Eq. (9.30) at this point to show explicitly the interconnection between the strands in the image chain. Equation 9.30 shows us how completely the radiometric and spatial strands of the image chain are interwoven and reminds us that if for convenience we treat them separately, we must recombine the strands or sub chains to understand any subtle interaction effects that can be introduced. These interactions are particularly important when we consider efforts at making precise radiometric measurements on objects with high spatial frequency (i.e., where the MTF deviates significantly from unity).

9.5 SUMMARY OF IMAGE CHAIN CONCEPTS

In this chapter, we have looked at some of the characteristic limitations placed on imaging systems by spatial, radiometric, and spectral resolution. In an ideal system, we would like to simultaneously extract information about fine spatial structure, subtle brightness changes, and detailed spectral character. However, one of the fundamental axioms of remote sensing is that there are never enough photons to let us image everything we want. To improve signal-to-noise for improved radiometric precision, we need more signal (more photons). This can be accomplished with larger detectors, longer dwell times, or bigger optics. For a fixed optical system, this generally means sacrificing spatial resolution, since larger detectors and/or increased dwell time will generally degrade resolution. Similarly, when we image in narrower spectral channels to improve spectral resolution, the number of photons available to be detected is reduced, decreasing signal-to-noise. Again, the spatial resolution must usually be sacrificed to maintain sufficient signal-to-noise to discern radiometric differences in the spectral image. Thus, we can think of spatial, spectral, and radiometric resolution as comprising a three-dimensional tradeoff space.

What is not well defined is how to perform tradeoff studies in this three space. Ideally, we would like to be able to determine uniform utility surfaces in this tradeoff space that would tell us how one combination of system parameters relates to another (e.g., is a three-meter, one-band data set with a signal-to-noise of approximately 150 "better or worse" than a nine-meter, three-band data set with similar signal-to-noise?). One tempting approach to addressing this question is to use classical information theory as developed by Shannon (1949). This approach, which is commonly used in channel capacity calculations, uses entropy as a measure of information. In the simplest case, the entropy can be expressed as

$$H = \sum p(\mathrm{DC}) \ln_2 (p(\mathrm{DC})) \tag{9.33}$$

where H is the entropy expressed in bits per pixel, and p(DC) is the probability of a digital count occurring.

Pratt (1978) discusses how higher-order entropy expressions can be developed that take into account the correlation effects between neighboring pixels or different spectral bands using joint probabilities. These higher-order entropy calculations provide a method for comparing the minimum information storage or channel capacity. However, because they rely on intensive calculations and are not necessarily directly related to effective measures of information from an application standpoint, this approach has not been extensively pursued. At present there is no generally accepted rigorous theoretical solution to this question. This is because the "effective information content" of an image is a function of the algorithms and analytical tools that are available for extracting the information from the image. For example, classifier-based tradeoffs have been used to select optimum spectral bands and/or texture metrics for use in land cover mapping (cf. Sec. 7.2; i.e., pick the bands that yield the best classification as described in Sec. 9.3.1). Because monochrome and multispectral algorithms are often very different, even application-specific comparisons can be very difficult (i.e., a change in algorithms at a later time could change the decision about the most appropriate spatial-spectral-radiometric trade). At present we must be satisfied with recognizing the importance of these tradeoffs and accept that design decisions are still often made on the basis of application-specific algorithms, often using ad hoc metrics.

9.6 REFERENCES

Beland, R.R. (1993). "Propagation through atmospheric optical turbulence," Chap. 2. Smith, F.G., ed., Vol. 2, *The IR and EO Handbook.* SPIE, Bellingham, WA.

Coltman, J.W. (1954). The specification of imaging properties by response to a sine wave input. *Journal of the Optical Society of America.* Vol. 44 No. 6, pp. 468-471.

Duggan, M.J., Israel, S.A., Whitehead, V.S., Myors, J.S., & Robertson, D.R. (1989). "Use of polarization methods in earth resources investigations." Proceedings of the SPIE, Vol. 1166, pp. 11-22.

Flittner, S.E., & Slater, P.N. (1991). Stability of narrow-band filter radiometers in the solar-reflective range. *Photogrammetric Engineering and Remote Sensing*, Vol. 57, No. 2, pp. 165-171.

Foos, D., & Fintel, W. (1990). Eastman Kodak Company, Personal communication with the author.

Gaskill, J.D. (1978). *Linear Systems, Fourier Transforms, and Optics.* Wiley, NY.

Gonzalez, R.C., & Wintz, P. (1987). *Digital Image Processing*, 2d ed., Addison-Wesley, Reading, MA.

Goodman, J.W. (1968). *Introduction to Fourier Optics.* McGraw-Hill, NY.

Granger, E.M., & Cupery, K.N. (1972). An optical merit function (SQF) which correlates with subjective image judgements.*Photographic Science and Engineering*, Vol. 15, No. 3, pp. 221-230.

Green, E.O., Conel, J.E., Margolis, J., & Chrien, T.G. (1994). "Spectral calibration of an imaging spectrometer inflight using solar and atmospheric absorption bands," IGARSS'94, surface and atmospheric remote sensing: technologies, data analysis and interpretation, California Institute of Technology, Pasadena, CA.

Hall, C.F., & Hall, E.L. (1977). A non-linear model of the spatial characteristics of the human visual system. *IEEE Transactions on System, Man, and Cybernetics*, Vol. SMC-7, No. 3, pp. 161-170.

Hecht, E. (1987). *Optics*, 2d ed. Addison-Wesley, Reading, MA.

Holst, G.C. (1993). "Infrared imaging system testing," Chap. 4,. In Dudzik, M.C. ed., Electro-optical system design, analysis, and testing, Vol. 4, *IR and EO System Handbook*, SPIE Press, Bellingham, WA.

Kaufman, Y.J. (1982). Solution of the equation of radiative transfer for remote sensing over nonuniform surface reflectivity. *Journal of Geophysical Research*, Vol. 87, No. C6, pp. 4137-4147.

Mausel, P.W., Kramber, W.J., & Lee, J.K. (1990). Optimum band selection for supervised classification of multispectral data. *Photogrammetric Engineering and Remote Sensing*, Vol. 56, No. 1, pp. 55-60.

Oppenheim, A.V., & Schafer, R.W. (1975). *Digital Signal Processing*. Prentice-Hall, Engelwood, NJ.

Pearce, W.A. (1977). "A study of the effects of the atmosphere on thematic mapper observations." Final Report under NASA Contract NAS5-23639.

Piech, K.R., & Schott, J.R. (1975). "Evaluation of Skylab earth laser beacon imagery." Calspan Report #KL-5552-M-1, prepared for NASA.

Pratt, W.K. (1991). *Digital Image Processing*, 2d ed. Wiley, NY.

Rabanni, M., & Jones, D.W. (1991). Digital image compression techniques. *SPIE Optical Engineering Press*, Vol. TT7.

Rees, W.G. (1990). *Physical Principles of Remote Sensing*. Cambridge University Press, Cambridge, NY.

Rosenblum, W. (1990). "Optimal selection of textural and spectral features for scene segmentation." Masters thesis, Center for Imaging Science, Rochester Institute of Technology, Rochester, NY.

Schowengerdt, R.H., Archwamety, & C. Wrigley, R.C. (1985). Landsat thematic mapper image derived MTF. *Photogrammetric Engineering and Remote Sensing*, Vol. 51, No. 9, pp. 1395-06.

Shannon, C.E. (1949). *The Mathematical Theory of Communication*. University of Illinois Press, Urbana, IL.

Slater, P.N. (1980). *Remote Sensing, Optics, and Optical Systems*. Addison-Wesley, Reading, MA.

Swain, P.H. (1978). "Fundamentals of pattern recognition in remote sensing," In P.H. Swain and S.M. Davis, eds. *Remote Sensing: The Quantitative Approach*, McGraw-Hill, New York, NY.

Tatian, B. (1965). Method for obtaining the transfer function from the edge response function. *Journal of the Optical Society of America*, Vol. 55, pp. 1014-19.

Travis, L.D. (1992). "Remote sensing of aerosols with the earth observing scanning polarimeter," Proceedings of the SPIE, Vol. 1747, pp. 154-163.

Warnick, J.S. (1990). "A quantitative analysis of a self-emitting thermal IR scene simulation system." M.S. thesis, Center for Imaging Science, Rochester Institute of Technology, NY.

Watkins, W.R., Crow, S.B., & Kantrowitz, F.T. (1991). "Characterizing atmospheric effects on target contrast." *Optical Engineering*, Vol. 30, No. 10, pp. 1563-75.

Whitehead, V.S. (1992). "A summary of observations performed and preliminary findings in the space shuttle polarization experiment." Proceedings of the SPIE, Vol. 1747, pp. 104-108.

CHAPTER

10

Image Modeling

In this final chapter, we will look at the process of generating synthetic images. This subject is of interest to remote sensing in its own right, but is particularly interesting from the image chain perspective. The process of effectively modeling nature to generate synthetic images that mimic real images requires a detailed knowledge and representation of the image chain. In this chapter, we will look at methods for generating synthetic images and how in many ways they represent a computer simulation of the image chain. From this image chain perspective, synthetic image generation (SIG) models become a powerful tool for the study of imaging systems and an aid in system and image analysis.

10.1 SIMULATION ISSUES

There are a host of reasons why synthetic image generation is rapidly becoming a powerful tool in remote sensing. The demand for this tool has come from many different sectors. Sensor designers have long looked to this approach to evaluate tradeoffs between the types of image fidelity parameters discussed in Section 9.4. Synthetic images can be produced over a range of spatial, spectral, and radiometric performance specifications that new sensors might produce. These images can then be evaluated in terms of application-specific performance metrics to determine the utility of the sensor in a given application. These performance metrics may range from visual assessment through performance using fully automated or manually assisted algorithms. This can also include sensor field of view and view angle studies to determine what range of look angles can be used to satisfactorily answer questions and to perform tradeoffs between synoptic perspective, resolution, and image acquisition costs.

A second group interested in synthetic image generation (SIG) are system operators. Here the task is to simulate a specific sensor and see how it can be used most effectively to image a phenomena of interest. For example, with a system that can image at different days of the year, times of the day, and look angles; what combination is optimal for observing a phenomena (e.g., crop stress), what combination is tolerable, and when will this sensor fail. Having delimited these types of variables, we could also ask when is the next acceptable acquisition and what are we likely to see in the image (i.e., is it worth the cost and effort to acquire the image?).

A third group with a great deal of interest in synthetic images is algorithm developers. Here the interest is in developing and testing algorithms on scenes that contain the target of interest in a variety of forms and over a range of acquisition conditions. This is motivated in some cases by a lack of real imagery, because the algorithms are for a sensor under development or to help motivate the development of a new sensor. However, far more often it is to supplement data from an actual sensor where a robust data set does not exist. Another group of algorithm developers are looking to synthetic image modeling, not just for the images but to incorporate the models into algorithms to assist in machine-aided analysis. In many advanced algorithms, there is a process of hypothesizing about what is present at a location in an image or about the condition of something that is presumed to be at a location in an image (e.g., is there a ship located at this dock; is it a freighter or a war ship; is it under power?). The algorithm could have the model produce images of several types of ships at the dock from the view angle of the image, under the tidal conditions at the time of image acquisition, and then extract features from the synthetic image for comparison with features from the actual image. The features would then be used in the hypothesis testing to determine if a ship is present and, if so, what it is and what its status is. For clarity, we have simplified this process, which would often involve many tests of the feature extraction algorithm over slightly different models to develop a statistical representation of the features for comparison with the actual image features. In many cases, some or all of these model runs are performed in advance to improve algorithm performance times.

This concept of hypothesis testing leads us to a fourth group of individuals interested in synthetic image generation. These are image analysts whose interest is also in hypothesis testing. An analyst might speculate about what he or she believes is in an image. By simulating what an image would look like based on that speculation and comparing the synthetic to the actual image, the analyst can often accept or reject the hypothesis. Alternatively, the analyst can use synthetic images to determine under what imaging conditions a hypothesis could be more definitively addressed (i.e., I can't tell if the crop in question is stressed, but if the image were acquired at 10 a.m. looking at 25° to 35° to the west, the probability of isolating the stressed vegetation will be much greater both in terms of spectral signatures and texture metrics).

A fifth group interested in synthetic imaging wants to use the images and the computer modeling processes associated with SIG for training purposes. Under the training umbrella there are at least two distinct groups. One group wants to use the images as a backdrop or context for training, where the images provide an environment but are not of fundamental interest in and of themselves. The use of SIG in flight simulators would be a good example of this category. The image fidelity required for these applications can range from very crude terrain structure for general purpose training to very location-specific detailed scenes where a landing in a particularly difficult location under adverse conditions is of interest. The other group interested in training are those individuals charged with educating image analysts or applications specialists in what information is contained in images and how various conditions impact the appearance and information content of an image. For this group, the ability of the SIG approach to reveal the impact of various phenomena on the final image is often as important as the image itself, since understanding these phenomena and their interplay in image formation is critical for effective image analysis.

This brings us to a final group interested in SIG, and, in the context of this volume, the group that would use the SIG process to study the image chain. These are the systems engineers who, either in designing a new system or analyzing an existing system's potentials and limitations for a particular task, are interested in using the SIG approach to combine the many complex interactions that go into the final image formation. If these interactions are well modeled by the SIG process, then the weak links in the image chain can be identified, including those links where the weakness is the result of some combination of interactions that might be difficult to isolate without an end-to-end simulation of the entire image formation process. The SIG approach also allows for the testing of alternative designs or improvements to the system before such changes are actually implemented.

These various interest groups place a broad range of demands on the SIG process. There are differing levels of fidelity, speed, interaction, etc., required for different groups and even the definition of fidelity is often different depending on the application. For our purposes here, we are interested in approaches that can be an aid to remote sensing image and systems analysts. This would include algorithm developers. This pushes us toward an interest in higher-fidelity models with a reduced concern about the types of real-time or near real time issues that flight simulation might require. Furthermore, from the standpoint of the underlying science of remote sensing and image analysis, we are interested in models that are based on the physical principles we have discussed throughout this book. SIG models based on these principles can help us assess the extent to which we understand the imaging process and point to where our understanding is flawed or oversimplified. If we simulate an image and compare it to an actual image in a controlled experiment, the discrepancies between the images can point us to limitations in our understanding of the imaging process. They may also point us to limitations in the implementation of the process in the SIG model, and care must be taken to differentiate between these two cases (i.e., when is the model inadequate and when is our understanding inadequate?). From this standpoint, the SIG process becomes a very powerful tool for the remote sensing scientist in pointing out where our understanding is inadequate and where more research and development effort is warranted. It helps us ascertain whether simplifying assumptions that were made are acceptable, or if we have cut too many corners. It can also point out the importance of a parameter or interaction mechanism we had completely overlooked. Conversely, a mechanism we could have spent a great deal of time trying to understand and include in our image analysis may be shown to be of little importance.

10.2 APPROACHES TO SIMULATION

In this section, we will briefly review some of the approaches to SIG and some of the methods that have been employed to simulate various phenomena that impact final image formation. In Section 10.3, we will review the basic elements of a SIG model and present some examples of how SIG can be used to analyze and understand certain phenomena.

10.2.1 Physical Models

One approach to SIG is to start with three-dimensional scale models of the scene of interest. These physical models can include terrain, land cover, structures, vehicles, etc., with the degree of detail a function of the resolution of the sensor to be simulated. This approach is described by Francis et al. (1993) for simulations in the reflective region. The model builder in this case includes all the texture and structure into the model that the eye would discern. The scene is illuminated with a collimated beam to simulate the sun and many diffuse sources to simulate the sky (cf. Fig. 10.1). The sunlight-to-skylight ratio and solar location are controlled to simulate the desired conditions. Path radiance is added either through the use of a beam splitter in front of the sensor or in postprocessing for digital simulations. The photographic or EO sensors are located above the target model and the optics adjusted to simulate

Figure 10.1 Simulation facility and synthetic image produced using physical models. (Courtesy of Itek Corporation.)

the desired field of view. The resolution of the imaging system in the simulation facility is usually constrained to be significantly better than the system to be simulated. The PSF of the system to be simulated can then be cascaded with the digital image acquired by the test system. The image in this case is designed to represent the radiance field at the sensor. This approach is attractive in that many of the target-background interactions (e.g., shape factor effects) and within-target texture are taken care of by the physical model. It is also easy to change camera and sun angles for multiple images of the same scene. It has the disadvantage that a skilled model builder is required to construct and shade the targets such that the right reflectance and reflectance variations are included in the scene. This problem becomes acute when this approach is used to simulate multispectral scenes. This can be done to some extent by color balancing the sunlight and skylight sources and carefully selecting paints for the scene elements. However, the process quickly becomes intractable over wide wavelength ranges. Maver and Scarff (1993) describe a hybrid approach to simulate multispectral scenes where the physical models and illumination system are used to generate templates that are then processed using computer models. For example, by just illuminating the scenes with the skylight lamps, an image can be obtained that can be segmented into material types (this can be simplified by the selection of paint shades in the model construction). The variations within a class can be used to generate maps of skylight irradiance levels, including shape factor effects. By using just the sunlight lamp, an image can be obtained that maps just the solar irradiance levels on each target including slope-azimuth effects and shadows. The material maps can then be used to select reflectance spectra from a spectral reflectance database and the scene radiometry reconstructed using radiation propagation models, such as LOWTRAN. The latter stages of this process resemble the steps used in the fully synthetic process discussed in the next subsection (10.2.2) and will not be treated here. The advantage of this approach is that some of the spatial variation can still be included by the model builder, and certain material interactions are also automatically included [e.g., shape factor (F)]. The disadvantage is that the scene segmentation of complex scenes can become difficult, requiring considerable manual editing. Also specular effects (BRDF) are very difficult to simulate. Model elements need to be quite diffuse in order to avoid specular glints from one of the many skylight illumination sources. If a more truly diffuse sky were produced, there would still be the problem of having the model builder properly building BRDF effects into the model. Finally, this method cannot effectively simulate thermal phenomena, so that alternative methods are required at longer wavelengths. While the physical-model approach has significant drawbacks, it has a great advantage when within class spatial detail is important to approximate the visual appearance of an image. This can be seen in Figure 10.1 where a test chamber and model board are shown along with a synthetic image produced by this approach.

10.2.2 Fully Computerized Models

An alternative to the use of physical models and the method that is almost universally used for SIG that includes the thermal infrared spectral region is a fully synthetic approach. The scene elements, radiation propagation, and sensor effects are simulated using computer models. This approach is attractive and widely used because it allows for essentially endless variation in the scene and interaction processes. On the other hand, the computational complexities of this approach in terms of the scientific issues, the coding, and the run time are disadvantages. Because we are interested in a method that covers the EM spectrum from 0.4 to 20 μm, and where all of the interaction mechanisms can be addressed independently, we will concentrate on the fully synthetic approach. Many of these SIG models rely heavily on what is referred to as a *first principles approach.* To the extent practical, they attempt to accurately model the physical processes taking place in the image chain. The net result is a rather high-fidelity model that can provide insight into the process of image formation, as well as the synthetic image itself.

The fully synthetic SIG models are characterized by at least the following components, which are treated in more detail in the next section. They all have some means of creating objects and land cover, which we will refer to as the *object data*, and of assembling these data into a three-dimensional representation of the world, which we will call the *synthetic scene*. Both the object database and the synthetic scene are usually created using some form of computer-aided design (CAD) software. The objects may be facetized surface models, or they may be comprised of combinations of solids. The solids are described by mathematical functions combined using Boolean logic (this process is referred to as *constructive solid geometry,* CSG, cf. Foley et al., 1990). In some cases, the objects will have internal structure, and the surface and internal structural elements will be connected according to thermodynamic linkages. In all cases, every element making up the scene must have associated with it a set of material properties which we will refer to as the *material database*. This database contains information on the optical and, where applicable, the thermodynamic properties of the materials.

The models operating in the thermal region also have to have knowledge of the temperature of each scene element. In some cases, these temperatures are assigned externally; however, in the more comprehensive models, the temperatures are computed by the SIG model itself. In this instance, a meteorological data base must exist to assist in temperature calculations performed by a thermodynamic model capable of computing the surface temperature of all scene elements. At least three types of thermodynamic models are in use: slab models that compute the one-dimensional heat flow perpendicular to a surface and may include internal heat sources or sinks; two-dimensional models that include lateral conduction between adjacent surface facets; and three-dimensional thermal models that include thermal interactions of surface facets with each other and with internal structures. All three types of models may or may not contain radiational heat-exchange interactions with other elements in the scene, depending on their level of sophistication.

A radiometry model is used to generate an estimate of the radiance field toward the sensor. This radiometry model is often coupled with a radiation propagation model such as LOWTRAN (cf. Sec. 6.3.3 and 6.4.2) to compute the level of irradiance onto the surface. The radiation propagation model also computes the effect of the atmosphere along the target sensor path to yield the radiance field at the sensor. The atmospheric propagation models will often have a database of atmospheric conditions that are required as input. Finally, a sensor model must be available to characterize the sensor location, view geometry, field of view, resolution, and spectral and radiometric response.

Figure 10.2 shows a block diagram of the flow and interactions associated with a generic SIG model. Any given SIG model might not be divided along the exact lines shown in Figure 10.2 and may contain more or fewer interactions than shown. Because of varying applications, computational approaches, and simplification schemes, the various SIG models include a rather diverse set of approaches. A detailed treatment of this diversity is far beyond the scope of this text. Rather, in the next subsection we will look in more depth at one particular model to see how the image chain is modeled within the SIG process. However, before beginning this more in-depth treatment, we want to mention some alternative models and highlight some particularly interesting features of the models. All of these models are under active development, so we will not provide any detailed information, since changes are occurring too rapidly for it to be relevant.

Many of the models use rather primitive background models to simplify calculations. Kornfeld and Penn (1993), on the other hand, discuss the use of extremely detailed background objects (trees in particular) for use in the modeling done at the Army Research Lab (cf. Fig. 10.3). This work, which emphasizes ground vehicle combat scenes, is also characterized by detailed modeling of the impact of specific sensor distortion effects on FLIR

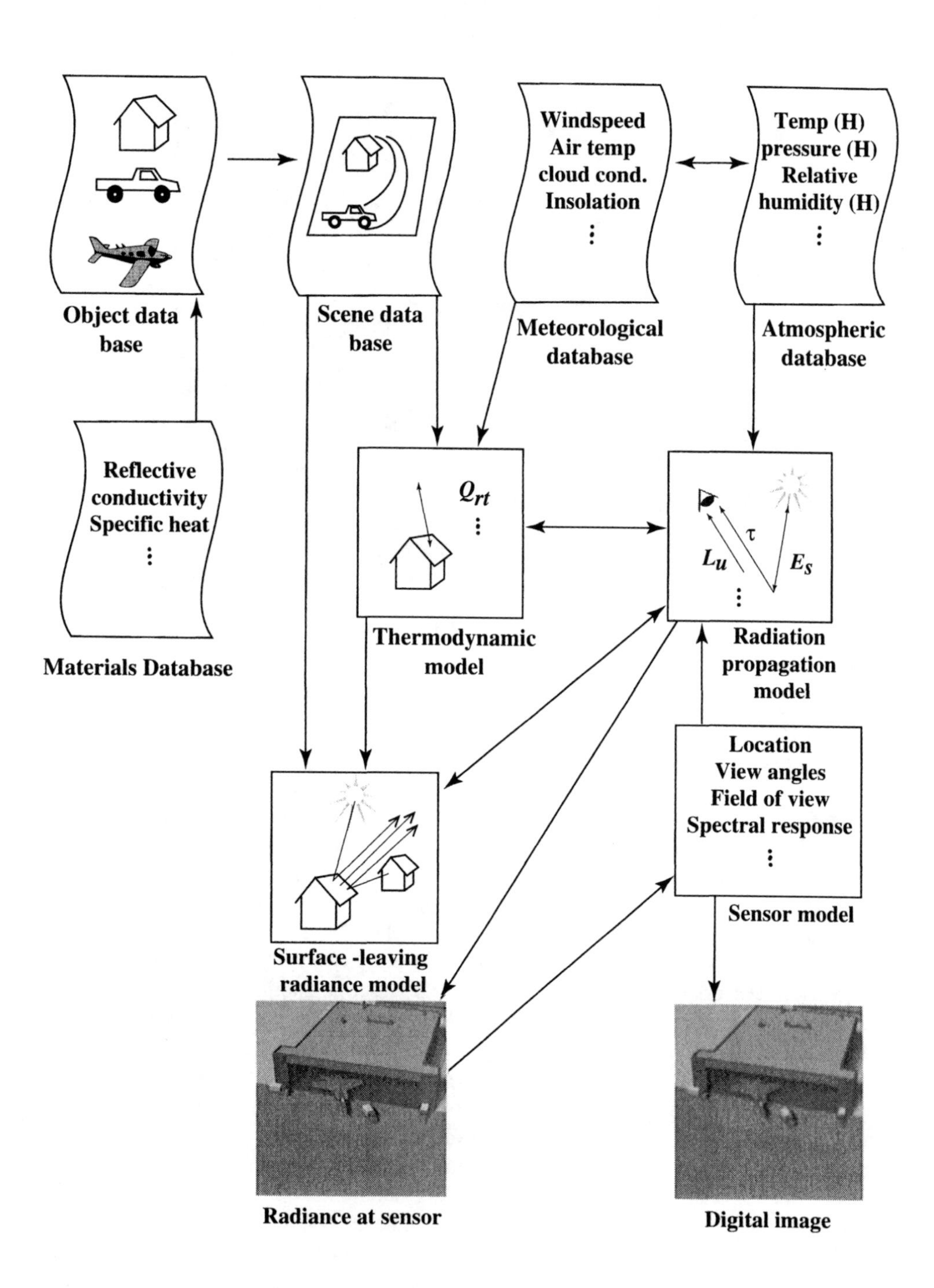

Figure 10.2 Diagram of conceptual data flow and interaction mechanisms in a generic SIG model.

Figure 10.3 LWIR synthetic image of targets in forested area. (Image courtesy of the Army Night Vision Lab.)

imagers. Similarly, many SIG models use simplified approximations to many of the thermodynamic computations. In part, this avoids the complexity of rigorous computations, but perhaps more importantly, it avoids the difficulties associated with laying out and assembling the detailed CAD figures needed to support higher-order thermodynamic calculations. However, Johnson et al. (1993) describe the physically reasonable infrared signature model (PRISM) designed to support very extensive thermodynamic models of targets, including detailed internal thermodynamic interactions. This level of detail is emphasized because one major application is to study the thermal signatures from ground vehicles imaged with very high-resolution sensors (cf. Fig. 10.4). Stewart et al. describe another interesting approach to the thermodynamic modeling problem used in the simulated infrared image model (SIRIM) code. This model divides the objects up into volume elements (voxels) and performs thermodynamic calculations on the interactions between voxels to provide a method to approximate three-dimensional heat flow within objects. These thermal calculations can then be coupled to background simulations (e.g., sea surface structure) to produce final image representations, including target and background interactions (cf. Fig. 10.5). Cathcart et al. (1993) and Sheffer et al. (1993) point out that background modeling can be as important as target modeling and discuss procedures that have been developed for use in the generation of land and sea backgrounds, respectively (cf. Fig. 10.6). All of the SIG models have particular strengths and weaknesses, and the user is cautioned to investigate their capabilities carefully to determine which approach is best suited to the user's particular requirements.

Figure 10.4 Synthetic LWIR image of a tank. (Image courtesy of Michigan Technological University, Keweenaw Research Center.)

10.3 A MODELING EXAMPLE

In this section, we will look in more detail at the components of a SIG model. In particular, since the SIG process is essentially an effort to produce a computer model of the image chain, we want to look at how SIG can be used to break out elements along the image chain that would help us to understand and analyze images. To do this, we will use one SIG model as a point of reference to see in a common framework how the image chain can be modeled. As discussed in the previous section, there are significant variations in the approach to SIG from one model to another. In addition, since even the model we will use as a baseline is under constant upgrade, we will tend to avoid the details that are dynamic and emphasize the conceptual issues. The model we will use as a baseline is the Digital Imaging and Remote Sensing Image Generation (DIRSIG) model described by Schott et al. (1992). It was obviously chosen in part because of the author's familiarity with the model, but more importantly, because it was developed along the image chain principles described throughout this treatment. Furthermore, it uses the same governing radiometry developed in Chapter 4, which will simplify the presentation.

In order to understand how the SIG process can help us to understand the image chain, and, conversely, how the image chain approach can be used in generating SIG images, we will walk through the components of the SIG process illustrated in Figure 10.2. We will look in more detail at how each step represents links and interaction mechanisms along the image chain and at how these links are implemented in at least one SIG model.

The first step in the SIG process is to create a scene. In order to do this, the ground coverage of the sensor and the flight path of interest must be determined (i.e., are we looking at a single frame, sequential frames along a track, an approach sequence, a strip beneath a push

Figure 10.5 Synthetic image of a ship on modeled ocean. (Image courtesy of ERIM.)

Figure 10.6 Synthetic LWIR image showing a heliocopter over simulated water. (Image courtesy of Georgia Tech Research Institute.)

broom, etc.?). We also need to know the approximate resolution of the sensor in terms of GIFOV. This helps determine the degree of detail needed in the CAD models. For example, to simulate a sensor with an 80-m GIFOV, it doesn't make sense to show the detailed structure on the roof of a building. On the other hand, to simulate an aerial system with a 0.1-m GIFOV, this level of detail may be important. The level of detail will also often be a function of the particular application. For example, if we are interested in studying stress in a particular crop, we might have very detailed models of the structure of that crop and yet include only the most primitive structure in adjacent crops. (N.B.: Depending on the approach taken and the resolution, this could be a 3-D spatial model of the structure with optical properties of each element or a detailed BRDF model of the canopy treated as a unit.)

The objects in a scene are created using CAD models such as shown in Figure 10.7. DIRSIG currently uses facetized surface models that may contain one or more internal heat sources (sinks) for use in simulations where self-emitted flux is relevant. Each facet is assigned a material ID, which is associated with the material database. The terrain can also be created using the CAD approach. However, if digital elevation models (DEM's) are available, they can be converted into facetized surface models in the same format as the object data (the DEM data can also be interpolated to finer sample centers and structure added if the data are too course for the required resolution). Material types can then be assigned to the facetized terrain using the CAD process. For large regions, this can be a very tedious process. An alternative approach that has been implemented involves using image data that is geometrically registered to the DEM. If this image data can be classified by land cover or material type, the material ID associated with that land cover can be automatically assigned to the corresponding facets derived from the DEM. In either case, the final scene is produced by using the CAD process to place objects where required in the scene, and a geographic location is assigned to the scene coordinates. Specialized objects or objects with particular features are also included at this point using the CAD process and control functions. For example, objects that move as a function of time (clouds, vehicles) are located at various points in the scene and the time sequencing assigned (i.e., when is the object where?).

Very closely linked to the scene database, whose construction we just described, is the material database. Each material in the scene is a pointer to this database. The database contains the optical and thermodynamic properties of each material type. The optical properties consist primarily of reflectivity (emissivity) as a function of wavelength and view angle from the nadir over the wavelength range of interest or spectral BRDF data where appropriate (due to limited databases, most objects are modeled as a linear combination of specular and diffuse components). For translucent objects, such as clouds, this will also include the spectral extinction coefficient. The thermodynamic parameters include the solar absorption coefficient, specific heat, conductivity, thickness, and broad-band emissivity, and may include information concerning the magnitude, operating times, and operating levels of internal heat sources (sinks).

In order to access the scene data, the DIRSIG model turns to the sensor model. The sensor model includes as a function of time the location of the sensor in three-dimensional space, the orientation of the primary optical axis, the sensor field of view, and sampling information. For example, for a simple framing sensor, the sensor model would include the FOV with respect to the sensor's x and y directions and the number of pixels in x and y. The sensor model also includes the number of spectral bands and the spectral response function of each band. In order to produce a final digital image, we want to first generate a representation of the image that contains spatial, spectral, and radiometric resolution, equal to or greater than the final image so that any degradation effects can be properly treated. This is accomplished in DIRSIG by producing a radiance image that represents the radiance in each band (spectral radiance weighted by the sensor response) at higher spatial resolution than the sensor and before any sensor-induced radiometric noise or quantization. To generate the required spatial resolution in the radiance image, 4 to 9 times as many pixels are generated

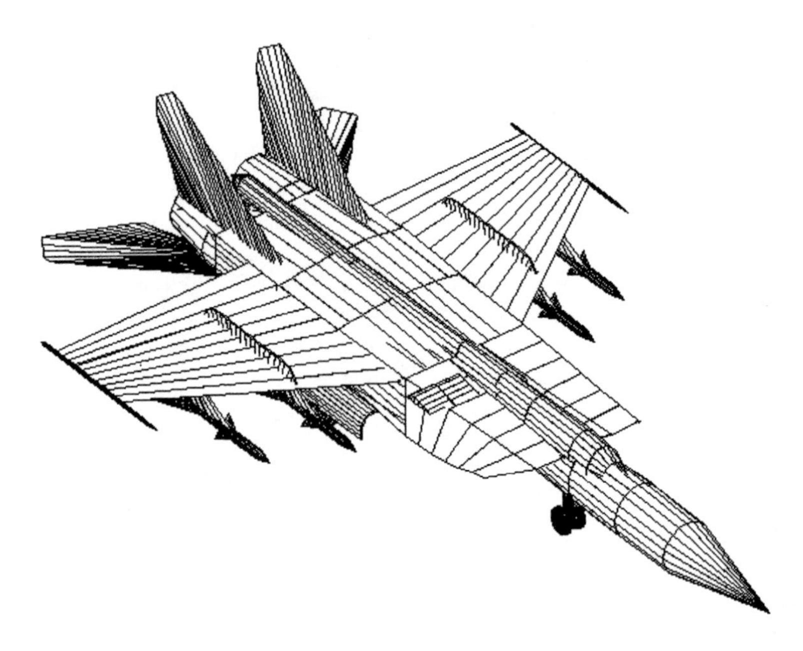

Figure 10.7 Wire frame of an object used in the SIG process. The object is produced using CAD software, and material types are assigned to each facet during the construction process.

in the radiance image as will be present in the final synthesized image.

A ray tracer is used to access the scene data and integrate the entire scene generation process. To simplify our treatment, we will assume, we are imaging with a pinhole camera and that for each "pixel" in the radiance image we trace a ray from the focal point of the optical system through the pixel center and into the scene, as illustrated in Figure 10.8. The ray tracer's function is to trace backward along the photon paths to determine what interactions took place that would affect the flux on the sensor at the point of interest. The ray tracer locates the point on a facet where the ray hits the scene and begins to gather data. First the material database is accessed to determine the material type, as well as the slope and aspect of the material. If the material is translucent (e.g., a cloud), the ray trace is continued until an opaque object or the sky is hit. For translucent objects, the distance the ray travels through the object must be computed so, the extinction can be determined. The ray tracer also computes a number of geometric factors such as; the view angle (θ), the slant path to the sensor (for transmission and path radiance calculations), angle from the surface normal to the sensor, and direction of specular reflection. It then casts out rays to ascertain three additional sets of data. First, it casts a ray toward the sun to see if the target is in sun or shade (assuming it's a daytime image), and, if running in the thermal region, it casts rays to where the sun has been every k minutes for the previous N hours (N is usually 12 or more hours; k is usually 15 or fewer minutes) to generate a sun-shadow history for the target (cf. Fig. 10.9). Second, rays are cast into a hemisphere above the target plane to determine the shape factor and the radiation environment of the target surround (cf. Fig. 10.10). In cases where full BRDF data are available, these rays can gather the temperature, reflectance, and sun-shade condition of adjacent surfaces to determine directional radiance loads for BRDF calculations. Third, a ray is cast in the specular bounce direction to determine if an object (including cloud) is hit (if so, its temperate, reflectance, and sun-shadow condition are determined), or if no object is hit, the ray is cast to the sky to determine where in the sky the specular ray came from. This is important because the sky is not isotropic, and the variations in the sky

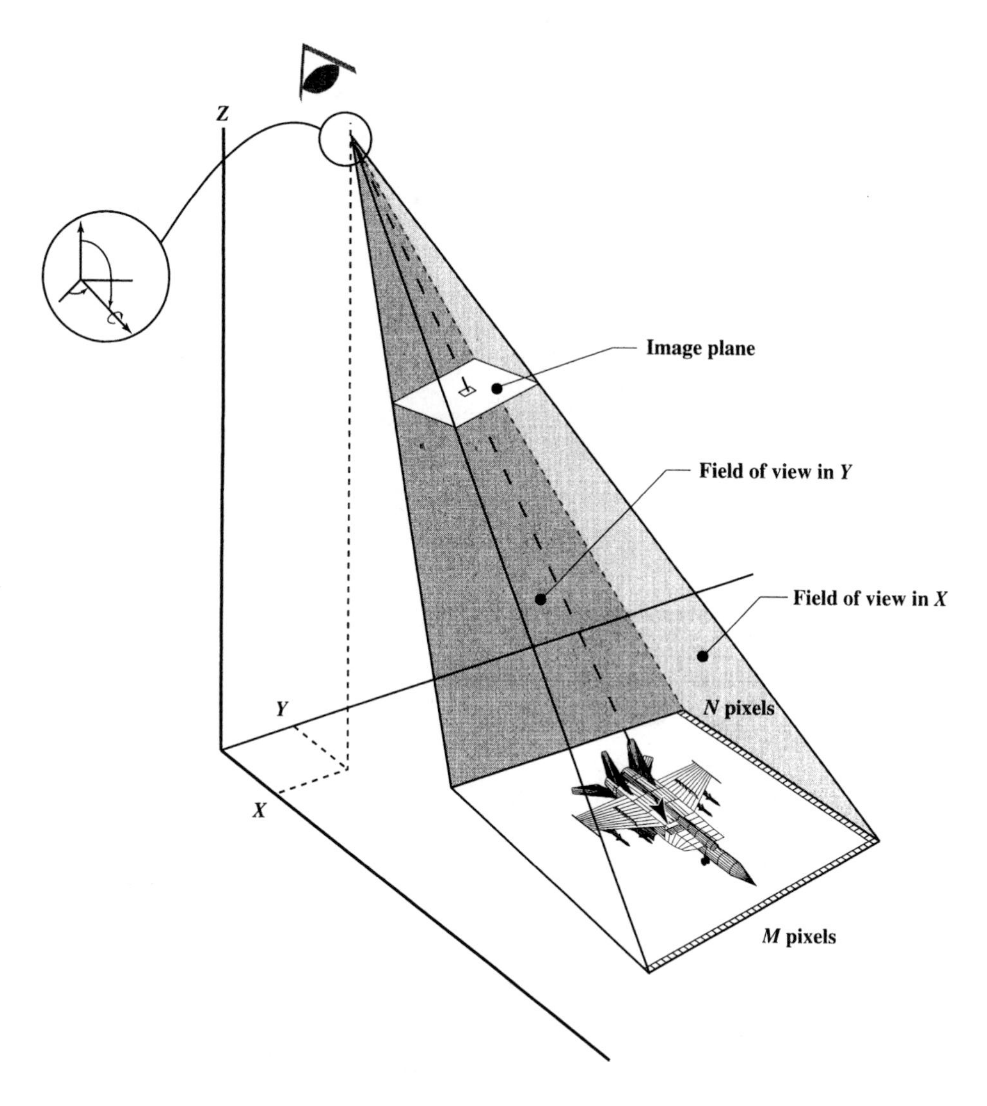

Figure 10.8 Illustration of the ray-tracing process for a simple framing camera. To generate an $N \times M$ radiance array, rays are traced from the focal point through each pixel center in an $N \times M$ image plane.

can greatly impact the appearance of specular objects.

The data the ray tracer has gathered is sent to the thermal model if self-emitted photons are relevant in the bands being simulated. The temperature of most objects is a function of the object's thermal inertia and environmental driving factors. The SIG model acquires data about thermal inertia from the materials database (specific heat, conductivity, density, thickness), but a meteorological history must also be available. This history consists of direct and diffuse insolation levels, air temperature, relative humidity, cloud type and coverage, precipitation rates (and temperature), air pressure, and wind speed as a function of time. This history may be available from experimental data or may be predicted based on standard meteorological forecast data along with geographic data and a forecast date. The insolation levels are adjusted based on the sun-shadow history information, and the radiational exchange is modified by the object's solar absorptivity, the shape factor term, and the temperature of adjacent objects. The temperature is calculated by a thermal model that uses all the available data

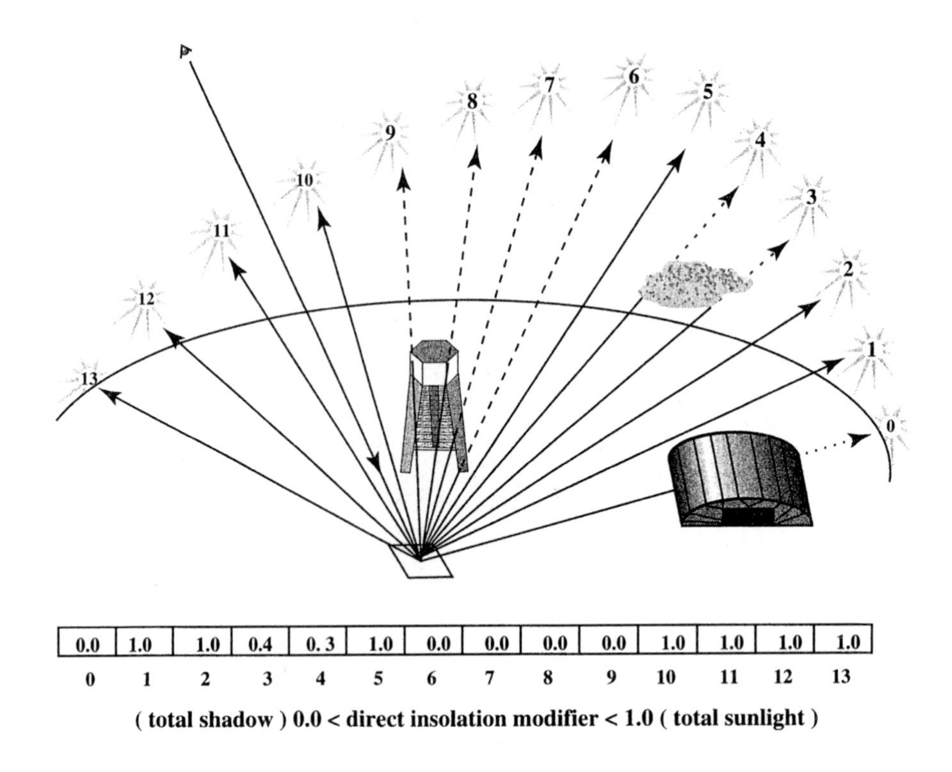

Figure 10.9 Sun shadow history.

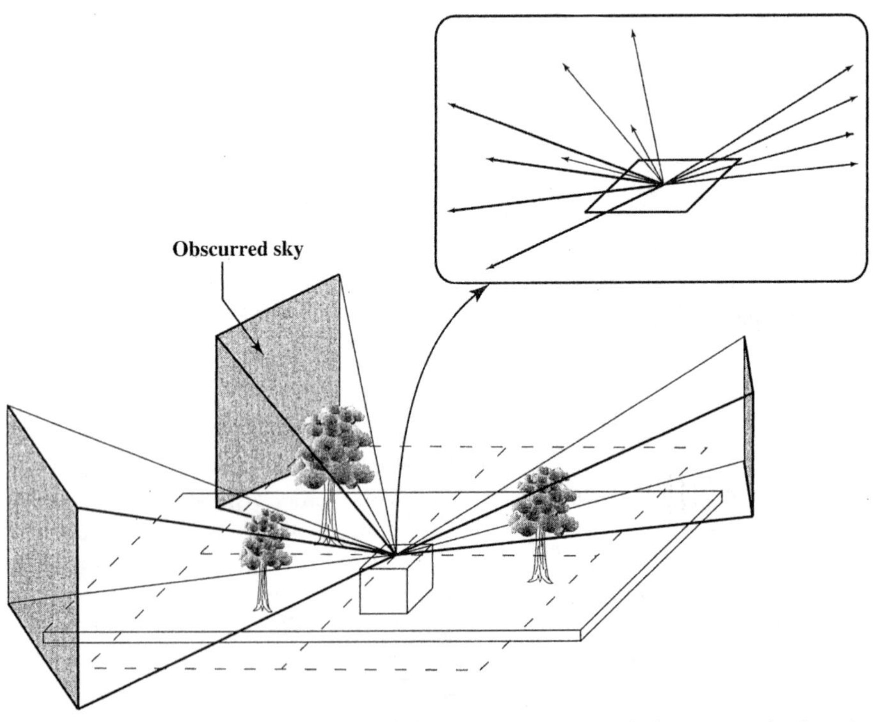

Figure 10.10 Rays are cast into the hemisphere above the plane of the target pixel to compute the shape factor for use in the thermodynamic modeling of radiational exchange and in the radiometric image formation model.

(including information on internal heat sources) to solve a forward chaining differential model. DIRSIG uses a modified version of the THERM thermal model, DCS Corporation (1991). This type of model is very effective for passive objects; however, more exotic models may be used where internal heating effects become complex. The target temperature is then added to the other data the ray tracer has collected, and the data are sent to the radiometry model.

The DIRSIG radiometry model essentially consists of the spectral version of the *big equation* of Chapter 4 [cf. Eq (4.61)] without most of the simplifying assumptions other than the use of numerical approximations for most of the integrations. The radiation propagation terms E_s, τ_1, τ_2, L_u, and L_d are all derived from MODTRAN (cf. Berk et al., 1989) and vary as a function of wavelength, slant range, and view angle. The atmospheric makeup is controlled by the standard user-supplied inputs to the radiosonde profiles in MODTRAN, with the lowest layers modified by the surface meteorology from the DIRSIG weather data files. The radiometry model solves for the spectral radiance reaching the sensor and then computes the effective radiance in the band using the spectral responsivity for each of the sensor's spectral bands. At this point, the radiance values are inserted into the radiance image arrays for each band. In addition, essentially all of the data that went into the computation of the final radiance can also be inserted into an image array for training, diagnostic, or advanced algorithm development purposes. From our perspective, we can think of these diagnostic images as a history of the radiometric image chain for each pixel in the radiance image. For example, the within-band transmission image consists of the $\tau_{2\text{avg}}$ value for each pixel in the radiance image. The radiance image can be thought of as the result of sampling the radiance field reaching the sensor with an array of delta functions located at the radiance image pixel centers. The radiance field must then be convolved with the PSF of the sensor system to generate the effective radiance values associated with the final image pixels. This is accomplished either in the spatial or frequency domain using the procedures discussed in Chapter 9 and illustrated in Figure 10.11. The radiance values are then converted to electronic values according to the response function of the system being simulated. Both random and structured noise, as discussed in Section 9.3, can be added to the signal at this point. Finally, the image is quantized with a quantizer that simulates the quantization levels in the sensor. Note that for simplicity we have neglected sensor platform motion effects, treatment of more exotic sensors, and most of the details of the SIG process. The reader is referred to the reference literature and the model developers for more detailed information.

A radiance field image is shown in Figure 10.12 along with some of the diagnostic images associated with its image chain. In Figure 10.13, a radiance image is shown along with some of the diagnostic images associated with the latter stages of the image chain.

The SIG images that result from this process are generally too flat because of naturally varying texture in many material classes. For example, the model builder will often not change the density or structure of grass or include wear marks in pavement unless these are considered critical elements (e.g., crop structure might be built into the canopy of a particular crop under study). It is, of course, possible to build models to the level of detail where structure is included automatically in the SIG process. However, the level of effort involved in the building of the models and the negative impact on run time is often not justified. An alternative procedure incorporates spatial structure into the latter stages of the SIG process. This procedure leaves the mean-level radiance values in a region unchanged and introduces variations about the mean by spatially varying the reflectance as a function of material type. Schott et al. (1995) describe how this approach can be implemented to preserve the spatial and spectral correlation within and between spectral bands for generation of multispectral SIG images. A limitation of this approach is that the spatial structure of the land cover or material in at least one spectral band must be known at the scale of the radiance field image. This structure data can be generated mathematically as described by Haruyama and Barsky (1984) and Pentland (1984) or extracted from actual imagery. The resultant SIG images con-

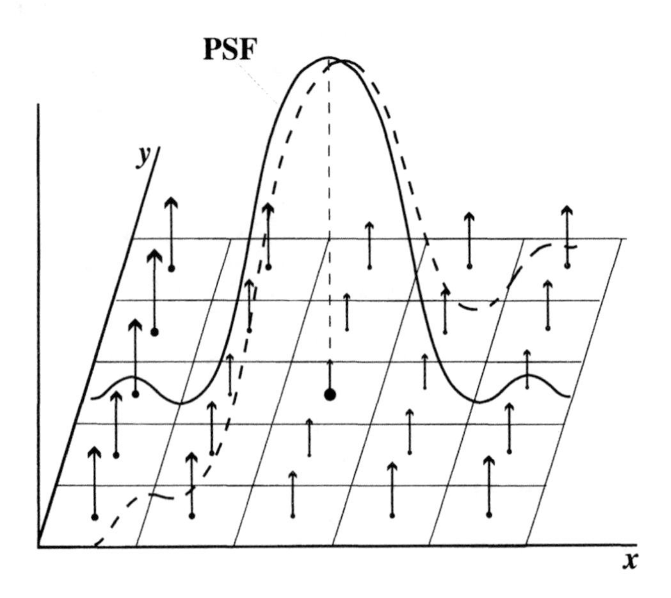

Figure 10.11 The final synthetic image is obtained by convolving the radiance image (indicated by the pixel heights in the figure) with the point spread function (PSF) of the system and sampling the resulting image.

tain a good approximation of the spatial, spectral, and radiometric characteristics of actual images. Figure 10.14 shows an example of an image where texture has been included.

10.3.2 Application of SIG Models

In this subsection, we will examine just a few of the ways in which SIG can be used to help us better understand and analyze certain remote sensing phenomena. These examples are not intended to be comprehensive, as they merely scratch the surface of what SIG is currently capable of doing . Ongoing improvements to SIG will make it even more valuable in understanding and extracting information from images.

Figure 10.15 shows an example of how radiometrically correct SIG can be used to characterize the impact of atmospheric effects on remotely sensed images. In this case, the scene was assumed to be a uniform, flat Lambertian reflector with a mean reflectance equal to the local albedo. The sensor modeled is a line scanner flying a north-south heading. Variations in the resultant image show how the atmosphere would be expected to influence a wide field of view sensor. An actual image has variations due to both atmospheric and BDRF variations in scene elements. These combined effects are shown in the Advanced Solid-State Array Spectroradiometer (ASAS) images shown in Figure 10.16. The ASAS sensor is designed to study BRDF effects by taking images of the same spot at several angles ahead of and behind a sensor. The images show the dramatic variations in how the same site appears when viewed at 19 different geometries (cf. Ranson et al., 1994). The SIG models can be used to assist in understanding both the relative shape and magnitude of these phenomena. If sufficient atmospheric data are available, the SIG data can even be used in generating spatially dependent atmospheric correction values for use in radiometric image analysis (i.e., atmospheric bias and gain "images" could be made and applied to the actual images through subtraction and division). This is just one example of how SIG can be used, both to understand a phenomena and potentially to generate analytical corrections to assist in image analysis.

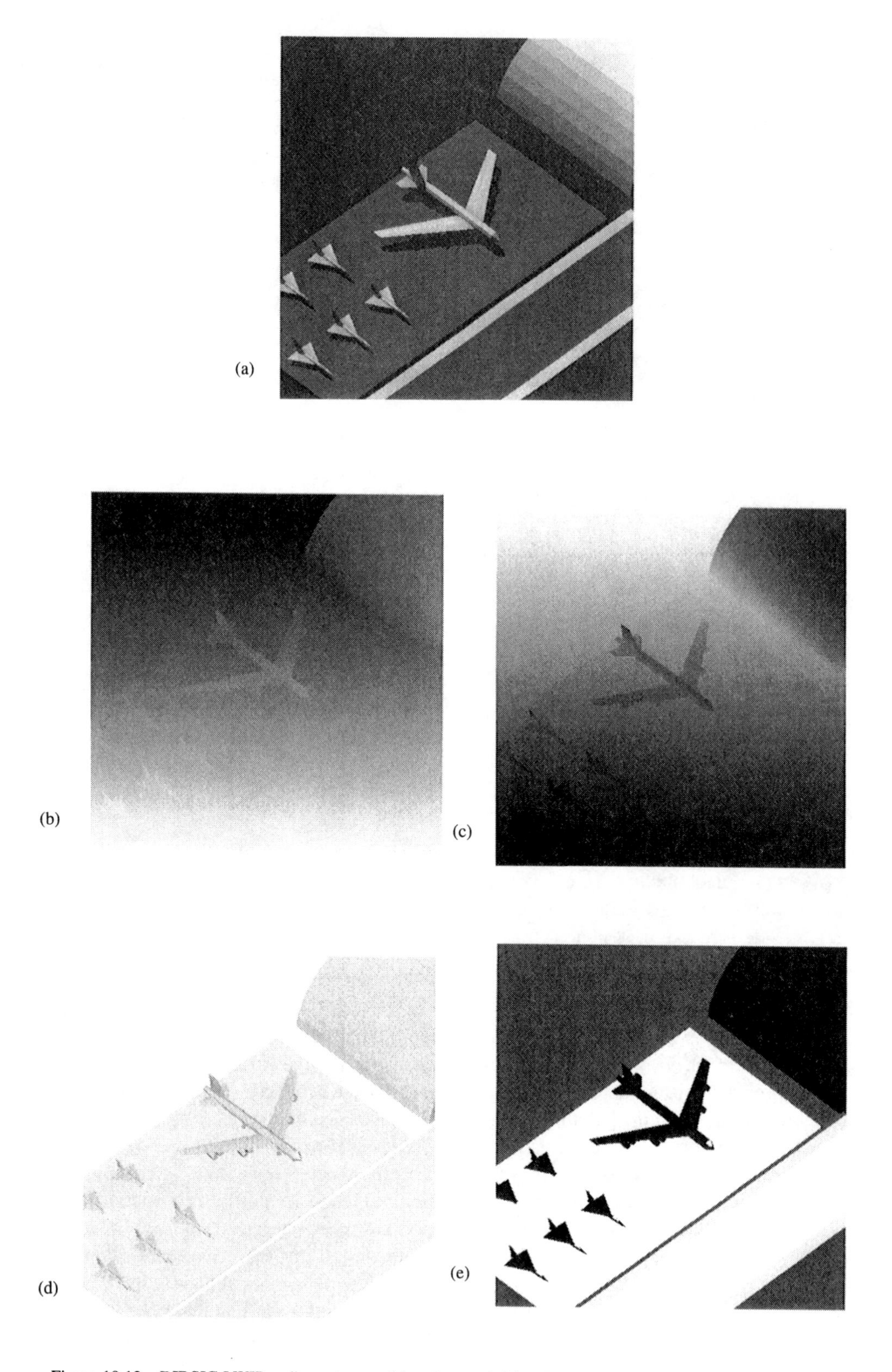

Figure 10.12 DIRSIG LWIR radiance image: (a) and several debug images, (b) transmission along line of site, (c) path radiance, (d) target emissivity, and (e) target material map.

(a) 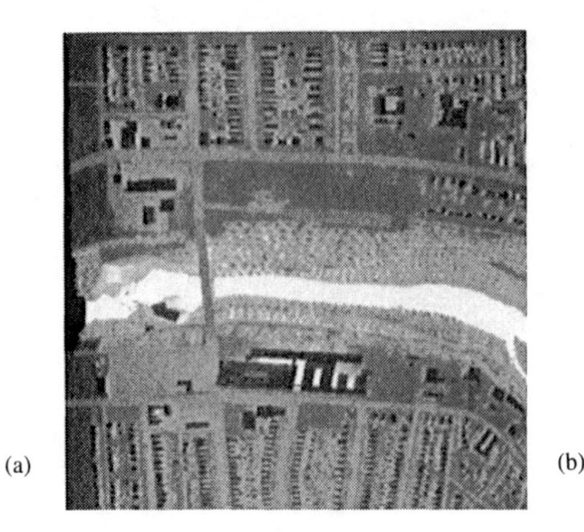 (b) 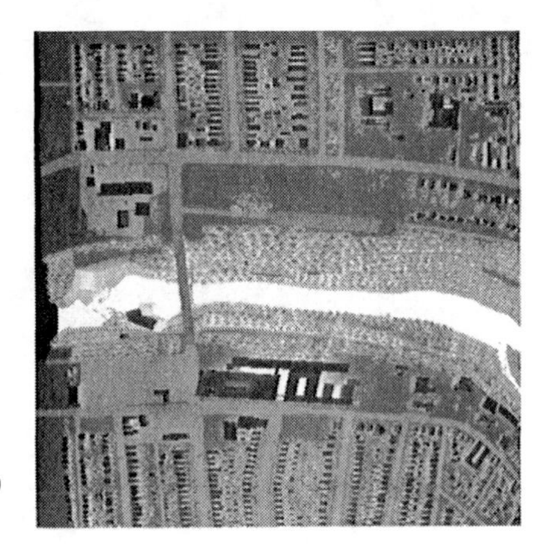

(c) 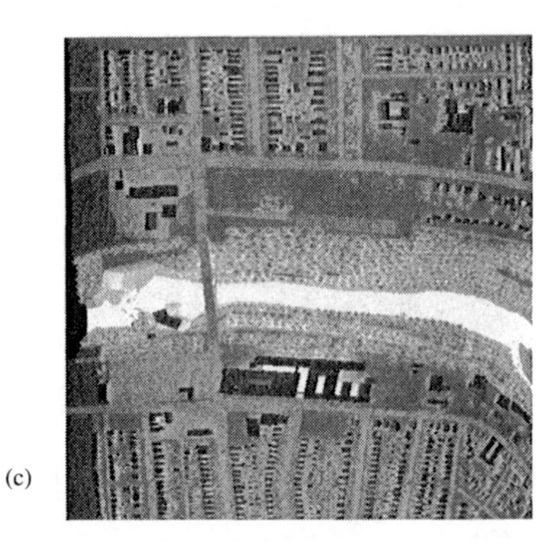 (d) 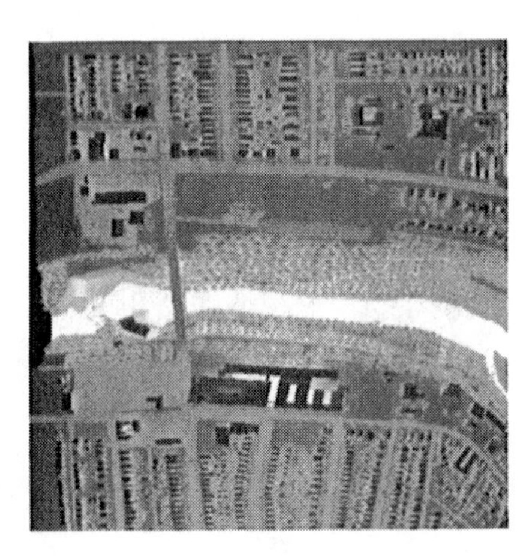

Figure 10.13 A final DIRSIG image (a) and debug images showing later stages of the image chain, (b) radiance image, (c) radiance image with noise effects, and (d) radiance image with MTF effects. Final image includes noise, MTF, and sampling effects.

An application of the ability of SIG to model thermal effects as a function of time is shown in Figure 10.17(a). These LWIR images show an airfield from which aircraft have left every 30 minutes for the past 3.5 hours (seven aircraft have departed). The differential temperature resulting from the length of time shadows have been cast on the high thermal inertia parking apron provides a potential signature to assess these departure (or arrival times). However, Figure 10.17(b), shows the sensitivity of this type of signature to environmental conditions. In this image the same scene is modeled, but the wind speed, which was assumed to have been constant at 3 mph in the first scene, has now been raised to 10 mph (wind speed in both scenes was held at a constant level for the previous 12 hours). A comparison of the images shows how thermal signatures are a function of environmental variables. In this case, the increased wind speed tends to drive all object temperature toward ambient air temperature and "scrubs" thermal contrast from the scene. The complex interaction of optical and thermal effects in MWIR and LWIR images make detailed interpretation of the images very difficult without the aid of tools such as SIG.

The importance of understanding the directional reflectance properties of surfaces is

Figure 10.14 A false-color synthetic scene produced by DIRSIG showing texture effects. See color plate 10.14.

Figure 10.15 Contrast-enhanced blue spectral band DIRSIG line scanner image of a uniform Lambertian ground model. Plane is flying north (up) with the sun east (right). The brightness variations in the image are strictly due to atmospheric transmission and scattering effects.

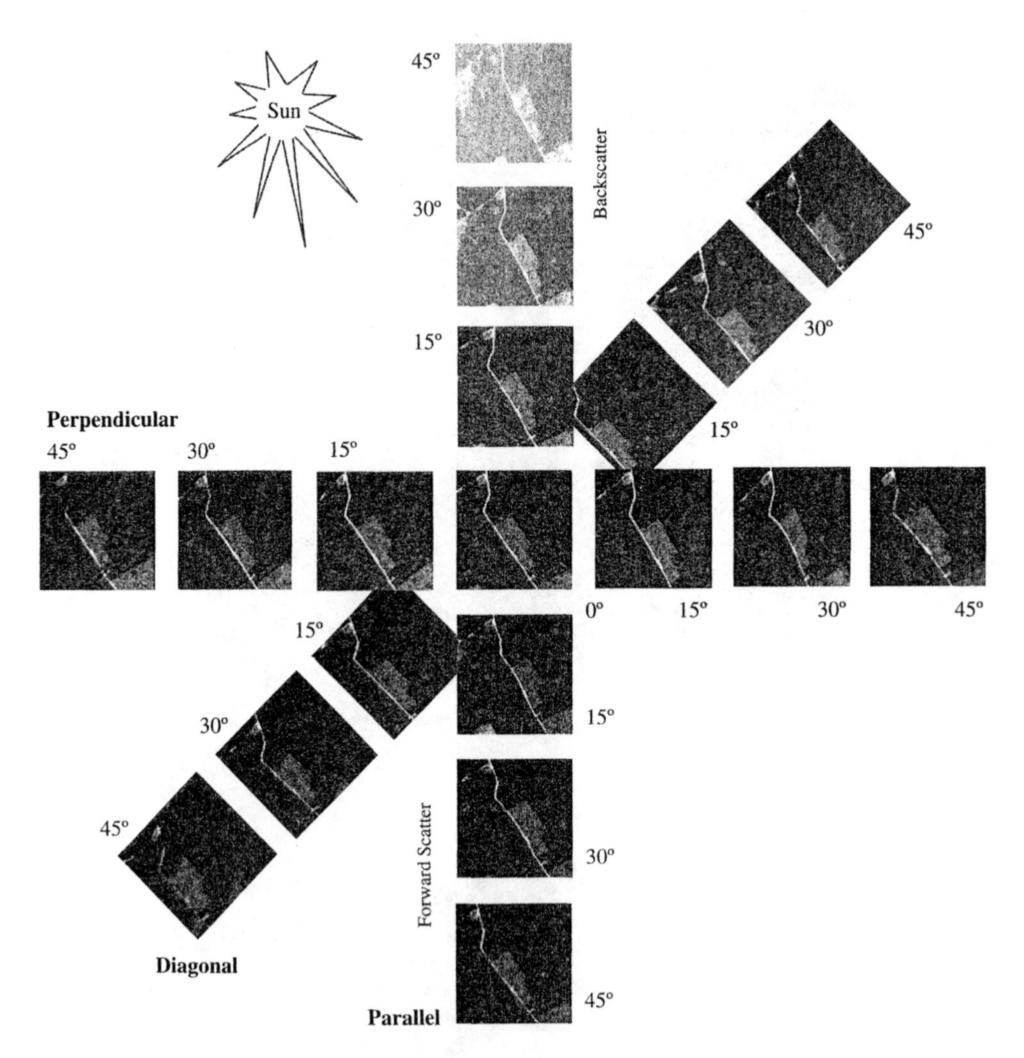

Figure 10.16 Color IR version of ASAS images from three flight lines showing combined effects of atmospheric and bidirectional reflectance variation. See color plate 10.16. (Image courtesy of NASA Goddard.)

shown in Figure 10.18. Here the reflectance of the parking apron is modeled first as highly specular, then as a more realistic surface that is largely diffuse with a small specular component. In the highly specular case, we see that the aircraft appears to cast two shadows. The first shadow is due to blocked sunlight. The second shadow is due to the plane obscuring the downwelled radiance from the sky in the specular direction. The effect is largely eliminated when the surface is modeled as a nearly Lambertian reflector that weights the downwelled radiance from all directions and reflects it toward the sensor. The actual surface has some specular component at the wavelength modeled and will actually be slightly darker at the specular angle as modeled by the case with the slightly specular surface.

The SIG modeling process can be used to study the impact of various sensor designs on imagery. One of the more fundamental differences is shown in Figure 10.19 where a scene is imaged as it would be seen by a framing camera and a line scanner. Note, in particular, how the frame camera image exhibits radial relief displacement, while the relief displacement in the scanner image is only in the across-track direction. Sensor platform effects are

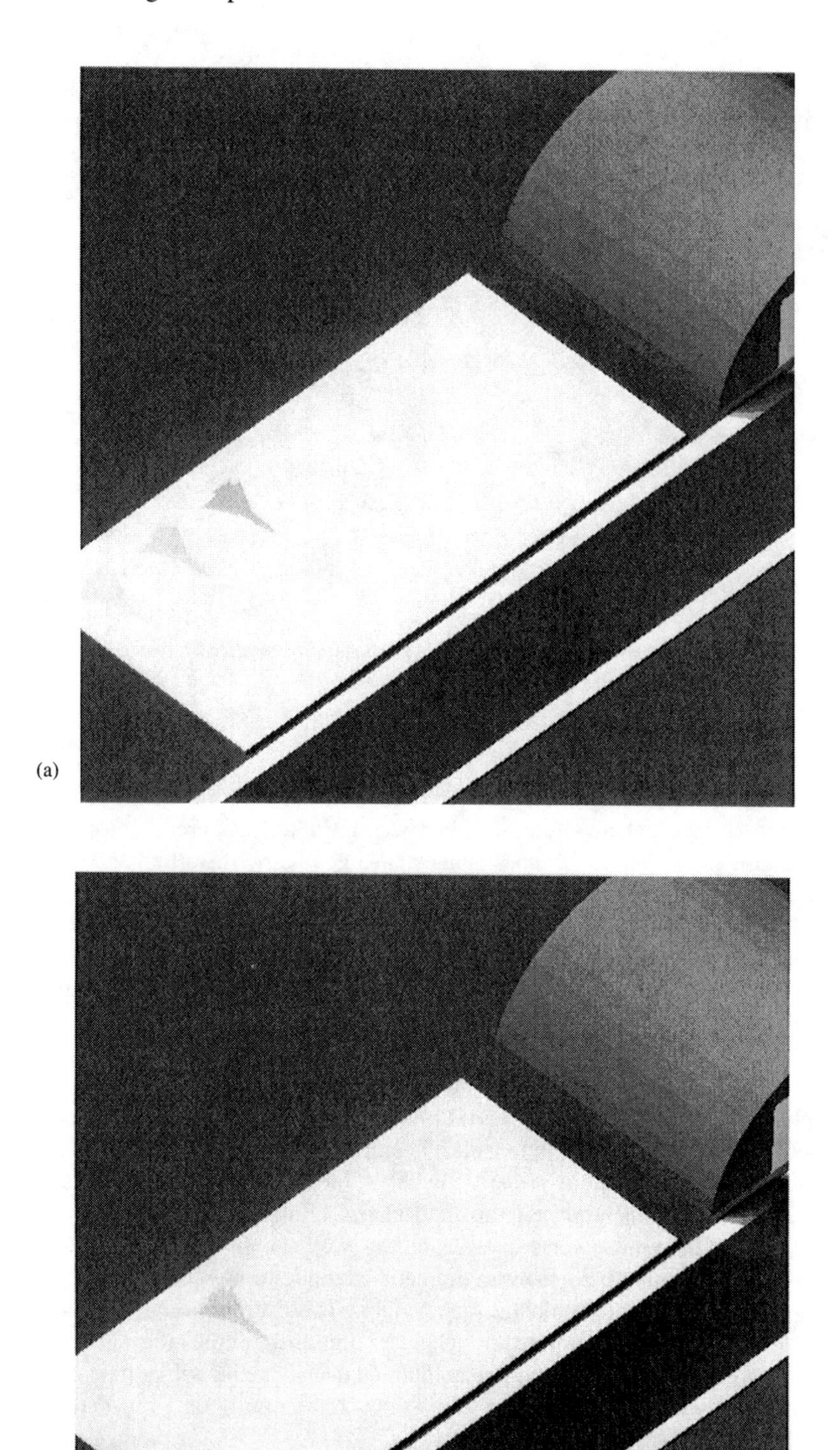

Figure 10.17 LWIR DIRSIG image showing airfield at different wind speeds (see text).

100% specular

10% specular

Figure 10.18 Sample image subsections produced by DIRSIG displaying the effects of specularity. Note the "double shadow" in the image modeled with 100% specular concrete.

also important, since the scanner will not be stable over the period of image acquisition. This is seen in Figure 10.19(c) where aircraft roll has been introduced. Careful analysis of these images would also show that the scanner images have geometric distortions due to angular sampling effects (i.e., tangent distortion and resolution degradation with angle). It is important to realize that these variations in the image characteristics are critical not only in design considerations. The performance of image processing and analysis algorithms will vary significantly with sensor characteristics, and SIG can play a crucial role in helping to develop and evaluate image analysis algorithms appropriate for different sensors or acquisition conditions.

One of the most powerful uses of SIG is to help us visualize and evaluate the importance of various phenomena on the remotely sensed image. Throughout the earlier chapters, we have repeatedly emphasized the importance of the atmosphere on the radiance levels reaching the sensor. If the atmosphere is properly modeled, the SIG process can be an extremely powerful tool in evaluating the impact of the atmosphere on image characteristics. It can also be used in reverse engineering studies to evaluate how well an approach works in removing atmospheric effects. Figure 10.20 shows a dramatic example of how the radiance from the sky will vary as a function of location and wavelength. These variations are reflected (literally and figuratively) in a muted form by the specular component of the reflectance of objects in the scene. Analysis of synthetic images can help us determine how important any particular phenomena is to our image analysis, as well as how to remove or take advantage of these phenomena that are significant.

One important issue regarding SIG has to do with the quality or fidelity of the models. How good are the SIG representations? Of course the answer is usually dependent on the application. Some applications need only to determine the relative contrast between a target and a background. Others need to properly reproduce mean radiance levels and all the within-object variations. Some need to reproduce only the statistical behavior of backgrounds and certain features of a critical target set. One way to put this is that a SIG image that correctly mimics a real image for those parameters we look at, or measure, is acceptable. Clearly, a measure based on these criteria would not be robust, since a slight change in what

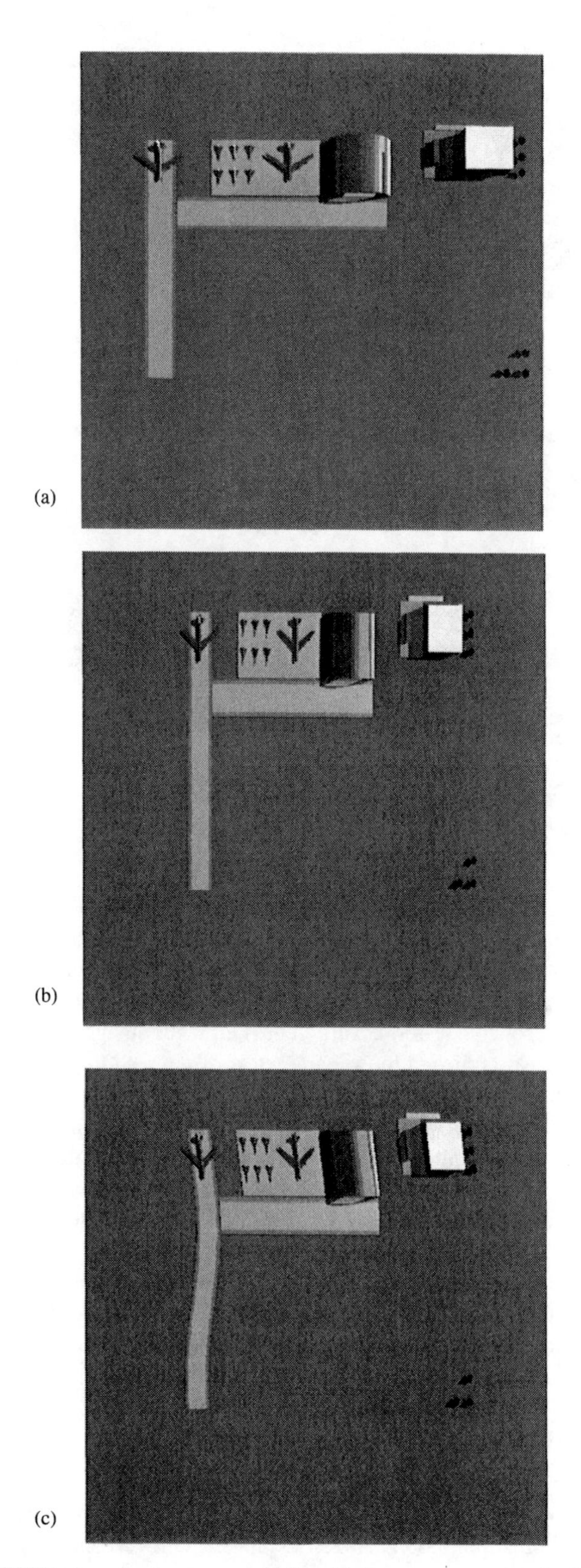

Figure 10.19 DIRSIG images showing sensor modeling: (a) frame camera model, (b), line scanner model including V/H and tangent errors, and (c) line scanner model with aircraft roll included.

Figure 10.20 A synthetic sunset produced by DIRSIG. This effect is possible due to the extensive spectral modeling DIRSIG incorporates with the help of MODTRAN. See color plate 10.20.

we look at (e.g., addition of a new segmentation feature) could cause a catastrophic failure in the effective quality of the SIG representation. To date, there has been only limited development of procedures and metrics for assessment of the quality of SIG images. The state of the art has been that reasonableness was sufficient (i.e., if the image looked reasonably like an actual image or at least mimicked those parameters of interest, that was sufficient). With improvements in the SIG modeling process, especially in the area of radiometric modeling, we are asking more and more of the SIG tools and, therefore, need more rigorous methods for evaluating their capabilities and limitations. Rankin et al. (1992) describe an experimental design and test metrics for evaluation of the mean-level thermodynamic and radiometric aspects of thermal IR SIG images. Mason et al. (1994a,b) provide some additional metrics for relative contrast assessment using rank order statistics. However, rather than answer the question of how to evaluate the total SIG process, these authors only offer some possible metrics for addressing certain issues. The more fundamental question of SIG image quality will undoubtedly be as ubiquitous as the general question of how to characterize image quality.

10.4 SIG MODELING AND THE IMAGE CHAIN APPROACH

We have chosen to close this text with a treatment of synthetic-image generation because it can be a powerful tool in helping to visualize the image chain. We emphasized sensor modeling, since that's where the initial emphasis on SIG has been, and modeling the rest of the image chain (processing and output) is a logical extension. From the image chain perspective, SIG offers a particular value in that if properly implemented, it merges the radiometric, spatial, and spectral strands of the image chain in a way that closely resembles the actual process. The subtleties of the interactions within and between these strands of the image chain can be very difficult to study and analyze without a tool like SIG. However, with all its pluses, it should not be perceived as a panacea. The ultimate goal of SIG is to mirror all the relevant aspects of the world being imaged. This is, for the foreseeable future, an impossible task. For many applications, SIG may already do an acceptable job and may fulfill more and more requirements as modeling tools are improved. However, the complexities of the world and the imaging process are such that many fine points are currently, and will continue to be, missed by the modeling process. Thus the models should be treated as tools that approximate the process but do not fully represent it. Such tools are very valuable in understanding and analyzing images, but at best they are only as good as our fundamental understanding of the process we are studying. SIG can help us in designing, analyzing, and often reducing the extent of field studies and collection programs, but it cannot replace them. In short, if you want to know what the world looks like, don't forget to look out the window.

10.5 REFERENCES

Berk, A., Bernstein, L.S., & Robertson, D.C. (1989). "MODTRAN: a moderate resolution model for LOWTRAN 7." GL-TR-89-0122, Spectral Sciences Inc., Burlington, MA.

Cathcart, J.M., Faust, N.L., Sheffer, A.D. Jr., & Rodriquez, L.J. (1993). "Background clutter models for scene simulation." Proceedings of the SPIE, Vol. 1938, No. 37, pp. 325-336.

DCS Corporation (1991). "AIRSIM thermal signature prediction and analysis tool model assumptions and analytical foundations." DCS Technical Note 9090-002-001.

Foley, J.D., vanDam, A., Feiner, S.K., & Hughes, J.F. (1990). *Computing Graphics: Principles and Practices*, 2d ed., Addison-Wesley, Reading, MA.

Francis, J., Maver, L., & Schott, J.R. (1993). "Comparison of physically and computer generated imagery." Proceedings of the SPIE, Vol. 1904, pp. 20-23.

Haruyama, S., & Barsky, B.A. (1984). Using stochastic modeling for texture generation. *IEEE Computer Graphics and Applications*, Vol. 4, pp. 7-19.

Johnson, K.R., Curran, A.R., & Gondo, T.G. (1993). "Development of a signature super code," Proceedings of the SPIE, Vol. 1938, No. 36, pp. 317-304.

Kornfeld, G.H., & Penn, J. (1993). "Various FLIR sensor effects applied to synthetic thermal imagery." Proceedings of the SPIE, Vol. 38, No. 39, pp. 350-367.

Mason, J.E., Schott, J.R., & Rankin-Parobek, D. (1994). "Validation analysis of the thermal and radiometric integrity of RIT's synthetic image generation model, DIRSIG." Proceedings SPIE, Vol. 2223, pp. 474-487.

Mason, J.E., Schott, J.R., Salvaggio, C., & Sirianni, J.D. (1994). "Validation of contrast and phenomenology in the digital imaging and remote sensing (DIRS) lab's image generation (DIRSIG) model." Proceedings of SPIE, Vol. 2269, pp. 622-633.

Maver, L., & Scarff, L. (1993). "Multispectral image simulation." Proceedings of the SPIE, Vol. 1904, pp. 144-160.

Pentland, A.P. (1984). Fractal-based description of natural scenes. *IEEE Transactions on Pattern Analysis and Machine Intelligence*, Vol. PAMI-6, No. 6.

Rankin, D., Salvaggio, C. Gallagher, T., & Schott, J.R. (1992). "Instrumentation and procedures for validation of synthetic infrared image generation (SIG) models." Proceedings SPIE, Infrared Technology XVIII, Vol. 1762, pp. 584-600.

Ranson, K.J., Irons, J.R., & Williams, D.L. (1994). Multispectral bidirectional reflectance of northern forest canopies with the advanced solid-state array spectroradiometer (ASAS). *Remote Sensing of Environment*, Vol. 47, pp. 276-289.

Salvaggio, C, Sirianni, J.D., & Schott, J.R. (1993). "Use of LOWTRAN-derived atmospheric parameters in synthetic image generation models." Proceedings of the SPIE, Vol. 1938, No. 34, pp. 294-307.

Schott, J.R., Raqueño, R., & Salvaggio, C. (1992). Incorporation of time-dependent thermodynamic model and a radiation propagation model into infrared three-dimensional synthetic image generation. *Optical Engineering*, Vol. 31, No. 7, pp. 1505-16.

Schott, J.R., Salvaggio, C., Brown, S.D., & Rose, R. (1995). "Incorporation of texture in multispectral synethetic image generation models." Presented at SPIE Target and Backgrounds: Characterization and Representation Conference, Orlando, FL.

Sheffer, A.D., Jr., Cathcart, J.M., & Stewart, J.M. (1993). "Ocean background model for scene simulation." Proceedings of the SPIE, Vol. 1938, No. 38, pp. 337-349.

Stewart, S.R., Lyons, J.T., & Horvath, R. Simulated infrared imaging (SIRIM): "user's tool for simulating target signatures (U)." Environmental Research Institute of Michigan.

INDEX